PAPERS PRESENTED AT THE

SIXTH INTERNATIONAL SYMPOSIUM ON

JET CUTTING TECHNOLOGY

HELD AT THE

UNIVERSITY OF SURREY, U.K.

6–8, APRIL, 1982

Conference sponsored and organised by

BHRA FLUID ENGINEERING

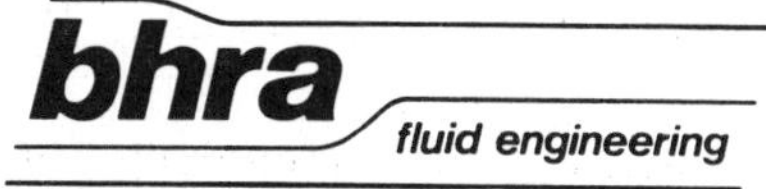

Editors: H.S. Stephens
E.B. Davies

When citing from this volume the following references should be used: Title, Author, Paper No., Pages,
6th International Symposium on Jet Cutting Technology,
Guildford, U.K.
BHRA Fluid Engineering, Cranfield, Bedford MK43 0AJ,
U.K. 6-8 April, 1982.

Printed and published by

BHRA Fluid Engineering
Cranfield, Bedford MK43 0AJ, U.K.

ISBN 0-906085-67-5

ACKNOWLEDGEMENTS

The valuable assistance of the Organising Committee and panel of referees is gratefully acknowledged.

ORGANISING COMMITTEE

E.B. Davies	BHRA Fluid Engineering
Dr. D.G. Edwards	University of Surrey
Dr. A. Lichtarowicz	University of Nottingham
D. Odds	F.A. Hughes & Co. Ltd
D.H. Saunders	BHRA Fluid Engineering
H.S. Stephens	BHRA Fluid Engineering
D. Walton	The Walton Mole Co. Ltd

OVERSEAS CORRESPONDING MEMBERS

N.C. Franz	British Columbia University, Canada
Dr. J. Henneke	Mannesmann Demag, FRG
Dr. K. Hoshino	Hoshino Research Inc., Japan
Dr. H. Louis	Hanover University, FRG
Dr. J.H. Olsen	Flow Research Inc., USA
Dr. D.A. Summers	Missouri-Rolla University, USA

PAPERS PRESENTED AT THE 6TH INTERNATIONAL SYMPOSIUM ON JET CUTTING TECHNOLOGY

CONTENTS

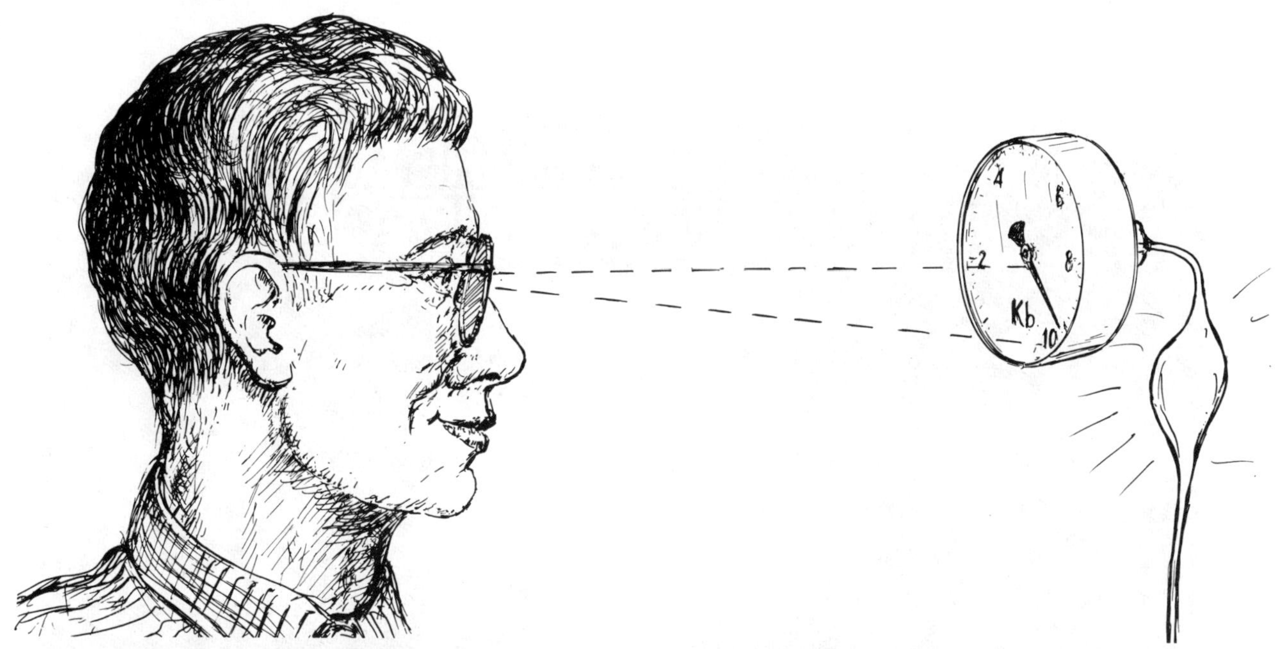

WE DO NOT LOOK AT HIGH PRESSURE THROUGH ROSE COLOURED GLASSES

because we know high pressure can be dangerous and the means of obtaining it often unreliable; so in the light of this knowledge and because of the increasing demand for high pressure pumps suitable for industrial applications we are guided by axioms learned through long experience. Confining highly stressed components to the slow cycling members of a system is of paramount importance when considering high pressure dynamic seal life and is just one part of Stansted Fluid Power design philosophy. Nevertheless, high pressure pumps do not run for ever without maintenance and for this reason ease of access to wearing components is another design feature on all Stansted equipment. This feature coupled with reasonably priced spares and service facilities available direct from the manufacturer ensures reduction in production time lost through mechanical breakdown.

TC 100/850 Pump/intensifier Systems and available for pressure up to 150.000 psi (10 kbar) for liquid and gas service.

Jet Cutting Technology

A review and bibliography

Editor: Robin Brown (BHRA Fluid Engineering)
Reviews by: R.D. Lee and A. Lichtarowicz
(Nottingham University)
R.E.P. Barton and D.H. Saunders

Vol. 9 in the BHRA Fluid Engineering Series

This bibliography, which updates the one published by BHRA in 1973, will be of particular interest to workers in this field, since it primarily records those articles, papers etc. on jet cutting technology that have been published in the period 1971-1980. The bibliography is divided into several sections; theory of jet cutting, including fluid mechanics of jets, and effect of jet impact on target materials; equipment used to generate the jets; and the application of jet cutting to various materials.

The review is similarly divided, and indicates some of the more important publications which may be consulted by readers interested in particular aspects of the subject.

A total of 806 citations are given, with full bibliographic details and informative abstracts. Subject and author indexes are also included.

Publication March 1982

ISBN 0 906085 64 0
188 Pages 806 Refs. A4. Softbound

Published by

Cranfield Bedford MK43 0AJ England

6th International Symposium on
Jet Cutting Technology
6-8, April, 1982

PAPER A1

SELF-RESONATING CAVITATING JETS

V.E. Johnson, A.F. Conn, W.T. Lindenmuth
G.L. Chahine and G.S. Frederick
Hydronautics, Incorporated, U.S.A.

Summary

Cavitation has been shown to provide substantial enhancement to the performance of jets used for cutting and cleaning applications. A further improvement to such jets can be obtained if the flow is caused to pulsate at frequencies corresponding to the predominant frequency of the structures in the jet shear layer. Several passive methods to enable these jets to resonate are now being developed. It should be emphasized that no external means or moving parts are utilized. The pulsations are achieved by self-sustaining oscillations within specially designed nozzles and supply piping. One design, the "PULSER CAVIJET®", consists of tandem orifices, with a resonating chamber in between. Another concept, the "ORGAN-PIPE CAVIJET®", causes pressure amplifications by means of standing waves in the pipe leading up to the nozzle.

The research leading to the development of these new types of cavitating jets is discussed, including descriptions of the experimental methods used to study their performance. Comparative testing has shown that self-resonating cavitating jets can cut rock and perform underwater cleaning more effectively and faster than either conventional jets or nonresonating cavitating jets.

Held at the University of Surrey, U.K.
Symposium organised and sponsored by
BHRA Fluid Engineering

NOMENCLATURE

Symbol	Description	Units
c	speed of sound in fluid	L/T
d	nozzle diameter	L
D	diameter of pipe	L
f	frequency	T^{-1}
K_n	mode parameter	ND
ℓ, L	length	L
M	Mach number, V/c	ND
n	mode number	ND
N	integer	ND
p	pressure	F/L^2
Δp	nozzle pressure drop	F/L^2
r	radius	L
R_e	Reynolds number, Vd/ν	ND
S_d	Strouhal number, $S_d = fd/V$	ND
t	time	T
u'	excited fluctuating velocity	L/T
V	jet velocity	L/T
∀	volume	L^3
X	distance from orifice	L
y	distance from surface	L
Γ	circulation	F^2/T
δ	boundary layer thickness	L
λ	acoustic wave length	L
ν	kinematic viscosity	L^2/T
ρ	fluid density	FT^2/L^4
σ	cavitation number	ND

SUBSCRIPTS

a	ambient
c	vortex core; centerbody
d	see Figure 9
i	incipient, induced
f	feed-pipe
o	initial value, jet exiting value
p	pipe
s	source pipe
t	see Figure 8
v	vortex, vapor pressure

SUPERSCRIPTS

'	rms of fluctuating component
*	critical Strouhal number, for optimum jet structuring

1. INTRODUCTION

1.1 Motivation

During the past two years our studies of cavitating jets have concentrated primarily on new ways to improve their cutting and cleaning action. The emphasis of these efforts has been on submerged jets, operating at large cavitation numbers, σ; typically: $\sigma \simeq 0.5$ to 2 (where: $\sigma \simeq p_a/\Delta p$, p_a: ambient pressure, and Δp: pressure drop across the nozzle). The main motivation for this work is a desire to develop jets capable of effectively enhancing the performance of mechanical drill-bits at greater depths than now possible with our conventional CAVIJET® cavitating jets.

Several papers have described our earlier efforts with what are now being called "conventional" cavitating jets (Refs. 1 to 6), to distinguish them from the "self-resonating" cavitating jets which are the topic of this paper. Results from laboratory and field trials have shown that the cavitation erosion will weaken or remove rock, thus facilitating the action of the mechanical bits. Our initial studies involved either "plain" CAVIJET nozzles, i.e., merely a nozzle-body designed to enhance cavitation with no inserts, or a "centerbody" CAVIJET, where the centerbody insert was usually a flat-ended circular cylinder having a diameter equal to one-half the diameter of the nozzle orifice.

To be effective at greater depths, the self-resonating cavitating jet nozzle should have a larger cavitation number at inception, σ_i, relative to that of the conventional jet, $\sigma_{i,o}$. If, for example, a nozzle is operated at a fixed Δp, then p_a must decrease below a limit value, $\sigma_i \Delta p$, for cavitation to begin to occur. In deep hole drilling this means that, depending on the density of the surrounding fluid, the depth has to be smaller than a certain value. Thus, for an operating $\sigma < \sigma_i$ (shallower depths) there will be cavitation; for $\sigma > \sigma_i$ (greater depths), cavitation will be suppressed. It has been demonstrated that a nozzle having a larger σ_i will remain erosive to a greater depth (or p_a), and at any depth will be more erosive than a nozzle with a smaller σ_i. By causing the jet flow to pulsate or oscillate, higher instantaneous driving pressures are available to collapse cavities — thus intensifying the erosive process.

In deep-hole drilling, rock cuttings fractured by the mechanical tool action must be rapidly cleared from the bottom, or they will be reground by subsequent cutters. Without removal of these chips (see, for instance, Maurer, Ref. 7), the drilling capability of a deep-hole bit is greatly reduced. The localized pressure fluctuations created by submerged, self-resonating jets provide an additional mechanism for increasing drilling rates, by serving to improve the process by which rock cuttings are lifted from the hole bottom. The differential pressures tending to hold the chips against the hole bottom, if overcome by the oscillating pressures of these new cavitating jets, would thus allow the washing action of the flow to more readily carry the chips away.

1.2 Background

Many investigators have examined phenomena related to fluctuating jets, either self-resonating or with an external means for stimulating or driving the pulsations in the jet flow. Schematics of several concepts available for pulsing jets are shown in Fig. 1. Crow and Champagne (8) examined structuring of an air jet, using a loudspeaker for excitation (Fig. 1a). When pulsing was provided at a frequency, f, which corresponds to a critical Strouhal number* of 0.3 (or a multiple of 0.3), discrete ring vortices were created and maximum jet velocity oscillations were observed. Moore (9) and Kibens (10) have conducted similar research. Morel (11) examined the passive oscillations of a Helmholtz tandem-orifice resonant air jet (Fig. 1e) at low Mach numbers (<0.1). The advantages of driving a pulsed jet at acoustic resonance frequencies was examined theoretically by Wylie (12), and more detailed discussions can be found in Ref. 13.

A number of studies, using various mechanical valves (Fig. 1b) or other devices to physically interrupt a jet, have shown the efficacy of delivering a series of

* The Strouhal number, S_d, based on nozzle diameter, d, is : $S_d = fd/V_o$, where V_o is the jet exiting velocity, and f is the frequency of oscillation.

waterhammer stresses to a surface (see, for instance Refs. 14, 15, 16) for cleaning or cutting purposes. None of these approaches, however, are operable at the frequencies (see the next section) required to achieve the jet structuring which was the objective of our investigation.

Galle and Woods (17) describe several methods for oscillating the pressure at the bottom of a drilled hole. However, the various fluid amplifier concepts they proposed (see Fig. 1c, for example) are only capable of frequencies several orders of magnitude lower than required to structure a jet at a practical value of V_o. Fülöp and Rókár (18) suggested a "fluid whistle" concept, for incorporation in deep-hole drills, to cause rock-weakening by creating pressure fluctuations at the natural frequency of the rock being drilled. No considerations of jet-structuring were cited by Fülöp and Rókár; however, a passive device (see Fig. 1e) of this general type has been examined and is discussed below. Indeed, such "bird whistles", or "Helmholtz resonators" have been discussed and understood for over a century (19, 20, 21), and were recently reviewed, along with other related self-sustaining oscillating flows, by Rockwell and Naudascher (22).

In summation, a variety of concepts for causing jet fluctuations have been examined, both for air jets and for liquid jets in air or submerged, and much of the basic phenomena of jet resonance have been recognized. None of these previous investigations, however, were seeking the objective pursued in the present study — namely the creation of explicit nozzle design procedures for high velocity submerged fluid jets where passive resonance could be exploited to increase the cavitation erosion intensity of such jets. Although not within the scope of this paper, the potential of self-resonance to create pulsed water jets for in-air cutting and cleaning purposes is also being examined.

2. JET-STRUCTURING MECHANISMS

A schematic suggesting the mechanism leading to a series of idealized ring vortices in the shear zone of a submerged jet is shown in Fig. 2. The flow leaves the nozzle, with orifice diameter, d, at a velocity, V. This velocity is assumed to be uniform over the nozzle exit plane except in the boundary layer region, which is of characteristic thickness, δ. In the upper portion of Fig. 2, an ideal shear zone, having no mixing with the surrounding fluid, is shown. In a real flow, exterior fluid is entrained, and vortices form as shown in the lower portion of this figure. If we assume that the cores of these vortices are made up of segments of the boundary fluid, spaced a distance, λ, apart, where λ is the distance between two vortices, then the cross-sectional area of each core is approximately $\lambda\delta$. If the cores are assumed circular, then the core radii, r_c, are given by:

$$r_c^2 \simeq \frac{\lambda\delta}{\pi} \quad . \qquad [1]$$

With these assumptions, the circulation, Γ:

$$\Gamma \equiv \oint v \, ds \quad ,$$

of the vortex is equal to that circulation in the segment of shear layer of length λ:

$$\Gamma = \lambda V_o \quad . \qquad [2]$$

If we use a Rankine model for the vortex, it can be shown that the pressure in the center of the vortex, p_{min}, is given by:

$$\frac{p_a - p_{min}}{\frac{1}{2}\rho V_o^2} = \frac{\Gamma^2}{2\pi^2 V_o^2 r_c^2} \quad . \qquad [3]^*$$

Inception occurs when $p_{min} \simeq p_v$ (where p_v is the vapor pressure of the fluid), then:

$$\sigma_i \simeq \frac{\Gamma^2}{2\pi^2 V_o^2 r_c^2} \qquad [4]$$

and combining [1], [2], and [4]:

$$\sigma_i \text{ (at vortex center)} \simeq \frac{\lambda}{2\pi\delta} \qquad [5]$$

* See a standard hydrodynamics text for properties of Rankine vortices, e.g., Ref. 23.

To cause increased cavitation and erosion, it is desirable to have σ_i as large as possible. This can be achieved, as suggested by Eq. [5], either by increasing λ, by decreasing δ, or both. For a given nozzle, jet fluid, and speed, the value of δ will be fixed. The spacing, λ, for a structured, resonating jet has been found to be of the order of the jet diameter, d, as shown schematically in Fig. 3a. In contrast a distribution of turbulent eddy sizes will be formed for an unstructured jet, and the mean scale is more nearly of the order of δ than of d (Fig. 3b).

The foregoing discussion did not examine the effects of a boundary upon which the jet is impinging, i.e., the surface being cleaned or eroded. The flows for an unstructured and a structured jet, striking a surface which is located at a standoff distance, X, from each nozzle, are contrasted in Fig. 4. As the ring vortex nears the surface, the ring radius, r_v, will increase. This "stretching" of a vortex is known to cause a decrease in the core size, r_c. Thus, from Eq. [4], this vortex core decrease is seen to create an increase in σ_i. Therefore, as the boundary is moved near to a cavitating jet, cavitation is first observed adjacent to the boundary. A combination of these mechanisms has been shown experimentally (see Sect. 5.1) to provide inception numbers for structured jets which are two to four times those of the same jet when resonance is not present.

3. EXPERIMENTAL FACILITIES

Two facilities were developed to study the characteristics of self-resonating jets. An air facility allows examinations of jet performance with the capability of rapid and economical changes in a design iteration that are not so readily made in the high-pressure water jet facility. Although the existing air facility has been operated at Reynolds numbers which are too low to allow complete scaling to the water results, valuable insights which have guided and minimized the more tedious tests in water were gained by studies of resonating air jets.

3.1 Air Facility

A schematic of the components contained in this facility are shown in Fig. 5. The rectangular plenum supplied air to the nozzle being tested. Pressure in the plenum was controlled by a needle valve in the 0.2 MPa supply line from the air compressor, and this pressure was monitored by a U-tube manometer. Both free jet and impinging jet tests were run, with a movable target plate placed normal to the jet axis at various locations. Perturbations in the jet axial velocity were surveyed with a 25μm diameter hot film sensor. This probe was mounted on a motor driven rig, allowing a continuous survey along either the axial or radial directions. A nearby microphone monitored acoustic signatures from the nozzle.

The electronic signal from the hot film sensor was conveyed to a Thermo-Systems Model 1050 hot-film anemometer bridge. The output from this bridge was then fed directly to an RMS voltmeter, a spectrum analyzer (Unigon Model 256), and an X-Y plotter. By also sending the RMS voltmeter output to the plotter, side-by-side comparisons could be graphed of the mean, V, and fluctuating, u', velocities. Output from the spectrum analyzer was viewed on an oscilloscope, to allow identification of the resonant frequency peaks in the fluctuating velocity. The microphone output was also fed to an RMS voltmeter and to the spectrum analyzer, thus providing a monitor of the frequency and intensity of sound pressure levels. These components were also used to process the signals from the pressure probes used in the water tests.

3.2 High Pressure Cell (HPC)

The test chamber used to observe the behavior of submerged cavitating jets is shown in Fig. 6. This cell can contain an internal pressure of up to 20.7 MPa (3,000 psi). This cell ambient pressure is controlled by a "choke", a floating double-ended piston valve, which is balanced by nitrogen gas. This floating action allows for the escape of particles during rock cutting trials. Rock specimens, typically cubes 15.2 cm (6 in.) on a side, can be rotated beneath the jet in the HPC, at rates up to 66 rpm by means of the variable speed drive.

The high pressure fluid for this cell is usually provided by either a three or a five-plunger pump; if a test requires higher flows then both pumps are run in parallel. The triplex pump, which is also rated for drilling mud, will deliver 303 ℓ/m (80 gpm) at 13.8 MPa (2,000 psi). The special quintuplex pump, with changeable plungers and head, can be run over a range; highest flow: 341 ℓ/m (90 gpm) at 17.2 MPa (2,500 psi), or highest pressure: 68.9 MPa (10,000 psi) with a 76 ℓ/m (20 gpm) flow capacity.

For flow visualization and photography, the HPC has been fitted with three circular viewing ports, each 3.8 cm (1.5 in.) in diameter. A typical photograph of a self-resonating cavitating jet taken in this cell is shown in Fig. 7. This photograph was taken by darkening the room, opening the camera shutter, and releasing a single flash from a stroboscopic light source. When observing the jets by eye, the strobe is operated at a suitable submultiple of the jet resonance frequency so as to reasonably "stop" the motion of the vortices. In this manner, jet structure can easily be monitored while operating parameters are varied.

Two piezoelectric pressure transducers (PCB Piezotronics Model No. 101A04; 6.4 mm dia.) are mounted in the HPC. One is placed in the inlet supply pipe feeding the nozzle; the other is in the wall of the main pressure vessel. As noted above, the same instrumentation used for the air tests serves to process the signals from these pressure transducers.

4. SELF-RESONATING JET CONCEPTS

The self-resonating submerged cavitating jets which have been studied can be grouped into three types. These have been given the descriptive names: "PULSER CAVIJET®," "PULSER-FED CAVIJET®," and "ORGAN-PIPE CAVIJET®," and the concept for each type will now be described.

4.1 PULSER CAVIJET®

The configuration for this self-resonating jet concept, a tandem-orifice Helmholtz resonator, is shown in Fig. 8a. The steady, high-pressure flow enters through the feed line, diameter D_f, and passes through an entrance section of length, L_s, and diameter, D_s. The flow then contracts through the first orifice, d_1, passes through the chamber (volume: ∀; length, L; diameter: D_t) and exits through the second orifice, d_2.

If operated at its optimum Strouhal number, $S_d = fd_1/V_1$, discrete ring vortices will be formed in the jet issuing from orifice d_1. When a vortex arrives at the second orifice, d_2, a distance L away from d_1, a pressure signal will be transmitted upstream, arriving back at d_1 after a time: $t_L = L/c$, where c is the speed of sound in the fluid. If the length, L, is selected as:

$$L = N\lambda - t_L V_c \quad , \qquad [6]$$

where: N is an integer number of vortices, and
V_c is the vortex convection velocity, $V_c = f\lambda$,
then the pressure signal will arrive at d_1 at exactly the time required to excite a new vortex. Using the expressions for S_d and t_L, and introducing the Mach number: $M = V_1/c$, Eq. [6] can be rewritten nondimensionally as:

$$\frac{L}{d_1} = \frac{N(V_c/V\ )}{S_d(1 + MV_c/V_1)} \quad . \qquad [7]$$

For applications where a space limitation exists, the so-called "LAID-BACK" PULSER CAVIJET, shown in Fig. 8b, might be used. In this concept, part of the necessary resonant chamber volume is contained in the region which surrounds the main flow passage leading to the first nozzle.

4.2 PULSER-FED CAVIJET®

A self-resonating nozzle concept, which has been shown to be capable of providing several advantages relative to the PULSER CAVIJET, is shown in Fig. 9. Shown are a basic design, Fig. 9a, and two designs (9b, 9c) with alternative diffusion chambers. In this concept the exit nozzle, d_3, is fed a fluctuating excitation by the same tandem-orifice with intervening resonant chamber configuration as described in Sect. 4.1.

The advantages of the PULSER-FED CAVIJET concept have been shown to be:

a. The jet formed by the exit nozzle, d_3, has a more uniform velocity distribution, and the vortices formed here are more cleanly defined.

b. The PULSER (d_1) nozzle can be selected to operate at a Strouhal number higher than that of the exit (d_3) nozzle. This implies that the resonant chamber pressure can be higher than p_a, the ambient pressure surrounding the jet exiting from d_3, and the velocity in the chamber less than V_o. Therefore, the cavitation number in the chamber will be much higher than that of the exiting jet, and cavitation in this chamber can be avoided.

c. The diffusion chamber (L_d, D_d) may be designed to enhance the amplitude of modulation provided by the PULSER chamber.

Disadvantages of the PULSER-FED concept include:

a. A more complex mechanical configuration, and

b. The overall energy loss, caused by losses in the diffusion chamber, is greater than for the PULSER CAVIJET concepts.
The latter problem can be minimized by using the alternative diffusion chambers shown in Figs. 9b and 9c.

4.3 ORGAN-PIPE CAVIJET®

A self-resonating jet concept which offers the simplest design, and seems to have considerable promise for being adaptable to existing deep-hole drill bits is shown in Fig. 10. This concept should achieve peak acoustic resonance when a standing wave forms in the "organ-pipe" section (length: L_p, diameter: D). This section is created by the upstream contraction, $(D_s/D)^2$ and the nozzle contraction, $(D/d)^2$. Peak resonance will occur when the frequency of the organ-pipe wave is near the critical jet structuring frequency, as given by the nozzle Strouhal number, S_d, but the exact resonance is dependent on the contractions at each end of the organ-pipe. For instance, if both $(D_s/D)^2$ and $(D/d)^2$ are large, then the first mode resonance in the pipe will occur when the sound wave length in the fluid is approximately four times L_p. Acoustic analyses and experimentation have led to the following approximation, useful for estimating the length of the organ-pipe:

$$\frac{L_p}{d} \simeq \frac{K_n}{MS_d^*} \quad , \qquad [8]$$

where the "mode parameter", K_n is given by:

$$K_n = \text{func.}\left\{n, \left(\frac{D_s}{D}\right)^2, \left(\frac{D}{d}\right)^2\right\} \simeq \frac{2n-1}{4} \quad \text{for} \left(\frac{D_s}{D}\right)^2 \text{ and } \left(\frac{D}{d}\right)^2 >> 1 \qquad [9a]$$

$$\simeq \frac{n}{2} \quad \text{for} \left(\frac{D_s}{D}\right)^2 >> 1, \text{ but } \left(\frac{D}{d}\right) \gtrsim 4 \qquad [9b]$$

In these expressions:

n = mode number of the organ-pipe,
S_d^* = critical Strouhal number, $fd/V \simeq 0.3$
M = Mach number, V/c.

Other examples of ORGAN-PIPE CAVIJET concepts are shown in Fig. 11, where single diameter, "two-step", and "three-step" diameter versions are indicated.

When the contractions are as designated in Eq. [9b], then an empirical relation found useful for designing an ORGAN-PIPE CAVIJET is:

$$M \simeq \frac{n}{2S_d^*}\left[\frac{d}{L_p} - 0.86\left(\frac{d}{L_p}\right)^2\right] \qquad [10]$$

In Fig. 12, comparisons are shown between the curves given by Eq. [10] and experimental measurements of jet structuring conducted in the air facility described in Sect. 3.1. It can be seen that the experimental data tend to fall within a band of Strouhal numbers, at each mode, of about: $0.4 \leqq S_d^* \leqq 0.5$.

5. SOME EXPERIMENTAL RESULTS

5.1 Cavitation Inception

As discussed in Sect. 1.1, one of the motivations for this study of self-resonating cavitating jets was to determine whether new nozzles could be developed which would exhibit larger cavitation numbers at inception, σ_i. Comparisons of the inception for three nonresonating nozzle types are shown in Fig. 13, where the trends of σ_i versus Reynolds number are plotted. It should be emphasized that this figure is intended to indicate only general trends; σ_i observations are highly dependent on details in nozzle design and fabrication, boundary layer characteristics, nuclei distributions and gas content in the fluid. For this reason, we prefer to plot (as in Fig. 14) relative curves of $\sigma_i/\sigma_{i,o}$, where $\sigma_{i,o}$ represents a base-line measurement with a standard nozzle under the same test conditions.

The curves in Fig. 14 summarize the results of cavitation inception measurements for each of the three self-resonating concepts. A curve for a conventional, centerbody CAVIJET is also given for comparison. Note that the downward trend of σ_i seen with respect to Reynolds number for the ORGAN-PIPE CAVIJET is accentuated by the way these data are normalized. That is, because the trend for $\sigma_{i,o}$ (inception for the plain CAVIJET nozzle, see Fig. 13) is steeply upward with increasing Reynolds number, dividing by these $\sigma_{i,o}$ values tends to create an exaggerated downward tendency for $\sigma_i/\sigma_{i,o}$.

The inception curve for the PULSER-FED CAVIJET nozzle has a well-defined peak within the range of Reynolds numbers in Fig. 14. The resonance in this nozzle shows the strong effect which jet structuring can have on inception of cavitation. Comparable peaks for the ORGAN-PIPE and PULSER nozzles did not occur in the range of Reynolds numbers (or Mach numbers) shown here. However, such peaks in σ_i, associated with critical resonant frequencies, have been observed for each type of self-resonating cavitating jet which has been examined. As seen in Fig. 14, each of the three types of self-resonating concepts have substantially higher inception numbers for cavitation than either the plain or centerbody versions of a "conventional" cavitating jet.

5.2 Rock Cutting

A typical comparison is shown in Fig. 15 between a self-resonating cavitating nozzle and a conventional ("SMITH") nozzle* supplied for use in a roller-cone deep-hole drill bit. Seen is a top view of one face of a 15.2 cm (6 in.) cube specimen of Indiana limestone. To allow a half-circle cut to be made by each nozzle, one half of the rock face was protected by a metal plate. The single pass was then made at a translation velocity of 6.4 mm/s — starting and stopping the rotation while the jet was impinging on the plate. The nozzle was then changed, the plate shifted to protect the other half of the cube-face, and the same conditions were run again. In this way, variations due to rock properties could be minimized. Under these conditions, the mean slot depth cut by the ORGAN-PIPE CAVIJET was 4.3 times deeper than that cut by the SMITH nozzle.

Similar slot-cutting comparisons are shown in Table 1 at several cavitation numbers. Note that, for the single pass tests, as σ is increased the performance of the ORGAN-PIPE CAVIJET improves relative to the SMITH nozzle. The cavitation inception number for this CAVIJET nozzle was more than twice that for the SMITH nozzle. This result is consistently observed, that is, the jet having the higher inception cavitation number is found to be more erosive than a jet of the same size, operated under identical conditions, but having a lower σ_i.

5.3 Underwater Cleaning

In a field trial of an early PULSER CAVIJET design, cleaning rates more than twice those achieved by a conventional CAVIJET of the same size were reported (24). This trial occurred during underwater cleaning of the "keel block areas" on the hull of an aircraft carrier, the USS LEXINGTON (CVT-16) which had just come out of drydock. The hull had been cleaned and repainted in drydock, except for those areas which had rested on the keel blocks. Navy divers, using underwater CAVIJET tools (25) developed by SEACO and HYDRONAUTICS, then proceeded to remove the old paint and corrosion products from the keel block areas while the LEXINGTON floated next to the dock in Pensacola, Florida. These divers reported, albeit subjectively since precise time trials were

---

* Manufactured by SMITH TOOL, Division of Smith International, Inc., Irvine, California, U.S.A.

not taken, that a PULSER nozzle, with d = 2.5 mm (0.10 in.) (see Fig. 8a) prepared the required white metal steel surface suitable for underwater repainting at a rate more than twice that of a conventional CAVIJET nozzle with the same orifice diameter. This trial involved limited use of the PULSER — most of the cleaning was with the conventional CAVIJET since only the latter nozzle has received an official Approval for Navy Use (ANU). We will be seeking an ANU for one or more of the self-resonating cavitating jet nozzles after further evaluation to determine the best designs for this underwater cleaning application.

6. CONCLUDING REMARKS

This paper has summarized some of the early results from an ongoing effort to develop new cavitating jets which might perform more effectively, particularly at high cavitation numbers (i.e., under deep submergence). Additional details can be found in the report, recently released by Sandia National Laboratories, which describes the feasibility study phase of this program (26). From this study, the following conclusions have been drawn:

a. Higher inception cavitation numbers can be achieved in a jet which is structured into discrete ring vortices at preferred resonant frequencies.

b. Several concepts have been demonstrated to be capable of producing a passive self-oscillation in a submerged cavitating jet. Three of these self-resonating nozzle types, the PULSER, PULSER-FED, and ORGAN-PIPE CAVIJET designs, have been found to have higher inceptions for cavitation than conventional (nonresonating) CAVIJETS and typical drill bit nozzles.

c. Limited rock cutting and cleaning trials have demonstrated a correlation between higher cavitation inception numbers and greater erosivity in submerged cavitating jets.

7. ACKNOWLEDGEMENTS

This study has been supported by the U.S. Department of Energy, Division of Geothermal Energy, under Sandia Laboratories Contract Nos. 13-5111 and 13-5129, with cost-sharing by NL/Hycalog, Houston, Texas, U.S.A.

8. REFERENCES

1. Johnson, V. E., Jr., et al.: "Tunnelling, fracturing, drilling and mining with high speed water jets utilizing cavitation damage". Proc. First Int'l. Sympos. on Jet Cutting Technology (Coventry, U.K.: April 5-7, 1972) Cranfield, U.K., BHRA Fluid Engineering, 1972, Paper A3, 19 pp.

2. Conn, A. F. and Johnson, V. E., Jr.: "Further applications of the CAVIJET™ (cavitating water jet) method". Proc. Second Int'l. Sympos. on Jet Cutting Technology (Cambridge, U.K.: April 2-4, 1974) Cranfield, U.K., BHRA Fluid Engineering, 1974, Paper D2, 14 pp.

3. Conn, A. F. and Rudy, S. L.: "Cutting coal with the CAVIJET™ cavitating water jet method". Proc. Third Int'l Sympos. on Jet Cutting Technology (Chicago, U.S.A.: May 11-13, 1976) Cranfield, U.K., BHRA Fluid Engineering, 1976, Paper D8, 9 pp.

4. Conn, A. F., Rudy, S. L., and Mehta, G. D.: "Development of a CAVIJET™ system for removing marine fouling and rust". Proc. Third Int'l Sympos. on Jet Cutting Technology (Chicago, U.S.A.: May 11-13, 1976) Cranfield, U.K., BHRA Fluid Engineering, 1976, Paper G4, 14 pp.

5. Conn, A. F. and Rudy, S. L.: "Conservation and extraction of energy with the CAVIJET™". Proc. Fourth Int'l. Sympos. on Jet Cutting Technology (Canterbury, U.K.: April 12-14, 1978) Cranfield, U.K., BHRA Fluid Engineering, 1978, Vol. 1, Paper H2, 20 pp.

6. Conn, A. F. and Johnson, V. E., Jr.: "The fluid dynamics of submerged cavitating jet cutting". Proc. Fifth Int'l Sympos. on Jet Cutting Technology (Hanover, F.G.R.: June 2-4, 1980), Cranfield, U.K., BHRA Fluid Engineering, 1980, Paper A1. pp. 1-14.

7. Maurer, W. C.: "The 'perfect cleaning' theory of rotary drilling". Journal of Petroleum Technology, Vol. 14, 1962, pp. 97-101.

8. Crow, S. C. and Champagne, F. H.: "Orderly structure in jet turbulence". Journal of Fluid Mechanics, Vol. 48, Part 3, August 1971, pp. 547-591.

9. Moore, C. J.: "The role of shear layer instability waves in jet exhaust noise". Journal of Fluid Mechanics, Vol. 80, 1977, pp. 321-367.

10. Kibens, V.: "Discrete noise spectrum generated by an acoustically excited jet". Paper 79-0592, AIAA Fifth Aeroacoustics Conference, March 12-14, 1979.

11. Morel, T.: "Experimental study of a jet driven Helmholtz oscillator". ASME, Journal of Fluids Engineering, Vol. 101, September 1979, pp. 383-390.

12. Wylie, E. B.: "Pipeline dynamics and the pulsed jet". Proc., First Int'l Sympos. on Jet Cutting Technology (Coventry, U.K.: April 5-7, 1972) Cranfield, U.K., BHRA Fluid Engineering, 1972, Paper A5, 12 pp.

13. Streeter, V. L. and Wylie, E. B.: Hydraulic Transients, McGraw Hill, 1967.

14. Nebeker, E. B. and Rodriguez, S. E.: "Percussive water jets for rock cutting". Proc. Third Int'l. Sympos. on Jet Cutting Technology (Chicago, U.S.A.: May 11-13, 1976) Cranfield, U.K., BHRA Fluid Engineering, 1976, Paper B1, 9 pp.

15. Lichtarowicz, A. and Nwachukwu, G.: "Erosion by an interrupted jet". Proc. Fourth Int'l. Sympos. on Jet Cutting Technology (Canterbury, U.K.: April 12-14, 1978) Cranfield, U.K., BHRA Fluid Engineering, 1978, Vol. 1, Paper B2, 6 pp.

16. Erdmann-Jesnitzer, F., Louis, H., and Schikorr, W.: "Cleaning, drilling and cutting by interrupted jets". Proc. Fifth Int'l Sympos. on Jet Cutting Technology (Hanover, F.G.R.: June 2-4, 1980) Cranfield, U.K., BHRA Fluid Engineering, 1980, Paper B1, pp. 45-55.

17. Galle, E. M. and Woods, H. B.: United States Patent Number 3,405,770, October 15, 1968.

18. Fülöp, M. and Rókár, G.: United States Patent Number 4,071,097, January 31, 1978.

19. Sondhauss, C.: Annual Physics, Volume 97, pg. 1854.

20. Helmholtz, Philosophical Magazine, Volume 36, 1868, pg. 337.

21. Rayleigh, J.W.S.: Theory of Sound, Chapter XXI, 2nd Edition, MacMillan (Republished 1945, Dover Publications), pp. 1825 ff.

22. Rockwell, D. and Naudascher, E.: "Self sustained oscillations of impinging free shear layers". Annual Review of Fluid Mechanics, Volume 11, 1979.

23. Milne-Thomson, L. M.: Theoretical Hydrodynamics, MacMillan, New York, 1969, pp. 351-354.

24. May, Robert A., SEACO, Incorporated, CAVICO Division, private communication, September, 1981.

25. Conn, A. F., Johnson, V. E., Jr., Lindenmuth, W. T., and Frederick, G. S.: "Some industrial applications of CAVIJET® cavitating fluid jets". Proc. 1st U.S. Water Jet Sympos., Golden, Colorado, April 1981.

26. Johnson, V. E., Jr., Lindenmuth, W. T., Conn, A. F., and Frederick, G. F.: "Feasibility study of tuned-resonator, pulsating cavitating water jet for deep-hole drilling". Sandia Laboratories Contractor Report SAND81-7126, August 1981 (originally: HYDRONAUTICS, Incorporated Technical Report 8001-1, May 1981), 142 pp.

Table 1 - Slot Cutting Comparisons

NOZZLES: Conventional SMITH, d = 4.3 mm (0.170 in.)
ORGAN-PIPE CAVIJET, d = 4.7 mm (0.185 in.)
ROCK: Indiana limestone; STANDOFF: X/d = 3
TRANSLATION VELOCITY: 6.4 mm/s (0.25 in./s)

Pressure, Nozzle Δp, MPa (psi)	Ambient Pressure, p_a, MPa (psi)	Cavitation Number, σ	Mean Slot Depth Ratio, CAVIJET / SMITH
SINGLE PASS			
17.4 (2,530)	1.2 (170)	0.067	1.16
17.2 (2,500)	4.1 (600)	0.24	2.54
17.2 (2,500)	8.3 (1,200)	0.48	2.33
TWENTY-FIVE PASSES			
8.4 (1,225)	2.2 (325)	0.27	1.90

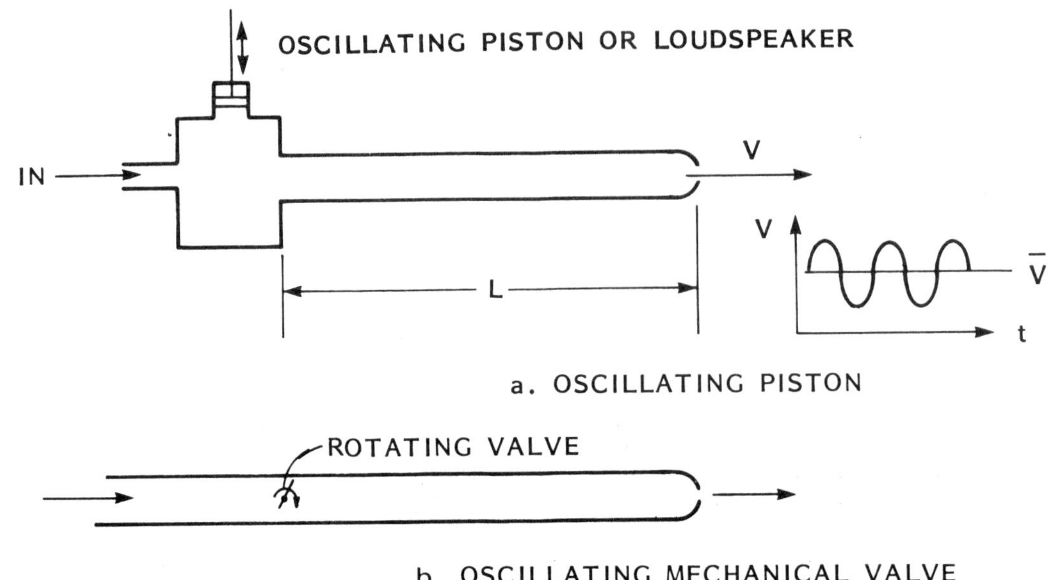

a. OSCILLATING PISTON

b. OSCILLATING MECHANICAL VALVE

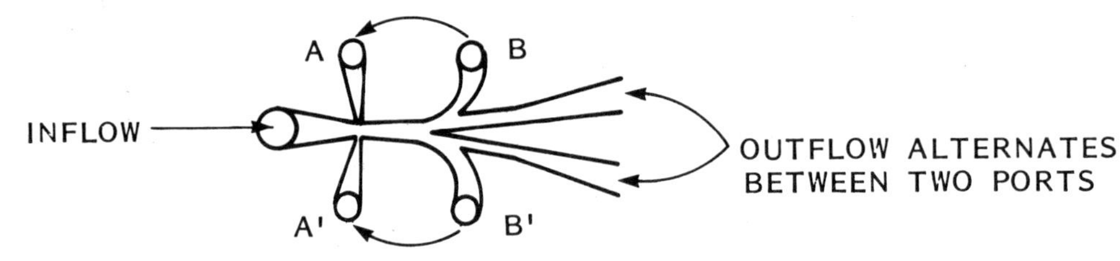

c. FLUID OSCILLATOR VALVE

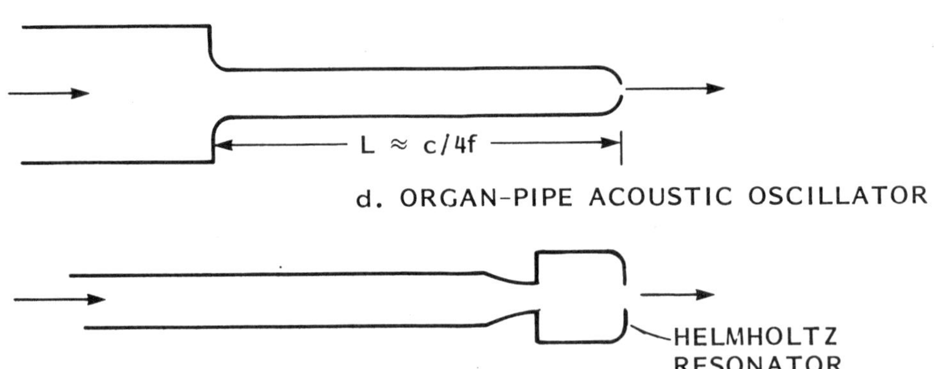

d. ORGAN-PIPE ACOUSTIC OSCILLATOR

e. HELMHOLTZ ACOUSTIC OSCILLATOR

FIGURE 1 - GENERAL CONCEPTS FOR PULSING JETS

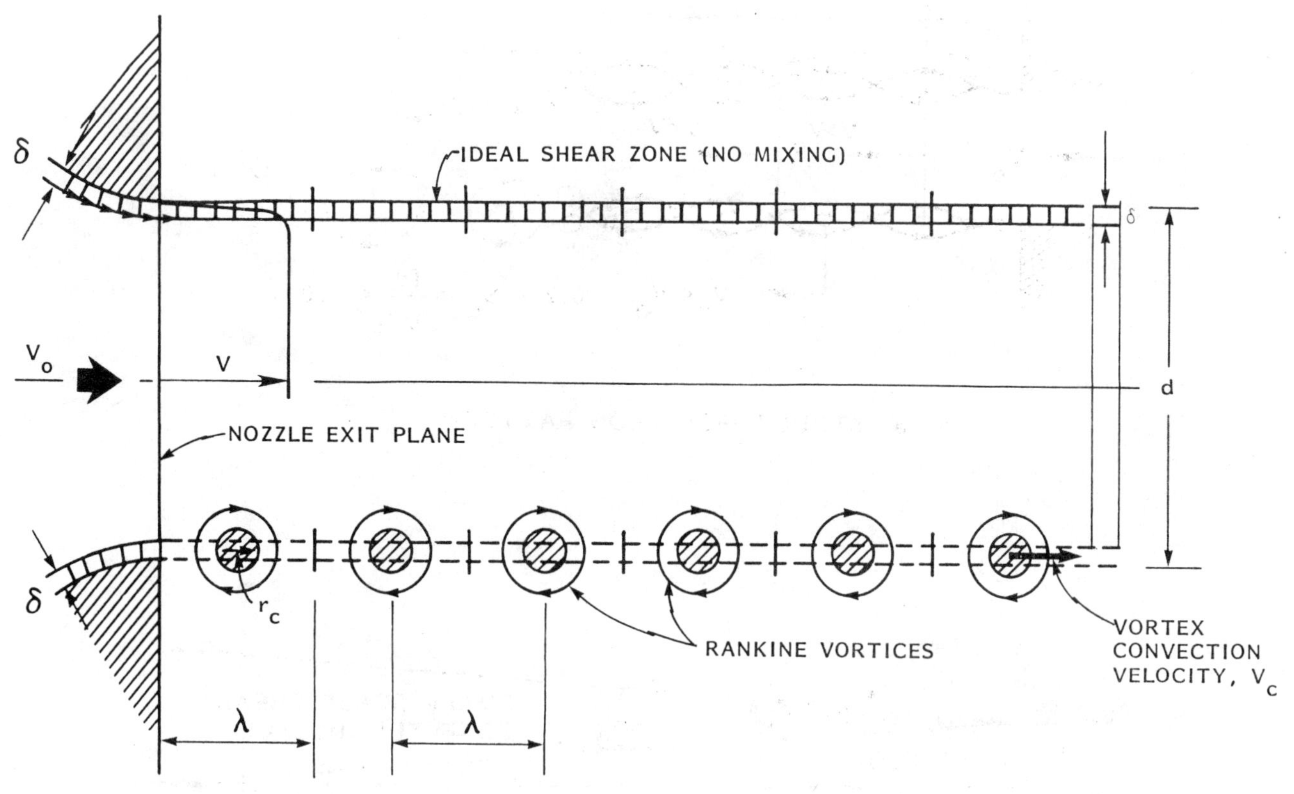

FIGURE 2 - SCHEMATIC OF MECHANISM FOR VORTEX ROLL-UP IN THE SHEAR ZONE OF A SUBMERGED JET

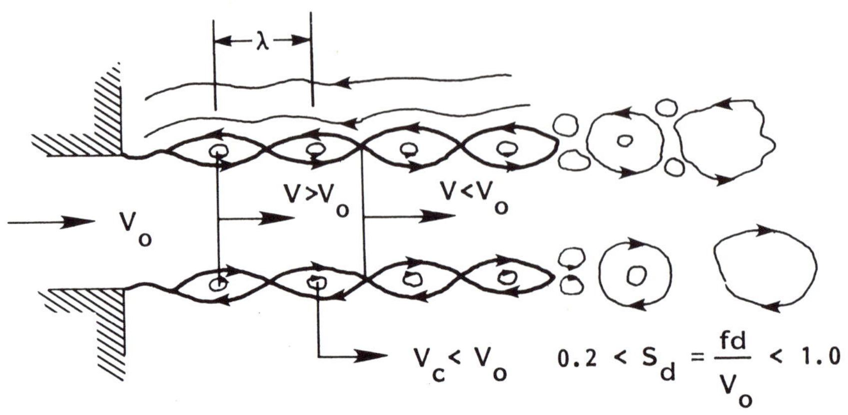

a. STRUCTURED FLOW PATTERN

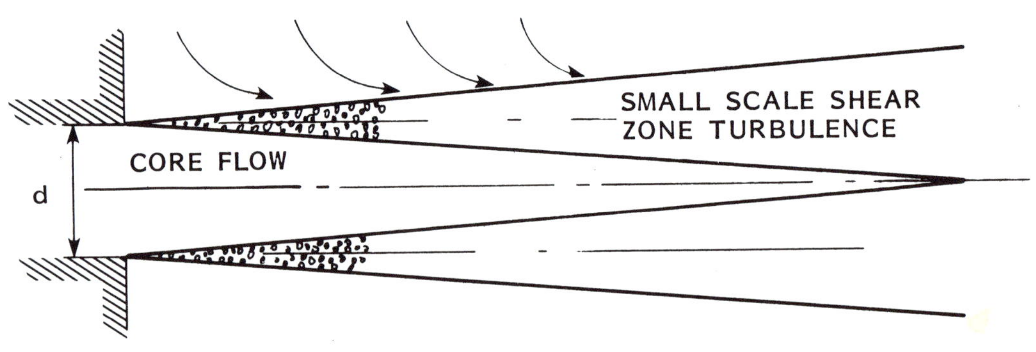

b. UNSTRUCTURED FLOW PATTERN

FIGURE 3 - FREE SUBMERGED JET FLOW PATTERN COMPARISON

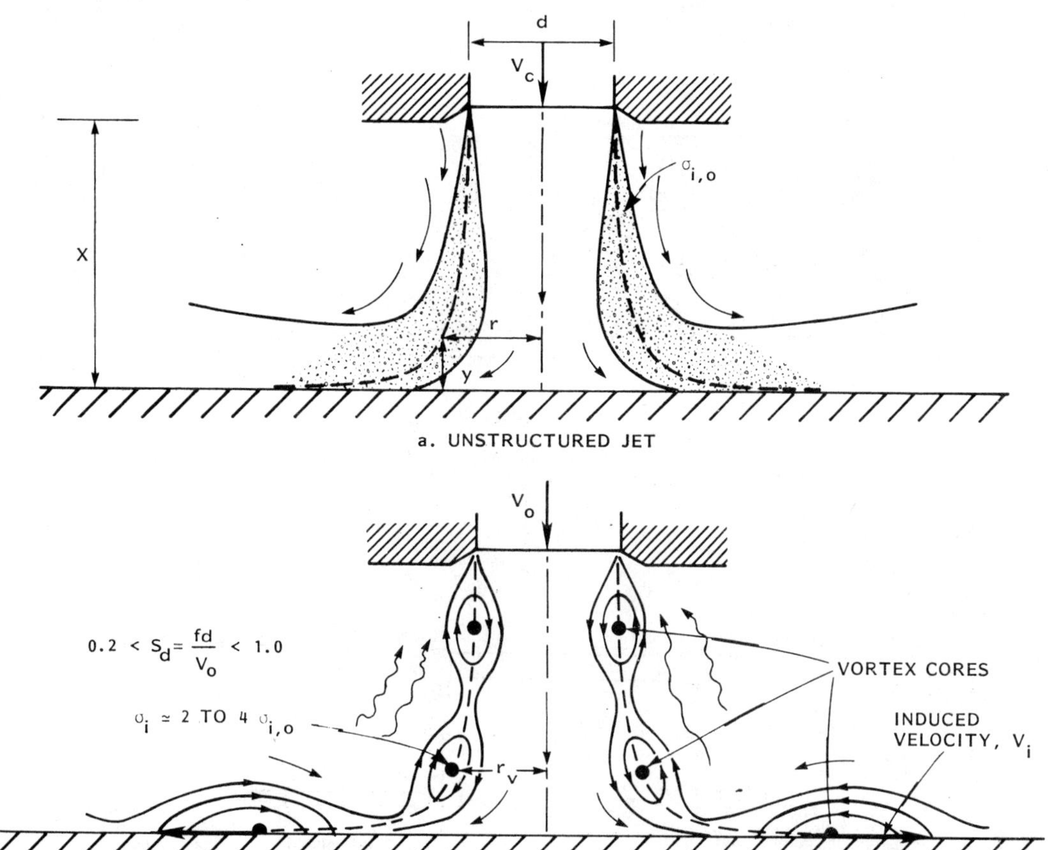

FIGURE 4 - FLOW COMPARISON FOR IMPINGING SUBMERGED CAVITATING JETS

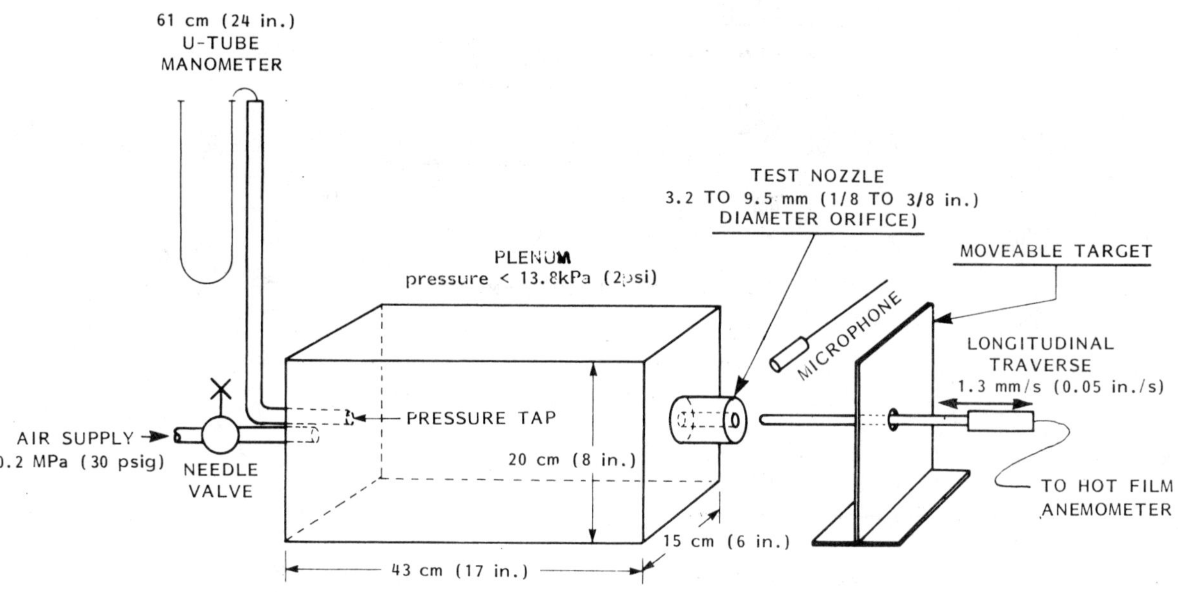

FIGURE 5 - APPARATUS FOR NOZZLE TESTS WITH AIR

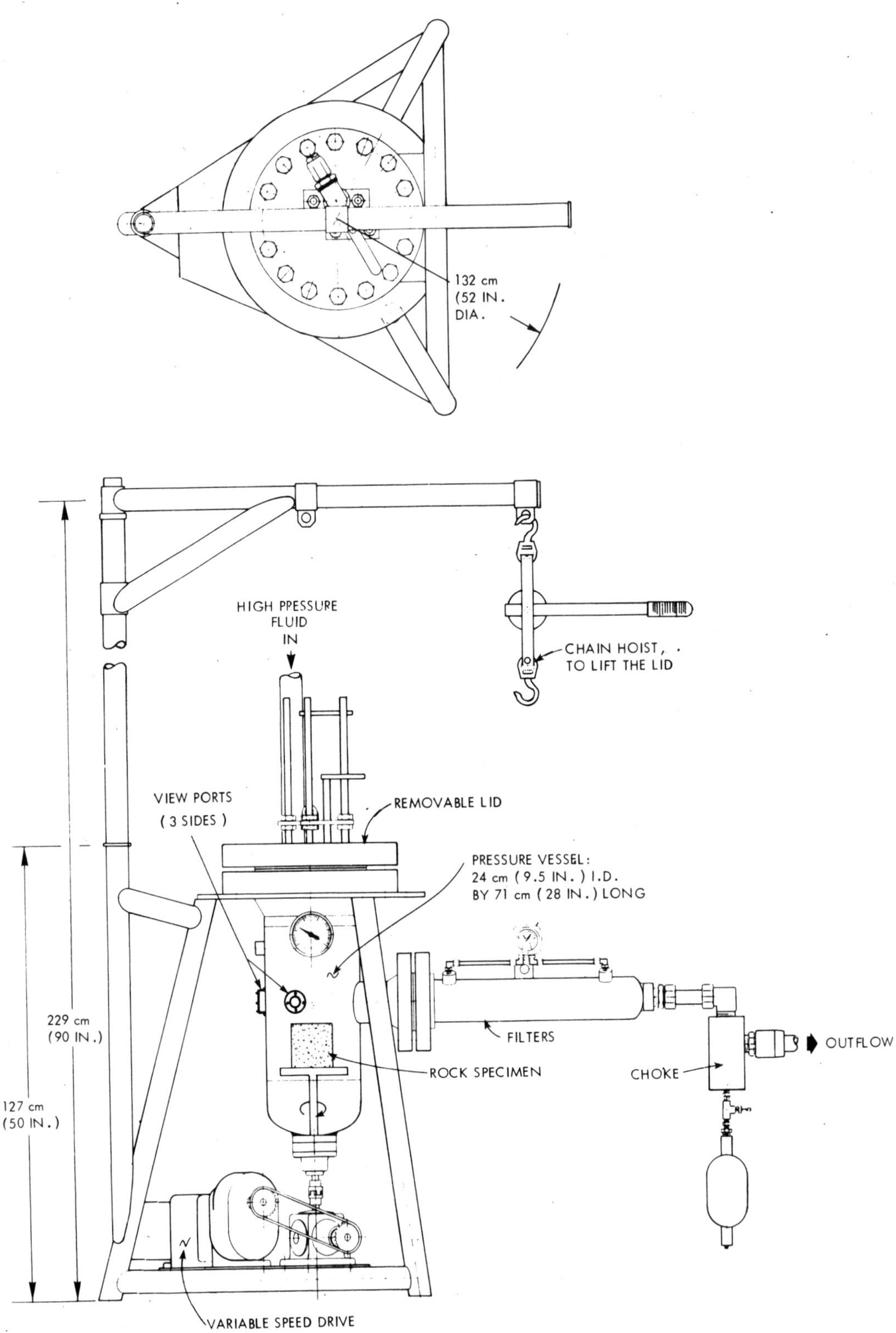

FIGURE 6 - HIGH PRESSURE CELL, HPC

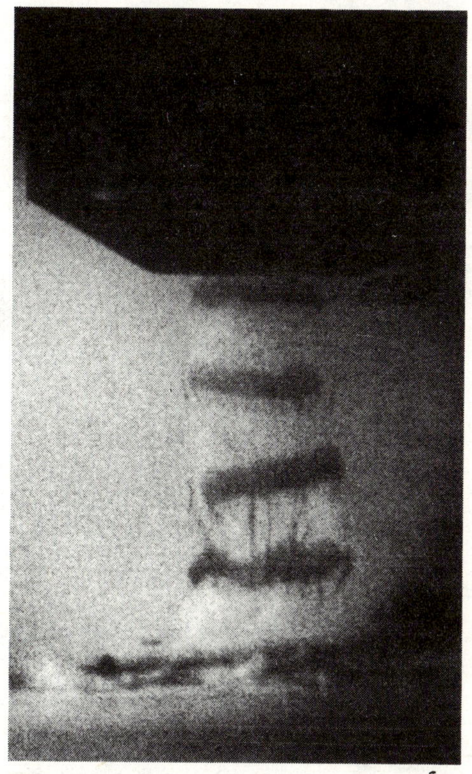

STANDOFF: $X = 12.7$ mm (0.5 in.); $X/d = 2.7$

NOZZLE PRESSURE DROP: $\Delta p = 17.2$ MPa (2,500 psi)

AMBIENT PRESSURE: $p = 3.9$ MPa (560 psi)

NOZZLE ORIFICE DIAMETER: $d = 4.7$ mm (0.186 in.)

RESONANCE FREQUENCY: 20.5 kHz

STROUHAL NUMBER: $S_d = 0.52$

CAVITATION NUMBER: $\sigma = 0.22$

NOZZLE: ORGAN-PIPE CAVIJET, with two-step pipe configuration. This nozzle is half-scale model of CAVIJET to be used in a three-cone roller bit.

FIGURE 7 - TYPICAL PHOTO OF RING VORTICES FROM SELF-RESONATING CAVIJET® NOZZLE. Note vortex spreading across surface of target plate.

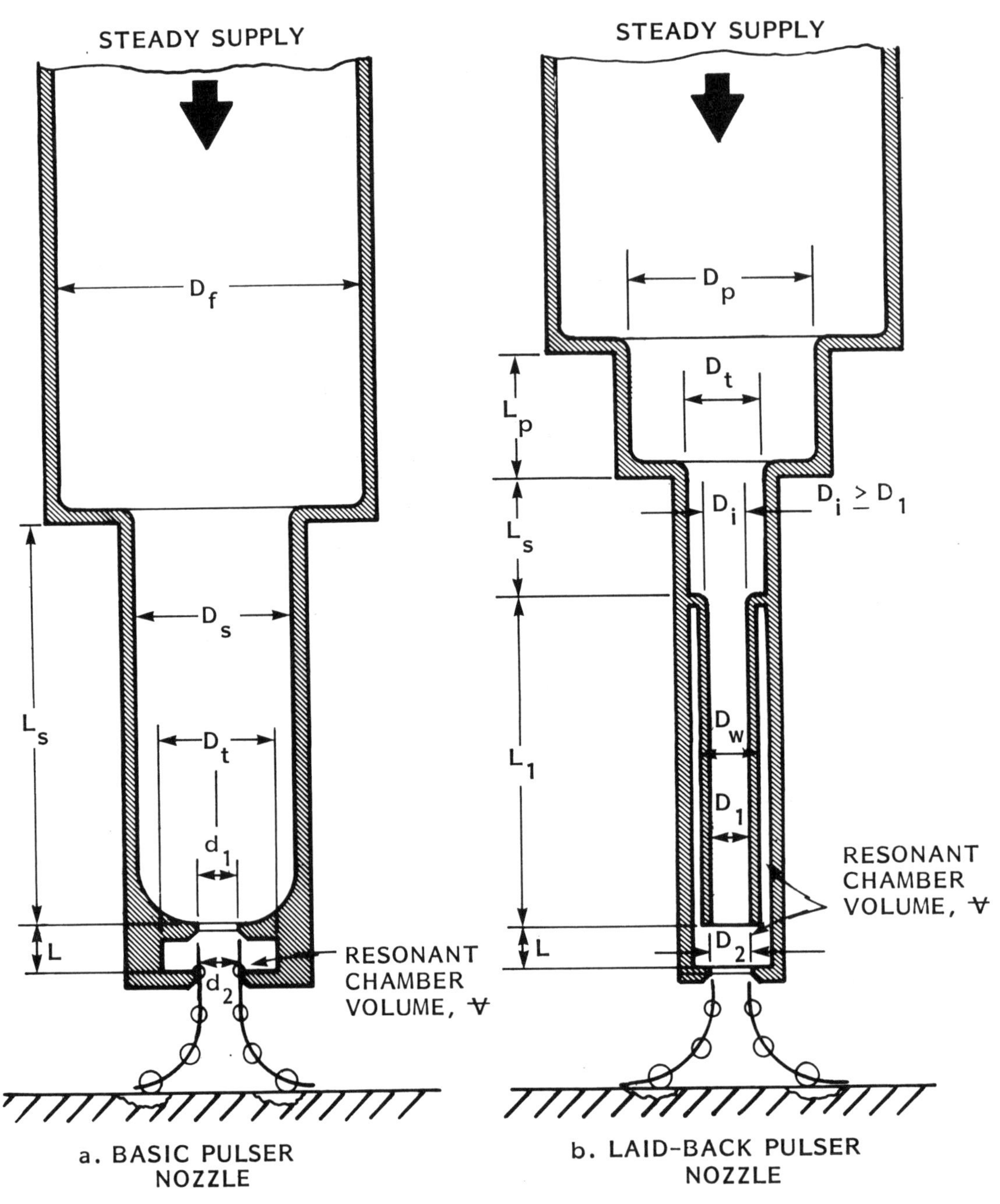

FIGURE 8 - CONCEPTS FOR PULSER CAVIJET® NOZZLES

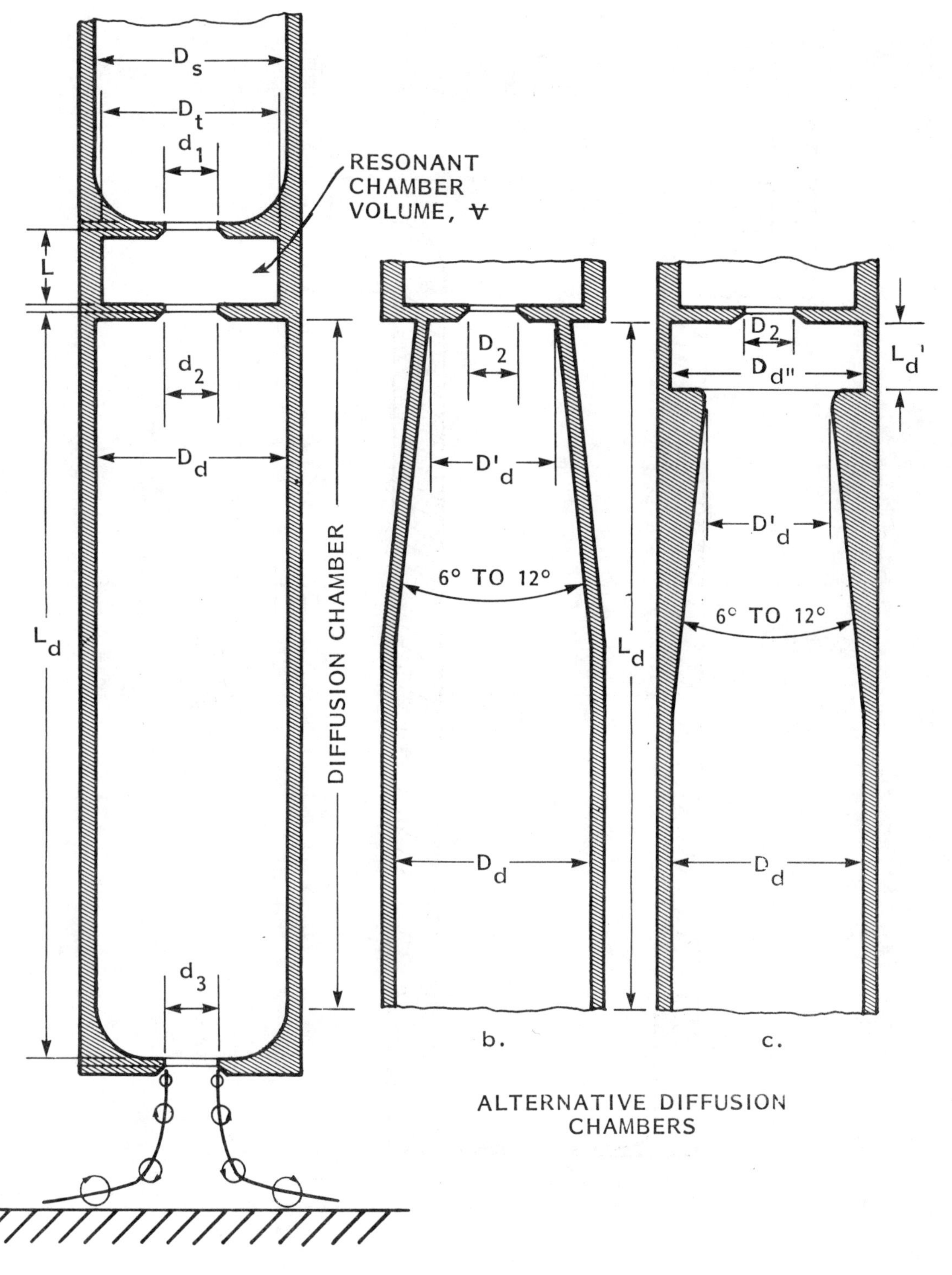

FIGURE 9 - CONCEPTS FOR PULSER-FED CAVIJET® NOZZLES

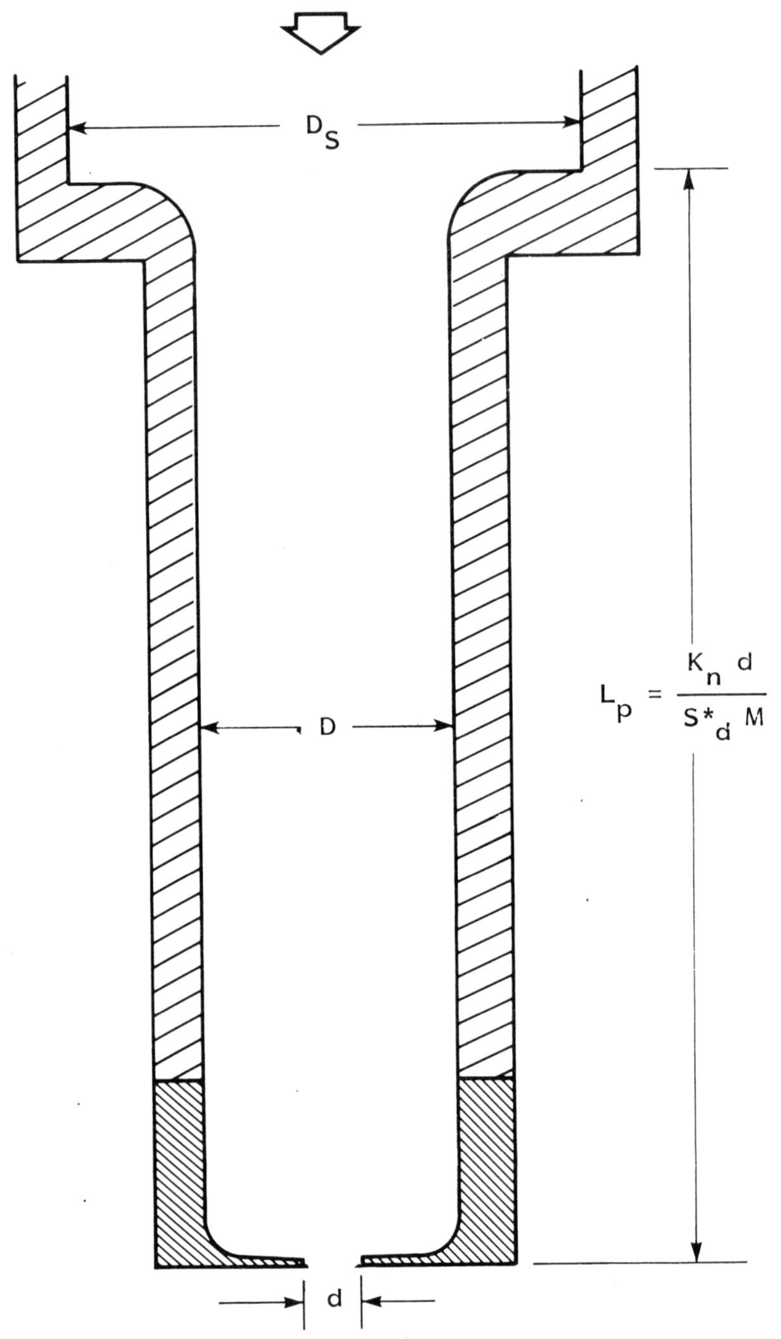

FIGURE 10 - ORGAN PIPE CAVIJET® NOZZLE CONCEPT

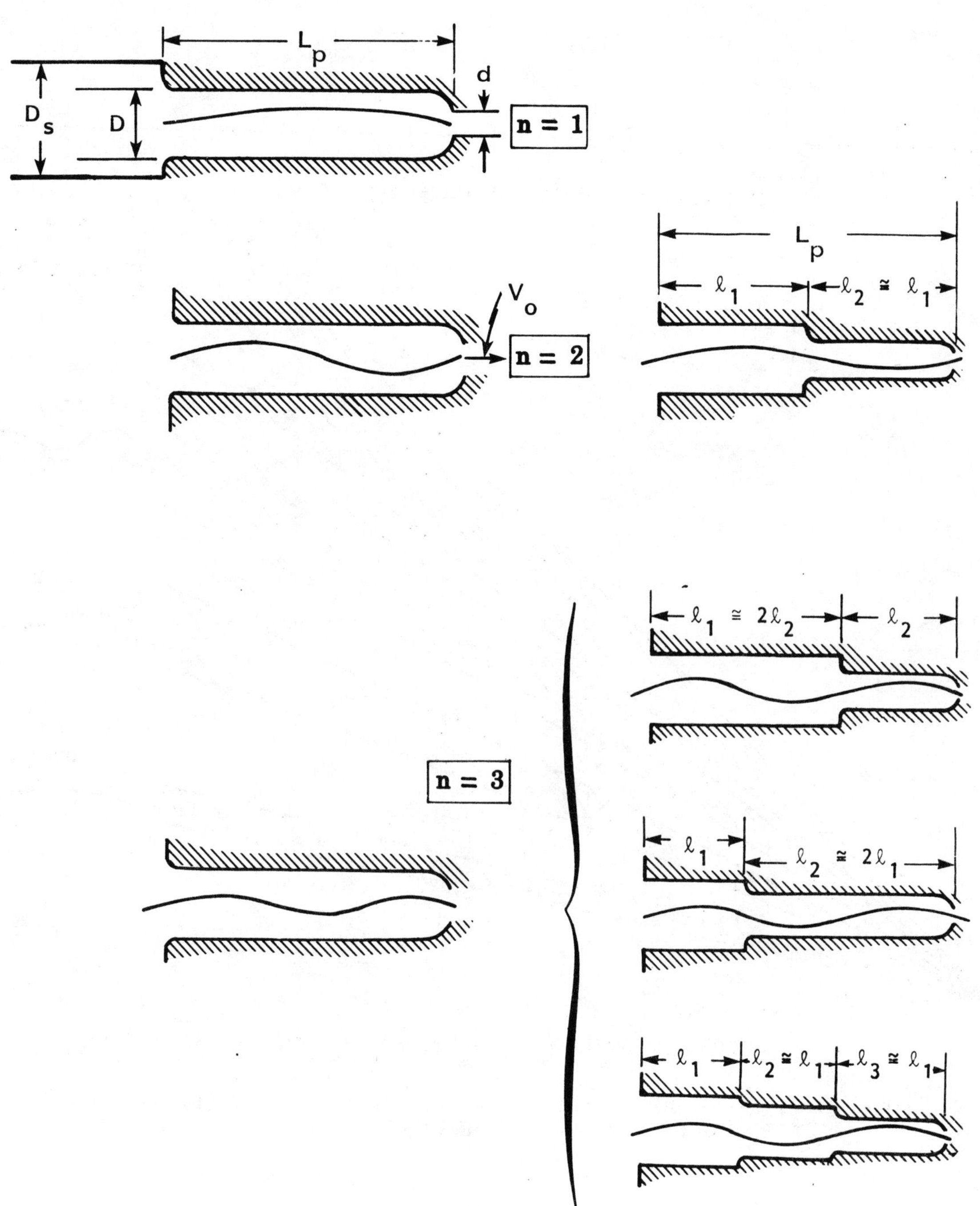

FIGURE 11 - MULTIPLE STEPPED ORGAN-PIPE CAVIJET® CONCEPTS

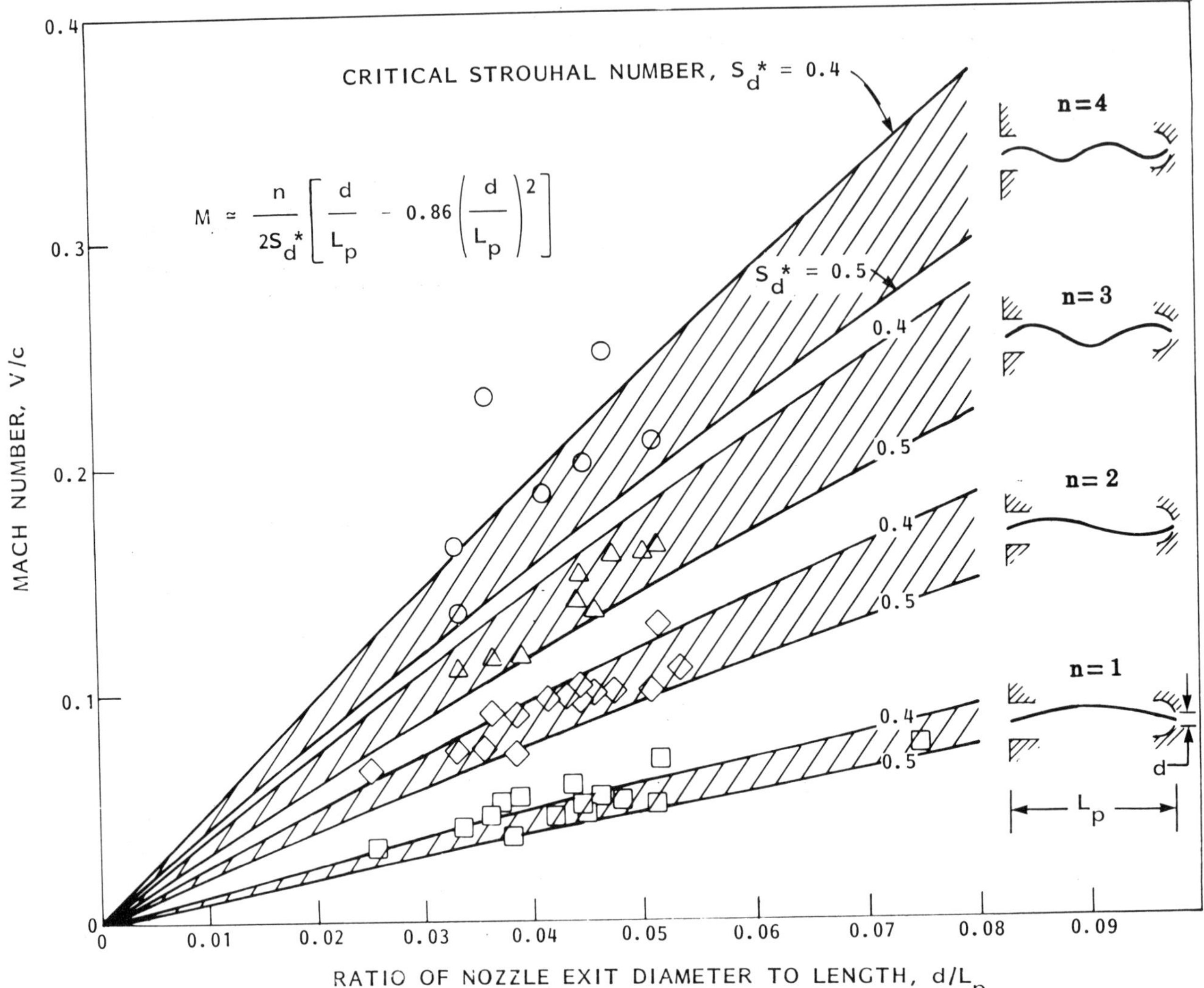

FIGURE 12 - RELATION BETWEEN MACH NUMBER AND NOZZLE LENGTH FOR ORGAN-PIPE CAVIJET® NOZZLES

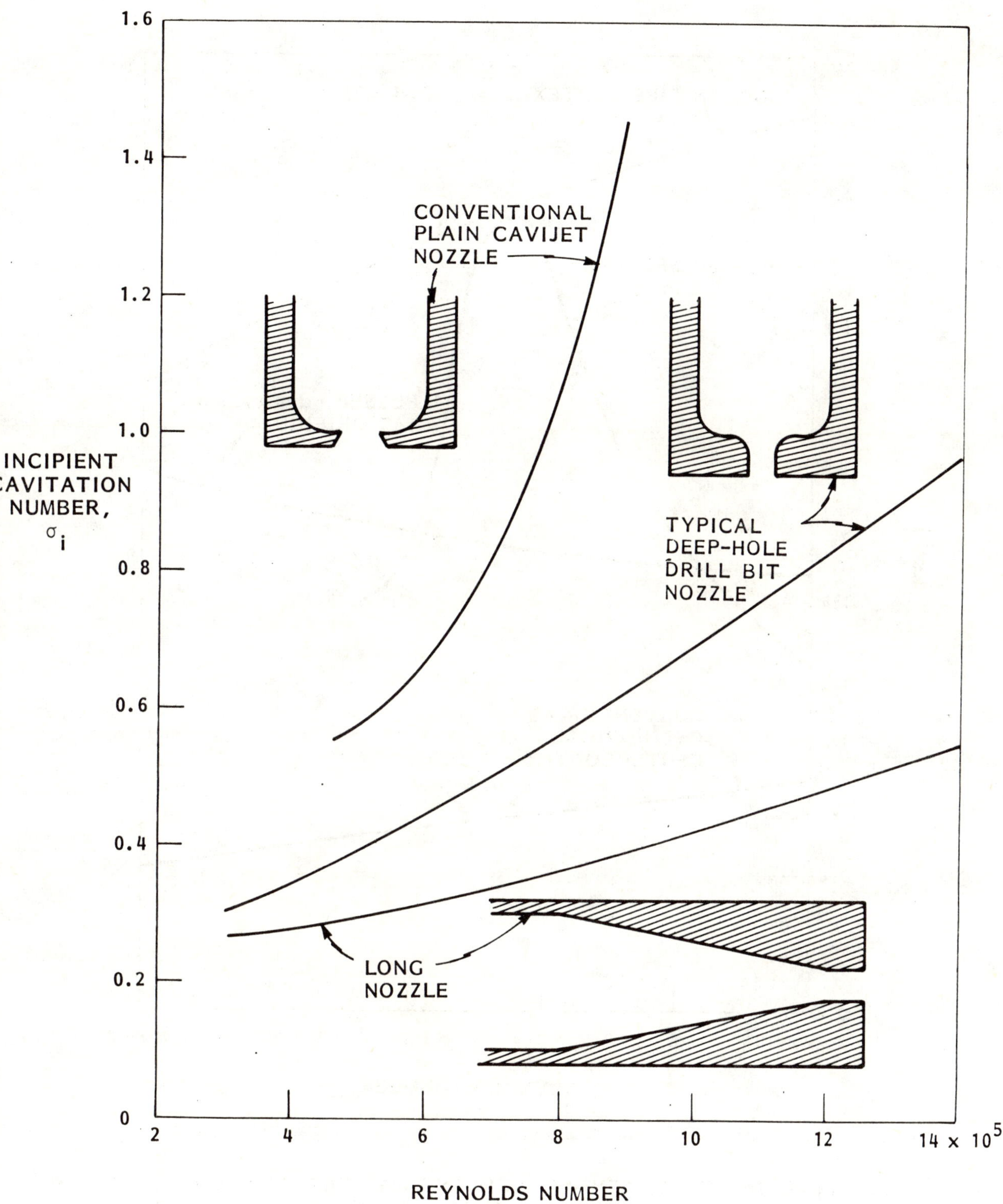

FIGURE 13 - APPROXIMATE INCIPIENT CAVITATION NUMBERS FOR CAVIJET® AND OTHER NOZZLE TYPES (ACOUSTIC DETECTION)

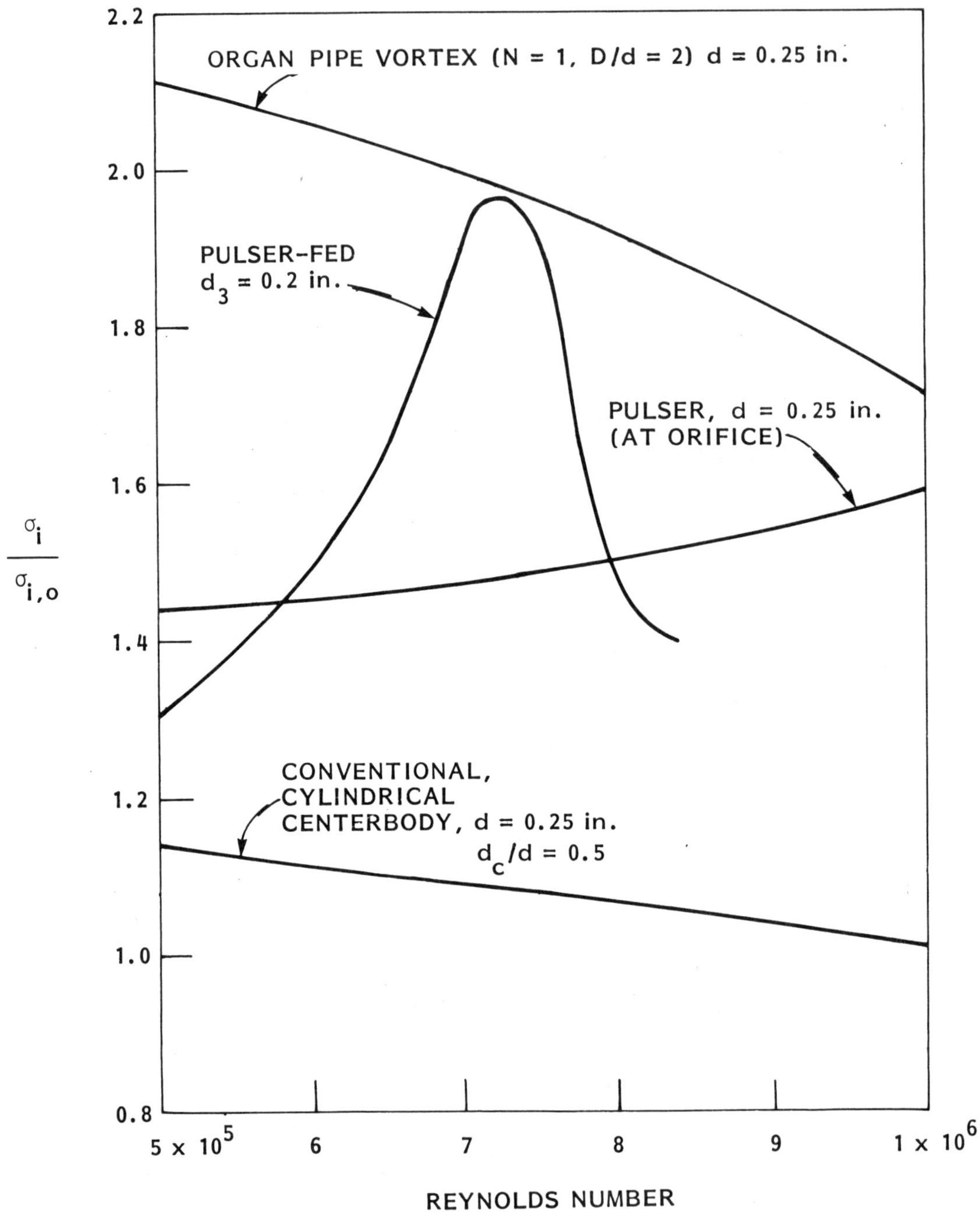

FIGURE 14 - EXPERIMENTAL INCIPIENT CAVITATION NUMBERS FOR VARIOUS NOZZLE TYPES (Relative to conventional CAVIJET®) $\sigma_{i,o} = \sigma_i$ (plain CAVIJET®)

ROCK: Indiana limestone, STANDOFF: X/d = 3

NOZZLE PRESSURE: Δp = 10.3 MPa (1,500 psi)

AMBIENT PRESSURE: p_a = 0.83 MPa (120 psi)

CAVITATION NUMBER: σ = 0.08

TEST MODE: Single pass in HPC

TRANSLATION VELOCITY: v = 6.4 mm/s (0.25 in./s)

NOZZLE ORIFICE DIAMETER: d = 6.4 mm (0.25 in.)

MEAN SLOT DEPTHS: SMITH 4.3 mm (0.17 in.)
CAVIJET: 18.3 mm (0.72 in.)

FIGURE 15 - COMPARISON OF SLOTS CUT BY SELF-RESONATING CAVITATING JET AND CONVENTIONAL JET

6th International Symposium on

Jet Cutting Technology

6-8, April, 1982

THE ENHANCEMENT OF CAVITATION DAMAGE AND ITS USE IN ROCK DISINTEGRATION

M. Mazurkiewicz and D. A. Summers

University of Missouri-Rolla, U.S.A.

Summary

A Lichtarowicz Cell is modified to allow travel of the rock sample under the cavitating jet. The effect of change in angle of impact and traverse velocity of the sample are determined. The impact of two converging cavitating jets on the target surface, and the effect of rapidly pulsating the chamber pressure, are both shown to be highly effective methods of increasing cavitation damage.

The usefulness of cavitation in disaggregating mineral bearing rock into its constituent grains, with an enhanced ease of cleaning is then discussed.

Held at the University of Surrey, U.K.
Symposium organised and sponsored by
BHRA Fluid Engineering

1. INTRODUCTION

The increasing societal demand for metals, taken with the limited available resources, will lead to the necessity of mining leaner and leaner deposits of ore in order to satisfy this burgeoning market. In the extraction of the valuable mineral from the ore body itself, it is frequently necessary, under current methods of rock separation, to pulverize the rock to a flour-like consistency, in order to separate the valuable mineral from the surrounding host rock. However, in many instances if the sample is examined before it is crushed, the valuable mineral can be seen as small discrete particles located within the host rock matrix.

It has been shown by investigators in the rock mechanics field (Ref. 1) that in many rocks the weakest phase of the rock lies at the interface between adjacent particles where, in most instances, small cracks already exist. It would therefore seem advantageous, were it possible, to grow these cracks to the point that the individual particles of rock were liberated from the mass. Under such circumstances the valuable mineral would be separated from the host rock and could be concentrated without the need to pulverize the entire volume of material. It was with that objective that this project was undertaken.

2. BASIC CONCEPT

It has been shown by work carried out at the University of Cambridge (Ref. 2) that where cavitation bubbles attack an established surface, they will preferentially collapse at pre-existing cracks on the rock lying within the cavitation zone. One thus has a method, were it possible to adapt it to current technology, of preferentially growing cracks in a material.

It was the initial objective of this program to see whether, in fact, by using cavitating water jet flows, it would be possible to generate this preferential crack growth so that mineral particles would be liberated from the solid. Further to determine if these discrete particles could, simply, be collected to separate the valuable ore from the virgin rock.

The project was undertaken in two phases, the first of which was to establish the veracity of the concept and as subsidiary part of this experiment to see if by parameterization of the fluid properties, it would be possible to improve the cavitation event and learn more about those parameters controlling the process to be undertaken. The second objective would then be to determine whether, once the concept was proven, it was either economic or practically viable.

3. EQUIPMENT DESIGN

The first step in the process was to develop a testing cell to determine whether cavitation could be made to work in the hoped for manner. Because the ultimate objective of the process was a procedure which could be fitted into a production cycle where material would be essentially flowing through a cavitation cell, it was determined that the test cell should allow translation of the sample under the cavitating jet nozzle. Existing cavitation cells built to test metals (Ref. 3) do not normally have this facility, nor do the cells constructed at the University of Missouri-Rolla, based on the design developed by Dr. Lichtarowicz of Nottingham University (Ref. 4). It was therefore decided to modify the cell design further from its original concept to allow translation of the sample.

The cell ultimately constructed (Fig. 1) sensibly comprises a small nozzle holder located approximately in the center of the upper surface of a hollow rectangular chamber. Within this chamber carved from a single piece of aluminum, a small brass carriage is located on the lower interior surface. The brass material was chosen so as to allow free sliding of the carriage across the floor of the chamber. A drive shaft was connected to this carriage and fed through the wall of the chamber and out to a sliding block arrangement in turn driven by a screw feed through a speed reduction gearing from the variable dc motor (Fig. 2). Viewing ports were located within the cell at the front and back surfaces, so that the jet could be illuminated and observed during a test (Fig. 3). Pressures in the cell were monitored by a small pressure gage which was attached to one port of the cell and two small holes were drilled in the bottom of the cell allowing fluid to bleed off from the chamber. These bleed-off holes fed initially into a chamber. In the chamber fluid flowed over a screen, which collected the particles from the comminution process, and the fluid then passed out, through a valve, to a dump circuit. By proper adjustment of this valve, it was possible to control the pressure in the cavitation chamber. Fluid flow to the cell was supplied by a 150 hp Flow Industries intensifier*.

4. INTRODUCTORY EXPERIMENTATION

The first experiments were carried out using relatively homogeneous samples of a local dolomite. One of the major advantages of using dolomite was that we have a copious amount of it available. Since the surrounding 100 sq km of Missouri consists of this rock, we are unlikely to run out of samples. Samples were prepared 2.5 cm x 5.0 x 2.5 cm in size and were weighed before and after testing.

The first experiment was to determine the speed at which the traverse under the nozzle would be most effective. The initial test conditions were at a jet pressure of 50 MPa, while the chamber pressure was kept at a level of .5 MPa, giving an approximate cavitation number of .01. Traverse velocities were set by assigning a set time for the jet to traverse the 5 cm sample length, and, typically, were in the range from 1 to 10 minutes.

An interesting phenomená was discerned during this period, seen by examining sequential photographs as the time of traverse is increased from 1 minute to 5 minutes. The growth of cracks in the sample surfaces under cavitation attack can clearly be seen. At the 1 minute traverse very small pitting of the surface is discerned (Fig. 4). At slower traverse speeds these pits coalesce into cracks and by the time a 5 minute traverse time has been tested, the cracks had coalesced to the point where fragments of dolomite are removed from the target surface (Fig. 5). This, in and of itself, seemed to indicate that the basic concept was viable.

However, to verify the concept in more detail, it was necessary to use a mineral ore.

5. ORE TESTING

Two ores were chosen for this particular demonstration, both of which are relatively local, and both of which contain lead. In one case the sample chosen was a sandstone in which a clearly discernible particle size difference between the sand and the galena. Samples were placed under the nozzle in a normal manner and tests run as previously. The results of the test clearly indicated, as can be seen by the figure (Fig. 6) that a size separation could be made between the galena and the sand, thereby allowing a separation of the lead from the sand without any great difficulty.

In the second case, however, a dolomite containing lead was used as the test sample. In this particular case the particle sizes of the dolomite lay very close to that of the galena (Fig. 7) and it did not prove possible to segregate the dolomite from the lead by simple sieving techniques. However, examination of the particles indicated that, below a 2.36 mm mesh screen, the particles of galena were clearly liberated from the dolomite (Fig. 8). It would therefore be relatively simple to separate the valuable material by a float and sink analysis, and thereby concentrate the ore without the necessity for complete material comminution, which would otherwise be the case.

The first objective of the program had therefore been undertaken, mainly to verify that cavitating jets are able to, as it were, disaggregate the rock into its component parts, such that a mineral concentration can occur without the need to totally comminute the rock. The current plan is to obtain samples of other ore in the near future and by examining a suite of different materials determine whether in fact this procedure is a viable one over a wide range of minerals.

---

*Commerically available equipment is named purely for identification purposes and does not constitute an endorsement of any product either by the University of Missouri or the Office of Surface Mining, or the Federal Government.

6. ENHANCED CAVITATION DAMAGE

During the course of this program it was recognized that the current pace of rock comminution is extremely slow. While it was possible to totally disintegrate the samples of galena in sandstone within the 5 minute test period, nevertheless this level of production is considerably below that of existing mining equipment and to be viable it must be recognized that the process needs to be optimized.

One early attempt to achieve this was to raise the pressure of the water jet, because this will intensify the pressures generated in the cavitation bubble collapse. In the event it is not considered that this was a completely successful move because the combined high pressure in the water jet and the intensified collapse pressure of the cavitation bubble led not only to a much more rapid erosion but also created a much more aggressive water jet which cut through the particles indiscriminately rather than, as was hoped, spreading out over the surface of the sample and selectively cutting into the crystal boundaries.

The next procedure therefore was to look at using two converging cavitation jets on the target surface. This has admittedly only been done in a somewhat superificial way at the present time. However, this procedure was carried out to compare the effect of a dual cavitating jet with a single orifice jet, the dual orifice jets were sized so that their combined area equated to that of the single orifice jets used in the earlier experiments, and thus flow volume and power should be approximately equivalent.

The result of this test indicates that at 50 MPa with a .5 MPa back pressure, improved cavitation results could be achieved. However, the results were not nearly as promising where the tests were carried out at either 62 MPa or 34.5 MPa. The particular nozzle design shown was at a convergent angle of 80 degrees, where smaller angles of convergence, 40 deg and 20 deg were used, the results obtained were simi-laraly much less dramatic (Fig. 9).

It must be recognized, however, that we are, as yet, children groping in the dark, in regards to the application and parameters controlling cavitation damage. At the present time all that one can report in this regard, is that under certain, perhaps, fortuitous circumstances, we were able to substantially improve cavitation erosion where under one specific condition a dual orifice converging cavitating jet was used rather than a single jet.

In similar manner a rotating disc was placed in the discharge line from the cell. By rotating this disc it was possible to open and partially close the flow line from the cell, thereby pulsating the chamber pressure. This was tried as a short experiment and again the results (Fig. 10) show that the technique is a highly effective one for improving cavitation damage. However, again as yet no parameters have been applied to this work to determine how best to make use of the phenomenon discovered.

7. CONCLUDING REMARKS

At the present time this study has really only just begun. However, we have clearly demonstrated that cavitation can be used to disaggregate rock and thereby potentially have a significant effect on the cost of comminuting rock and thereafter concentrating the valuable mineral without going to the very high intensive energy method of pulverization. Results from our tests have shown that it is possible to substantially improve cavitation damage over that which has been achieved to date, but there is as yet too little known about the parameters controlling cavitation within the test cell for us to be able to predict whether in fact the method will be ultimately economically viable. Such an evaluation we would hope will be made based on the studies that we will undertake during the course of the next year.

8. ACKNOWLEDGEMENTS

This work was funded under the Mining and Minerals Research and Resources Institute Program of the Office of Surface Mining of the Department of Interior. We are pleased to acknowledge this support and that of our students, Mr. R. Brandom and Mr. T. Fort III, and staff Mr. B. Hale and Mr. J. Tyler, who made this report possible.

9. REFERENCES

1. Brace, W.F., "Brittle Fracture of Rocks," State of Stress in the Earth's Crust, Elsevier, New York, 1964, pp 111-180.

2. Brunton, J.H., "Cavitation Damage," Paper 7.1 in Proc. 3rd Int. Conf. on Rain Erosion and Associated Phenomena, ed. A.A. Fyall and R.B. King, RAE, Farnborough, August 1970, pp 821-845.

3. Standard Method of Vibratory Cavitation Erosion Test, ASTM, Part 10, Designation G32-72, 1975, pp 724-728.

4. Lichtarowicz, A., "Cavitation Jet Apparatus for Cavitation Erosion Testing," in Erosion - Prevention and Useful Applications, W.F. Adler, ed., 1977, pp 530-549, ASTM STP 664.

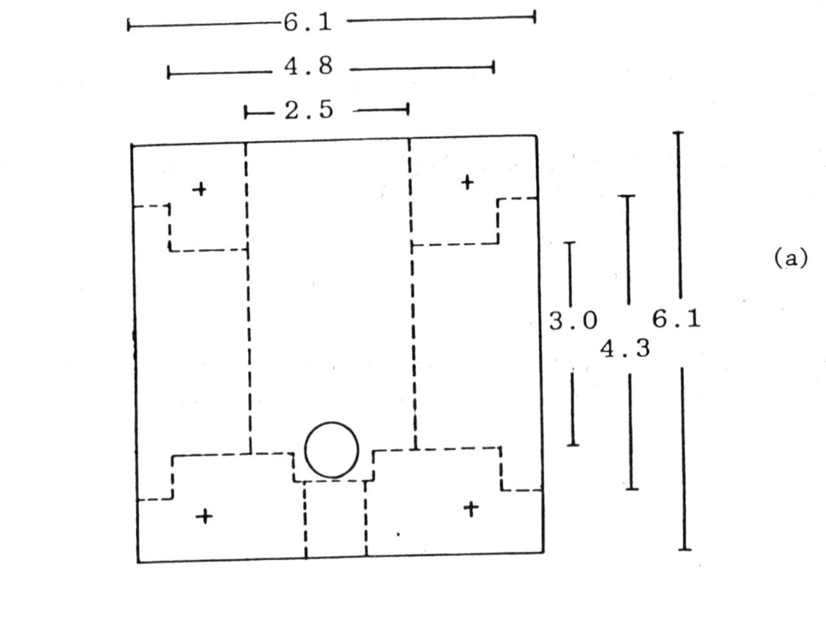

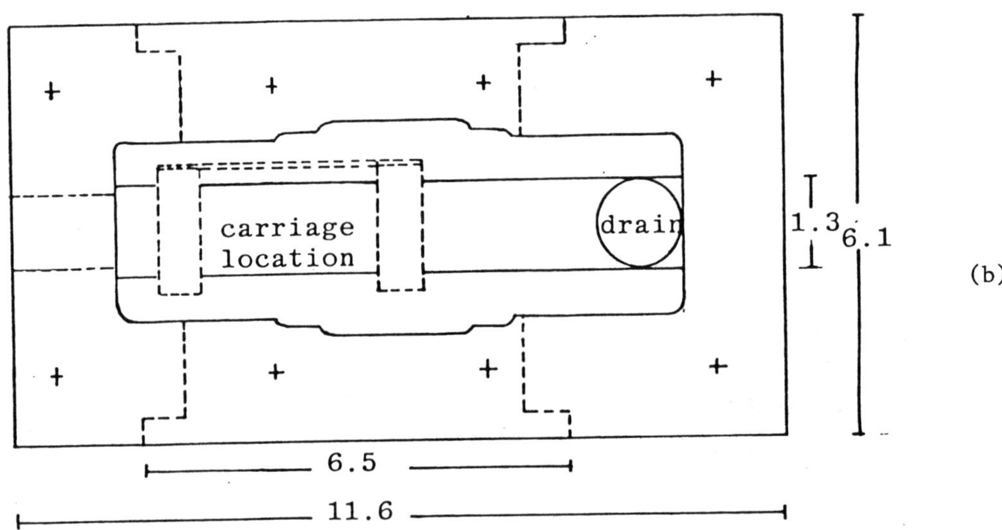

Figure 1. Half scale view from (a) side and (b) top of the new cavitation cell. (Dimensions in cm.) The + marks denote bolt locations.

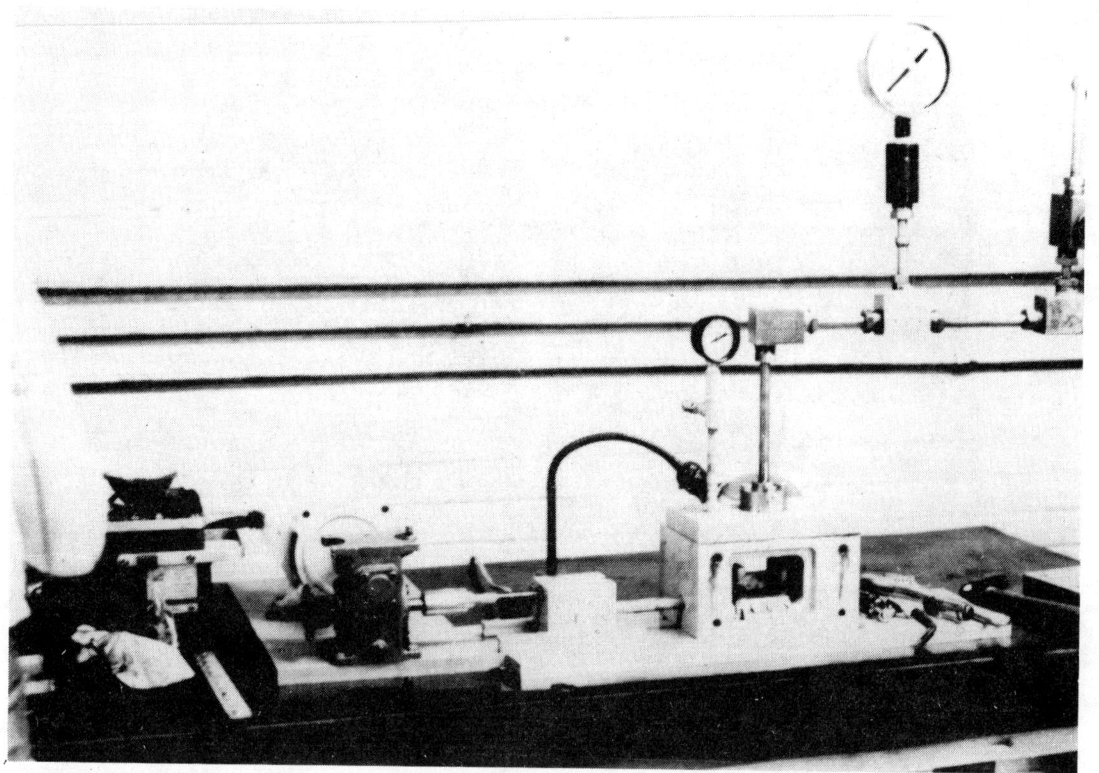

Figure 2. Overall view of test rig showing drive mechanism.

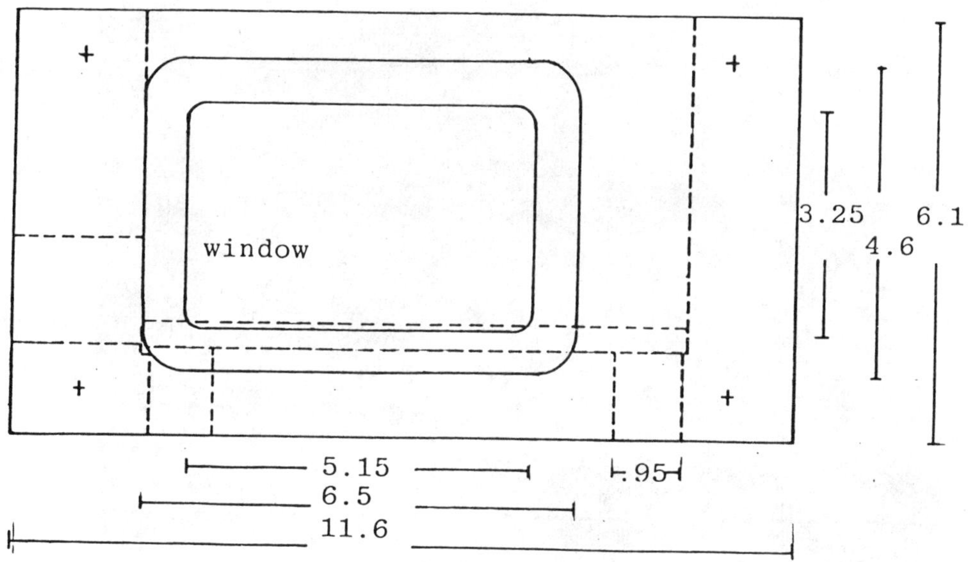

Figure 3. Half scale drawing of the side view of the chamber showing the location of the viewing ports.

Figure 4. "Fast" traverse showing initial pitting of the dolomite surface.

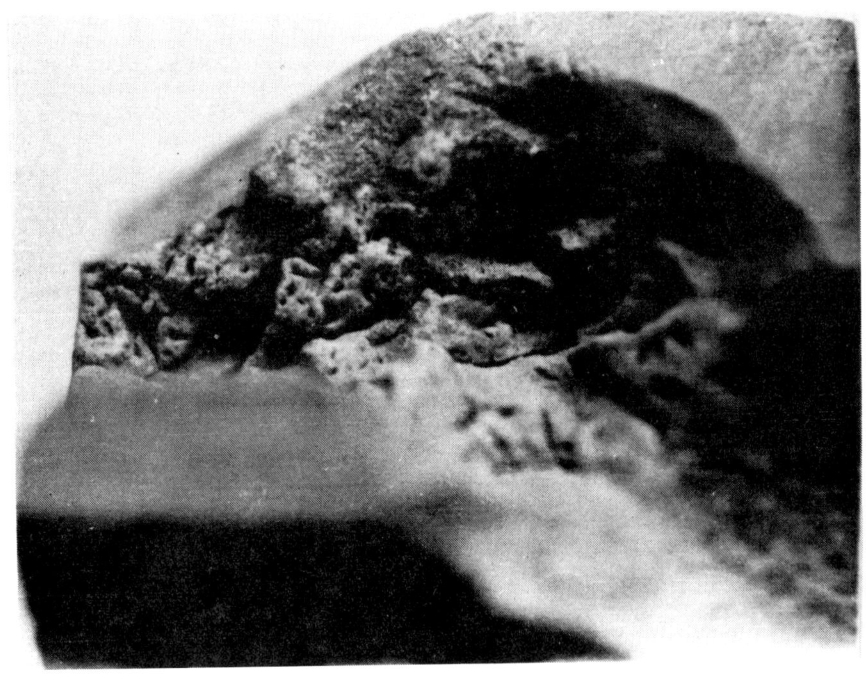

Figure 5. "Slow" traverse (5 min) showing the development of cracks and large scale material removal.

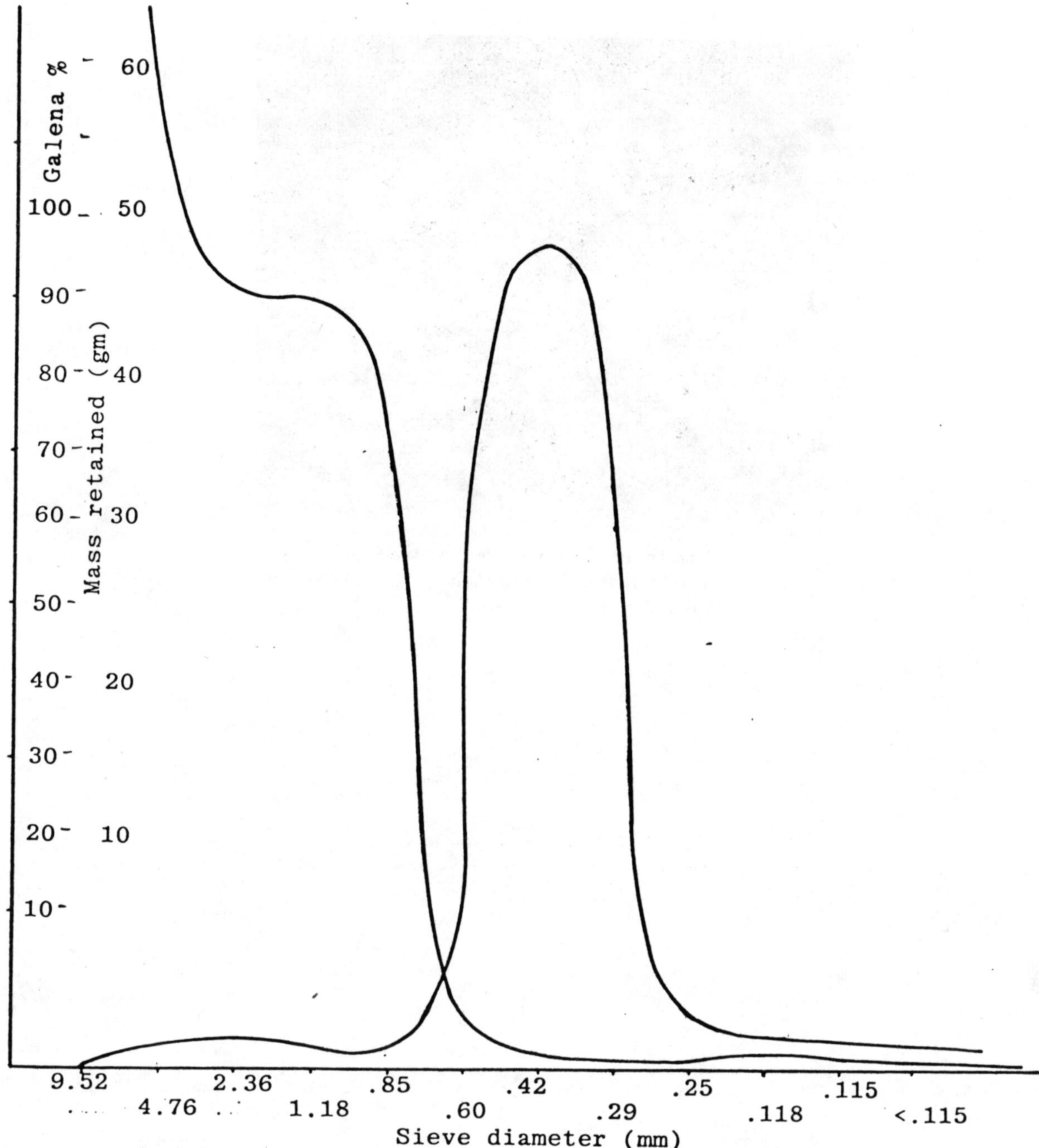

Figure 6. Galena distribution and sand particle size distribution for a sandstone ore after disaggregation.

Figure 7. Sample of galena ore before test. The dark material is galena, the light is dolomite.

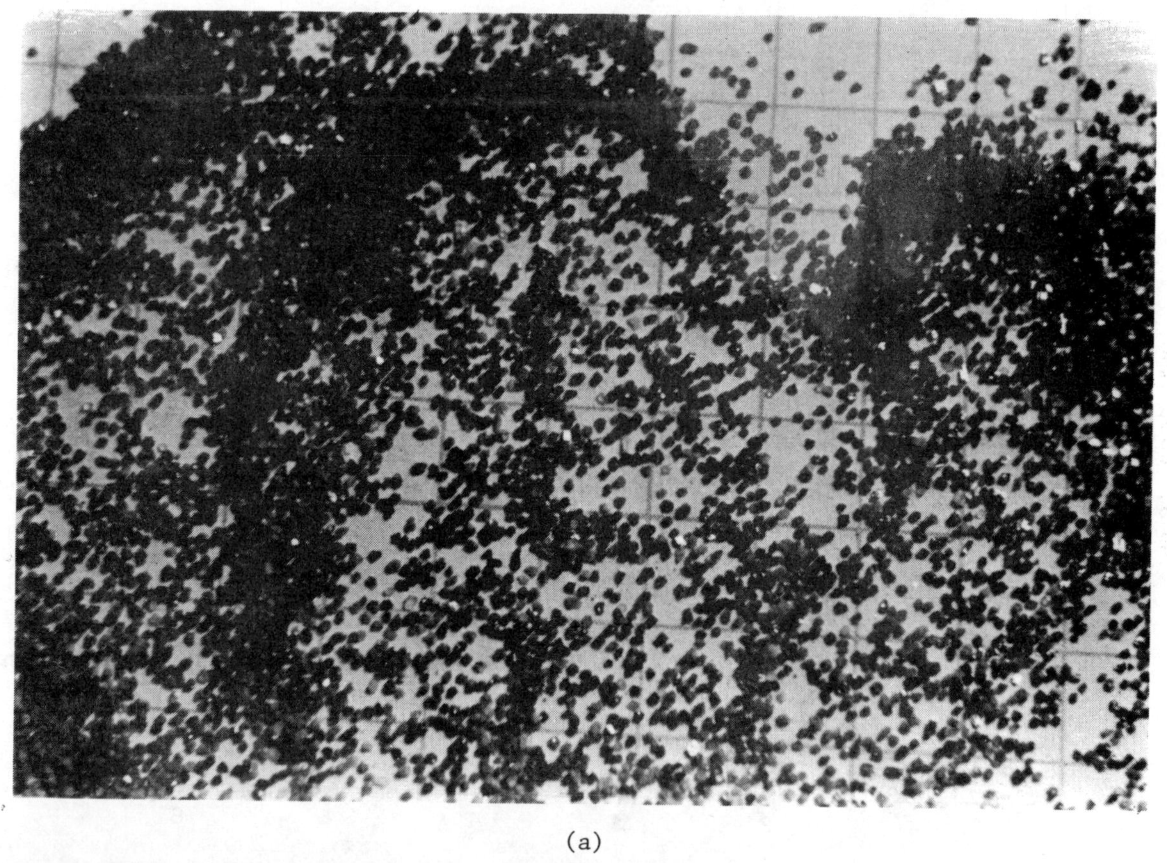

(a)

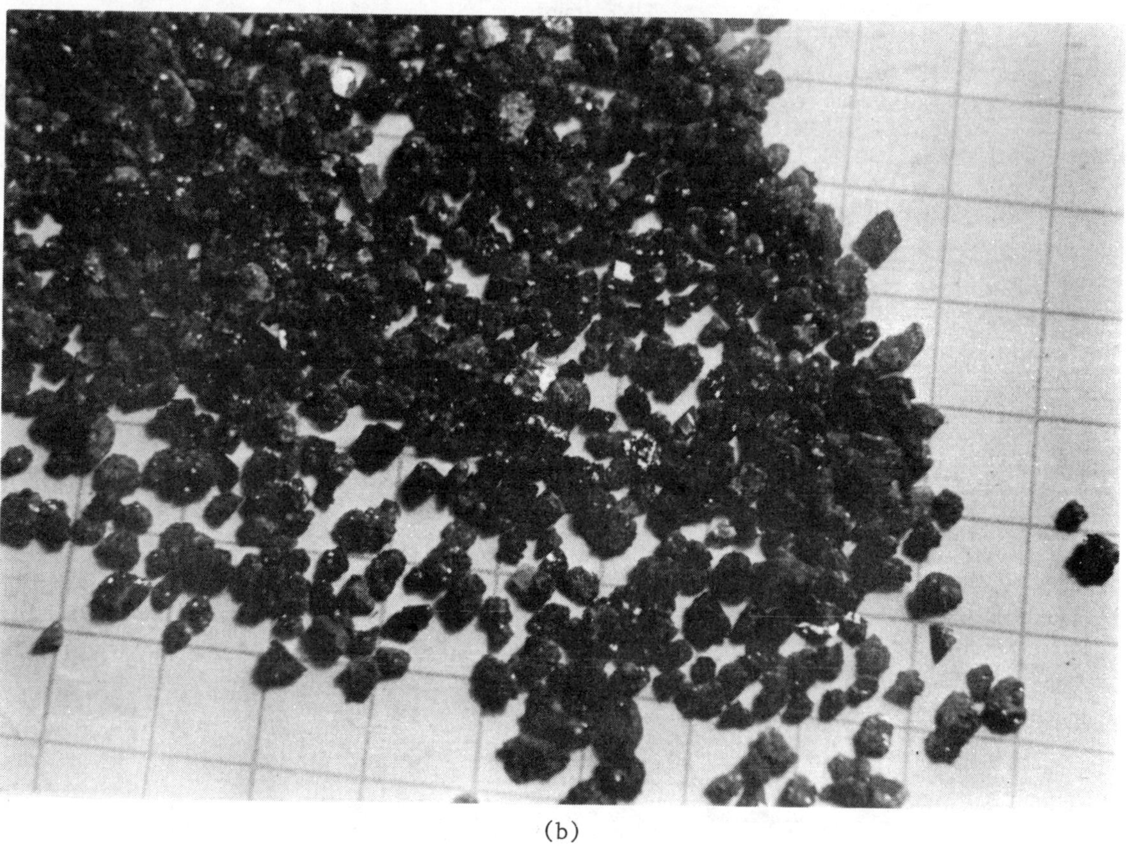

(b)

Figure 8. Showing separation of dolomite and galena grains (a) and (b).

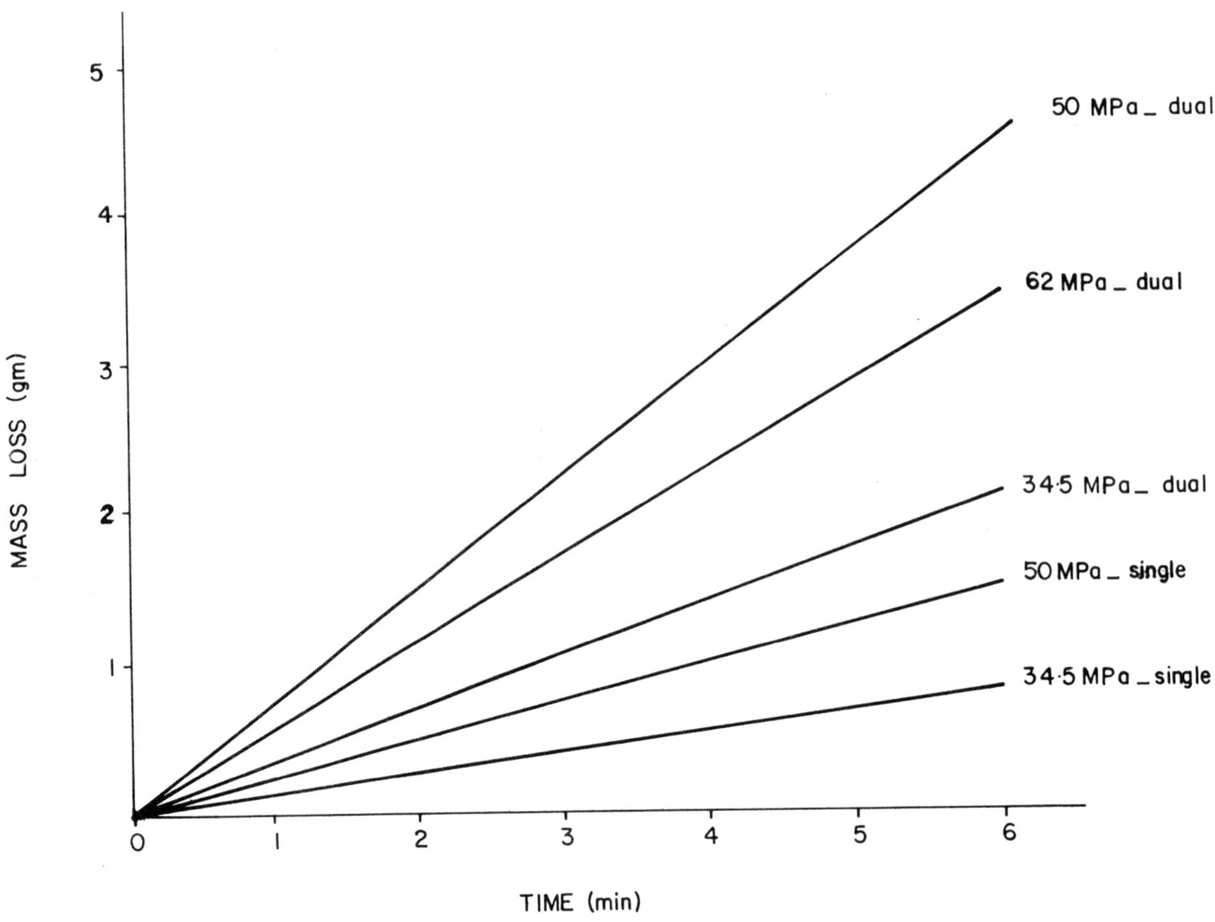

Figure 9. Relative mass loss for single and dual orifice converging nozzles, at 0.01 cavitation number and as a function of sample traverse time.

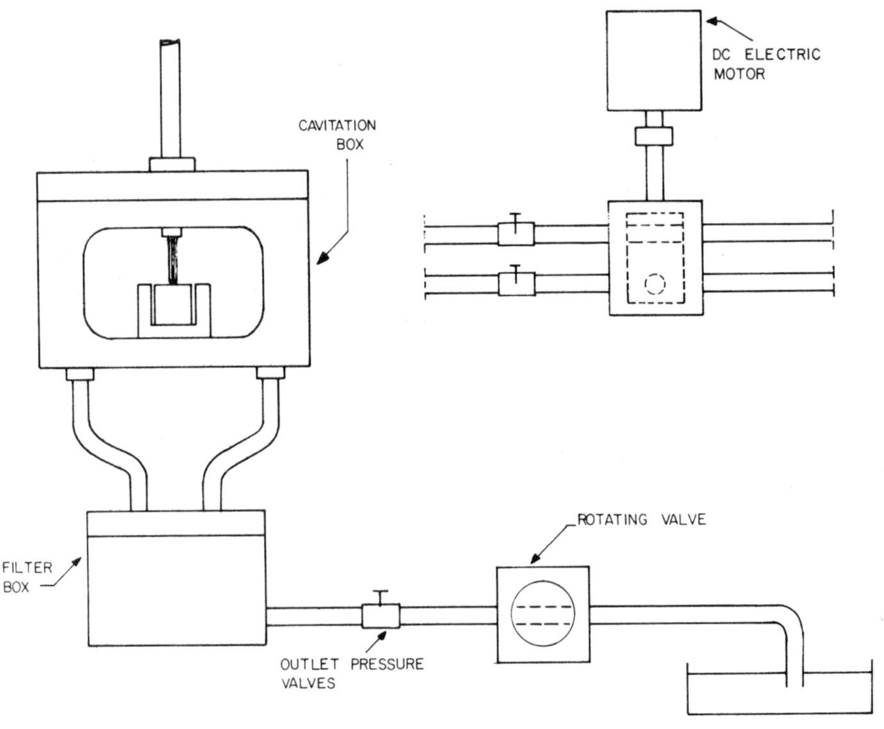

Figure 10. Layout for causing cavitation number variation.

6th International Symposium on
Jet Cutting Technology
6-8, April, 1982

ENDOSCOPIC JET-CUTTING
A NEW METHOD FOR STONE DESTRUCTION IN THE COMMON BILE DUCT

K. Jessen, J. Philipp and M. Classen
University of Frankfurt, F.R. Germany
W. Schikorr and H. Louis
University of Hannover, F.R. Germany

Summary

We present our preliminary in-vitro data using a jet-cutting system for destruction of human gallstones.

In our experimental test device we use a high pressure pump system. High pressure liquid jets are pressed through a flexible endoscopic teflon probe on human gallstones.

The concrements are fixed by a basket attached at the tip of the jet-probe.

The stones can be split in fragments by liquid jets.

This new method could become a successful nonoperative therapeutical procedure in high risk patients suffering from stones in the common bile duct.

Held at the University of Surrey, U.K.
Symposium organised and sponsored by
BHRA Fluid Engineering

Introduction

Gallstones are a common health problem in western countries. The incidence ranges between 15-20 percent and increases with age. In the USA 1 million new cases occur each year (Ref. 1). A recent Lancet editorial claimed that 1 percent of deaths in England are attributable to gallstone disease (Ref. 2).
In West-Germany about 5 million people have stones in the biliary tract (Ref. 3).
These data reflect the enormous medical and economical importance of this disease.
Surgery had been the therapy of choice up to 1973. Since then stones in the common bile duct can be removed by endoscopic papillotomy (Ref. 4). Two thirds of these concrements pass spontaneously into the duodenum. In 20 percent the calculi can be grasped and extracted with the aid of a basket. But in nearly 15 percent stone extraction is not possible (Ref. 5). For these cases other methods ie. endoscopic electrohydraulic lithotripsy or chemical stone dissolution have been performed (Ref. 5,6). Complications led our interest to new techniques with less serious side-effects.
In the last years technical experience was achieved using jet-cutting technology (Ref. 7+8). The excellent results in mining draw our attention to a practicable application of the jet-cutting technique to medical therapy, especially to endoscopic destruction of gallstones.

Material and Methods

1 Technical arrangement:

For generating a fluid-pressure up to 1000 bar the "Institut für Werkstoffkunde", University of Hannover, has a three-plunger-pump available. (WOMA-Atümat)
This system was used for the first experiments involving the cracking of gallstones. The fundamental possibility of realisation was checked. some observations were made to determine the parameters for the construction of a very small pump, perhaps an intensifier, for further in-vitro gallstone experiments and series in cadavers and in animals. The arrangement of the experiments is described: Fig. 1 shows a high pressure magnet-valve integrated in the tube from the pump to a nozzle of 1.8 mm diameter.Immediately behind the magnet-valve an adapter is fitted. Installed on the adapter a pressure gauge and a nozzle-pipe (300 mm long) joining the 0.2 mm nozzle and the modified Dormia basket (fig.2), necessary for fixing the stone just before the orifice of the nozzle.
A screen of PMMA (Plexiglas) prevents the lost of split parts of stones during or after the pressure pulse.
The 1.8 mm load-nozzle is necessary because the magnet-valve's velocity of closing depends on the flow-rate through the valve. The flow-rate through a 0.2 mm nozzle is too small for a fast closing of the valve.
Some experiments have shown that a nozzle diameter of 0.2 mm is necessary. The use of a 0.1 mm nozzle showed that it was not able to crack most of the gallstones. To minimize the flow-rate, the

nozzle-diameter should be made as small as possible.

2 Test-realisation:

The water-pressure is adjusted at the pump while the magnet-valve is opened, that means water is streaming through the 1.8 mm load-nozzle and the 0.2 mm main-nozzle.
The pressure is measured by using a pressure transducer, an instrumentation bridge amplifier with digital display (Hellborg) and a digital storage oscilloscope (Gould). The magnet-valve is opened and closed by a manget-valve driver (Hellborg) with variable programme. Using the storage oscilloscope the height and the duration of the pressure pulse near the nozzle can be controlled and adjusted on the programme-chart of the magnet-valve driver.
The duration of the pressure pulse was adjusted to 0.25 s. Before triggering a "shot" onto a stone, the system is triggered idle to be sure the pressure and the duration of pressure will not be different from the adjusted values.
The experiments were made with pressures ranging from 100 to 250 bar. The working pressure of 150 bar used normally was experienced in pre-experiments.
If a stone resists a pressure pulse of 150 bar and 0.25 s, a second pulse (up to ten pulses) follows.
If ten pulses show no result too, the stone is loaded by continuous pressure to estimate its resistance against the jet, and the time of destruction is measured.
Then a second stone is loaded in the same way with a higher pressure until the first pulse will be successfull.
Table 1 gives a view about the fluid-volume streaming through a nozzle of 0.2 mm diameter during a 0.25 s pulse. A nozzle efficiency of 1 is assumed for this calculation. The fluid volume is of importance for a later use of this method in the human body because the bile duct is not able to contain more than a few cm^3 of fluid.

3 Investigation of some gallstones using metallographic methods and scanning electron microscope (figs 13 & 14):

In General there are two different kinds of gallstones, pigment stones and cholesterol stones.

They differ in their response to water-jet loading.

While the dark coloured pigment stone may be split by a jet of 150 bar, the light coloured cholesterol stone demands a pressure of more than 200 bar.

To find out the reason for the different properties, two gall stones of ca. 10 mm diameter composed of different amounts of cholesterol, bile pigments, calcium, and bile acids were embedded in plastics and graphs of cross-sections were produced. The graphs (fig.3 and fig.4) show that there are already cracks in the unloaded stones.

The dark stone shows more and larger crack-areas than the light coloured stone. Both stones show a skin, but the light coloured stone a much stronger one.

The next step was an investigation of a light-coloured stone being split by a water jet of about 200 bar using a scanning electron microscope.

Fig. 5 shows the border between skin and the inner surface of a light coloured stone.

A crack is shown in fig.6 and in higher magnification in fig.7. Significant are the crystals on both sides of the cracks. Their direction lead to the conclusion that the water-pressure has opened or widened the crack.

The comparison of fig. 8 and fig. 9 shows that there are differences in the structure of skin and inner stone.

While the structure of the skin is very tiny grained, the structure of the inner stone shows coarse lamellas.

It is clear that if a jet has penetrated the skin, the interior of the stone will cleave at once because of the pressure build-up by the jet.

[Table 2: Experimental data using high pressure liquid jets for gallstone destruction (three-plunger-pump, Woma-Atumat).]

4 Endoscopic procedure:

A routine flexible side-viewing endoscope (fig.10) is introduced into the duodenum of a cadaver or animal.
When the papilla of Vater is visualized (region where pancreatic and biliary duct system is open into the duodenum), the papilla is cannulated first with a small catheter. Through the probe dye radiopaque is introduced into the biliary tree. The cholangiogramm demonstrates the installed gallstones (fig. 11 and 12).
The flexible jet-cutting probe combined with a Dormia basket at the tip, can pass the instrumental channel of the endoscope and reach the common bile duct.
The gallstone is grasped and fixed just before the orifice of the nozzle. The duration of the pressure pulse and the jet pressure are adjusted before triggering a "shot" onto the stone. When the pressure pulse splits the stone, fragments can easily be removed out of the duct. If a stone resists a single pressure pulse another "shot" with higher pressure can be performed without risk.

Conclusions

Our experimental studies have shown that the test device, high pressure-pump-system and flexible jet-cutting probe combined with a Dormia basket, permits the destruction of gallstones.
Further experiments in cadavers and animals are planned to confirm our preliminary results and to calculate possible risks of damaging the tissues of the biliary tract.
The future tests will prove if this new method will be available in human patients.

References

1. Schoenfield, L.J.:
"Diseases of the gallbladder and biliary system."
New York, London, Sydney, Toronto, John Wiley and Sons, 1977, p. 109

2. Editorial, Lancet, 1, p. 1061, 1976

3. Peters, H.:
"Epidemiologie der Gallenwegserkrankungen."
Internist 21, p. 559, 1980

4. Classen, M. and Demling, L.:
"Endoskopische Shinkterotomie der Papilla Vateri und Steinextraktion aus dem Ductus choledochus."
Dtsch. Med. Wochenschr. 99, p.496, 1974

5. Koch, H., Rösch, W., Walz, V.:
"Endoscopic lithotripsy in the common bile duct."
Gastrointestinal Endoscopy, 26, 1, p. 16-18, 1980

6. Schenk, J., Schnack, B., Riemann, J.F., Rösch, W.:
"Treatment of choledocholithiasis using the transpapillary perfusion technique."
Endoscopy 12, p. 224, 1980

7. Beutin, E.F., Erdmann-Jesnitzer, F., Louis, H.:
"Material behaviour in the case of high-speed liquid jet attacks."
Second international symposium on jet cutting technology, Cambridge, 1974.

8. Erdmann-Jesnitzer,F.,Louis,H.,Schikorr,W. : Cleaning , Drilling , and Cutting by Interrupted Jets .
Fifth international symposium on jet cutting technology .
Hannover ,June 1980.

Table 1 :

Relation between pressure and calculated fluid volume per 0.25S through a 0.2 mm nozzle (loss-free)

Pressure in (bar)	Volume (ml)	jet-velocity (m/s)
100	1.10	140
150	1.34	171
200	1.55	198
250	1.74	221
300	1.90	242

Table 2 :

Experimental data using high pressure liquid jets for gallstone destruction

No of stone (fig. 13 and 14)	Diameter (mm)	Weight (g)	Liquid volume (ml)	Water pressure (bar)	No. of Pulses	Duration of Impulse (sec)	Result
1	22,4	3,889	5x1,34	150	5	0,25	stone perforation no smashing
2	20,3	2,537	10x1,34	150	10	0,25	no smashing, little change on surface
3	18,6	1,805	10x1,34	150	10	0,25	minimal change on surface
4	16,4	0,959	1,34	150	1	0,25	3 fragments
5	15,3	0,841	1,34	150	1	0,25	smashed
6	12,4	0,526	1,34	150	1	0,25	smashed
7	21,9	4,721	1,34	150	1	0,25	two fragments
8	20,9	3,044	1,74	250	1	0,25	stone perforation two fragments
9	19,8	3,031	1,62	220	1	0,25	two fragments
10	18,1	2,397	2x1,34	150	2	0,25	two fragments
11	18,1	2,516	1,34	150	1	0,25	two fragments
12	13,9	0,853	5x1,34	150	5	0,25	only small lesions on surface

1
2
9
3
4
5
150
6
7
8
1 ballast - nozzle (1, 8 mm)
2 gall - stone
3 main nozzle (0, 2 mm)
4 pressure transducer
5 magnet - valve
6 digital storage oscilloscope
7 instrumentation - bridge amplifier
8 magnet - valve drive
9 dormia - basket
test assembly for experiments
in gall - stone cracking
three - plunger - pump

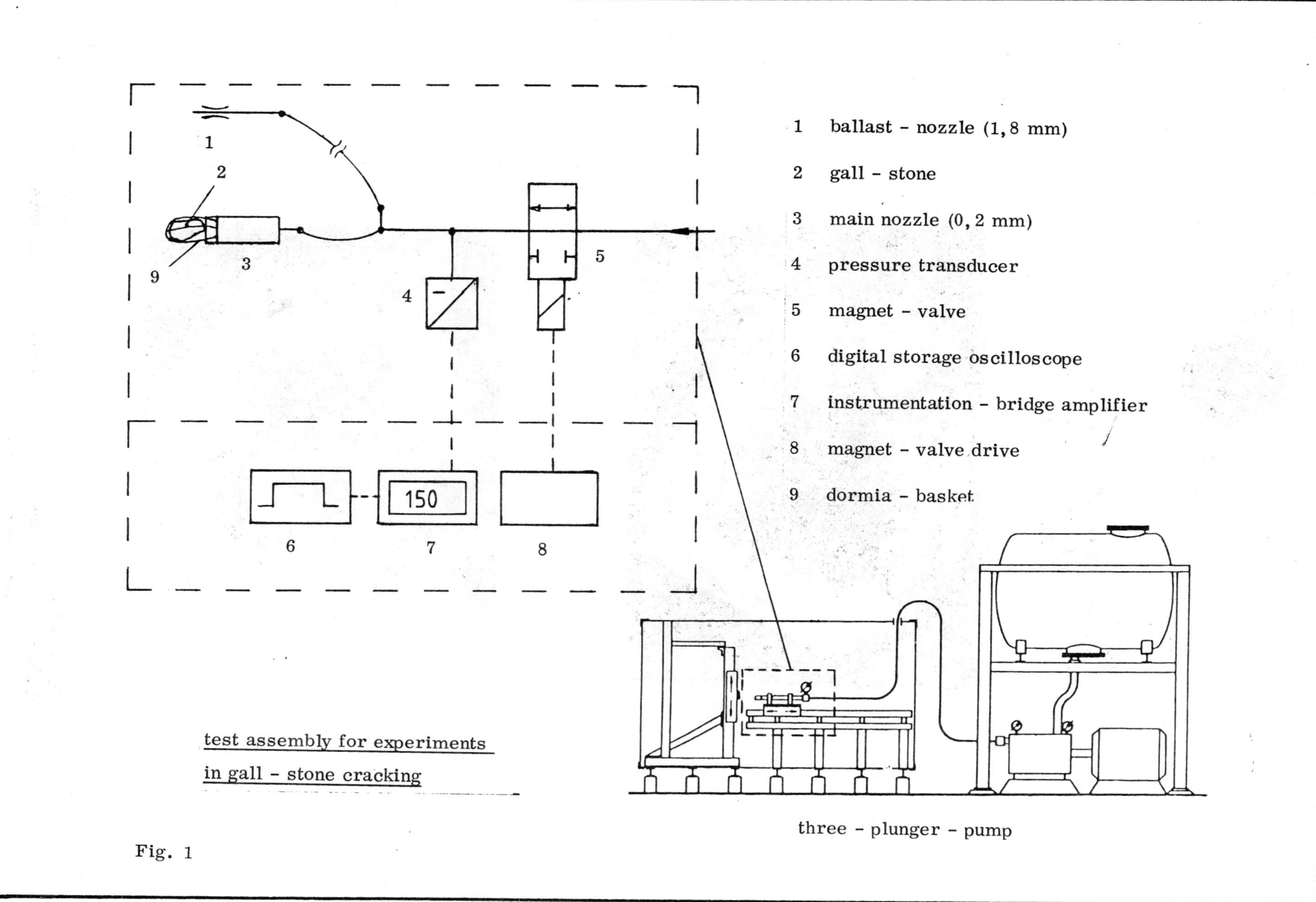

Fig. 1

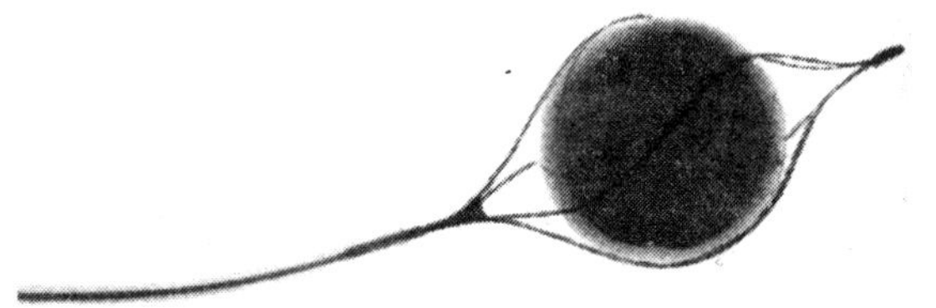

Fig. 2
Legend : Model of the jet-cutting probe with a Dormia basket at the tip. In the Dormia basket a gallstone.

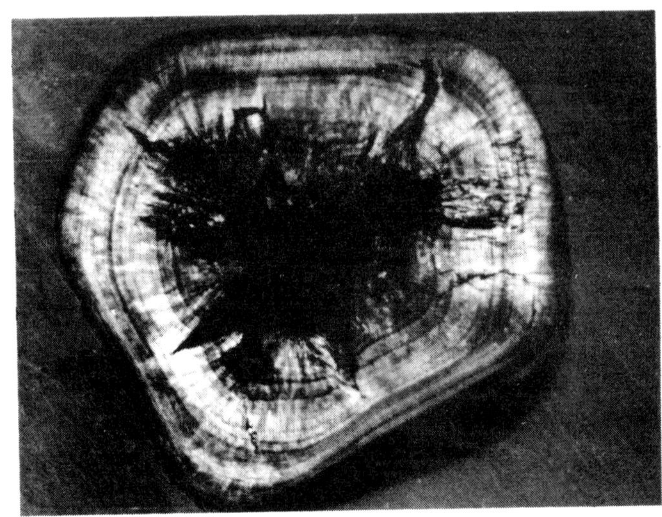

Fig. 3

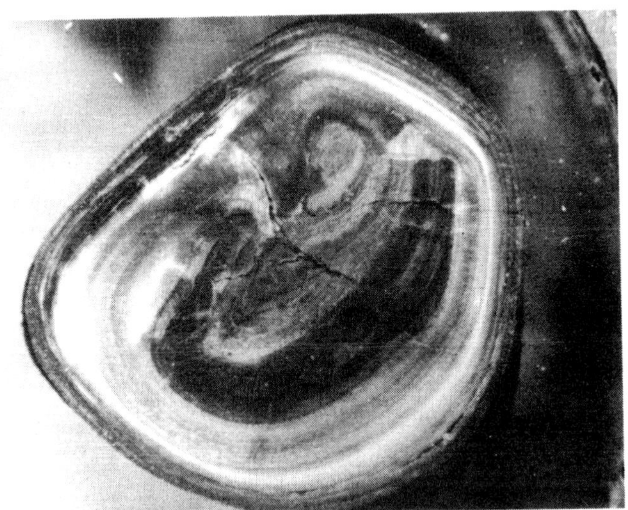

Fig. 4

Fig. 3 & Fig. 4
Legend : Two different gallstones embedded in plastics and cross-sectioned.

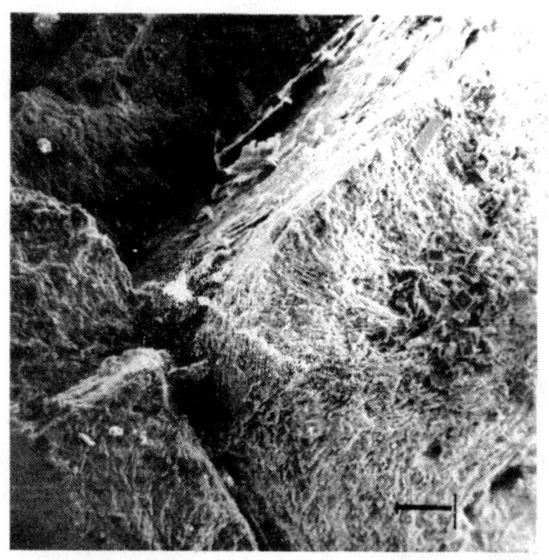

Fig. 5 Photograph : scanning electron microscope ⊣ 262
Gallstone, border between skin and inner surface of a light coloured stone,

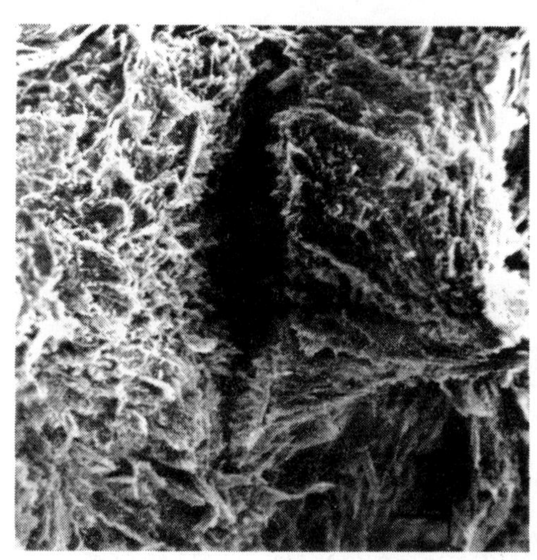

Fig. 6

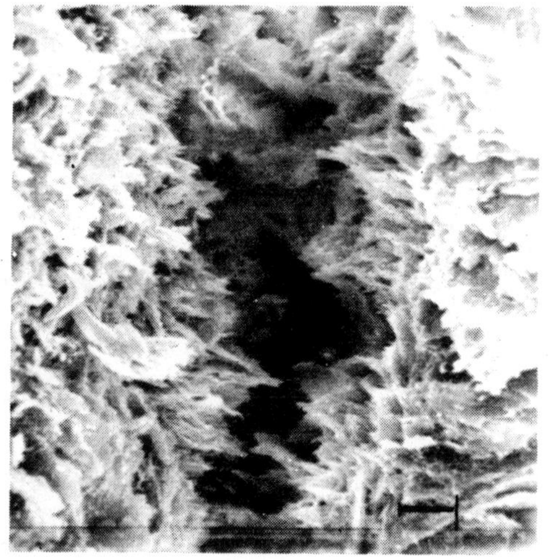

Fig. 7

Fig. 6 & Fig. 7
Legend : Scanning electron microscope ⊣ 160
Gallstone, exposed to water jets, look at the crack
(Fig. 6)
Magnification ⊣ 53
(Fig. 7)

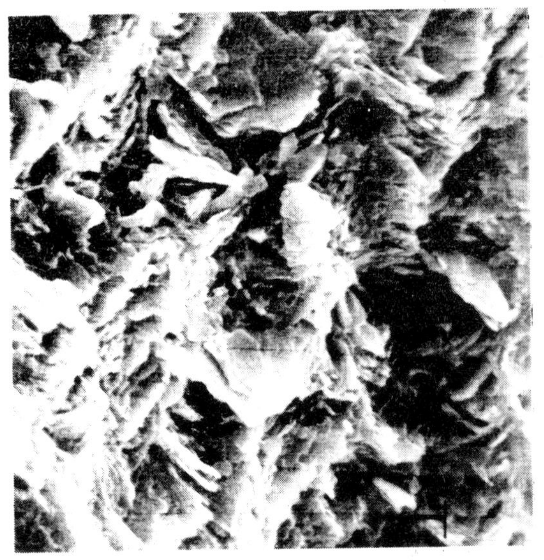

Fig. 8

Fig. 9

Fig. 8 & Fig. 9
Legend : Gallstone structure. Fig. 8 surface,
Fig. 9 inner stone
scanning electron microscope, 160

Fig. 10
Legend : Flexible side-viewing endoscope
Dormia basket in the instrumental channel.

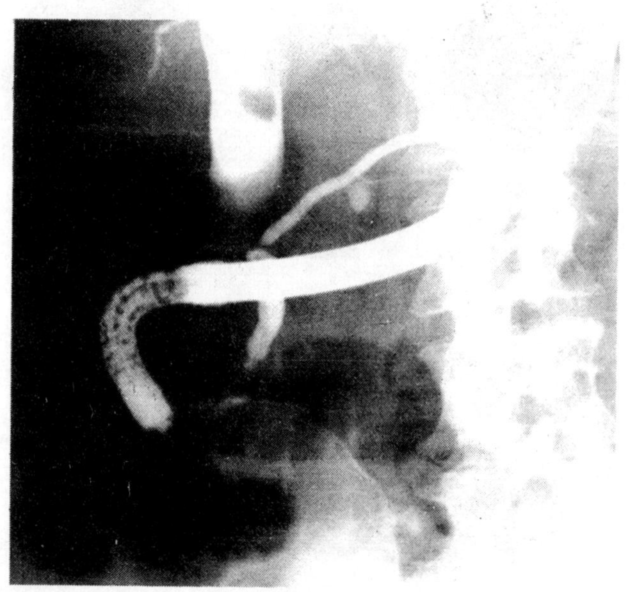

Fig. 11

Legend : Fig. 11
X-ray-photograph
Endoscope in situ, cholaniogramm
Demonstrates a gallstone in the common
bile duct.

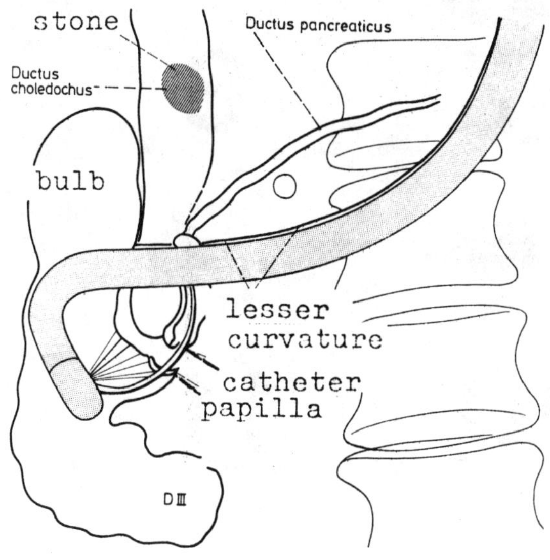

Fig. 12

Fig. 12
Diagram of the x-ray photograph

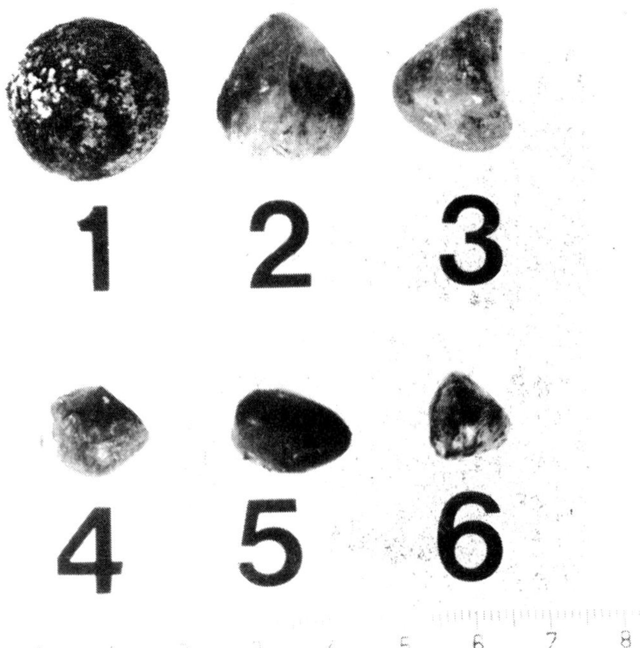

Fig. 13

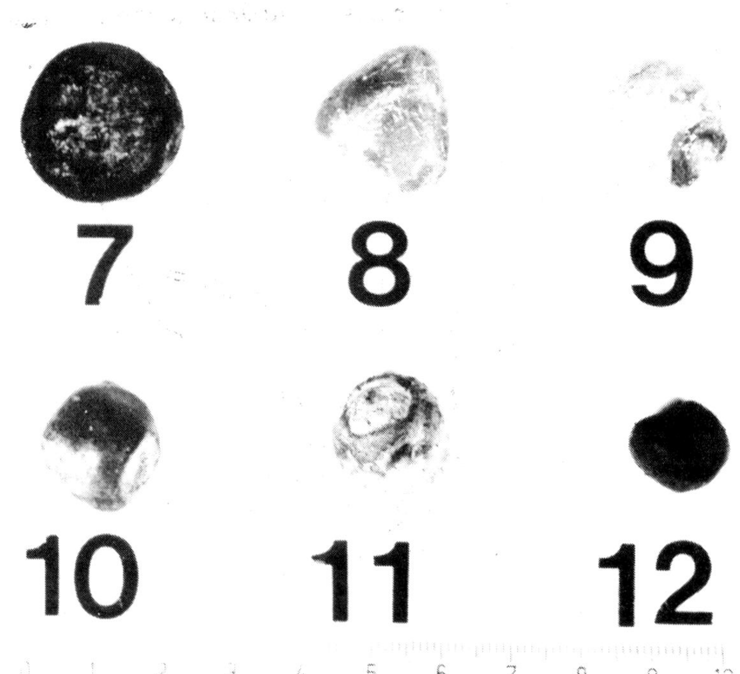

Fig. 14

Legend Fig. 13 and Fig. 14
Macroscopic photographs of different gallstone split by liquid jets.

6th International Symposium on
Jet Cutting Technology
6-8, April, 1982

SPACE-AGE JET CUTTING FOR UNDERWATER SALVAGE

J.J. Ridgeway

Aqua-Tech International Ltd., France

Summary

In this paper the limitations of present cutting methods are discussed and the need for a better method is presented. Design problems associated with explosive cutting are considered and a "cost-optimized" design is determined. Predictions are made of the role that a cost-effective cutter will play in the new world of underwater technology in support of the offshore petroleum industry.

Held at the University of Surrey, U.K.
Symposium organised and sponsored by
BHRA Fluid Engineering

NOMENCLATURE

W = core weight, gr./m

T = thickness of material cut, mils

H = brinell hardness

P = density, g./c.c.

K = a constant reflecting efficiency of LSC design and material.

1. INTRODUCTION

Little has been accomplished to improve the antiquated methods of cutting offshore structures and conductor pipe underwater, that is, until recently.

The purpose of this paper is to bring the reader up to date on new developments in the field of explosive cutting and the future that lies ahead.

2. THE NEED FOR A BETTER METHOD OF CUTTING

The amount of work which man can accomplish underwater is greatly limited, and even then is only accomplished at a premium in cost.

The time and cost involved to cut steel underwater with a conventional cutting torch or electrical arc is prohibitive, not to mention the potential hazard involved. A diver subjects himself to at least a limited amount of risk every time that he enters the water and the less time that he spends in the water to accomplish a task, reduces the risk involved.

It is natural for one not familiar with explosive technology to associate this type of energy with an element of danger and a hazard. However, if one were to compare the two methods of underwater steel cutting, the record will show that the converse is true.

The need for an economical application of explosive energy to perform the job of an underwater torch is not new. Nor are underwater explosives, however, the existance of a "cost-effective" underwater explosive cutter is.

3. DESIGN CONSIDERATIONS

In order to design a truly "cost-optimized" underwater cutter, the following design objectives were set:

(a) Design a linear shaped charge device which is legally "air-transportable" to serve the immediate needs of a customer regardless of where his offshore project might be located.

(b) Design a light weight device which will be easy to handle in the field and inexpensive to ship.

(c) Design a device which will be resistant to the corrosion of an offshore environment.

(d) Design a linear shaped charge device which is superior in performance to any existing similar explosive method.

(e) Design a device which is ultra-safe to handle; and,

(f) Design a device which is inexpensive in cost to fabricate so that the savings can be passed along to the customer.

It has only been recently that a new "two-component" field mix explosive has been developed, which is legally air transportable in its component form, has an exceptionally high detonation velocity, is safe to handle and impossible to detonate when unmixed. And when mixed, it is more resistant to shock than other commercially available explosives. It has a higher detonation velocity than nitroglycerine, pentaerythritoltetra nitrate (PETN), cyclotrimethylene trinitramine (RDX), or any other commercially available liquid or solid explosive. This binary energy system is known as "Astrolite".

Since penetration of steel by a Linear Shaped Charge (LSC) is a direct function of the detonation velocity, it is apparent that the efficiency of cutting is directly related to the explosive which is used with the LSC device.

Based on the attributes of Astrolite and the need for an inexpensive cutter, a development program was initiated to determine the optimum parameters for a high performance LSC. Said performance was evaluated according to the calculated efficiency of LSC designs and materials tested. The formula used was:

$$K = \frac{W}{T^2 \text{ (HP) } 0.6}$$

The parameters which most influenced "cut" were varied for the two most promising LSC configurations. Standoff is the air space between the target plate and the LSC, and is necessary for the shaped charge jet to form and cause "penetration". Varying standoff was used initially to determine the optimum standoff for maximum penetration. The liner angles and thicknesses were varied to determine their optimums, using various cost-effective liner materials.

Cut is made up of penetration (which is the actual displacement of the metal target due to melting) and "break" (which is the fracture beyond the point of penetration). The depth of penetration is a function of several parameters, but is varied easiest by standoff distance. At optimum standoff, penetration nearing 5 cm (a common pipe wall thickness offshore) was recorded in steel target plates with effective cuts of up to 6.5 cm.

Cost "tradeoffs" were made for all affected activities from design conception through to operations. Labor cost to fabricate, handle, install, operate and maintain the design product was the design criteria for this minimum-cost product. A low-cost, light-weight material which would not greatly reduce performance was also a goal. Several inexpensive and readily available aluminum alloys were then tested. What was lost in LSC performance due to low ductility and density of LSC material, was gained in cost savings elsewhere. Aluminum not only fulfilled the requirements for a light-weight device, but is resistant to corrosion and is very compatable with the corrosive liquid Astrolite explosive.

At the conclusion of the developmental test program, all of the foregoing listed design objectives were accomplished. Inexpensive, efficient, light-weight cutters were the result and patents were accordingly applied for and approved.

4. MOBILITY - SOMETHING NEW WITH EXPLOSIVES

Conventional explosives require special handling, licenses, escorts, special vehicles for transportation and special storage facilities. Long distance transportation of explosives is limited to surface transportation only and can be time consuming and expensive. Jobs utilizing such explosives need long-range planning and can be severly hampered if work does not progress exactly according to plan. As a result, distant work is considered a risk and as such, additional contingency costs are loaded into the job at the customer's expense.

The restrictions of conventional explosives are not applicable to Astrolite since it is not an explosive mixture until the actual time that it is needed. This also establishes a control of explosives in the field which was not possible heretofore.

The LSC has primarily emerged from the aerospace program where it was successfully used in space and on the moon to precisely and quickly cut away and separate "spent" missile stages in-flight. Now, such amazing technology has reached the commercial applications of underwater salvage work to benefit our offshore petroleum industry. LSC's utilize the "Monroe Effect" to direct and concentrate their explosive forces and were perhaps first recognized for their importance as an anti-tank weapon commonly known as the "bazooka".

No longer are remote projects in foreign lands limiting. With the advent of air-transportable "non-explosive" explosives like Astrolite and hardware designed to efficiently perform work utilizing this energy, worldwide service with instant response-time can be offered. Jobs requiring immediate resolution half way around the world are now, for the first time, possible to accomplish with new underwater tools.

5. FIELD APPLICATIONS

Linear shaped charges can quickly separate thick steel plate of sunken ships for salvage so that economical light surface lift equipment can remove derelicts in sections. Often, these sections can be reassembled later at dry dock.

Access ports can be easily made for entry into such sunken derelicts to facilitate cargo salvage as well.

The Corps of Engineers employed this technique to remove a sunken barge in the Arkansas River, which was a navigational hazard. And the author had the opportunity to demonstrate this system at Port Jefferson, New York, to cut the Integrated Tug Barge, Martha R. Ingram, in-two for return to the ship yard for refitting after an accident. This "slice" required less than one second of time to be performed, rather than weeks with conventional means.

LSC's can also be bent into a circle so that pipelines, conductors, offshore platform legs, and piling can be precisely cut in two. Such devices, called Circular Cutters, can be bent into practically any size diameter and are made to cut pipe from the inside as well as from the outside. With modification, they can be used to cut "windows" in pipe as well. One major use of these devises is to cut "stab piles" or "pinning piles" free from inside their offshore platform jacket legs for platform relocation. Great savings are possible by reducing derrick-barge and workover-rig standby time.

6. APPLICATIONS OF THE FUTURE

Chemical high energy from safe explosives will one day become as commonplace as electrical or gas energy is today; however, this will probably not occur until the public attitude towards explosives has been changed. The public must first be educated about the recent technological advances which have been made with such binary systems as Astrolite and subsequently their ultra-safe operation.

The use of plastics for LSC fabrication has recently been developed as a second generation of low-cost explosive cutting devices. This has good practical application in cutting or breaking of reinforced concrete, timber, and rock.

Modern two-component explosives will shortly be replaced by a new self-sterilizing mixture which becomes non-detonatable even after mixing, if not utilized within a short period of time.

Tomorrows construction requirements offshore will be for such things as deep water logistics support platforms, deep water drilling platforms, and pipelines at depths unimaginable only a few years ago. Maintenance, modification and salvage are areas which will need efficient low-cost tools to aid the diver or technician.

Whatever man can do on the surface of the earth he can do above and below it if the demand warrents it. And the use of explosive energy and modern devices will not only make his work easier underwater, but possible in many cases.

Man working in a hostile environment, where every unit of work is only accomplished at a premium, will need every advantage available. The advantages of explosive cutting is but one tool available to him; those of welding, bending, breaking and torquing explosively underwater will lessen his burdens. These are only the conventional construction tools of explosives, but their applications can be as far-reaching as the ingenious imagination of the inventor.

6th International Symposium on
Jet Cutting Technology
6-8, April, 1982

THE DEVELOPMENT OF ICE-BLASTING FOR SURFACE CLEANING

G. Galecki

Wroclaw Technical University, Poland

G.W. Vickers

The University of British Columbia, Canada

Summary

The entrainment of small ice particles within air jets (ice-blasting) to clean and abrade surfaces is considered as an alternative to grit-blasting, water-jets or other mechanical cleaning methods. The approach utilises the hardness and cutting potential of ice particles, with the benefit that once the ice has performed its cleaning duty it melts and drains away.

An assessment is made of the important parameters of the ice-blasting process and the cleaning effectiveness and efficiency is considered in relation to other cleaning methods.

Held at the University of Surrey, U.K.
Symposium organised and sponsored by
BHRA Fluid Engineering

1. INTRODUCTION

There are a variety of methods available for cleaning surfaces in industrial situations. The most common and effective method for cleaning large areas of difficult materials, such as rust, paint or barnicles, is grit-blasting. Here grit (or sand) is projected on to the target surface by a jet of air. The difficulties encountered, however, are the grit residue left after cleaning (900-1200 kg/hour) and the cost of grit (0.05-0.09 $/kg). Indeed, there are many applications where the grit residue is totally unacceptable and in these cases recourse is frequently made to the more laborious mechanical cleaning methods (such as percussive needle guns or rotating scrub brushes) or to water jets.

In this work a new approach is investigated utilizing the hardness and cutting potential of ice particles entrained within air jets. The process is much akin to grit-blasting, with the benefit that once the ice has performed its cleaning duty, it melts and drains away.

The equipment for producing ice particles, of approximately 3 mm diameter at a rate of up to 225 kg/hour, is described together with the nozzles and other experimental equipment. From the testing program an assessment is made of the important variables of the process, such as ice temperature, ice particle size, rate of ice mass-flow, standoff distance and air supply pressure. The cleaning potential of ice-blasting is given for a series of painted surfaces and shown to be very effective. The economics of ice production, in relation to grit costs are also most favourable.

2. EQUIPMENT

A schematic drawing of the testing facility is given in Fig 1 and a general view given in Fig. 2.

Ice pellets are produced by crushing 3 cm blocks of cube ice which have been cooled in liquid nitrogen. The ice pellets are drawn into an air jet through a venturi orifice in the nozzle housing, see Fig. 3. The nozzle mixes the ice particles with the air jet and directs the jet to the target surface.

The ice crusher, shown in Fig. 4, consists of two rotating multiblade drive shafts, the teeth of which mesh with a serrated plate. The two shafts are mounted in an aluminium housing which also holds the blocks of cube ice. Rotation of the drive shafts cause the cube ice to be forced through the 3 mm gaps in the serrated plate. A styrofoam/plywood enclosure surrounds the ice crusher so that liquid nitrogen can be held between the two containers. The temperature of the ice crusher is thereby kept at a level which prevents it from unduly heating the outer layers of the ice particles. This in turn gives a more consistent ice particle size and prevents the particles from sticking together.

With a variable speed motor drive on the ice crusher the rate of ice mass-flow can be regulated from 0-225 kg/hr. Fig.5 shows the quantity of pellet ice produced against drive shaft speed. Above 3000 rpm the quality of ice deteriorates as larger amounts of snow and flake ice are produced.

3. TESTING PROCEDURE

Assessment of the cleaning efficiency was done on the basis of the maximum area removal rate of underwater hull paint (grey aluminium-vinyl anticorrosive primer, red cuprous-oxide antifouling paint) from sheet metal specimen.

The nozzle, mounted as shown in Fig. 6 was traversed across the target at an increasing speed, until a clear path was no longer evident. The width of cleaned path at the maximum traverse speed gives the maximum rate of cleaning. The speed of travel of the nozzle across the specimen could be adjusted from 0-2½ m/min.

During testing the conditions of standoff distance, supply pressure, mass flow-rate of ice, ice particle size, ice temperature and angle of attack, were all systematically varied.

Cube ice, formed in a freezer unit at -24°C, was usually immersed in liquid nitrogen (-195°C) before being placed in the crusher (this is referred to as type A ice). Tests were also made with ice taken directly from the freezer unit (this is referred to as type B ice). Measurements taken of the ice pellets as they left the crusher showed that type A ice had attained a temperature of approximately -110°C and type B ice a temperature of approximately -15°C.

4. TEST RESULTS AND DISCUSSION

The variation of the maximum cleaning rate of underwater hull paint with standoff distance, ice temperature, air supply pressure, ice particle size, ice mass flow-rate, and angle of attack are given in Figs 7 - 16.

4.1 Standoff Distance

Figs. 7 - 12 show that the maximum rate of cleaning occurs at a standoff distance, between the nozzle and target surface, of approximately 7-15 cm. For all conditions considered, the cleaning rate increases moderately with standoff distance up to a maximum level and thereafter it decreases markedly.

This characteristic is probably due to a combination of the speed and hardness of the ice particles as they travel along the air jets. Ice particle velocity will likely increase during its initial travel in the jet, while hardness (and intensity of impact) will decrease as heat is transferred to it from the surrounding air.

It is noticeable that the optimum standoff distance increases with supply pressure (and faster particle speed) and with lower ice temperatures.

4.2 Ice Temperature

Figs 7 - 12 show that the colder, type 'A', ice gives 2 to 3 times higher cleaning rates than type 'B' ice. This is due to the increased hardness, and hence impact strength of ice at lower temperatures. The actual temperature of the ice during impact is not known, although estimates of -110°C and -15°C for type 'A' and 'B' ice respectively were made at a point where the ice enters the nozzle.

It should be noted that type 'A' ice is more difficult and expensive to produce than type 'B' ice particularly in a continuous on-line process and this has to be balanced against the improvement in cleaning rates.

4.3 Air Pressure

Figs 13 and 14 show that doubling the air pressure from 0-27 to 0.54 MPa makes approximately a fourfold increase in cleaning rate for types 'A' and 'B' ice.

This is due to the increased ice particle velocity with air pressure. The pressure of 0.55 MPa is typical of those used in grit blasting.

4.4 Ice Particle Size

Fig 15 shows that 3 mm diameter ice particles supplied at a rate of 68 kg/hour, can clean more effectively than 4½ mm diameter particles at a higher rate of 112 kg/hour

4.5 Ice Mass Flow-Rate

Figs 13 and 14 show that doubling the quantity of ice from 68 to 135 kg/hour increases the cleaning rate by 20 - 25 per cent for type 'A' ice and somewhat less for type 'B' ice.

This increase is due to the increased number of impacts per unit time but is obviously not so important as ice temperature or supply pressure. Also, the increase in ice cost has to be balanced against the marginal improvement in the cleaning rate.

4.6 Angle of Attack

Fig 16 shows that nozzle inclination of $\pm$ 30 degrees to the normal of the target surface has little effect on the cleaning rate. However, angles greater than this show a sharp drop off in cleaning rate.

5. COMPARISON OF CLEANING METHODS

The maximum cleaning rate, the cleaning efficiency and the area cleaned per unit cost are given, in Table 1, for ice-blasting, grit-blasting, water-jets and percussive-needle guns. The cleaning rates given for the four methods are those pertaining to the removal of underwater hull paint. Cleaning efficiency is the rate of paint removal per unit of power supplied. The area cleaned per unit cost includes assessment for consumable material costs (i.e. cost of ice or grit), power supply costs, and labour and overhead costs (at a rate of $30/hr). Thus the area cleaned per unit cost, $\bar{A}$, is given by

$$\bar{A} = \frac{\text{maximum cleaning rate (m}^2\text{/hour)}}{\text{consumable material and power supply costs (\$/hr) + 30 (\$/hr)}}$$

No account is taken of equipment and depreciation costs as these are essentially similar.

From Table 1, it can be seen that grit blasting gives the highest cleaning rate range of 2.8-41 m^2/hr, ice-blasting is next with a range of 0.6-2.1 m^2/hr followed by water jets at 0.5m^2/hr and percussive needle guns at 0.3m^2/hr. The area cleaned per unit cost and cleaning efficiency show a similar trend.

The best cleaning rates for ice-blasting were obtained with the colder, type A, ice at the higher mass flow-rates of 135 kg/hr. However, the additional ice costs at the higher mass flow-rate resulted in only marginally higher figures for area cleaned per unit cost.

While ice-blasting is not as efficient as grit blasting it is 1½ times more efficient than water jets and 2½ times more efficient than percussive needle guns. Thus, for the many cleaning applications where grit cannot be tolerated it is a very promising technique.

The equipment for producing ice pellets was designed as a laboratory tool for assessing the potential of the method. For on-site work a more compact freezing unit is required for producing ice pellets directly.

6. CONCLUSIONS

The ice-blasting technique, presented in this work, is shown to be more effective than either water-jets or percussive needle guns but not as effective as grit-blasting. However, for cleaning applications where grit cannot be tolerated ice-blasting is a very promising technique which merits further development.

ACKNOWLEDGEMENTS

The authors wish to thank Messrs. John Wiebe and John Hoar for their help with the project. The financial support provided by the Department of Nation Defense under Contract No.08SB 3290031 is gratefully acknowledged.

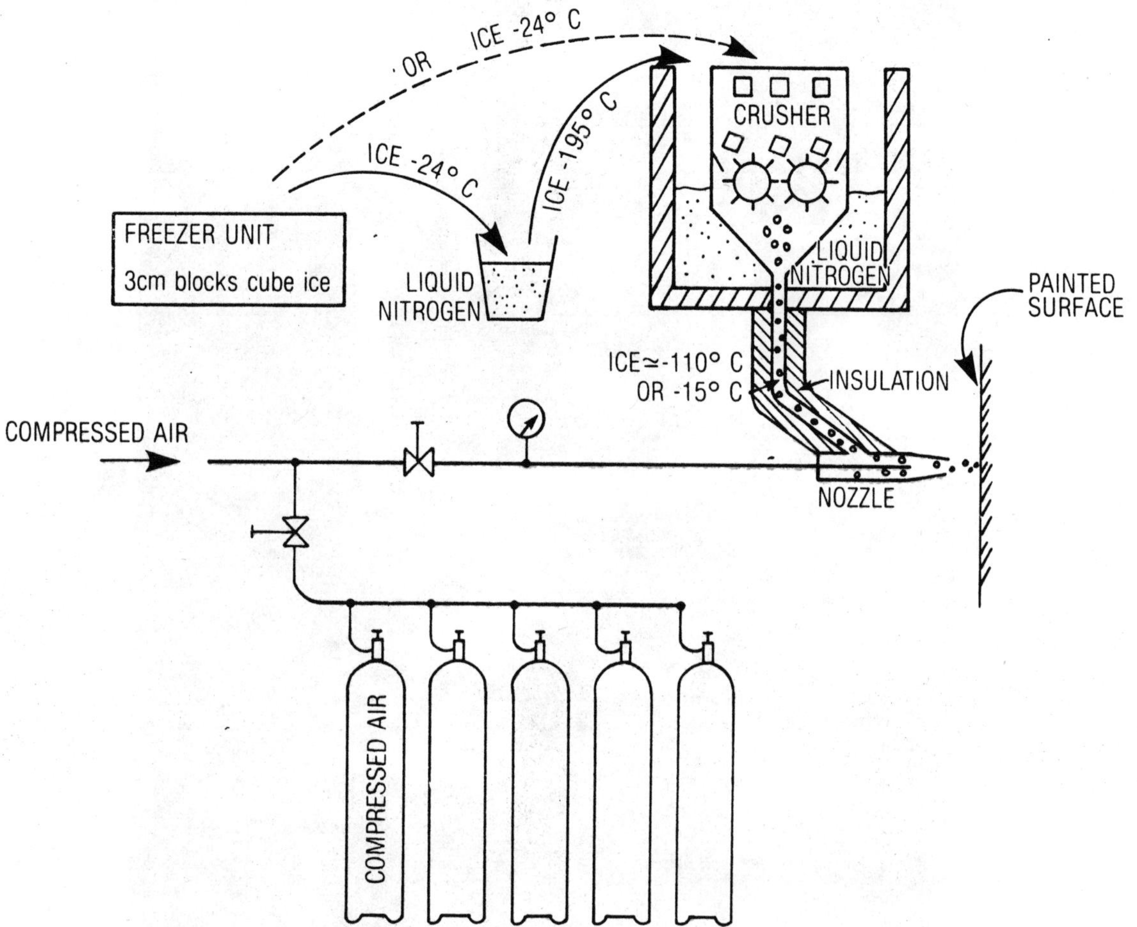

Figure 1. Schematic drawing of the testing facility

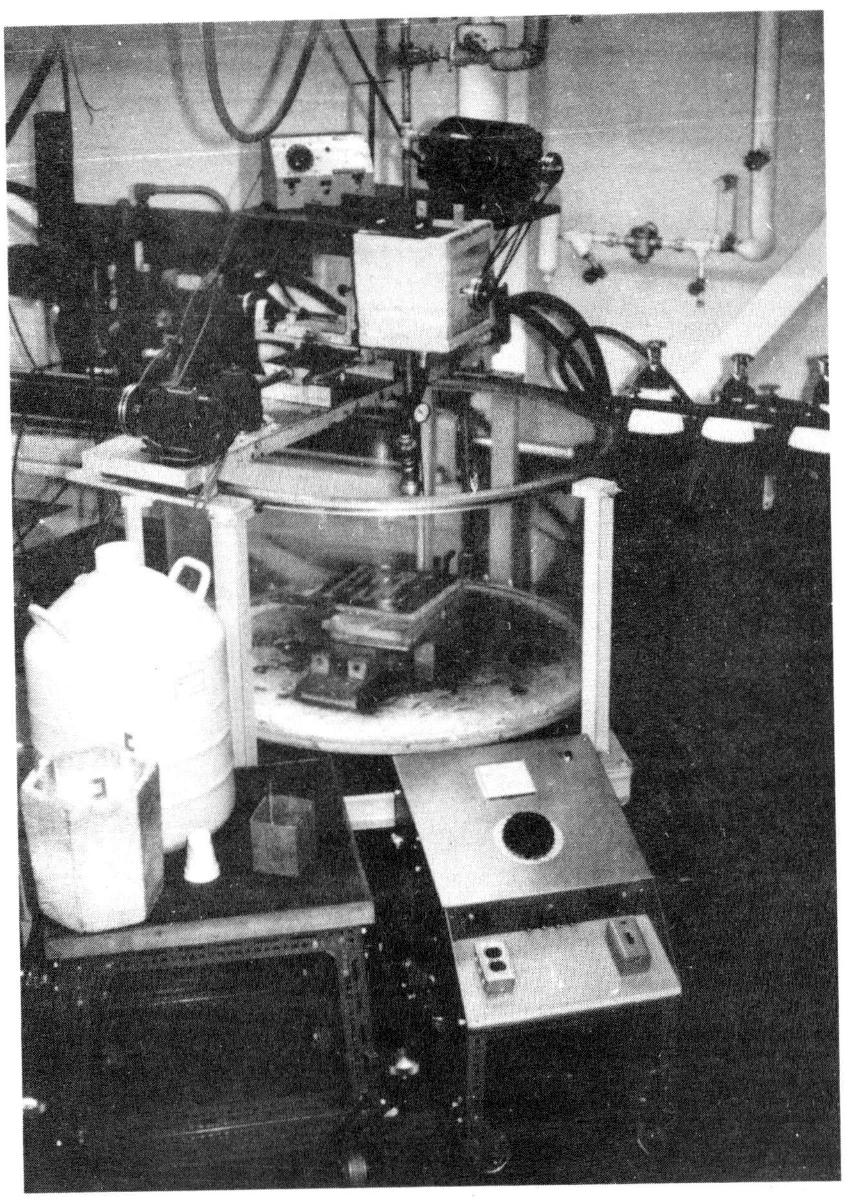

Figure 2. General view of testing facility showing tank, nozzle, speciment, ice-crusher and traversing arrangement.

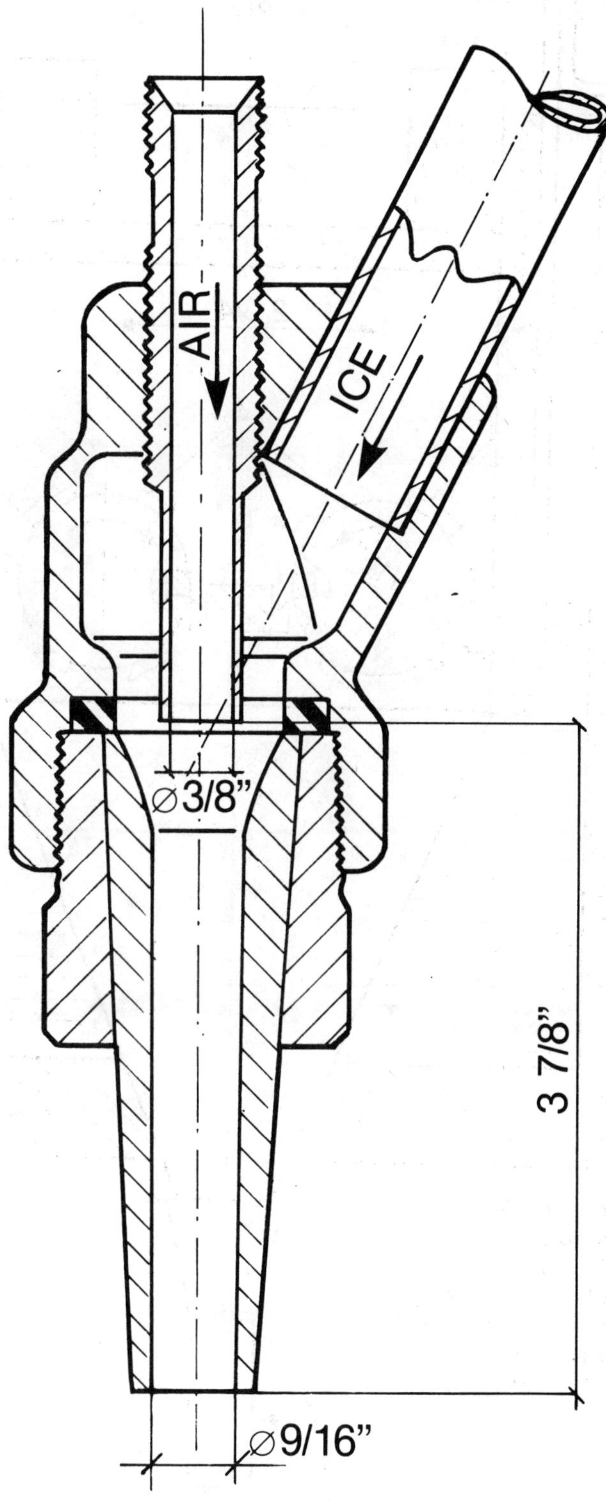

Figure 3. Air/ice venturi nozzle housing

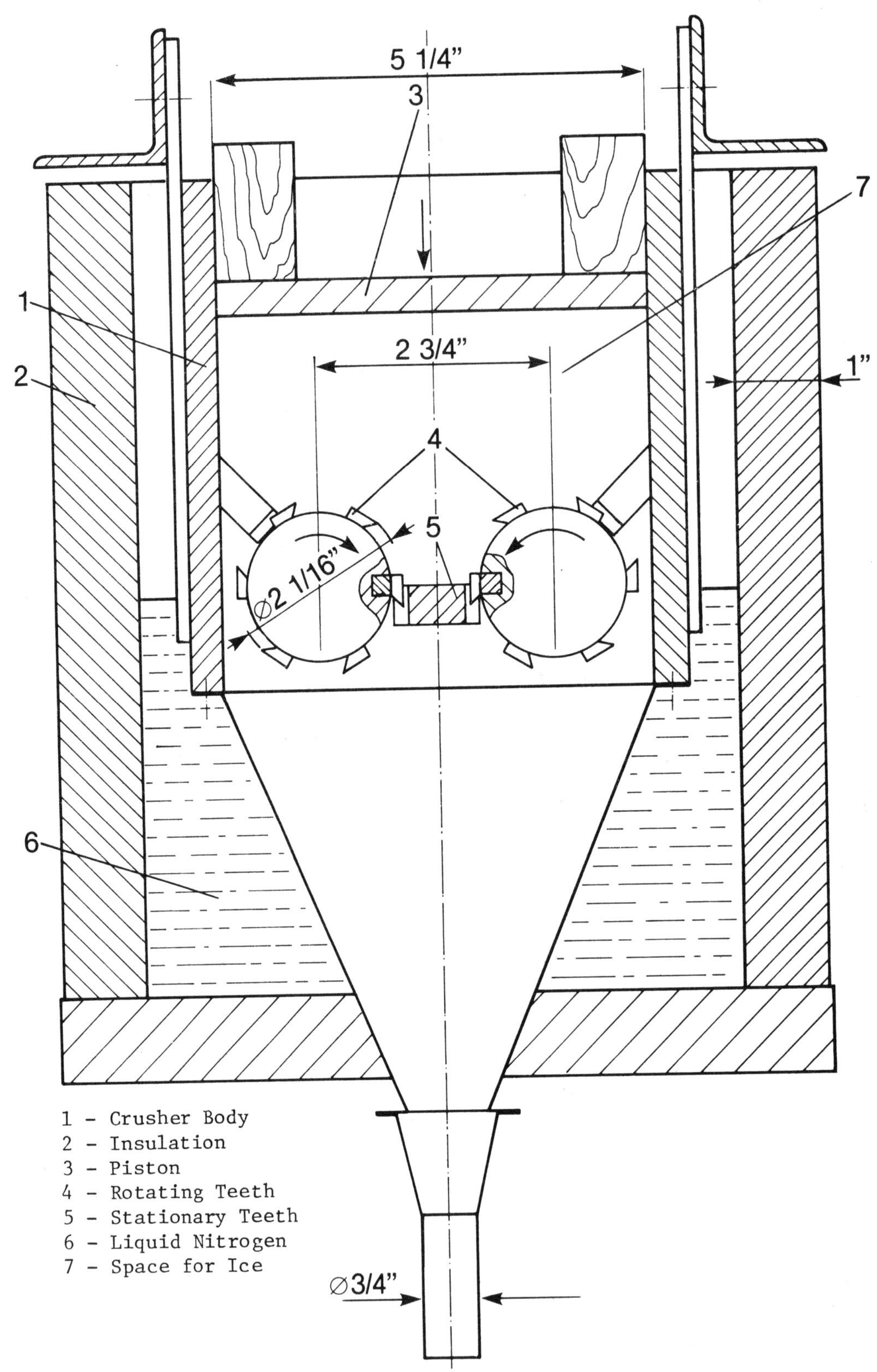

Fig. 4 Section diagram of the ice-crusher.

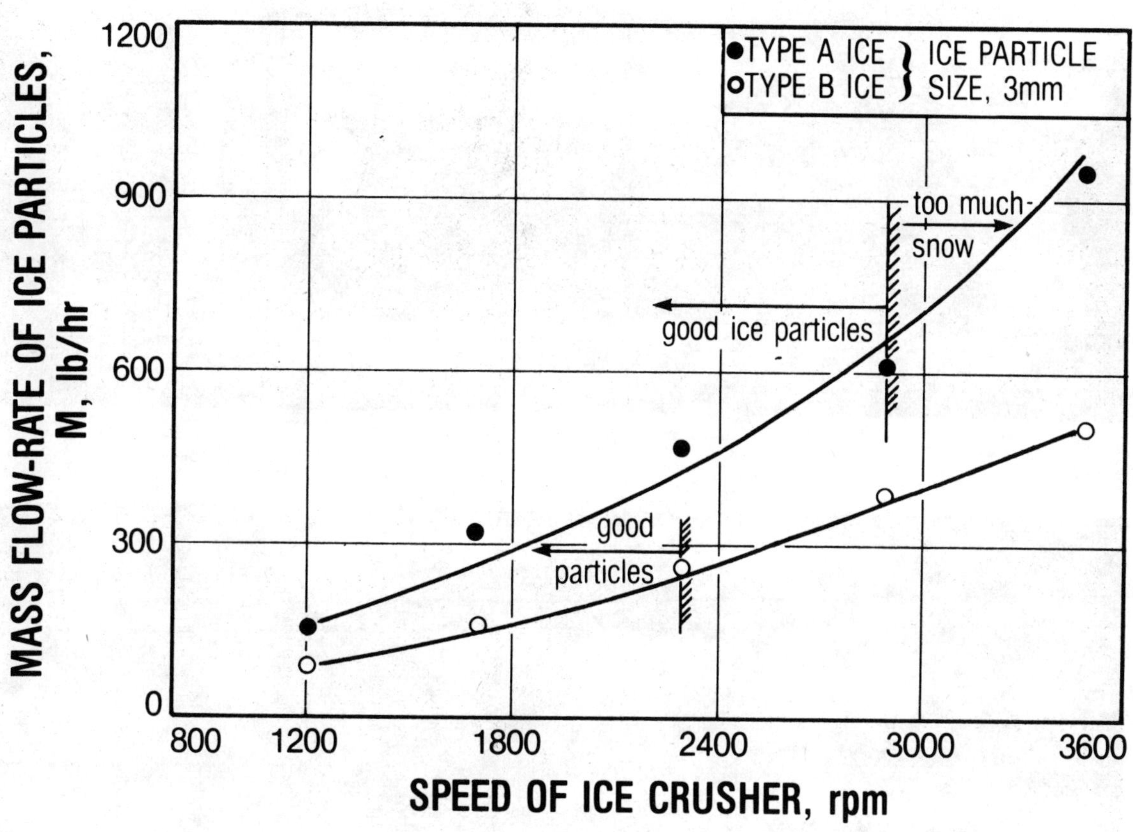

Figure 5. Quantity of pellet ice produced versus the rotational speed of the crusher drive shaft.

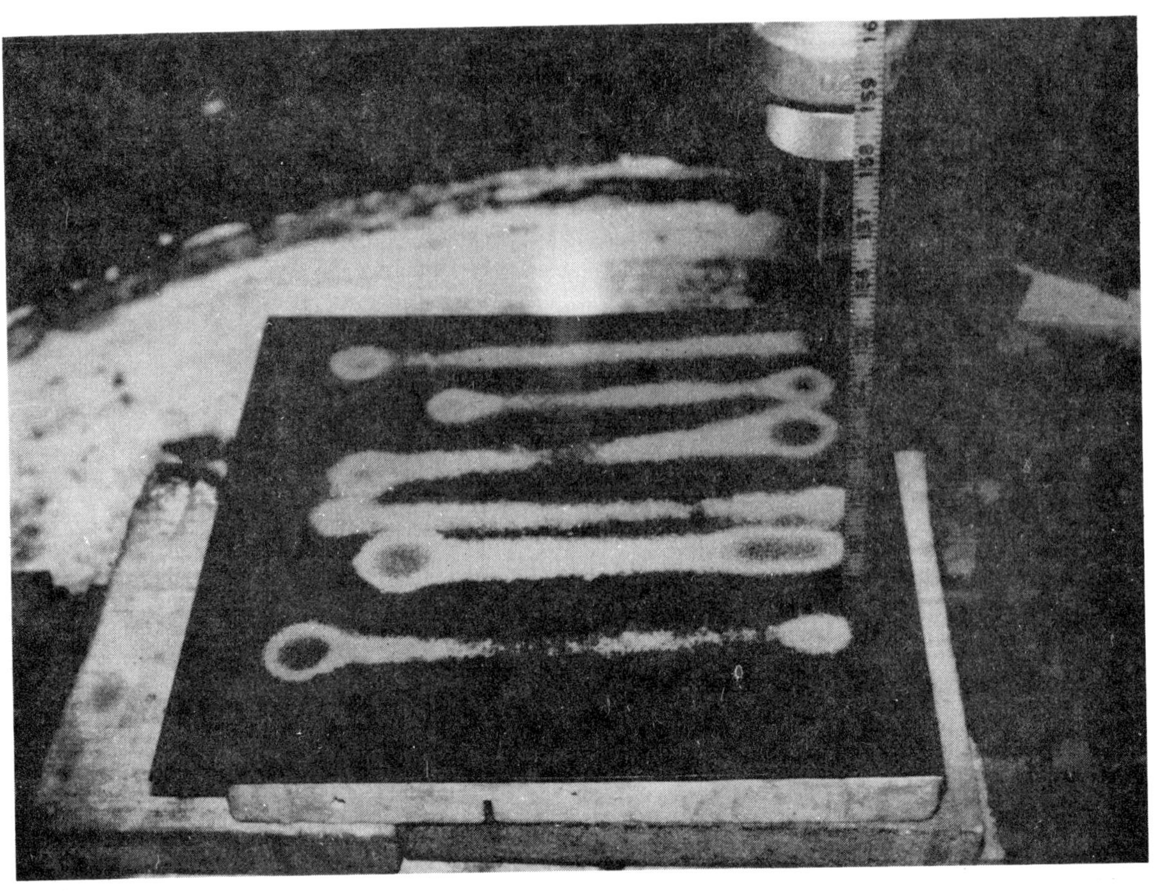

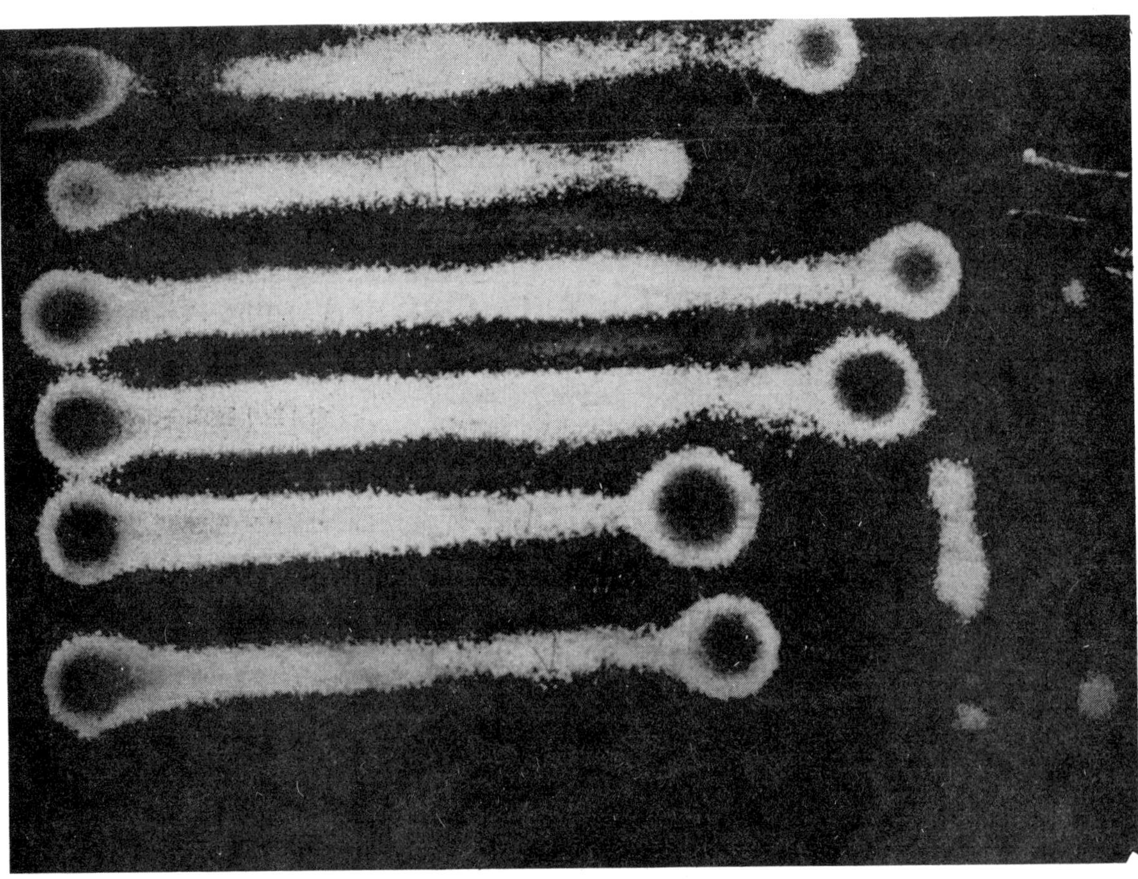

Figure 6. Test specimen - Grey aluminum vinyl anticorrosive primer, red cuprous oxide antifouling paint.

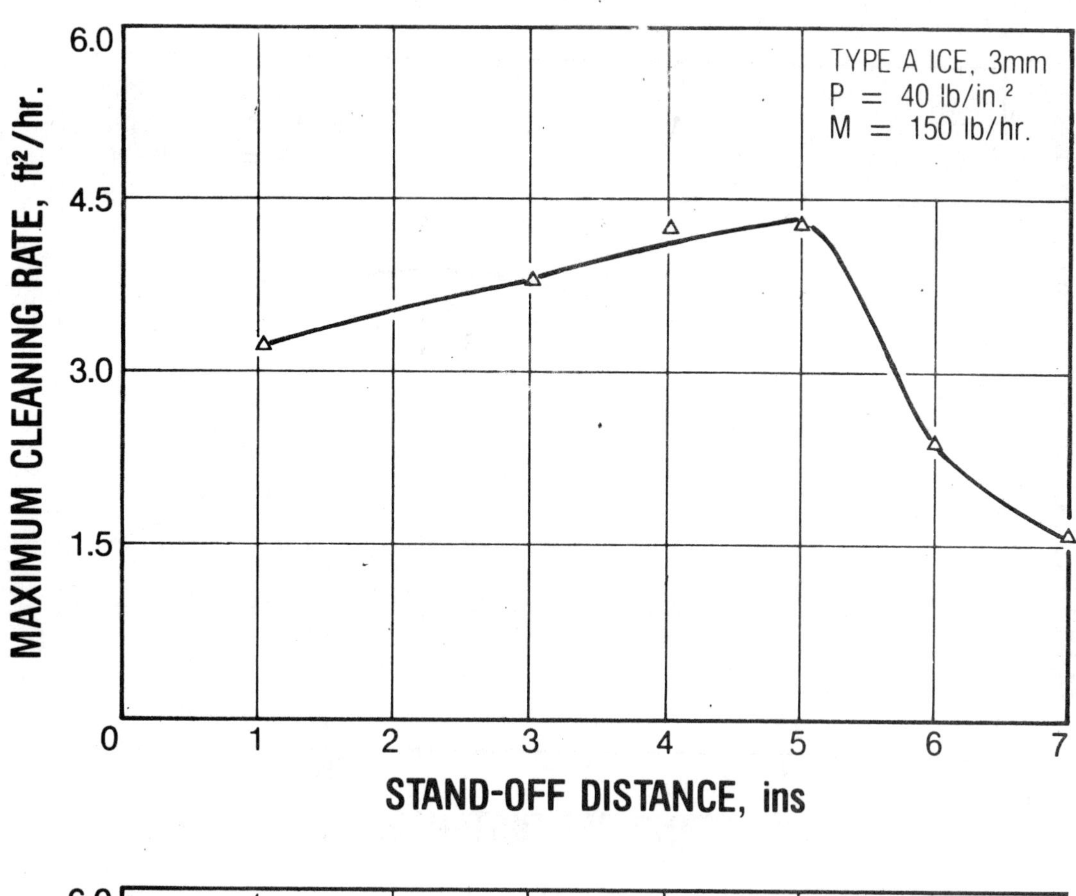

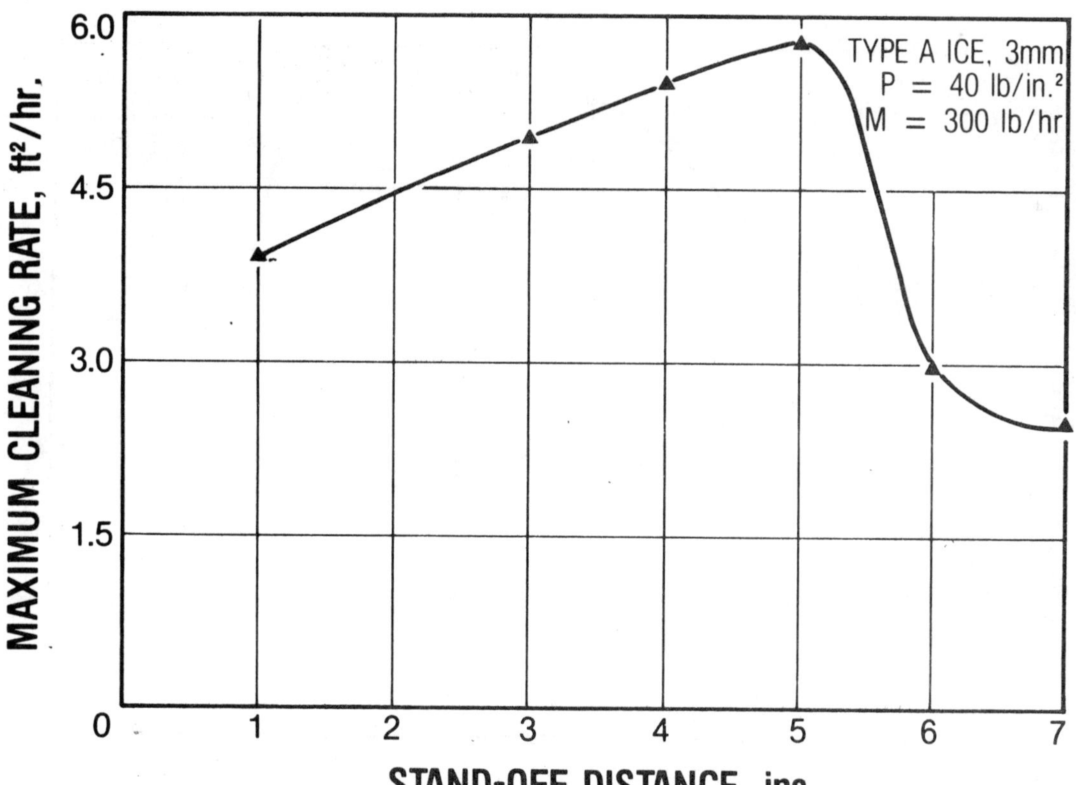

Figure 7. Maximum cleaning rate versus standoff distance for type "A" ice at an air supply pressure of 40 lb/in^2

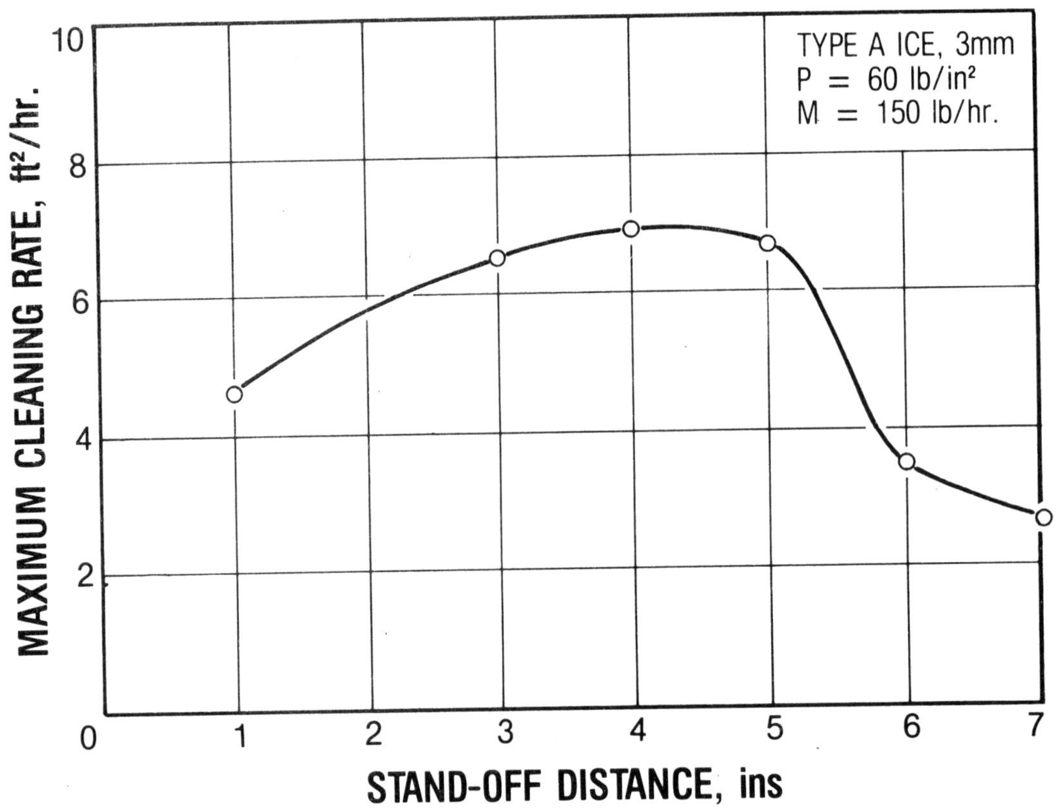

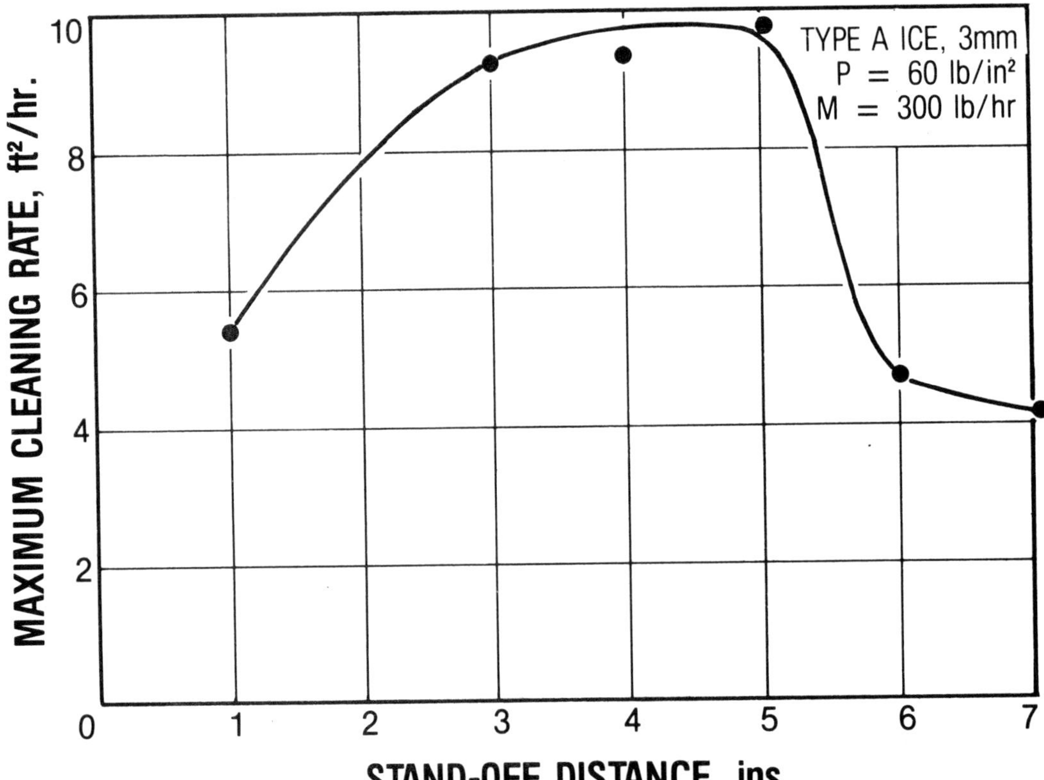

Figure 8. Maximum cleaning rate versus standoff distance for type "A" ice at an air supply pressure of 60 lb/in^2

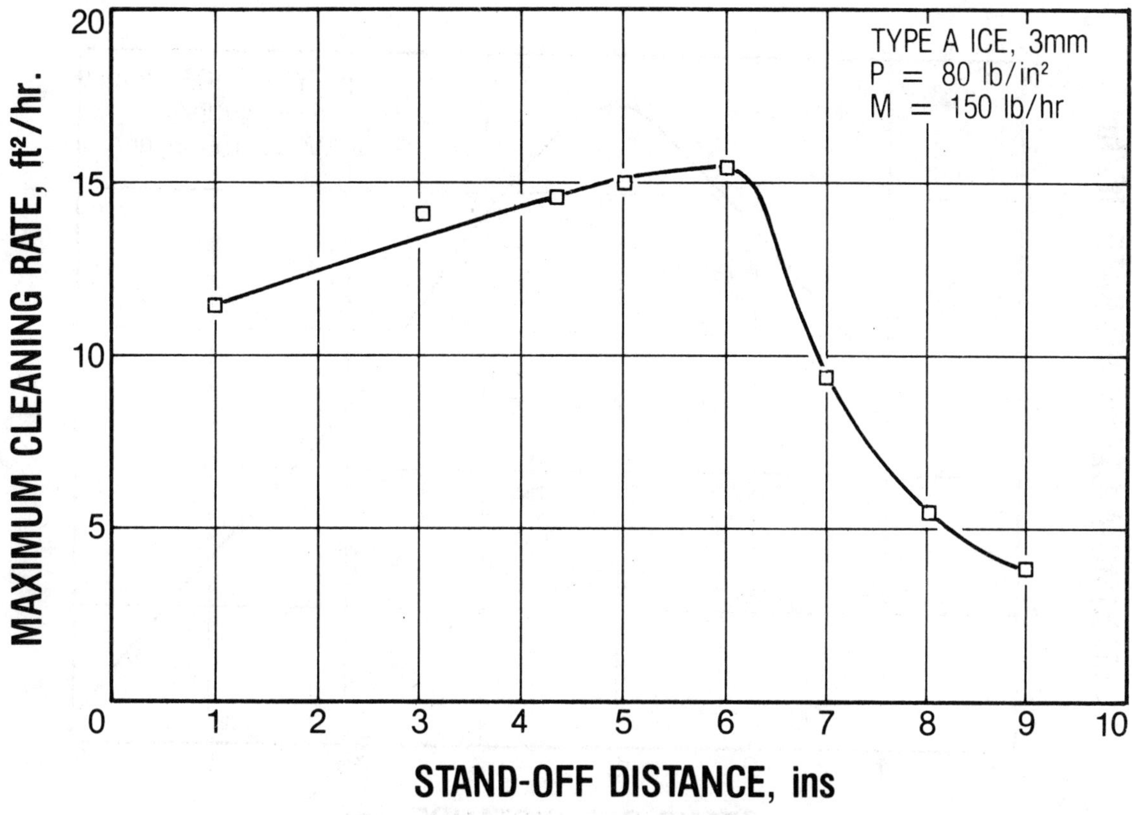

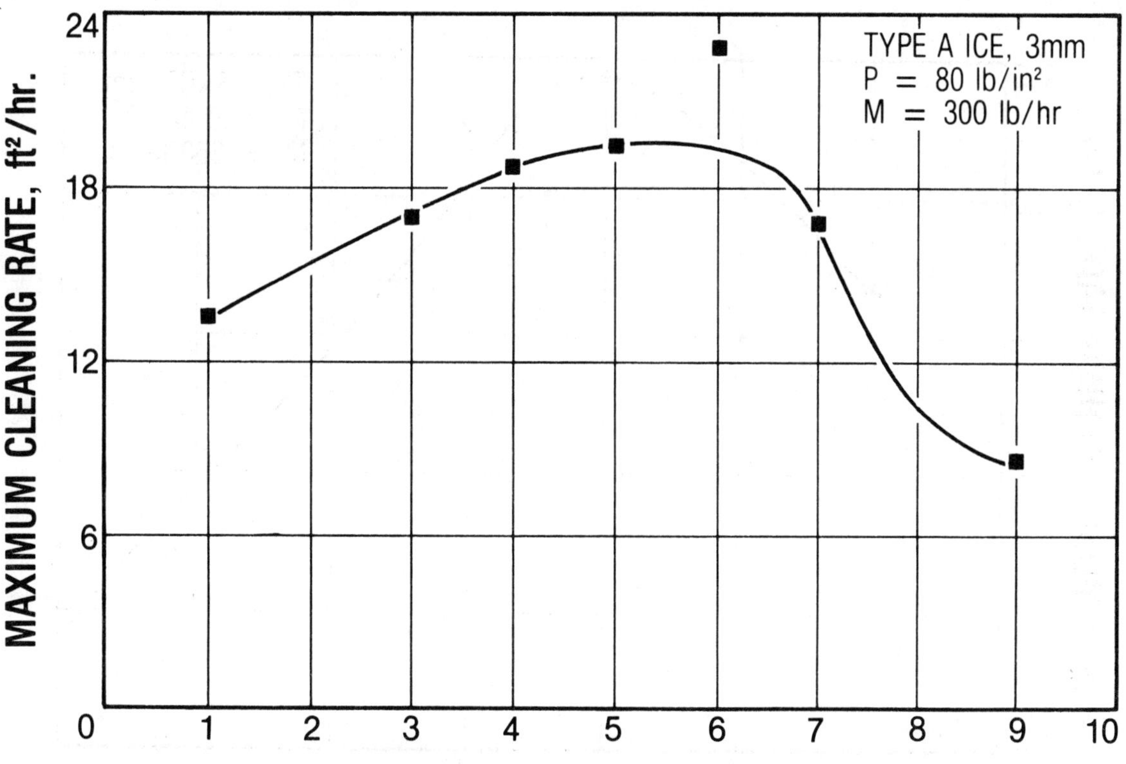

Figure 9. Maximum cleaning rate versus standoff distance for type "A" ice at an air supply pressure of 80 lb/in^2.

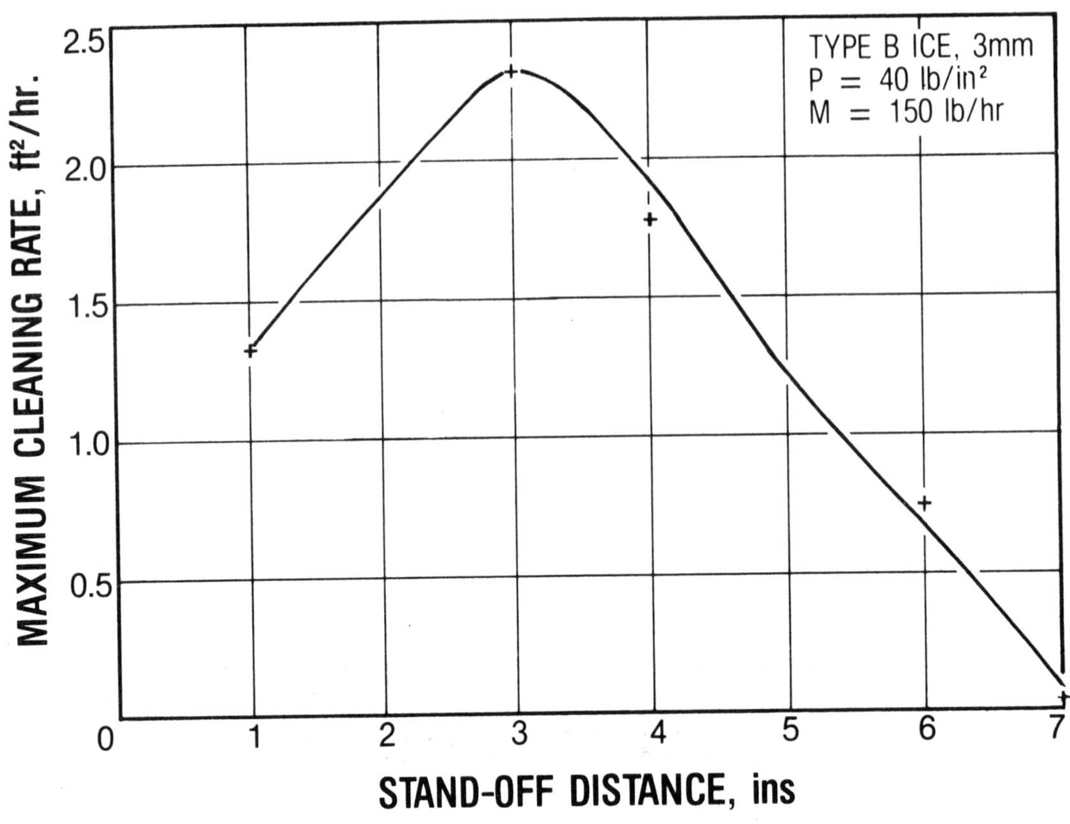

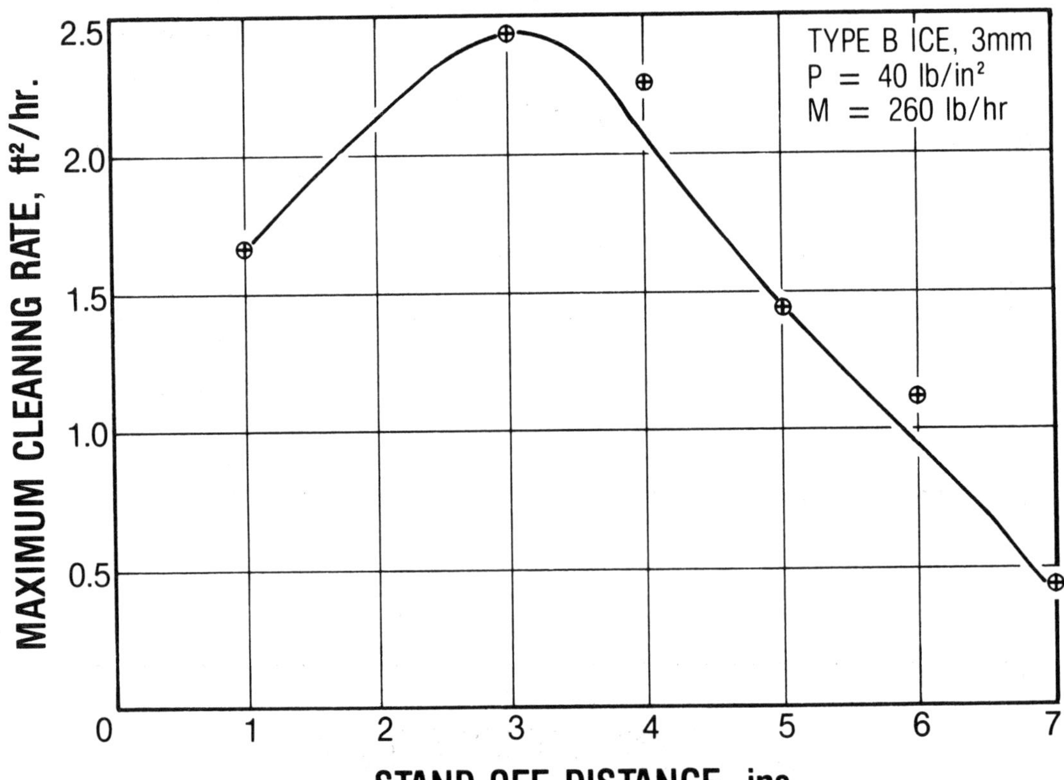

Figure 10. Maximum cleaning rate versus standoff distance for type "B" ice at an air supply pressure of 40 lb/in^2.

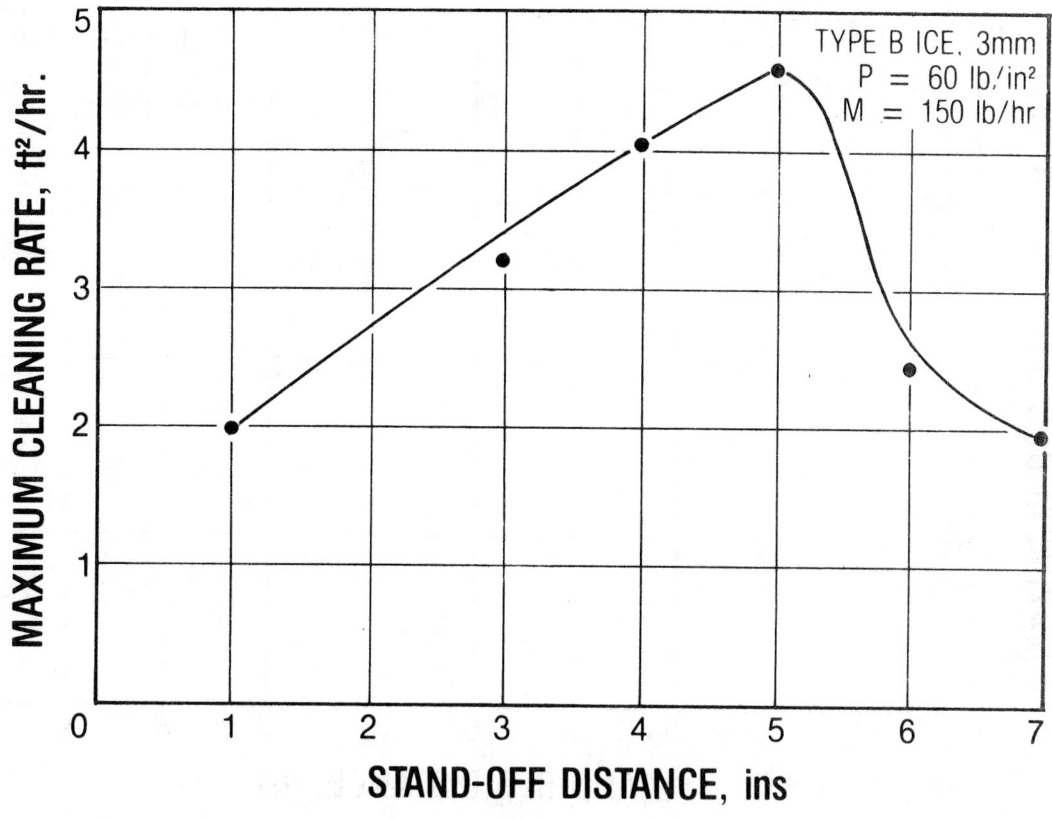

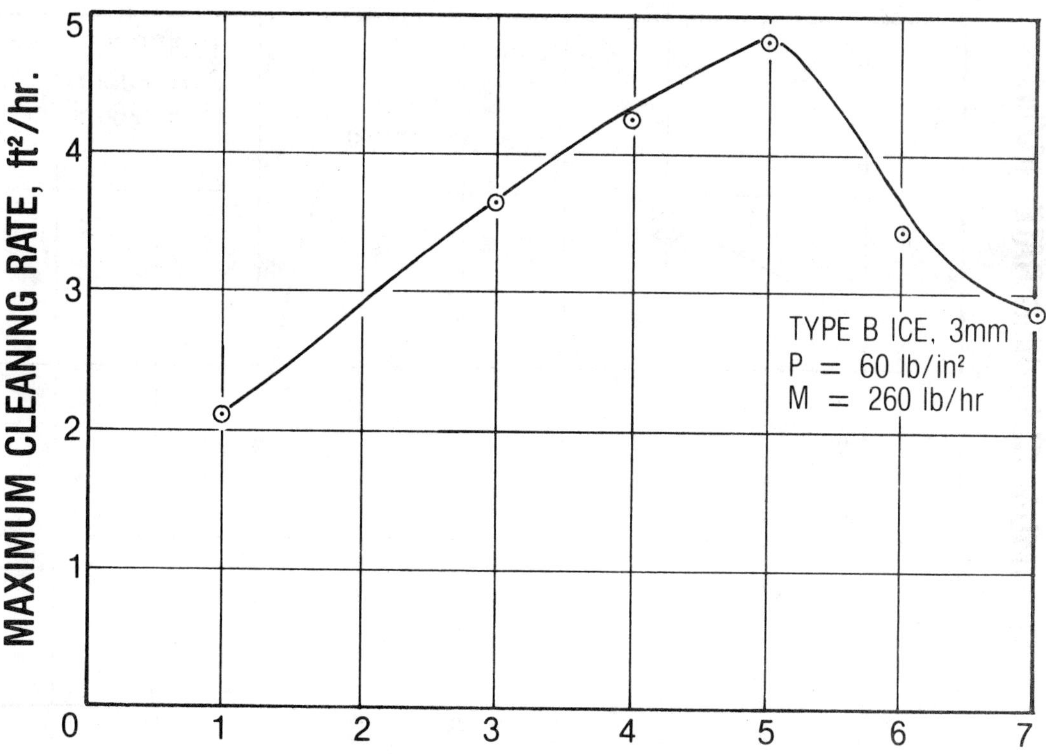

Figure 11. Maximum cleaning rate versus standoff distance for type "B" ice at an air supply pressure of 60 lb/in^2.

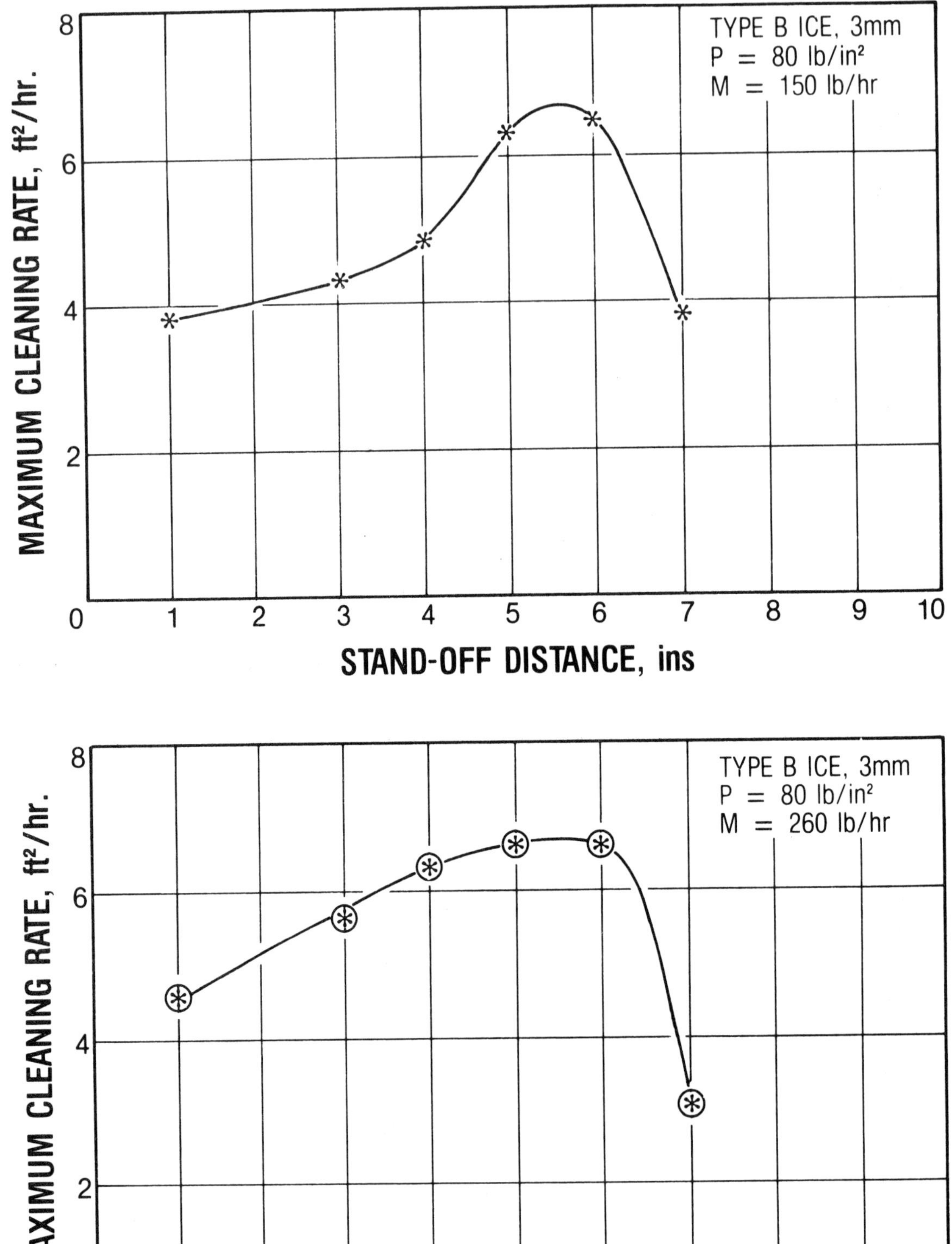

Figure 12. Maximum cleaning rate versus standoff distance for type "B" ice at an air supply pressure of 80 lb/in^2.

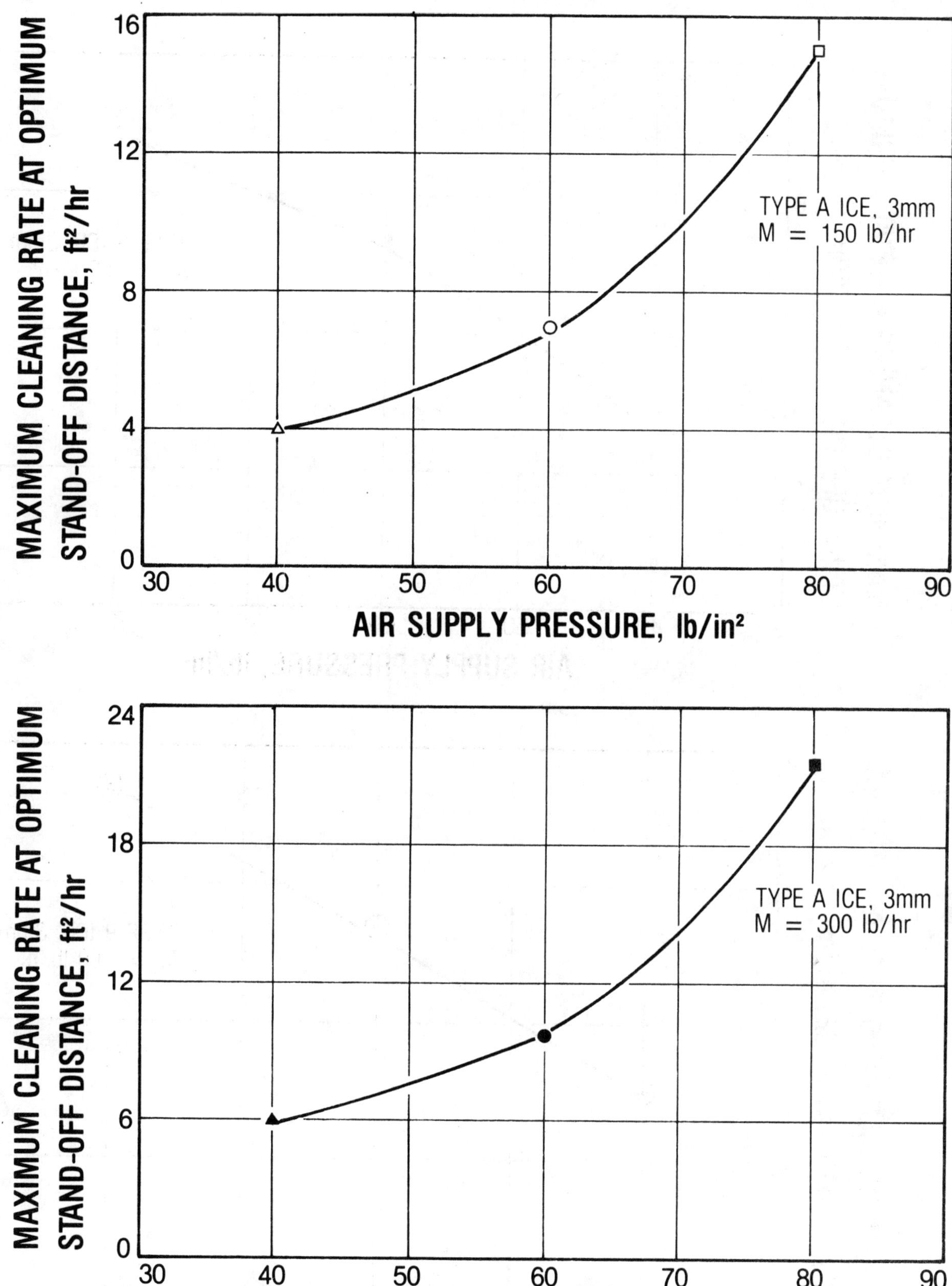

Figure 13. Maximum cleaning rate at optimum standoff distance versus air supply pressure for Type "A" ice.

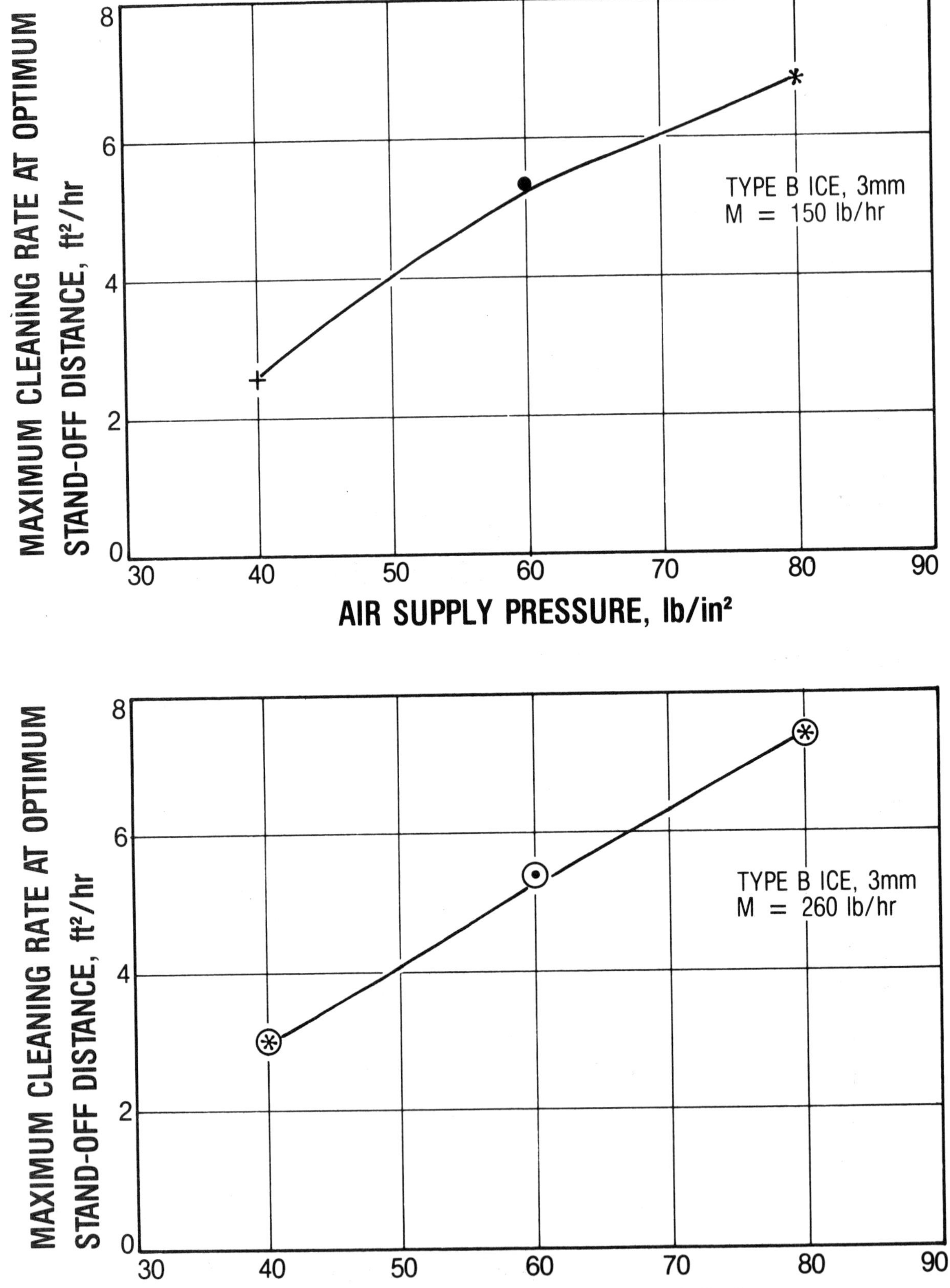

Figure 14. Maximum cleaning rate at optimum standoff distance versus air supply pressure for type "B" ice.

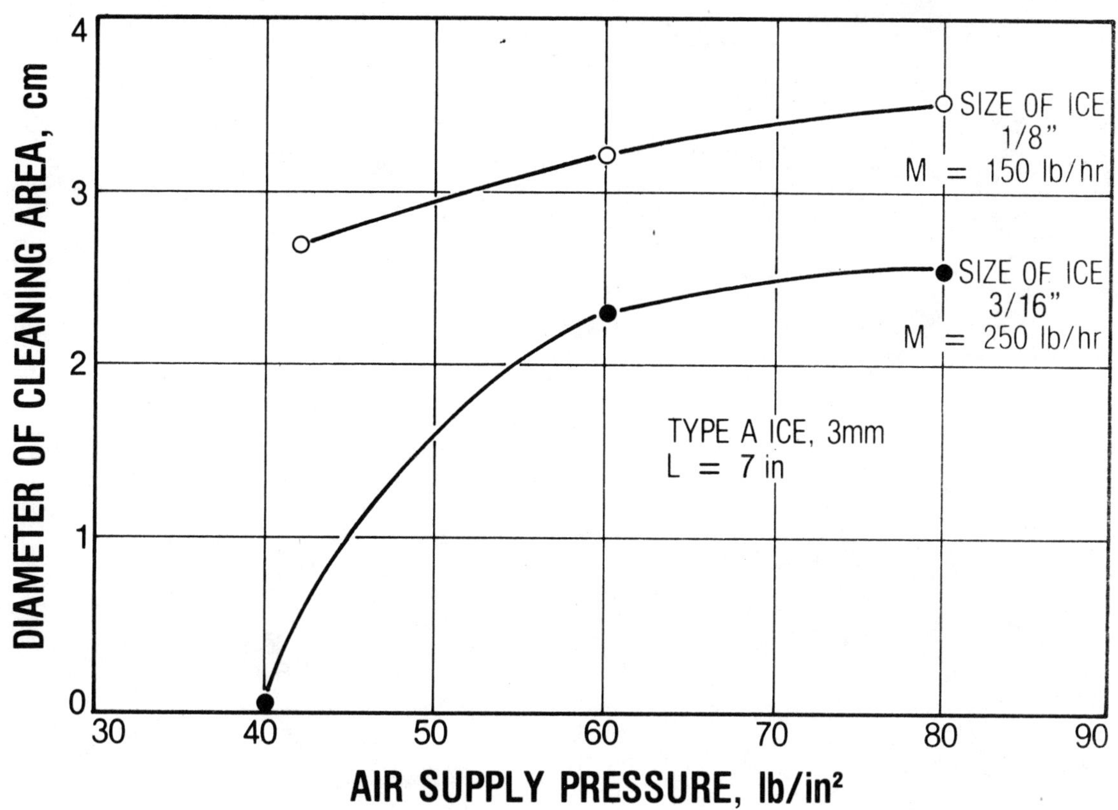

Figure 15. Cleaning effectiveness of 3 mm and 4 1/2 mm diameter ice particles

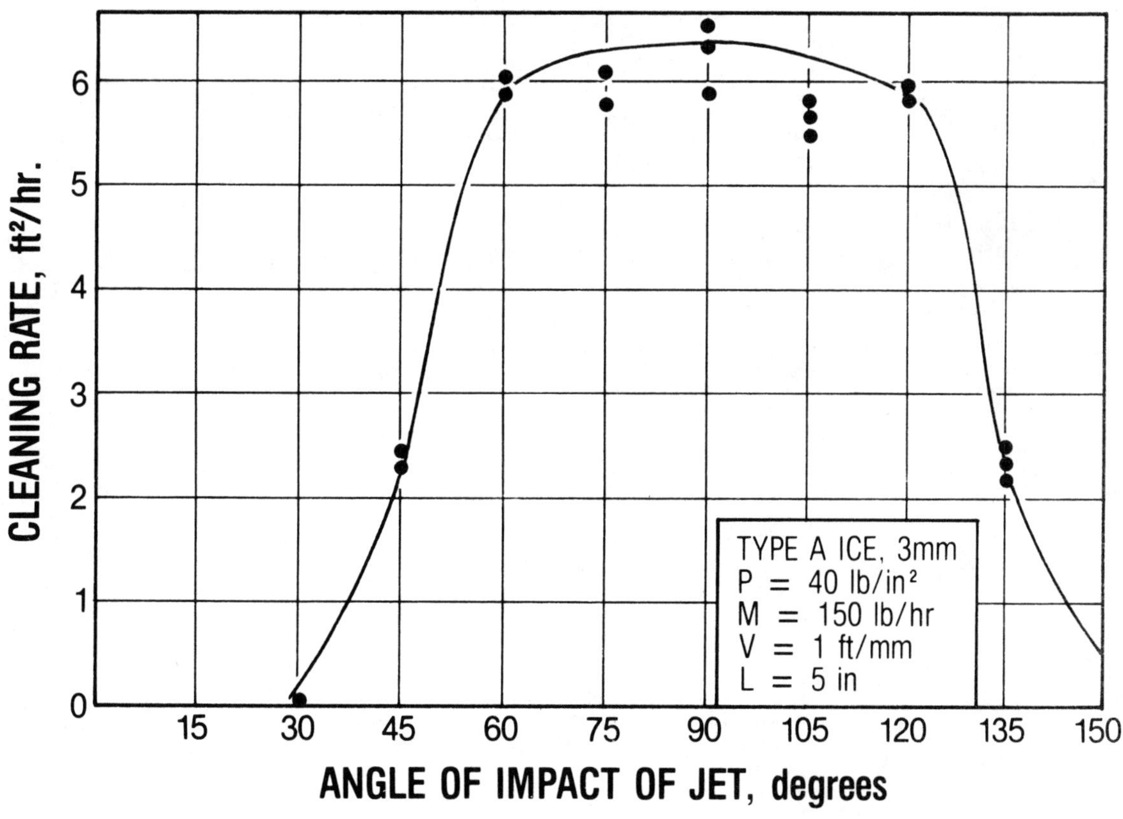

Figure 16. Cleaning rate versus nozzle inclination to target surface

6th International Symposium on
Jet Cutting Technology
6-8, April, 1982

PAPER C1

THE EFFECTIVENESS OF WATER JET SPRAYS IN CLEANING AND THE MECHANISMS FOR DISINTEGRATION

D.A. Summers
University of Missouri-Rolla, U.S.A.

Summary

Most water jet cleaning nozzles produce a fan-shaped jet. The limiting conditions under which this is effective are discussed, and illustrated. High speed photographs are used to show the very rapid disintegration of fan jets and the mechanism by which such disintegration occurs.

The addition of heat to the water, to enhance removal of oil:grease layers is examined, as is the introduction of polymers into the jet stream.

Held at the University of Surrey, U.K.
Symposium organised and sponsored by
BHRA Fluid Engineering

1. INTRODUCTION

High pressure water jets are increasingly being used for cleaning surfaces on an industrial basis. In many instances the use of high pressure water, in such applications as car cleaning, is carried out in combination with mechanical brushes. However, there are many instances, particularly in self service units and where large industrial vehicles, which do not fit into a standard carwash, are cleaned, where a hand held lance alone, is used for the cleaning task. Because of the manuverability of the lance and the relatively slow speed at which it is traversed over the surface, it is common practice to use a fan shaped nozzle in order to cover as large a surface area as possible as the wand is moved. Because of the nature of the operation and the geometry of the surface being cleaned, widely varying standoff distances can quite frequently occur between the nozzle and the surface. This can be of considerable importance since, as will be shown in this paper, there is a relatively rapid decrease in jet cleaning effectiveness with standoff distance.

The work described herein was to evaluate the effectiveness of such nozzles. A preliminary literature survey (for a large part of which I am indebted to BHRA) revealed that comparatively little technical information, other than that previously reported at these symposia, was available. A test program was therefore required.

2. TEST PROCEDURE

In order to evaluate the effectiveness of fan jets in cleaning different surfaces, the water jet spray was tested for a number of different nozzle designs and as a function of jet pressure and standoff distance from the nozzle. The ultimate result was then applied to the cleaning of painted metal surfaces which have been covered with various different dirts to verify the cleaning effectiveness achieved. These results will be discussed in the following section.

In evaluating the relative performance of fan jets as opposed to the more conventionally used round jets of water jet cutting technology, reference is made to work previously carried out by Pardey who had contrasted the relative performance of fan jets and round jets in cutting polystyrene (Ref. 1).

Pardey had shown that there was a very rapid drop off in the cutting effectiveness of a jet in cutting polystyrene up to distances of 30 cm from the nozzle and that thereby it was recommended that such nozzles only operate within the 30 cm standoff distance. Similar tests were therefore carried out in this program.

The individual nozzles to be tested were placed in a specially fabricated holder, designed to the request of one of our suppliers, with a 100 inlet nozzle diameter straight section, leading into the nozzle body itself. Nozzles were fitted to this holder in "as received" condition and high pressure was supplied to the system at either 6.9 or 13.8 MPa.

Fluid flow from the nozzle was directed at the chuck of a lathe (Fig. 1). The chuck was rotatable and a large thick piece of polystyrene was locked in the jaws of the lathe. By making a half revolution of the chuck, the water jet spray from the nozzle would thus intersect, and cut into, the leading surface of the polystyrene. In this manner a very rapid evaluation of the performance of different nozzles could be made. An examination of the results achieved from this study indicated a very disparate set of results for the wide range of nozzles examined in the program.

While the best nozzles could indent polystyrene up to a distance of 30 cm from the nozzle, some nozzles tested were incapable of creating an impression on the polystyrene if located more than 7.5 cm from the polystyrene surface (Fig. 2). Insofar as the basic flow supplied to these nozzles was the same and the water was supplied at pressures in excess of 6 MPa while it has been possible to cut polystyrene at a pressure of approximately 0.06 MPa, some explanation of this phenomena seemed required. The explanation was sought in two separate ways. The first procedure was to evaluate the pressure profile generated by high pressure sprays as they issue from the nozzle.

The procedure carried out was to locate the nozzle in a fixed holder located above the bed of a surface grinder (Fig. 3). It is particularly pertinent to point out that in both this trial and in the previous trials, nozzles were held fixed while the target was moved. The reason that this is stressed is that it has been the investigator's previous experience that relatively rapid oscillation, of the nozzle itself, will increase jet disintegration considerably, particularly in smaller nozzle diameters, and have a deleterious effect on jet structure. The results derived from this and the previous test are, in large measure, therefore, somewhat better than would be achieved, were the test to be carried out with the nozzle oscillating, as would normally be the case.

Pressure profiles were taken from representative nozzles at standoff distances of 7.5, 15, and 22.5 cm from the nozzle surface (Fig. 4). A pressure transducer was attached to the table of the surface grinder, positioned so that as the table moved, the head of the transducer would pass completely through the jet. The direction of travel and orientation of the nozzle were aligned with the long jet axis along the direction of greatest travel of the grinder table, and the feed was set up such that the table would be incremented over by a distance of .5 mm/increment while the signal from the pressure transducer was monitored through a small commercial computer at .5 mm increments. In this manner it proved possible to generate pressure profile across the body of the jet in the two orthogonal directions.

The first point that came out of the study with the extremely rapid decay in pressure away from the nozzle. For example, even with the very best nozzles at a distance of 7.5 cm from the nozzle at a supply pressure of 6.9 MPa pressures had already formed to below .6 MPa and at 22.5 cm from the nozzle, the pressures were typically on the order of 0.06 MPa.

At nozzle pressures of 13.8 MPa pressures at 7.5 cm from the nozzle were still only on the order of .7-.8 MPa and at 22.5 cm from the nozzle, the pressure range for the three best nozzles ranged from approximately .138-.26 MPa. Such very low pressures, even relatively close to the nozzle, and a very rapid method of disintegration led to a photographic study of the types of jets created by the various nozzle sizes in an attempt to understand why the jets break up and whether in fact this breakup could be delayed. The experimental setup for this program is shown in the attached figure and is based on many previous pieces of equipment very similar to that shown. I suspect that the initial equipment for this was developed at the Cavendish Laboratory in Cambridge, and certainly that is where we got the idea (Ref. 2).

The equipment consists of an air rifle attached to a bench and directed such that the slug that it fires will impact on a small metal plunger placed in the holder section of the apparatus (Fig. 5). This holder section sensibly comprises nothing but a threaded port into which the nozzle body under test is threaded; after which the nozzle assembly is filled with water. A small metal cylinder is then placed in the back of the port, such that when this cylinder is struck by the slug, it is pushed forward and water is forcibly ejected from the nozzle (Fig. 6).

This technique was used to compensate for the wide variety of different nozzle structures which was tested during the course of this program. As the slug passes from the rifle barrel to the impact point, it cuts through two light beams. These light beams serve a dual purpose, firstly they allow determination of the slug impact velocity and secondly they allow triggering of a stroboscopic flash source set up to illuminate the water jet produced by the nozzle. A camera was located on the jet side of the nozzle, set up, initially on the same side of but, ultimately on the opposing side to, the stoboscopic light source. A frosted plastic sheet was also placed over the stroboscopic light source to provide a diffused background illumination.

Two types of photographs were taken with the equipment. In the first of these (Fig. 7), the jet issuing from the nozzle was illuminated by the light source located on the same side as the camera and again contrasted against the dark background. In order to generate the scale, double exposure of film with a grid placed in the jet plane was used. This technique however did not give a sufficient illumination of the structure of the jet and accordingly the light source was moved to the opposing side of the jet for the remainder of the program.

The results obtained from the procedure were of considerable interest in that they allowed a clear discernment of the breakup of the characteristics of the jet stream and thereby allowed some analysis of why the results generated occurred. If one looks, for example, at a typical photograph (Fig. 8) taken by this technique, it should first of all be noticed that the marks at the bottom of the picture indicate 2.5 cm increments away from the nozzle. As the water jet exits from the jet, initially a thin sheet of fluid is created (as aside it may be mentioned that this angle exceeded by approximately 5 deg on average, the nominal angle spread designated by the nozzle manufacturer). As this sheet continues to spread it is disturbed by surface ripples creating relatively thin and thick segments of the jet (Fig. 9). At some point in the jet travel, as the sheet spreads, it universally becomes thinner and will reach a point where it perforates. This point typically is some 5 cm from the nozzle face, although it is a function of nozzle design pressure and, to a degree, fluid properties. When this perforation occurs surface tension virtually immediately causes a circular collapse of the spray in the immediate vicinity. Because of the simultaneously occurrence of this phenomenon at a number of adjacent points within the jet structure, the jet is broken up into a series of circular spaces, rapidly expanding until such time as they meet an adjacent collapsing space and a string of fluid is then left between the two adjacent collapse zones. This string of fluid in itself is not stable and breaks up relatively rapidly into large droplets (large is used here as a relative term only). These droplets in turn as they travel through the air appear to be relatively rapidly broken into a smaller and finer spray so that by the time that the jet has traveled to a distance of some 10 cm from the nozzle all that is left of the initial solid jet structure is a fine spray of droplets and these droplets in turn become further disrupted by passage through the airstream.

After due consideration, it was considered there is likely very little that one can do to change the characteristics of the jet to improve performance over the results obtained without such improving techniques. The next stage in the process however is to examine the effect of such phase jets on cleaning metal surfaces.

3. PLATE CLEANING

Particular question has arisen as to the benefits likely to be achieved where heat is added to the water jet surface and additionally it would be interesting to determine the effects likely to occur where long chain polymeric additives are mixed in the fluid. In order to best illustrate the effectiveness of cleaning the grease, small target plates were prepared and located at 7.5, 15, and 22.5 cm from the leading surface of the nozzle.

In this particular instance the nozzle was traversed across the surface of the plates rather than as has previously been the case, having the target moved. This did cause some reduction in the effectiveness of the jet particularly as the speed of jet traverse increased.

Under the type of oscillating speed which has been discerned in watching high pressure water jet operators clean equipment, single traverses were made over these target plates, a shutter mechanism was located in front of the target plate so that as the nozzle oscillated, only a single traverse of the jet over the surface could be achieved (Fig. 10).

The results of the tests were as follows: in all cases, increasing the temperature of the water jets was clearly beneficial over the case where a cold water test was carried out. The water droplets which remained on the test surface, after the test was completed indicated that they were wetting the surface, at a water jet temperature of 210 deg, which was not the case where cold water was used. The result was clearest where an oil or greasy layer covered the plate and, at 7.5 cm standoff even at 190 deg, the surface appeared clean.

At a distance of 15 cm from the nozzle, the cleaning effectiveness was much reduced. It will be remembered that from the photographic analysis that at this point, the jet was disintegrated into droplets and although the higher temperature fluid was still more effective than cold jets, nevertheless considerable dirt still existed on the target surface.

At a 22.5 deg standoff distance while the hot water gave a discernably cleaner target surfaces than was the case with cold water, very little material removal had occurred and effectively no cleaning of the surface was visible under any of the test conditions.

In order to determine whether adding polymer to the water jet was effective, three different concentrations of three polymers were added to the fluid at varying different concentrations. As a result of the test which was carried out, at 15, and 22.5 cm standoff distances, with cold water some improvement was found. At 15 cm standoff distance the grease layer on the surface was cleaned down to a thin film of grease over the surface; at 22.5 cm standoff distance and at a concentration of polymer around 3000 parts/million, the majority of grease was removed from the surface and only a surface film of grease remained. Thus, some improvement in cleaning performance was discerned when long chain polymers were added to the fluid although the fluid was not sufficiently improved that total surface cleaning could be achieved at 22.5 cm. Photographs of fluid from fan nozzles with polymers were then studied and these are shown in the attached photographs (Fig. 11).

4. FINAL CONCLUSIONS

It is concluded from this study that fan nozzles are only relatively effective in cleaning surfaces where the nozzle is held within 10 cm or less of the target surface. Where such is the case, then the use of high temperature water as opposed to cold water is more effective in removing grease.

5. ACKNOWLEDGEMENTS

This work was funded under contract DAAK11 80 C 0061 from the Chemical Systems Laboratory. Mr. Bruce Lewbart was technical monitor on the project. I am pleased to acknowledge this support and the assistance of Messers. J. Blaine, R. Robison, and J. Tyler of the Rock Mechanics staff and Messers. R. Brandom and T. Fort III as students who carried out, in large part, the work herein described.

6. REFERENCES

1. Pardey, P.H., "Underwater Applications of Continuous Water Jets," Paper G1, Proc. 3rd Int. Symp. Jet Cutting Technology, BHRA, 11-13 May 1976, Chicago, Ill., USA.

2. Rochester, M.C. and Brunton, J.H., "High Speed Impact of Liquid Jets on Solids," Paper A1, Proc. 1st Int. Symp. Jet Cutting Tech., BHRA, 5-7 April 1972, Coventry, UK.

Figure 1. Location of nozzle and test sample of polystyrene on a lathe.

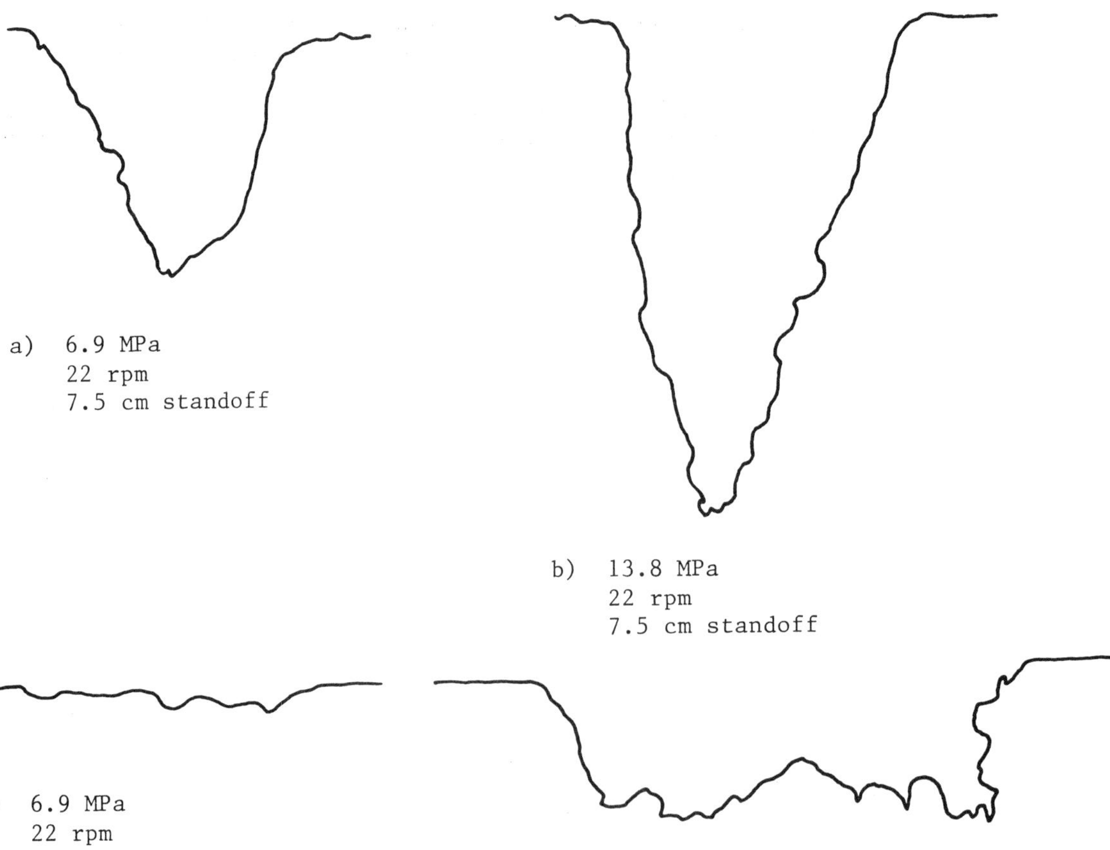

Figure 2. Profiles of the cuts made in polystyrene by (a) and (b) a good nozzle; and (c) and (c) a bad nozzle as a function of pressure and distance. The lathe rpm is also given.

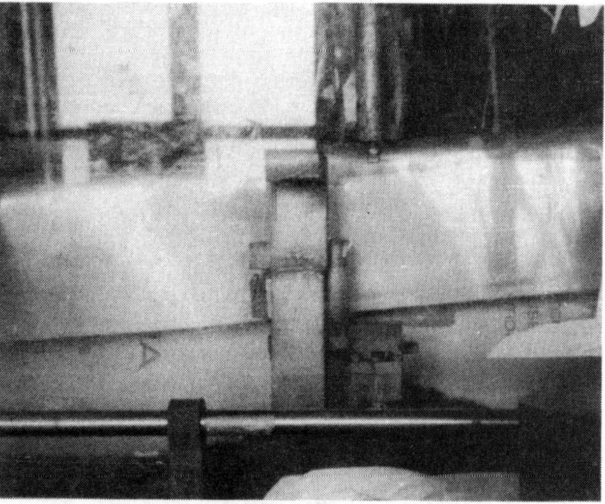

(a) (b)

Figure 3. Layout for pressure profiling of jets using a surface grinder as a traveling carriage (a) showing overall layout; (b) showing pressure transducer stations (with transducers located at the 7.5 and 15 cm stations) and nozzle location.

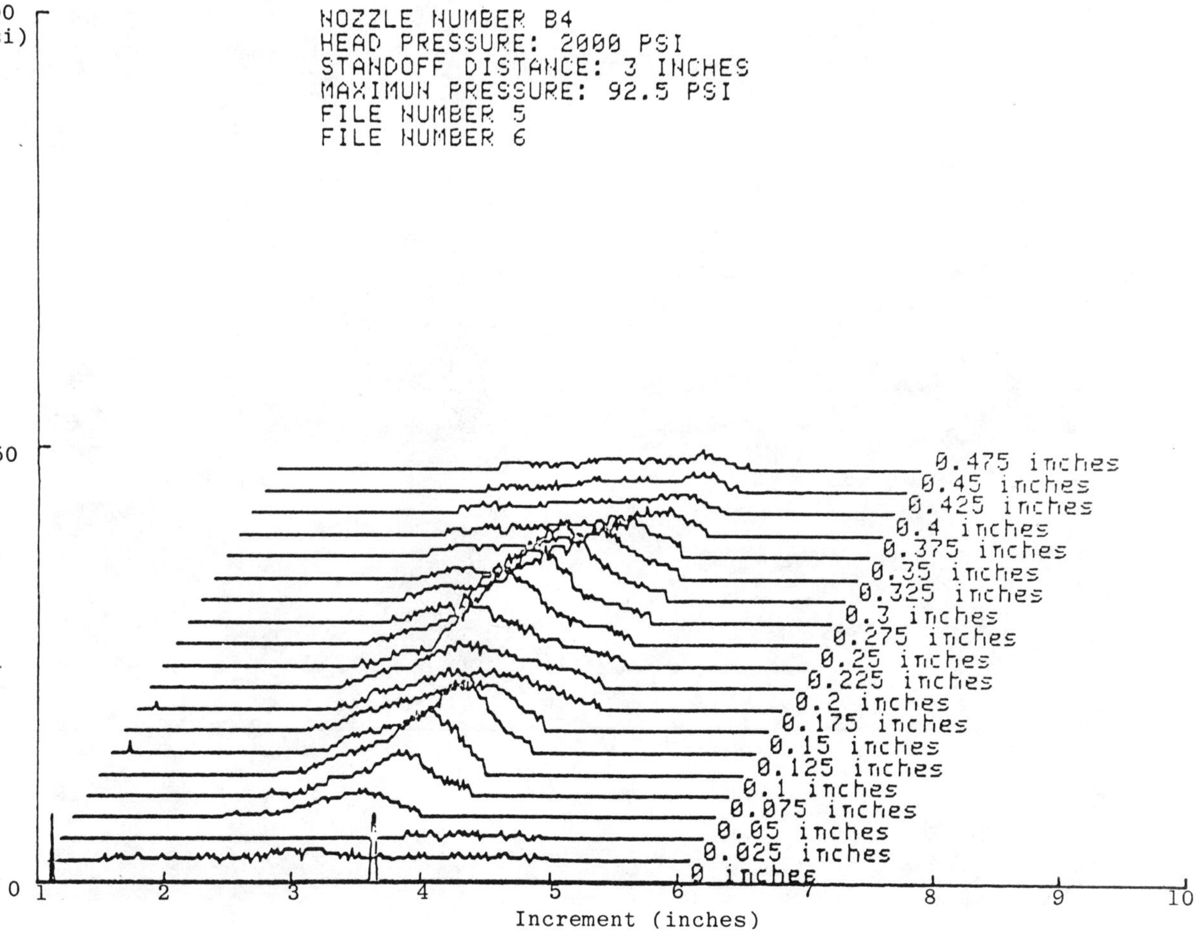

NOZZLE NUMBER B4
HEAD PRESSURE: 2000 PSI
STANDOFF DISTANCE: 3 INCHES
MAXIMUN PRESSURE: 92.5 PSI
FILE NUMBER 5
FILE NUMBER 6

500 (psi)

250

0

0.475 inches
0.45 inches
0.425 inches
0.4 inches
0.375 inches
0.35 inches
0.325 inches
0.3 inches
0.275 inches
0.25 inches
0.225 inches
0.2 inches
0.175 inches
0.15 inches
0.125 inches
0.1 inches
0.075 inches
0.05 inches
0.025 inches
0 inches

1 2 3 4 5 6 7 8 9 10

Increment (inches)

Figure 4. Pressure profiles through a typical jet.

Figure 5. Experimental layout for photography, showing the location of the photodiodes and the nozzle holder, the airgun in the cylinder to the right.

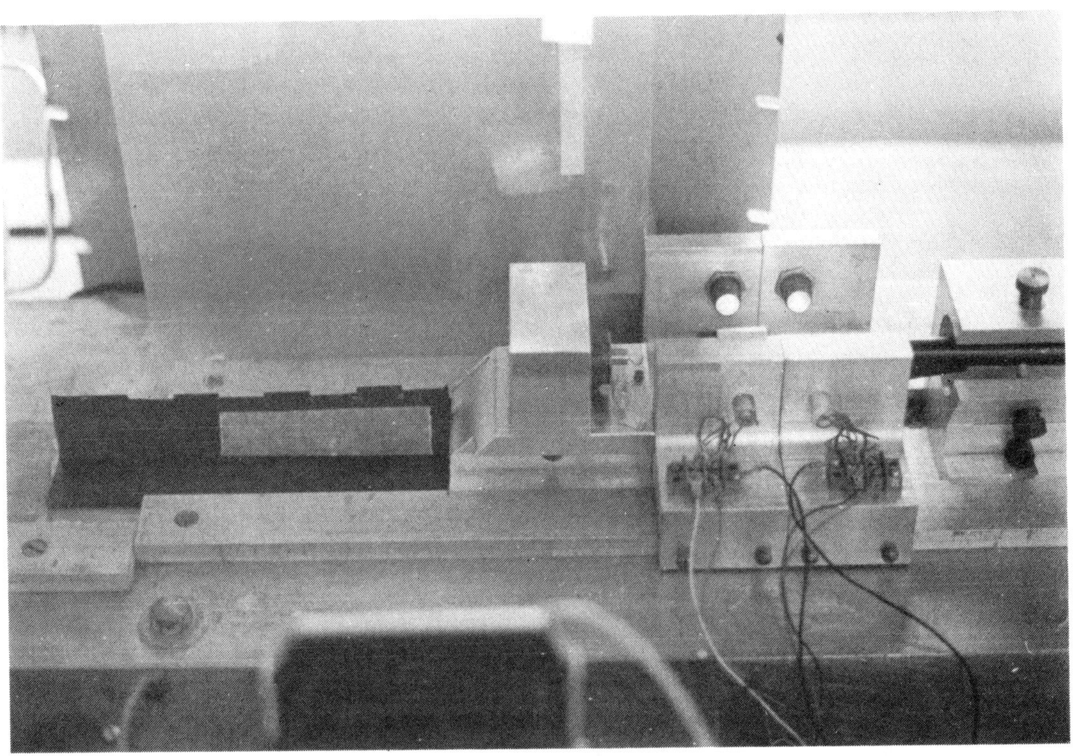

Figure 6. Experimental layout showing light screen behind the nozzle (the camera is the black block in the immediate foreground).

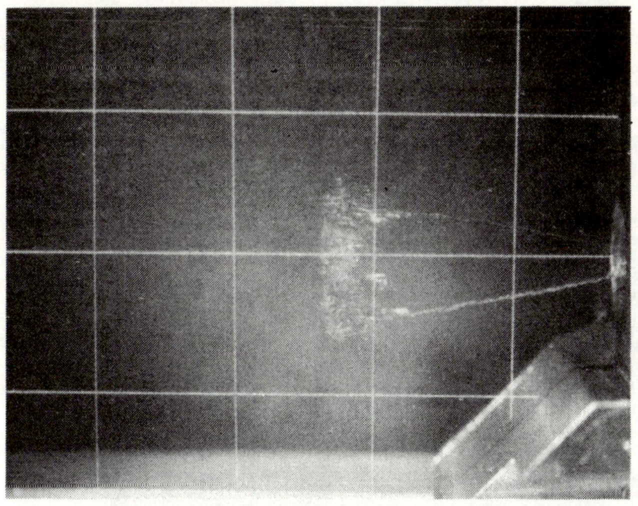

Figure 7. Jet from nozzle 14, travel time 0.54 ms, delay 2.4 ms.

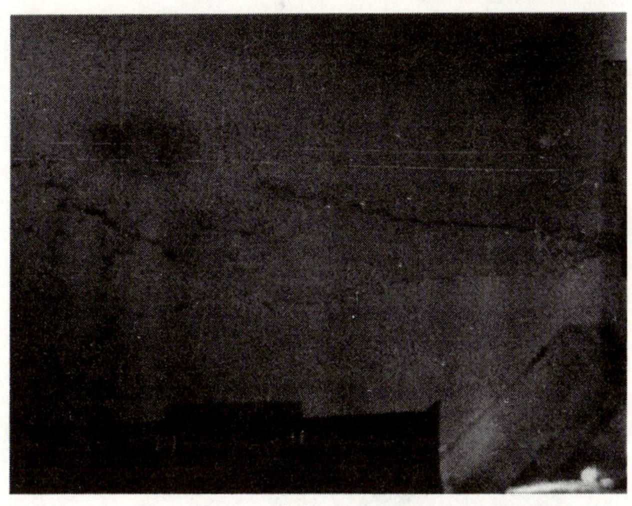

Figure 8. Side view of jet.

Figure 9. Nozzle (as in Fig. 8) turned 90 deg.

(a)

(b)

Figure 10. Showing plate location in front of nozzle. The shutter shown in (b) is laid on the nozzle and falls, after the nozzle has passed out from under it.

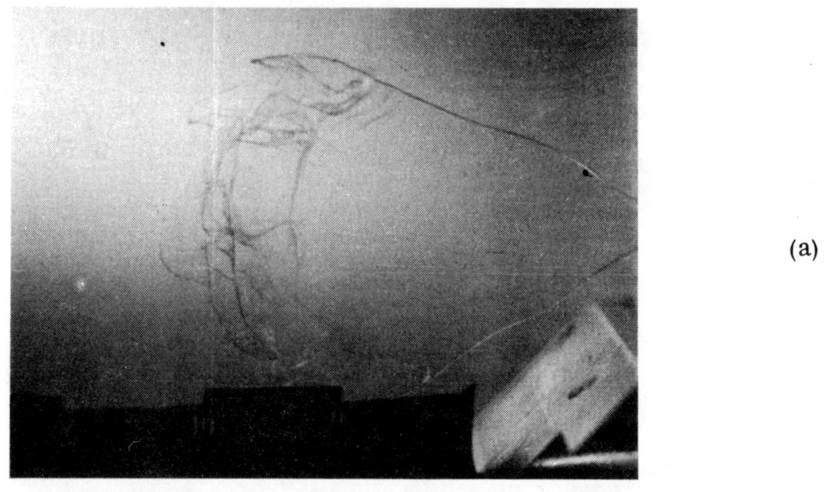

(a)

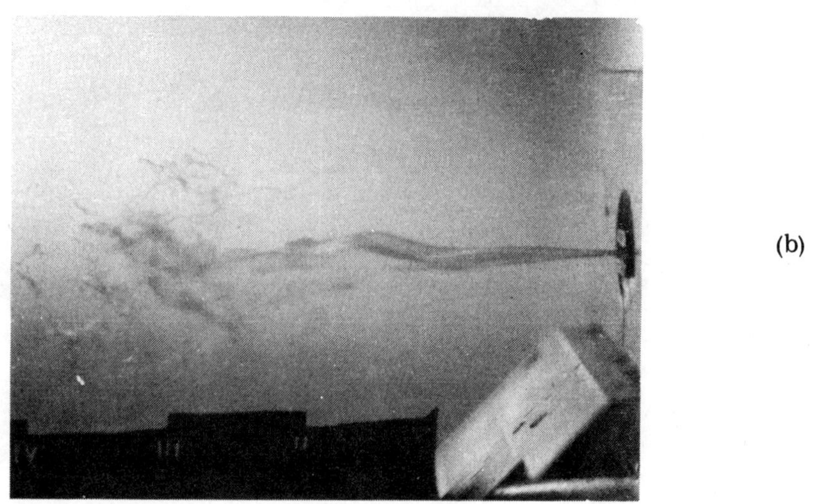

(b)

Figure 11. Front and side views of a jet containing 0.3% polymer solution.

6th International Symposium on
Jet Cutting Technology
6-8, April, 1982

PAPER C2

IMPACT PRESSURE CHARACTERISTICS OF A WATER JET

A.J. Watson, F.T. Williams, R.G. Brade

University of Sheffield, U.K.

Summary

This paper presents an experimental study of the characteristics of water jets propagated at velocities of 170 m/s to over 900 m/s and of the impact pressures produced in a solid target.

Jets in both liquid and gel form are produced using propellant powder in a water cannon. High speed cine photography and a simple velocity measuring device has been used to study the anatomy of the jet. This has been compared with the pressure produced in an instrumented impact test on a steel bar and also with the residual damage produced in steel and plastic targets.

Held at the University of Surrey, U.K.
Symposium organised and sponsored by
BHRA Fluid Engineering

1. INTRODUCTION

This paper presents an experimental study of the characteristics of a water jet propagated at velocities of 170 m/s to over 900 m/s and of the impact pressures produced in a solid target.

Such high velocity water jets are already in use, or under development, in the quarrying and mining industries for rock and coal cutting. They also have applications for cutting out high explosive filled stores without initiating detonation.

High velocity jets are produced by firing water from a barrel with a small diameter nozzle. Cartridges filled with propellant powder provide the driving energy and the water emerges from the nozzle as a reasonably coherent jet. In this work the characteristics of a jet produced in this way have been studied using high speed cine photography and a simple velocity measuring device. The impact pressures have been measured by firing the jet at the end of a Hopkinson bar which is a long steel bar instrumented with electrical resistance strain gauges. More coherent jets have been observed by using water in gel form.

When a water jet impinges upon any material a high initial pressure arises from the water hammer effect. This has been well documented and the theoretical approach well developed by many research workers (Ref. 1). High impact pressures arise during the initial stage of water jet impingement due to compressive deformation of the water before outward radial flow begins. If a hemispherical leading edge is assumed for the jet then this impact pressure lasts until the radial velocity of the circumference of the contact zone becomes less than the compression wave velocity for water. After this point the behaviour is more complex until the dynamic pressure reduces to that given by full imcompressible flow, defined by Bernouilli's equation and termed the stagnation pressure (Ref. 2).

The distribution of impact pressure depends on the shape of the water jet head and both hemispherical and Hertzian pressure distribution have been used for theoretical predictions. The Hertzian approach indicates maximum pressure near the edge of the contact surface and minimum pressure at the centre. This theory has been found most appropriate by many workers especially using large diameter jets (i.e. greater than 5mm). If the leading edge of the water jet is curved, measurement of edge pressure is difficult because the high pressure zone expands radially across the surface and total impact pressure will be a radial integration of the edge pressure.

It is apparent that with very high jet velocities the shock wave velocity rather than the isentropic sound velocity should be used to determine impact pressures. The exact relationship is complex and semi empirical approaches have been found sufficient by various workers to predict the high initial peak pressure over considerable ranges of velocity. This is well documented for both rigid and non rigid targets (Ref. 3 & 4).

The duration of the high initial peak pressure is theoretically given by the time taken for the compression wave in the water to reach the circumference of the impact area. Various researchers have shown the experimental values to be longer than that predicted theoretically. However the duration has been shown to be dependant on the shape of the water jet head and proportional to jet radius (Ref. 5).

2. EXPERIMENTAL METHOD

The experimental set up used in this investigation is shown in Fig. 1 and consists of a cannon which fires the water at a target placed 200mm from the nozzle mouth. In the series of tests described here, the target is either a steel plate; a steel Hopkinson bar 29mm diameter 1.0m long, colinear with the jet; a 380mm sided cube of concrete or a 300 x 300 x 50mm polymer slab. When using the concrete cube the jet was directed at the centre of one face. When using the polymer slab the jet was directed at the centre of a 300 x 50mm edge. Except for the steel bar, these targets were not instrumented and were chosen in an attempt to determine the features of the jet impact by the residual damage produced. This is reported later in the discussion.

High speed cine photographs were taken of the jet from its emergence from the nozzle of the cannon to impact on the target. Impact pressures were obtained from the Hopkinson bar which has electrical resistance strain gauges at 250mm and 750mm from the impact end, Fig. 2. The result from these gauges are recorded on electronic digital storage devices and then transferred to a microcomputer for subsequent storage and further analysis.

3. EQUIPMENT AND INSTRUMENTATION

3.1 Water cannon

The water cannon comprises three basic sections: nozzle, barrel and breech. In all tests the nozzle had an exit diameter of 19mm which tapered at an angle of 2° over 107mm from a barrel bore of 29mm. Initially the water was loaded to fill the barrel and nozzle but in later testing the water has been loaded solely into the barrel and retained with a thin acrylic diaphragm. With the latter method higher velocities are possible. Using carageenan gelatine instead of liquid water a more coherent jet was produced without the variations caused by the front diaphragm which could then be omitted.

The gelled water has to be placed in the cannon while warm and was poured in through the nozzle after a polypropelene piston had been placed at the back of the barrel. A concavity was then left at the front face by the cooling gel which was topped up by more gel to leave a flat front surface. There was some evidence that this might have left a concave discontinuity producing a shaped charge effect which may have ejected the conical section (that used to top up the concave surface) before the main jet.

The method of casting the water gel was therefore changed. With the new method a temporary acrylic diaphragm was inserted between the nozzle and barrel, then 100ml of the gel was poured in from the back of the barrel. The acrylic diaphragm was then removed to leave a flush flat front surface to the gel.

3.2 Water jet velocity measurement

The system initially adopted for velocity measurement was a framework, which supports two sensors each consisting of a pair of brass strips connected to a capacitor. As the water jet impinges upon the leading brass strip of each sensor, it deflects and makes contact with the back strip thus discharging the capacitor which had previously been charged to 27 volts. The collapse of potential on each sensor in turn is recorded on a storage oscilloscope. The time interval between each detection of the water jet head thus enables the average velocity between the sensors to be calculated. This system has been used successfully although the major drawback is that the jet shape and velocity may be affected as it perforates the brass strips.

Initial attempts to photograph the jet at successive intervals of 10.5μs were made with a Barr and Stroud CP5 Rotating mirror camera. These were unsuccessful due to the variability in the time which elapsed between the firing pulse and the ejection of the water and also because the film was fogged by light from the burning propellant gases which were ejected after the water jet.

A Hycam rotating prism cine camera has been used successfully to photograph the water jet at up to 8000 pps. The camera's ancillary equipment was modified to enable the propellant cartridge to be fired automatically by the footage switch using a 27 volt signal. The actual film speed is calculated from a pulsed neon light registered on the edge of the film and hence the water jet velocity can be determined. Reasonable results have been obtained by using 3 or 4 kW photoflood lights placed within 300mm of the water jet and 400 ASA Ilford HP5 film uprated to 1200 ASA during development.

The Hycam cine camera has been used for almost all the tests conducted during the period covered by this paper and in addition to the jet velocity, provides details of the jet anatomy. Later tests used white poster paint mixed in the 100ml of water for greater clarity in distinguishing the water jet.

It was found necessary to provide an adequate power supply because the power surge caused by the camera during the acceleration of the film up to running speed, in conjunction with more than 3kw of photoflood lighting caused blacking out and early triggering of the recording equipment used in the pressure measurement system.

To date, attempts to correlate pressure measurements with water jet photography have been only qualitative mainly because of the difficulties in obtaining a common time base for the high speed cine photographs and the pressure measurements. With the increasing use of the more coherent gelled water jet, consideration may be given to further use of velocity sensors to provide this common time base.

To date all velocity measurements have been taken from the leading edge of the water jet. Photographs from tests 046 to 063 have been examined for any relative velocity differences between the leading edge and the tail of the water jet. However accurate measurements are difficult on the water jet tail as it is dispersed by the gaseous products of the propellant. Tests 064 and 065 using gelled water show the tail of the jet more clearly and measurements on the tail give jet velocities of 230 m/s and 190 m/s respectively. This compares with velocities of 315 m/s and 265 m/s for the leading edge of jet in the same tests. These represent a lower relative velocity for the tail end of 70-75% compared with the jet front. However in these two tests pistons were still used at the propellant end of the water gel and it is possible that the tail end velocity is reduced by the inability of the piston to pass freely through the tapered nozzle. No dispersion of the jet is apparent from the photographs.

3.3 Water jet impact pressure measurement by Hopkinson bar

The majority of tests during the period covered by this report concerned impact pressure measurement of a water jet using the EN27 steel Hopkinson bar, Fig. 2. The series of tests 046 to 065 used charge weights in the ratio 2, 3, 4, 5, 6 and the jet impacted the steel anvil on the Hopkinson bar. The end of the anvil was flush with a steel plate which was to protect the gauges on the bar. Signals from the strain gauge stations (Fig. 2) were amplified and recorded digitally. The Hopkinson bar enables stress pulses to be measured without interference from reflections.

4. TEST RESULTS

All jet velocity measurements have been taken from high speed photographs of the leading edge of the jet and results are given in Table 1.

The high speed photographs have also revealed four principal types of liquid water jet anatomy. These are a) diffuse jet heads, b) precursor jets, c) narrow jets, d) wide spread jet fronts. Occasionally other anomalies appear which do not fit easily into any of these four types and it has been observed that the jet anatomy sometimes changes category during flight and more than one of these categories can exist simultaneously.

Examples of each type are described below and attempts have been made to interpret the pressures recorded on the steel bar with the observed anatomy.

4.1 Diffuse jet head

A diffuse jet head may be either a droplet zone or the actual jet head which has split into fingers. The jet head diameter may be approximately equal to or larger than the nozzle diameter but is of low mean density.

A diffuse jet head of diameter slightly greater than the nozzle is evident on the photographs taken from test 047, Fig. 3. The impact pressure recorded by the Hopkinson bar for this test is also shown in Fig. 3 and shows an ill defined peak lasting 100μs. The separate peaks are probably discrete packets of water impacting the anvil.

Test 050 produced a jet head with a diffuse zone spreading wider than the nozzle diameter, Fig. 4. The Hopkinson bar record gives a low initial peak impact pressure compared with the apparent stagnation pressure. The diffuse jet head may impact over a period of about 200μs with the main jet impact giving another minor peak pressure after that initial period.

4.2 Precursor jet

Precursors appear as narrow jets leading the main jet front and travelling at higher speeds than the main jet. These may be caused by local irregularities in the jet front such as a concavity giving rise to Munroe effects. It may also be a result of the manner in which the front diaphragm ruptured.

Test 049 showed one clearly developed precursor, which is evident as soon as the jet emerges from the nozzle, and a second precursor developing by the time the first impacts the Hopkinson bar, Fig. 5. The photograph also shows that the precursors are followed by a diffuse region.

The interframe time in this test is 136μs and impact of the first precursor occurred 136μs before the main jet front. However the Hopkinson bar recording in Fig. 5 shows two impact pressure peaks 26μs apart. It appears therefore that the second precursor impacts 26μs after the first and the diffuse main jet front gives rise to low pressure peaks.

The photographs in Fig. 6 shows a precursor which impacted the Hopkinson bar about 75μs before the main jet front, which appeared to have a velocity of about 400 M/s. The Hopkinson bar recording shown in Fig. 6 substantiates this as there is an initial double peaked impact pressure followed by a further pressure peak 75μs later. The initial double peaked impact could be caused by an irregular front to the spike.

It appears from other tests e.g. 062, that precursors may have velocities considerably in excess of the main body of the jet.

4.3 Narrow jet

This category is devoted to those jets which have a core of diameter much less than the nozzle diameter. The core is typically surrounded by a droplet zone which has a diameter at the most not much greater than the nozzle diameter. The peak impact pressure is typically higher than average although the velocity does not appear from photographs to be particularly high with respect to the higher than average impact pressure peak.

A droplet cloud sometimes shrouds a narrow jet which produces a high impact pressure peak evident on the Hopkinson bar recording. The photograph in Fig. 7, test 056 shows that the droplet cloud spreads out and a more coherent jet emerges. The Hopkinson bar recording shows an irregularity before the peak impact pressure, Fig. 7, which may be caused by the droplet cloud.

Multiple narrow jets sometimes appear. A double jet is evident in the photograph on Fig. 8 and from the Hopkinson bar recording the main jet appears to impact about 100μs after the first recorded peak impact pressure which is recorded as a single peak.

The narrow jets often have a high velocity (>900 m/s) and can show multiple initial impact pressure peaks but with other peaks following.

4.4 Widespread jet fronts

This category is devoted to jet anatomies that have a leading edge or edges that spread over a much wider area than the nozzle diameter. For instance photographs from test 062, Fig. 9, show a widespread jet developing. It appears to spread over a diameter of about 125mm but has also a precursor jet which is probably responsible for the high peak impact pressure in the Hopkinson bar recorded in Fig. 9 with the diffuse widespread jet head giving the lower impact pressures following the peak value.

Photographs from test 063, Fig. 10 show a widespread jet has developed and at impact the jet front spreads over a diameter of about 115mm. The Hopkinson bar recording in Fig. 10 shows three peaks spread over 250μs. The first impact pressure peak may be the result of a precursor jet while the other two may be caused by anomalies in the jet anatomy concealed by the droplet cloud.

5. DISCUSSION AND CONCLUSIONS

The variation in jet front velocity, jet anatomy, and impact pressures does not appear as yet to allow any rigorous mathematical analysis of the results. These experimentshave provided a photographic record of the water jet produced by a water cannon using propellant charges in the ratio 2 to 6 with impact pressures measured using the 29mm diameter EN27 steel Hopkinson bar. For each test 8-10 photographs of the jet have been examined to determine the anatomy of the water jet and calculate velocities. These photographs have then been compared with the resulting impact pressures recorded from the Hopkinson bar strain gauges.

The Hopkinson bar anvils have an impact surface 29mm in diameter and the nozzle is 19mm diameter. Those jets described as narrow jets or precursors have a diameter of less than 19mm and may well in some cases be less than 3mm diameter. Indentations in the anvils, Fig. 11, provide evidence that leading edges may be of this magnitude but not necessarily circular.

The impact pressure recorded from strain gauges on the 29mm diameter Hopkinson bar would therefore be reduced by the ratio of the area of the jet to the bar.

Droplet zones around the jet core have made an accurate assessment of jet diameter from photography difficult. However the polymer block in test 044 was a composite Polyol and Isocyanate Polymer with a Shore A hardness value of 70 and the surface of the sample was pitted by the impact in segments of an annulus of outside diameter about 30mm and width 5mm, Fig. 12. Tensile cracking occurred in the centre of the impacted area. The high speed cine photograph on Fig. 12 indicate that this jet is diffuse and the diameter scaled from the photograph appears to be in excess of 38mm.

When the water was gelled using carageenan the jet appeared to be more coherent and observations from a typical high speed cine photograph, test 064, show a parallel jet emerging from the nozzle and remaining parallel up to target impact. The jet head is well defined but appears to have a leading edge which is inclined at 45° to the vertical. The density of the jet also appears to vary with a dense jet head followed by a less dense region again followed by a dense coherent jet. The impact pressure recording shows multiple impact pressure peaks lasting about 100μs then a flat response stagnation pressure. It has been mentioned earlier that as the gel cooled in the water cannon it contracted leaving a slightly concave front surface. In a later test, 065, the concavity left by the cooling gel was topped up by more gel to leave a flat front surface. Photographs of this test again show a parallel sided jet but show a narrow low density jet impacting the Hopkinson bar some 800μs before the dense coherent jet which travels at a velocity of 266 m/s. No impact pressure recording is available for this later jet but from the Hopkinson bar recording the narrow diffuse jet appears to give a typical high impact pressure peak followed by a relatively constant apparent stagnation pressure.

ACKNOWLEDGEMENTS

This research is being carried out at the Department of Civil and Structural Engineering, University of Sheffield supported by Procurement Executive, Ministry of Defence. Thanks are due to MoD for the finance which allows the work to continue and to the technical staff of the Department for their interest and enthusiasm.

REFERENCES

1. Haskins, P.J., Private communication, MoD (RARDE), 1979.

2. Hancox, N.L. and Brunton, J.H., "Erosion of solids by the repeated impact of liquid drops", Phil. Trans. Roy. Soc. A. 260, 1966.

3. Rochester, M.C. and Brunton, J.H., "High speed impact of liquid jets on solids", 1st Int. Symp. on Jet Cutting Tech., 1972.

4. Heymann, F.J., "On the shock wave velocity and impact pressure in high speed liquid impact", Trans. ASME Jnl. of Bas. Eng., 1968.

5. Johnson, W. and Vickers, G.W., "Transient stress distribution caused by water jet impact", Jnl. Mech. Eng. Sc., Vol. 15, No. 4, 1973.

TABLE 1

TEST NO	CARTRIDGE RATIO	INSTRUMENTATION Velocity rig dimensions (mm) Camera speed, etc. Hopkinson bar dia = d(mm) Standoff distance = S(mm)	JET VELOCITY (ms^{-1})	RESULTS IMPACT PRESSURE (N/mm^2) Station No.	APPARENT STAGNATION PRESSURE (N/mm^2) (1) or (2)	COMMENTS
044	3	Polymer Block A, 70 Shore A hardness S = 200	max 460 red to 440	-	-	Slight erosion occurred over segments of annulus outer dia 25 mm, tensile crack in centre of impact area.
045	3	Polymer Block A, 70 Shore A hardness S = 200	max 370 red to 290	-	-	All tests from 045 onwards have used white poster paint dissolved in the water to improve the contrast on photographs.
047	3	Hopkinson bar EN27, d = 29, S = 200 Hycam @ 7660 pps	380	(1) 148 (2) 132	47	Jet head appears diffuse giving a series of shock impact pressures over ≃ 130 μs.
049	4	Hopkinson bar EN27, d = 29, S = 200 Hycam @ 7350 pps	varies 230/460	(1) 302 (2) 279	71	Photographs show two precursors leading a diffuse jet head. Precursors give two high shock pressures 26 μs apart.
050	4	Hopkinson bar EN27, d = 29, S = 200 Hycam @ 7410 pps	460	(1) 121 (2) 118	55	Diffuse jet head giving low shock pressure over about 220 μs.
056	5	Hopkinson bar EN27, d = 29, S = 200 Hycam @ 7500 pps	810	(1) 202 (2) 228	39	High velocity. Probable narrow jet shrouded by droplet cloud leading main jet.
057	3	Hopkinson bar EN27, d = 29, S = 200 Hycam @ 7500 pps	varies 170/410	(1) 68 (2) 79	32	Early diffuse jet followed by faster main jet.
058	3	Hopkinson bar EN27, d = 29, S = 200 Hycam @ 7710 pps	max 440 red to 390	(1) 78 (2) 85	48	Initial shock pressure followed by diffuse zone then main jet.
059	4	Hopkinson bar EN27, d = 29, S = 200 Hycam @ 7130 pps	max 470 red to 390	(1) 179 (2) 185	51	Precursor jet giving highest shock pressure followed by further impact pressure peaks.
060	5	Hopkinson bar EN27, d = 29, S = 200 Hycam @ 7530 pps	460	(1) 178 (2) 159	45	Narrow jet front with possibly two precursors giving multiple shock pressure peaks.
061	6	Hopkinson bar EN27, d = 29, S = 200 Hycam @ 7400 pps	>910	(1) 146 (2) 152	43	Very high velocity narrow jet front. Jet appears to become denser and more coherent after ≃ 500 μs. Low shock pressure.
062	6	Hopkinson bar EN27, d = 29, S = 200 Hycam @ 7630 pps	620 red to 290 see notes	(1) 293 (2) 250	61	Possible precursor jet at >700 m/s, giving multiple high shock impacts, followed by diffuse jet with wide spread.
063	6	Hopkinson bar EN27, d = 29, S = 200 Hycam @ 7510 pps	870	(1) 194 (2) 180	44	Precursor followed by droplet shroud shielding main jet. Two equal shock impacts 140 μs apart.
064	2	Hopkinson bar EN27, d = 29, S = 200 Hycam @ 7660 pps	310	(1) 48 (2) 48	20	Carageenan gelatine 440 used to gel water. Dense jet head followed by 75-100mm diffuse then dense jet.
065	2	Hopkinson bar EN27, d = 29, S = 200 Hycam @ 7320 pps	390-180 main jet 270	(1) 34 (2) 35	19	A narrow diffuse jet emerges ≃500 μs before main jet. Diffuse jet reduces in velocity before impact.

Fig. 1 WATER CANNON EXPERIMENTAL RIG

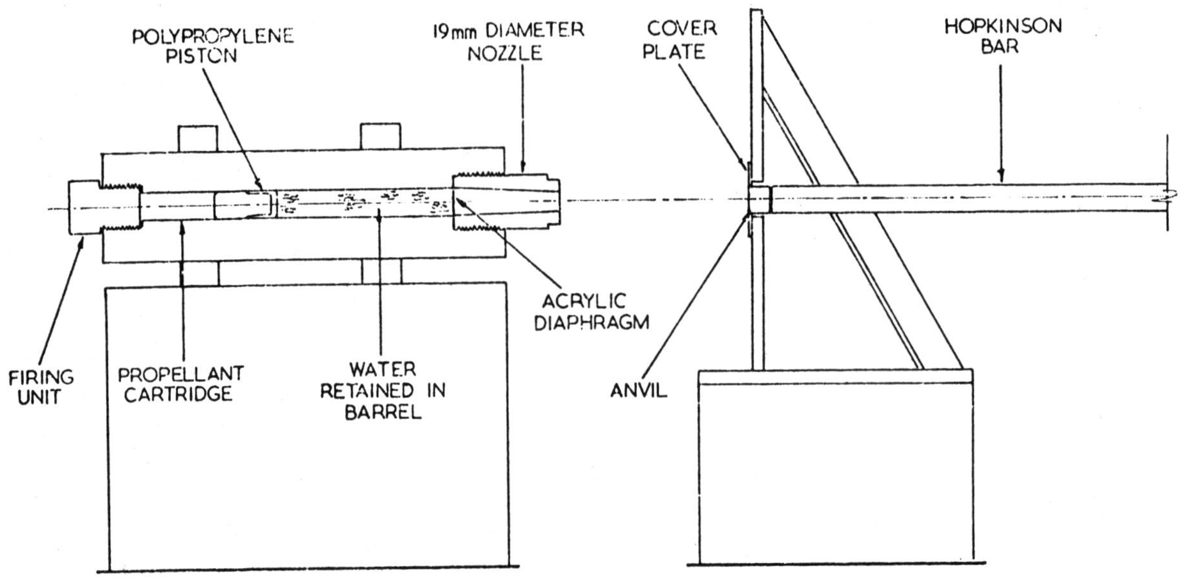

Fig. 2 EN27 HOPKINSON BAR - STRAIN GAUGE POSITIONS

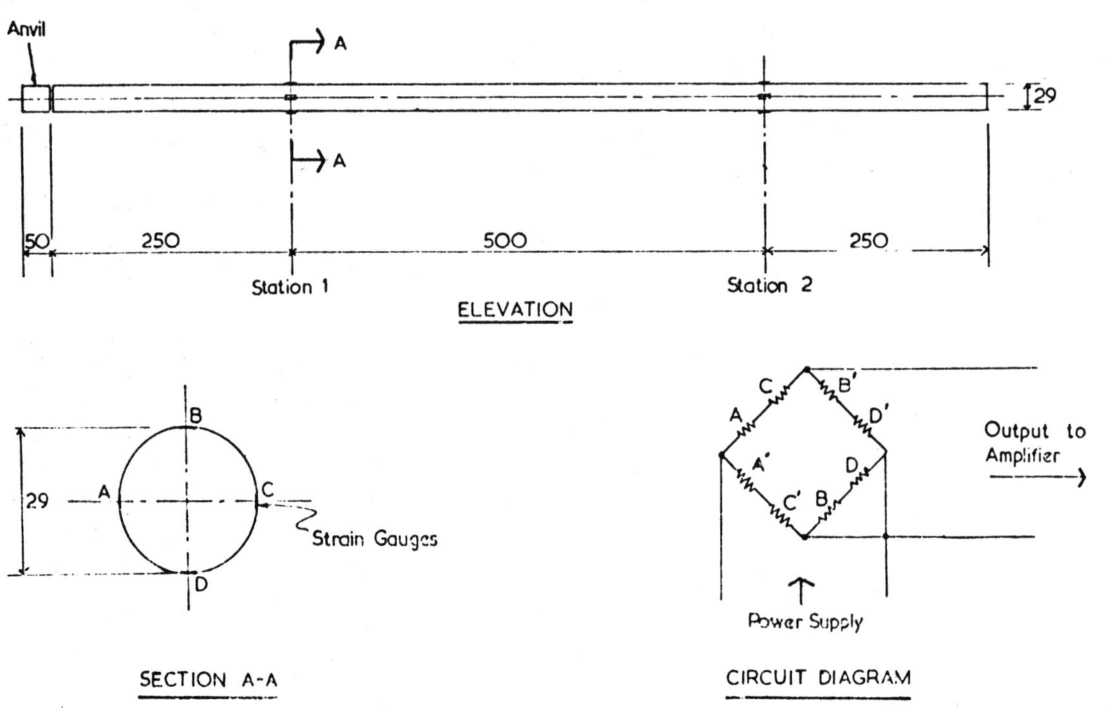

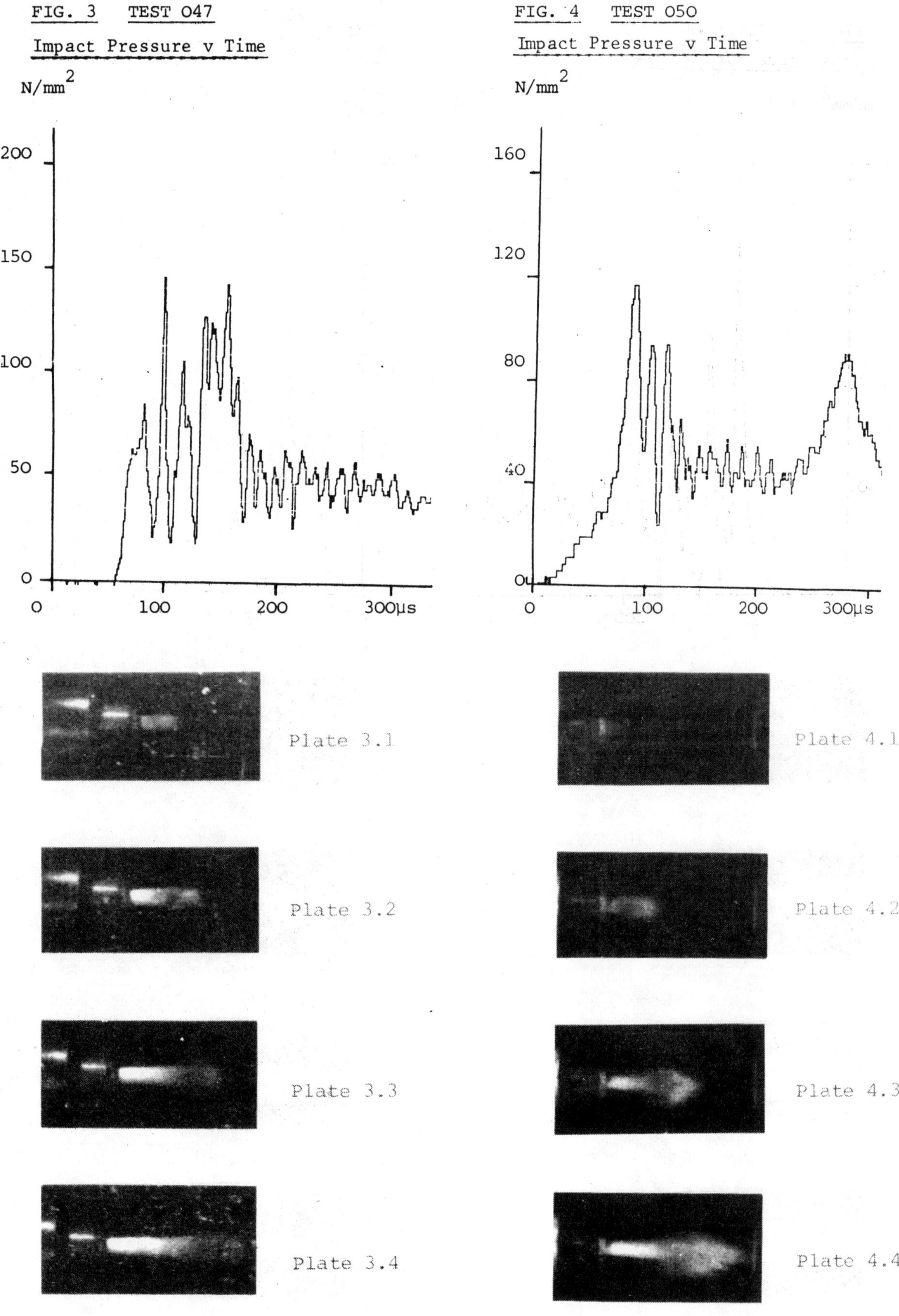

Interframe time 130μs

Interframe time 135μs

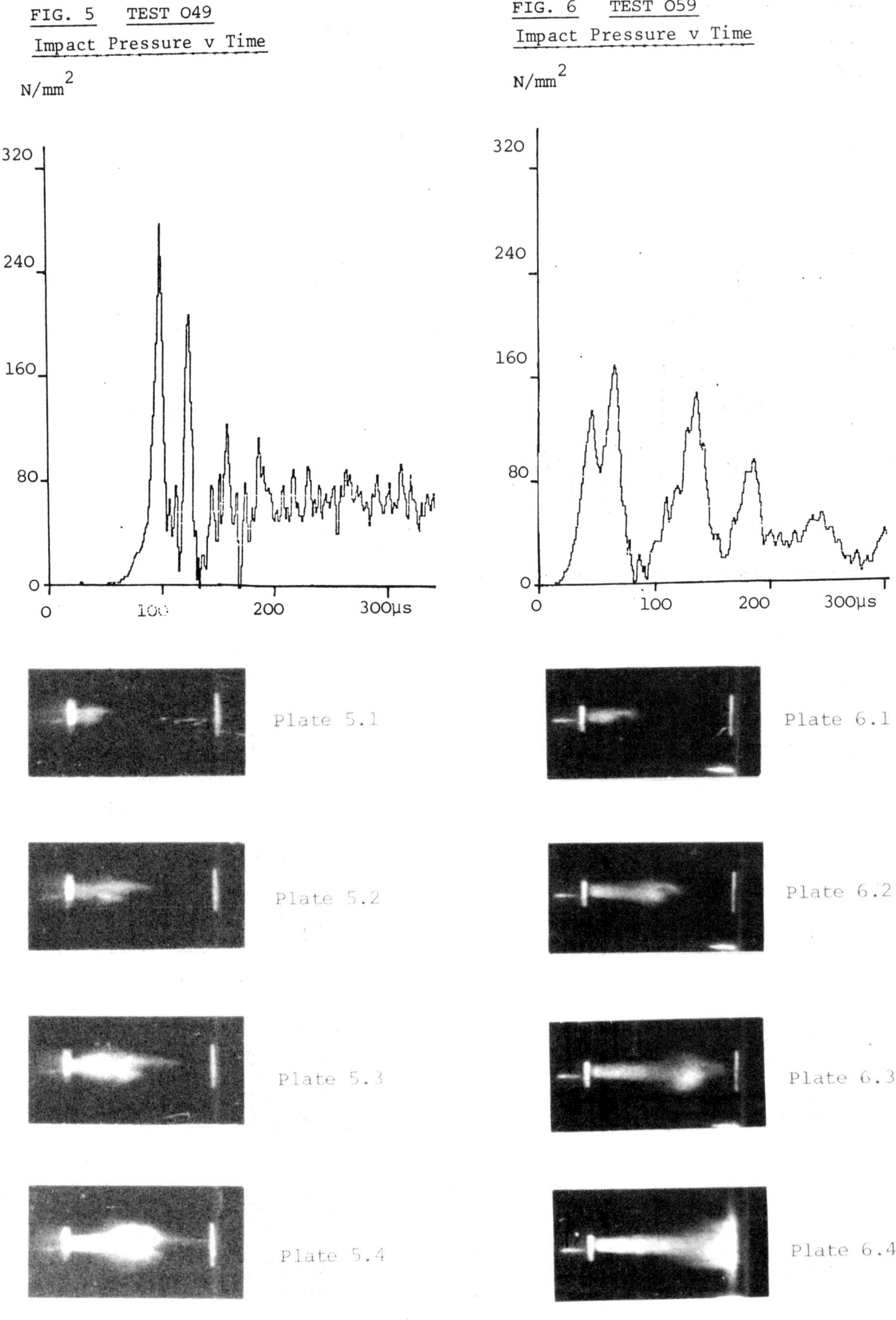

Interframe time 136µs

Interframe time 140µs

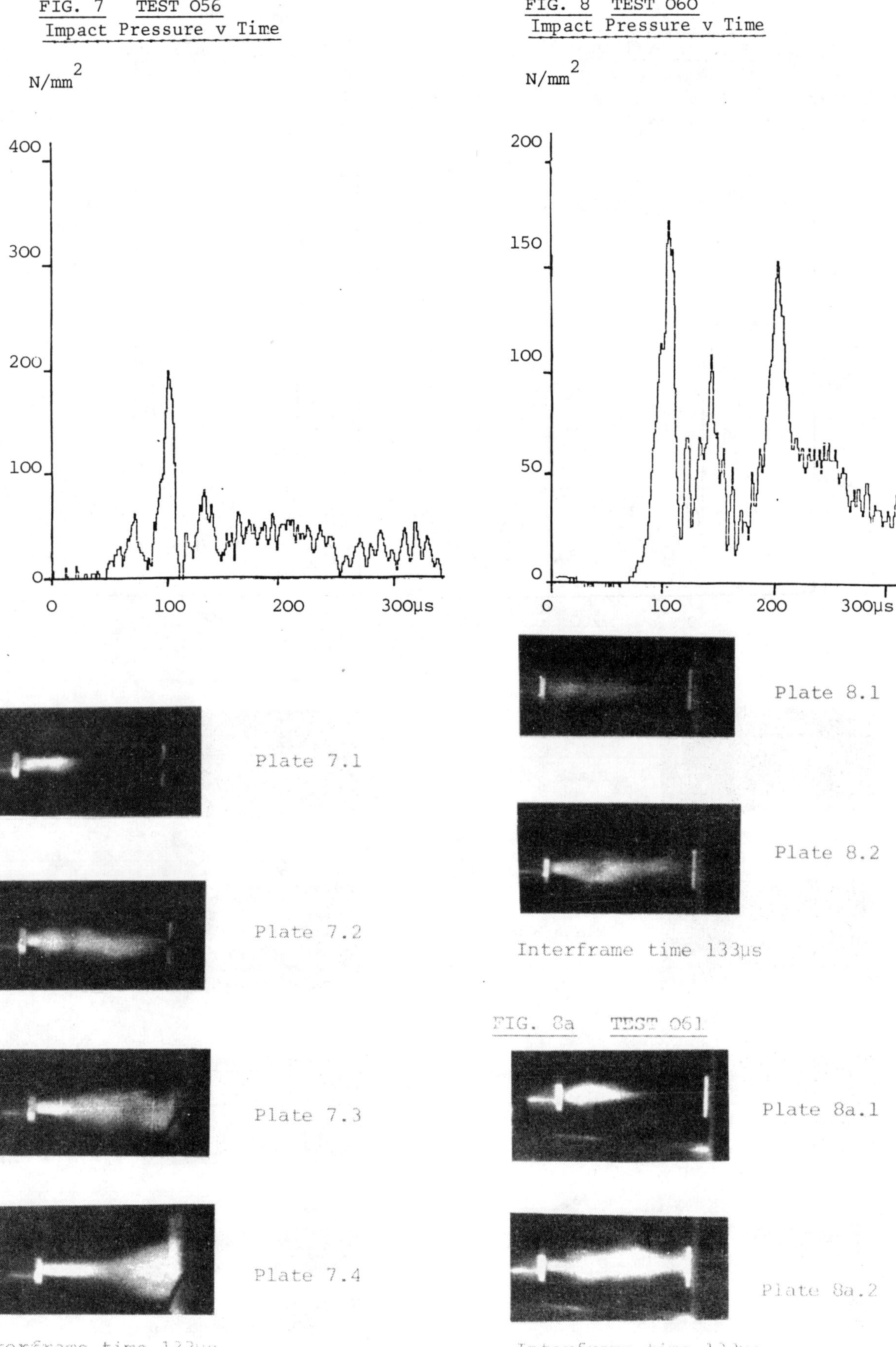
FIG. 7 TEST 056
Impact Pressure v Time
N/mm^2
400
300
200
100
0
0
100
200
300µs
FIG. 8 TEST 060
Impact Pressure v Time
N/mm^2
200
150
100
50
0
0
100
200
300µs
Plate 7.1
Plate 7.2
Plate 7.3
Plate 7.4
Interframe time 133µs
Plate 8.1
Plate 8.2
Interframe time 133µs
FIG. 8a TEST 061
Plate 8a.1
Plate 8a.2
Interframe time 133µs

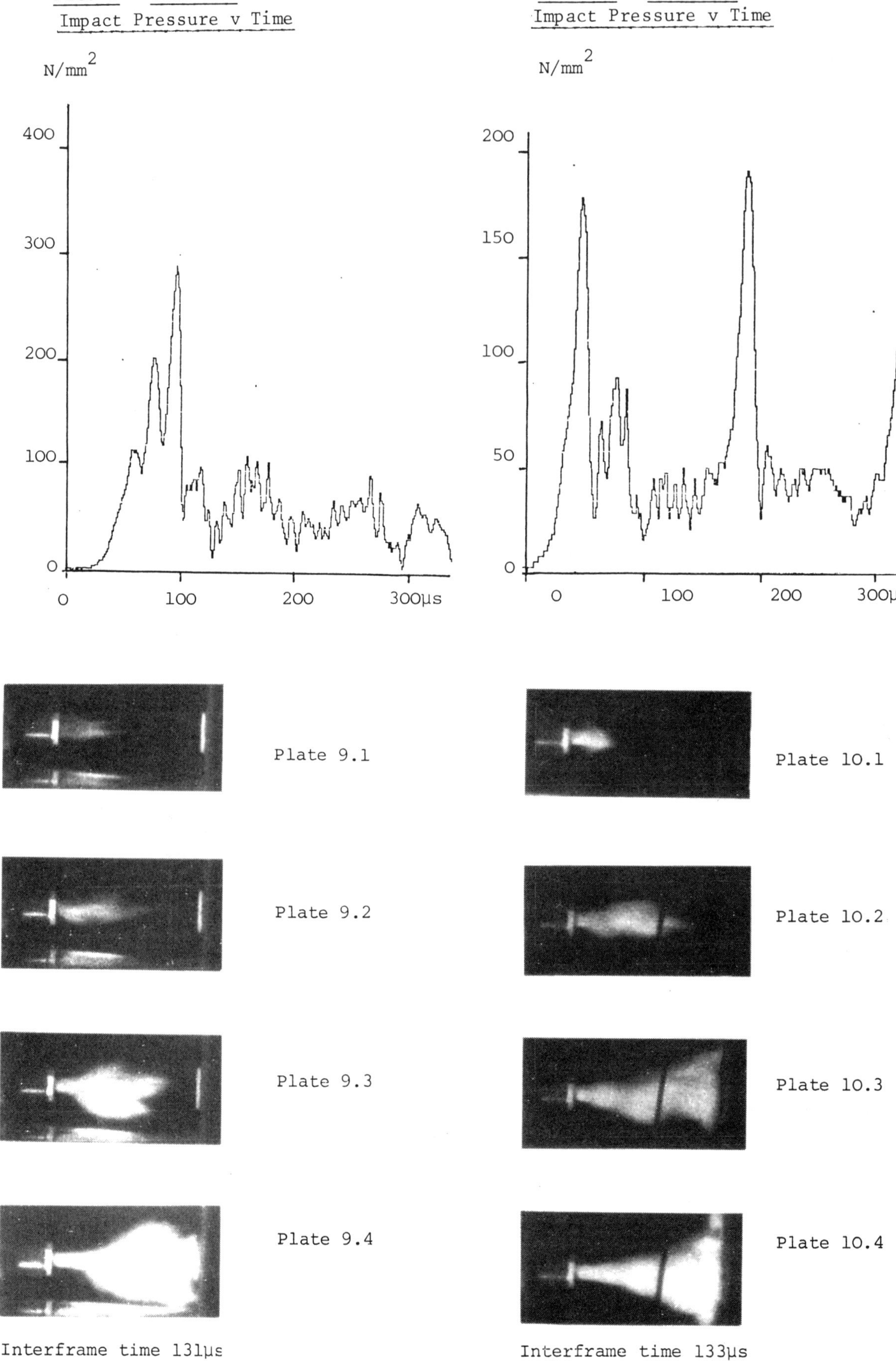
FIG. 9 TEST 062
Impact Pressure v Time
N/mm^2
400
300
200
100
0
0
100
200
300µs
FIG. 10 TEST 063
Impact Pressure v Time
N/mm^2
200
150
100
50
0
0
100
200
300µs
Plate 9.1
Plate 9.2
Plate 9.3
Plate 9.4
Interframe time 131µs
Plate 10.1
Plate 10.2
Plate 10.3
Plate 10.4
Interframe time 133µs

FIG. 11 EN27 STEEL HOPKINSON BAR ANVILS

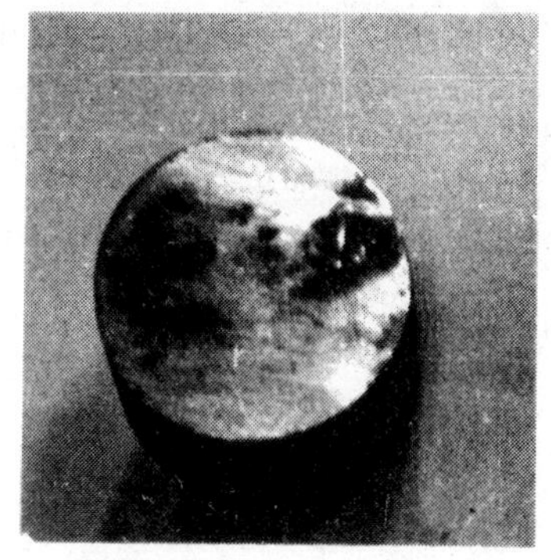

10mm

Plate 11.1 TEST 059

10mm

Plate 11.2 TEST 060

FIG. 12 WATER JET IMPACT ON POLYMER BLOCK

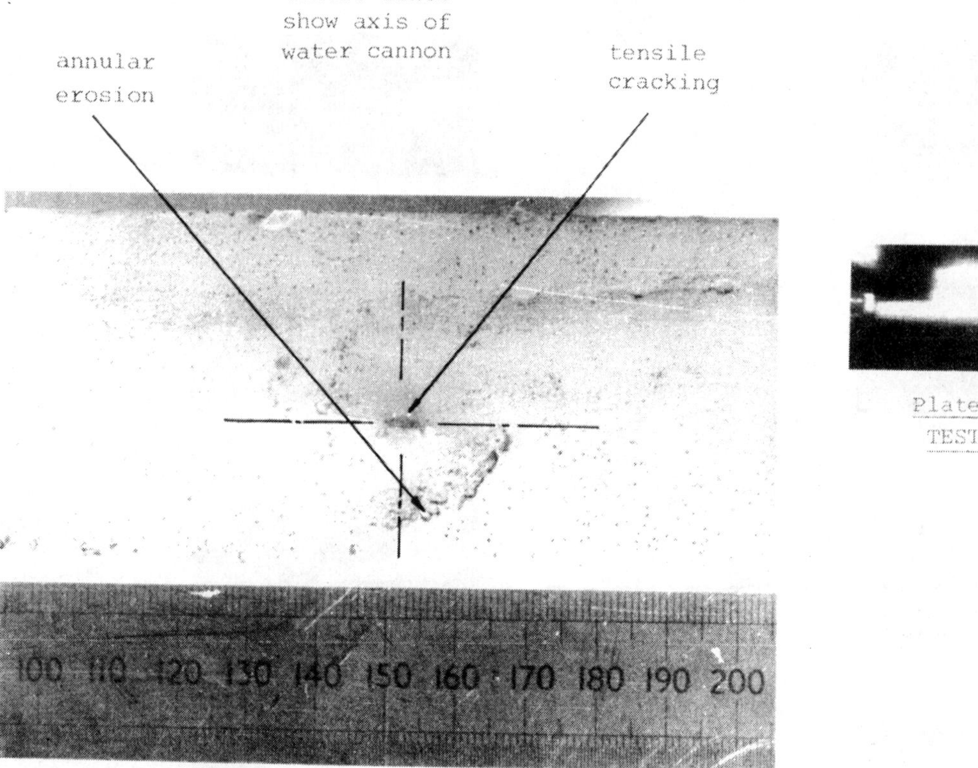

Plate 12.1 TEST 044

Plate 12.2 TEST 044

FIG. 13 TEST 064

Impact Pressure v Time

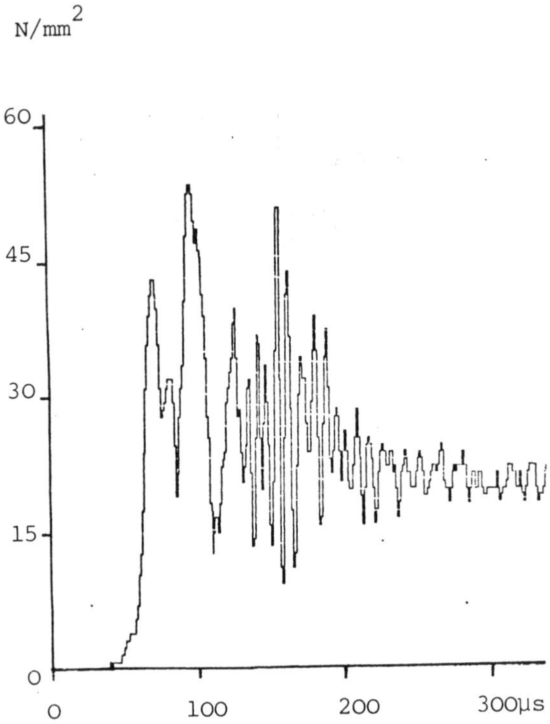

N/mm^2

60
45
30
15
0
0 100 200 300μs

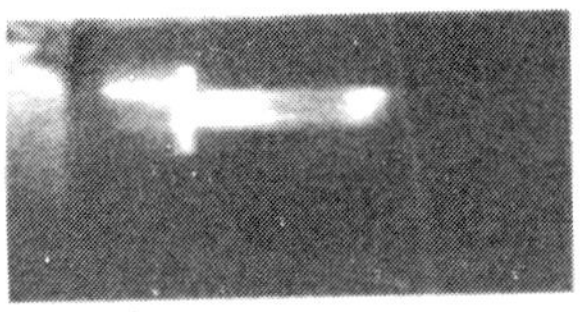

Plate 13.1

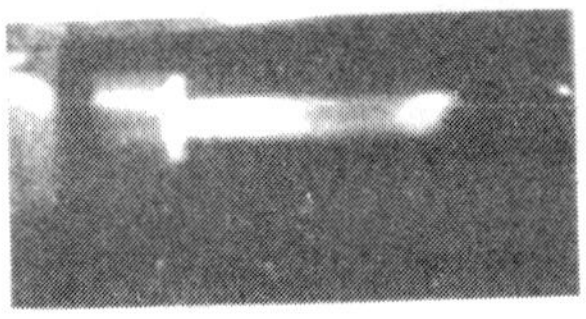

Plate 13.2

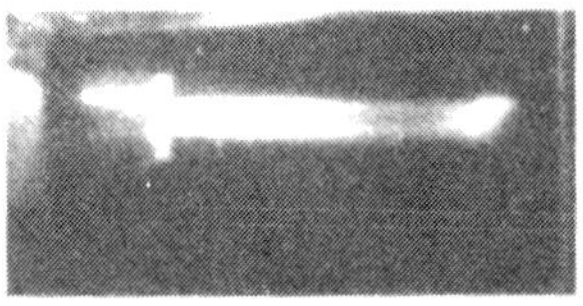

Plate 13.3

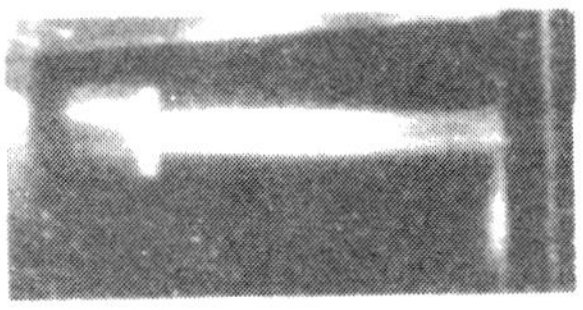

Plate 13.4

Interframe time 137μs

FIG. 14 TEST 065

Impact Pressure v Time

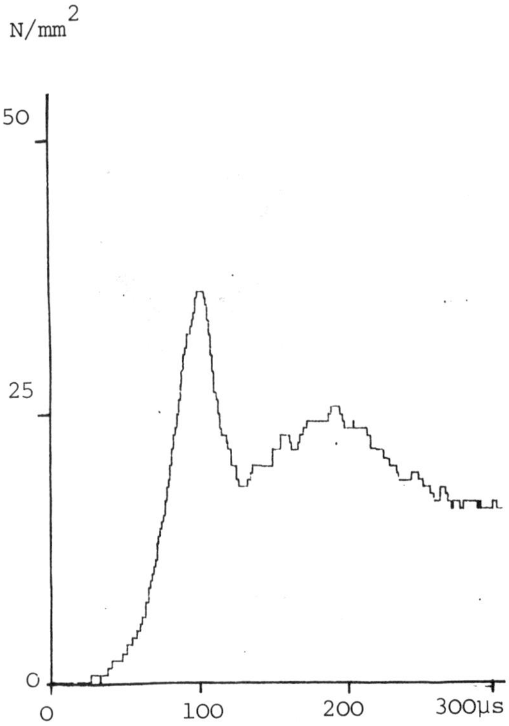

N/mm^2

50
25
0
0 100 200 300μs

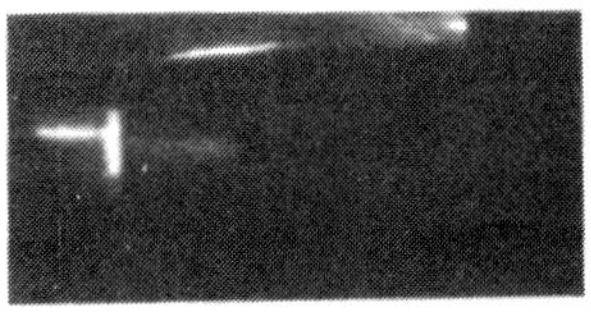

Plate 14.1

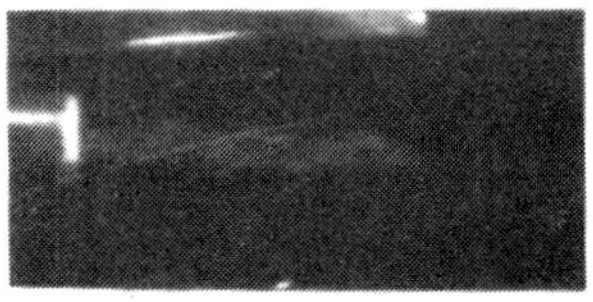

Plate 14.2

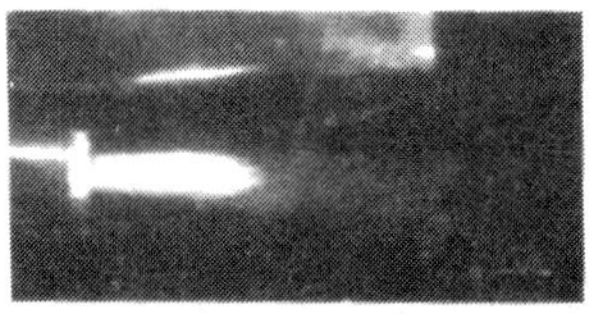

Plate 14.3

Plate 14.4

Interframe time 136μs

6th International Symposium on
Jet Cutting Technology
6-8, April, 1982

A NUMERICAL AND EXPERIMENTAL INVESTIGATION OF HIGH-VELOCITY JETS IMPINGING ON A FLAT PLATE

R.S. Amano and K.F. Neusen

University of Wisconsin-Milwaukee, U.S.A.

Summary

A numerical and experimental study is made of the characteristics of turbulent axisymmetric jets impinging on a flat plate. The purpose of the study is to obtain a better understanding of the behaviour of high-velocity jets used in jet cutting.

In the computations, a hybrid scheme of central and upwind finite differencing is used to solve the full Navier-Stokes equations. The turbulence model used in this computation is a two-equation turbulence model which provides a turbulent kinetic energy equation and a turbulent energy dissipation equation (k- ε model). For cells at the wall, the wall function is treated in a special way such that the turbulent kinetic energy has a parabolic profile in the viscous sublayer and has a linear profile in the fully turbulent region. This gives better physical meaning and accuracy, and also saves much computational time since it is not necessary to use very fine grids near the wall. The turbulence energy and the dissipation rate at the exit of the nozzle are obtained carefully, taking the nozzle design into account.

The experimental portion of the study was performed on a quasi-steady jet flow system. Each experiment was begun at a nozzle pressure of 100,000 psi (689 MN/m^2), with the flow velocity and pressure gradually decreasing to zero after the reservoir valve is opened. A pressure transducer and a force transducer were used to indicate the nozzle pressure and plate force respectively as the experiment progressed.

The computed results of pressure, force on the wall (plate), and velocities in the computational field are compared with the experimental data. The computations are made for stand-off distances ranging from 16 to 28 nozzle diameters and for nozzle Reynolds numbers ranging from approximately 10,000 to 30,000.

Held at the University of Surrey, U.K.
Symposium organised and sponsored by
BHRA Fluid Engineering

NOMENCLATURE

C_1, C_2, C_μ	=	coefficients in turbulence model
C_f	=	skin friction coefficient ($\tau_W / {}^1{}_2 \rho U^2{}_N$)
C_V	=	velocity coefficient
D	=	nozzle diameter
E, N, S, W	=	node points at east, north, south and west respectively
F	=	force
H	=	stand-off distance (nozzle-to-plate distance)
i	=	turbulence intensity
k	=	turbulent kinetic energy
l_m	=	mixing length
P	=	pressure
Re_D	=	nozzle Reynolds number
r	=	radial coordinate
r_N	=	nozzle radius
t	=	time
U	=	velocity in χ direction
U_N	=	nozzle exit velocity
V	=	velocity in r direction
χ	=	coordinate parallel to nozzle
y	=	distance from wall
Γ_ϕ	=	diffusivity of ϕ
ϵ	=	turbulence energy dissipation rate
κ	=	von Karman constant
λ	=	constant
μ_1	=	dynamic viscosity of fluid
μ_T	=	turbulent viscosity
ρ	=	density
$\sigma_k, \sigma_\epsilon$	=	turbulent Plandtl numbers for diffusion of k and ϵ respectively.

τ	=	shear stress
ϕ	=	variable

<u>Subscripts</u>

E,N,P,S,W	=	values at node points, E, N, P, S, and W.
eff	=	effective valve of diffusion coefficient, i.e. sum of molecular and turbulent contributions.
in	=	value at the inlet condition.
k	=	value pertaining to kinetic energy.
n	=	levels prevailing at outer edge of near-wall cell.
T	=	turbulent value quantity.
w	=	wall values
ϵ	=	value pertaining to energy dissipation rate.

1. INTRODUCTION

The fluid dynamics of a free jet impinging on a solid surface are basic to the study of jet cutting technology. As advancements are made in the area of analyzing turbulent high velocity free jets, the probabilities increase that a comprehensive model can be constructed. In the long run, the calculation schemes for jets will have to be linked to calculation schemes which describe the target material removal dynamics, if a comprehensive jet cutting model is to evolve.

In the present study, a numerical procedure is described which allows solution of the Navier-Stokes Equations in conjunction with a two-equation turbulence model. The procedure is then applied to jets with nozzle exit velocities up to 900 m/s. To complete the study, experiments were run over the same range of jet velocity and at two different target to nozzle stand-off distances.

It is, of course, possible to make some comparisons between the numerical and experimental results. Pressure distributions over the whole flat plate target were obtained from the calculations. These pressures were then integrated over the plate area to produce the normal force on the plate (in the axial direction). Thus enabling a comparision to be made with the experimentally determined forces. The calculations provide some additional features that are not easily measured (and were not measured in the experiments). These include: 1) velocity distribution in the free jet; 2) pressure distribution in the jet; 3) skin friction coefficient on the plate.

2. MATHEMATICAL FORMULATION

The following set of equations, incorporating the Boussinesq turbulent viscosity concept, describe the mean velocity fields in statistically stationary, turbulent flows of an incompressible jet impinging on a flat plate.

Continuity equation:

$$\frac{\partial}{\partial x}(r\rho U) + \frac{\partial}{\partial r}(r\rho V) = 0 \qquad (1)$$

Momentum equations:

$$\rho\frac{DU}{Dt} = -\frac{\partial P}{\partial x} + \frac{1}{r}\left[\frac{\partial}{\partial x}\left(r\mu_{eff}\frac{\partial U}{\partial x}\right) + \frac{\partial}{\partial r}\left(r\mu_{eff}\frac{\partial U}{\partial r}\right)\right] + S_u$$

$$S_u = \frac{\partial}{\partial x}\left(\mu_{eff}\frac{\partial U}{\partial x}\right) + \frac{1}{r}\frac{\partial}{\partial r}\left(r\mu_{eff}\frac{\partial V}{\partial x}\right) \qquad (2)$$

$$\rho\frac{DV}{Dt} = -\frac{\partial P}{\partial r} + \frac{1}{r}\left[\frac{\partial}{\partial x}\left(r\mu_{eff}\frac{\partial V}{\partial x}\right) + \frac{\partial}{\partial r}\left(r\mu_{eff}\frac{\partial V}{\partial r}\right)\right]$$

$$- \mu_{eff}\frac{V}{r^2} + S_v \qquad (3)$$

$$S_v = \frac{\partial}{\partial x}\left(\mu_{eff}\frac{\partial U}{\partial r}\right) + \frac{1}{r}\frac{\partial}{\partial r}\left(r\mu_{eff}\frac{\partial V}{\partial r}\right) - r\mu_{eff}\frac{V}{r^2}$$

The effective turbulent viscosity is obtained from the k~ε viscosity model of Jones and Launder (Ref. 1). The form employed here is the high-Reynolds-number form.

Viscosity:

$$\mu_{eff} = \mu_1 + \mu_T \tag{4}$$

$$\mu_T = C_\mu \rho k^2/\varepsilon \tag{5}$$

Turbulent kinetic energy equation:

$$\rho\frac{Dk}{Dt} = \frac{1}{r}\left\{\frac{\partial}{\partial x}\left[r\left(\mu_1 + \frac{\mu_T}{\sigma_k}\right)\frac{\partial k}{\partial x}\right] + \frac{\partial}{\partial r}\left[r\left(\mu_1 + \frac{\mu_T}{\sigma_k}\right)\frac{\partial k}{\partial r}\right]\right\}$$

$$+ G - \rho\varepsilon \tag{6}$$

Turbulent energy dissipation equation:

$$\rho\frac{D\varepsilon}{Dt} = \frac{1}{r}\left\{\frac{\partial}{\partial x}\left[r\left(\mu_1 + \frac{\mu_T}{\sigma_\varepsilon}\right)\frac{\partial \varepsilon}{\partial x}\right] + \frac{\partial}{\partial r}\left[r\left(\mu_1 + \frac{\mu_T}{\sigma_\varepsilon}\right)\frac{\partial \varepsilon}{\partial r}\right]\right\}$$

$$+ C_1\frac{\varepsilon}{k}G - C_2\rho\frac{\varepsilon^2}{k} \tag{7}$$

where

$$G = \mu_T\left\{\left(\frac{\partial U}{\partial r} + \frac{\partial V}{\partial x}\right)^2 + 2\left[\left(\frac{\partial U}{\partial x}\right)^2 + \left(\frac{\partial V}{\partial r}\right)^2 + \left(\frac{V}{r}\right)^2\right]\right\}$$

and where the asymptotic, high-Reynolds-number values of the coefficients are:

C_μ	C_1	C_2	σ_k	σ_ε
0.09	1.44	1.92	1.0	1.3

3. NUMERICAL SOLUTION PROCEDURE

The axisymmetric equations for a turbulent jet were presented above as equations (1) through (3), (6), and (7). These equations can be rewritten in general form as:

$$\frac{1}{r}\left[\frac{\partial}{\partial x}\left(\rho r U\phi\right) + \frac{\partial}{\partial r}\left(\rho r V\phi\right) - \frac{\partial}{\partial x}\left(r\Gamma_\phi\frac{\partial\phi}{\partial x}\right) - \frac{\partial}{\partial r}\left(r\Gamma_\phi\frac{\partial\phi}{\partial r}\right)\right]$$

$$- S_\phi = 0 \tag{8}$$

where

ϕ = U, V, k, ε, etc.
Γ = μ_{eff}, Γ_k, Γ_ε, etc.

Here S_ϕ represents the "sources" relevant to the transport of the variable ϕ. Equations of this form were solved by finite-difference means as described below.

Fig. 1 shows part of a finite-difference grid where the values of ϕ are assured known at the nodes P, N, S, E and W. Equation (8) may be integrated over the indicated control volume to give

$$\left[\int_{r-}^{r+} r\left(\rho U\phi - \Gamma_\phi \frac{\partial\phi}{\partial x}\right) dr\right]_{x-}^{x+} + \left[r\int_{x-}^{x+}\left(\rho V\phi - \Gamma_\phi \frac{\partial\phi}{\partial r}\right) dx\right]_{r-}^{r+}$$

$$= \int_{r-}^{r+}\int_{x-}^{x+} rS_\phi dxdr \tag{9}$$

The following finite-difference approximation may be made:

$$\int_{r-}^{r+} r\rho U\phi dr\Bigg|_{x+} \approx \frac{1}{2}\left(\phi_P+\phi_E\right) \quad (\rho U)_{x+} \int_{r-}^{r+} rdr = C_{x+}\left(\phi_P+\phi_E\right) \tag{10}$$

$$\int_{r-}^{r+} r\Gamma_\phi \frac{\partial\phi}{\partial x} dr\Bigg|_{x+} \approx \frac{\phi_E-\phi_P}{x_E-x_P}\left(\Gamma_\phi\right)_{x+} \int_{r-}^{r+} rdr = D_{x+}\left(\phi_E-\phi_P\right) \tag{11}$$

$$\int_{r-}^{r+}\int_{x-}^{x+} rS_\phi dxdr \approx (S_\phi)_P \int_{r-}^{r+}\int_{x-}^{x+} rdrdx = S_\phi dVol. \tag{12}$$

with $A_E = D_{x+} + C_{x+}$, $A_W = D_{x-} - C_{x-}$ etc., substituting (10)-(12) into (9) gives

$$\phi_P(A_N+A_S+A_E+A_W) = A_N\,\phi_N + A_S\,\phi_S + A_E\,\phi_E + A_W\,\phi_W + S_\phi\ dVol. \tag{13}$$

The finite-difference coefficients are modified according to a hybrid scheme (Ref. 2). If the pressure is known, then equations (13), written for each variable at each grid node, yields a closed set of algebraic equations. However, there is no guarantee that the resultant velocity field would satisfy the continuity relation (Ref. 1). The two problems of determining the pressure and satisfying continuity are overcome by adjusting the pressure field so as to satisfy continuity (Ref. 3).

4. BOUNDARY CONDITIONS

The boundary conditions for the jet impinging on a flat plate are shown in Fig. 2. The boundary AB is a symmetric axis, the boundary BC is a wall boundary, the boundaries CD and DE are inflow and outflow boundaries where the flows are assumed to be boundary layer flows.

In Fig. 3 the physical model for near wall conditions is shown. The transport equation for turbulent kinetic energy, k is integrated over the control volume near the wall assuming the diffusion of turbulent kinetic energy normal to the wall is negligible. Within the viscous sublayer, a parabolic variation of kinetic energy is assumed which corresponds to a linear increase of fluctuating velocity with distance from the wall. Since the parabolic profile of k has a zero gradient at the wall, there is no diffusional leakage of kinetic energy to the wall. The diffusion rate of turbulent kinetic energy out of the cell at its north, east, and west boundaries are handled by the same conservative differencing scheme that is employed over the remainder of the flow. The turbulent shear stress is zero within the viscous sublayer, undergoes an abrupt increase at the edge of the sublayer, and varies linearly over the remaider of the cell. The detailed procedure to obtain the mean rates of generation and dissipation are given in (Ref. 4).

The turbulent energy dissipation was fixed at the wall node point by using

$$\varepsilon_P = k_P^{3/2} / (\kappa \; y_P C_\mu^{3/4}) \tag{14}$$

The conditions in the nozzle are given in two ways; one is by assuming a uniform velocity profile, the other is by assuming a 5.5 power velocity profile for fully developed nozzle exit conditions.

When the velocity profile at the nozzle exit is uniform, the turbulent kinetic energy in the nozzle also must be given. The turbulent energy dissipation in the nozzle can be computed from the turbulent kinetic energy.
The conditions in the nozzle are given as:

$$U = U_{in}, \quad k = k_{in} = iU_{in}^2, \quad \varepsilon = \varepsilon_{in} = \frac{k_{in}^{3/2}}{\lambda r_N}$$

When the velocity profile at the nozzle exit is non-uniform, the turbulent kinetic energy and the turbulent energy dissipation need to be calculated.

Assuming fully developed pipe flow, the transport equation of k can be simplified by neglecting the convection and diffusion for the near wall flow as

$$\mu_T \left(\frac{\partial U}{\partial r}\right)^2 - \rho\varepsilon = 0 \tag{15}$$

Using equation (5), this expression becomes

$$\varepsilon = C_\mu^{1/2} k \left(\frac{\partial U}{\partial r}\right) \tag{16}$$

The shear stress can be expressed with equation (5) as

$$\tau = \rho C_\mu \left(\frac{k^2}{\varepsilon}\right) \left(\frac{\partial U}{\partial r}\right) \tag{17}$$

The shear stress is also expressed by using mixing length, l_m, as

$$\left(\frac{\tau}{\rho}\right)^{1/2} = l_m \left(\frac{\partial U}{\partial r}\right) \tag{18}$$

Then we obtain the turbulent kinetic energy from euqations (7) and (18) as

$$k = C_\mu^{-1/2} \; l_m^2 \left(\frac{\partial U}{\partial r}\right)^2 \tag{19}$$

Therefore, from a specification of the velocity profile and a knowledge of the mixing length, we can obtain ϵ and k at the nozzle exit. The detailed procedure to obtain the mixing length is given in ref. (5).

5. EXPERIMENTAL WORK

In conducting the experiments, the UWM jet cutting system was employed. The elements of the system are shown in the schematic, Fig. 4. This is a "blow-down" system in which the reservoir pressure is pumped up to a desired level and then allowed to gradually decrease as the flow exits through the nozzle. The flow leaving the nozzle and impinging on the target is quasisteady in nature. One advantage of the blow-down system is that a large range of nozzle velocity can be studied in a single experiment. Each experiment was begun at a reservoir pressure of 690 MPa (100,000 Psi). This pressure produced initial nozzle exit velocities of approximately 900 m/s.

A schematic of the arrangement used to attach the nozzle to the system tubing is shown in Fig. 5. A cone type fitting provides the seal between the nozzle and a tube fittings. The nozzle provides a smooth flow transition from the 1/8 inch (3.175 mm) nominal tubing diameter to the 0.0094 inch (0.239 mm) exit diameter. It should be noted that some pressure loss is incurred in the flow as it moves from the pressure transducer (where the nozzle pressure measurement is made) to the nozzle, because of the connecting tube and tube fittings. However, there was no way to measure this pressure loss, since no pressure port was located in the hardened steel nozzle. An estimate of the pressure losses was made by calculating the frictional and form losses produced in the supply pipe and nozzle combination. This pressure loss estimate was used as a guide in selecting values of the velocity coefficient.

Target force measurements were made using a strain-gage-type force transducer which was rigidly mounted behind a bearinged target support. The output from the force transducer and the pressure transducer were simultaneously recorded on a strip chart recorder.

Because of the small diameter of the jet and the high kinetic energy of the flow leaving the nozzle exit, no Pitot tube or other velocity measuring device was placed in the jet. Consequently, the exit velocity determined for the experiments was really a theoretical velocity based on the measured pressure. The frictionless Bernoulli equation for a compressible fluid was used to calculate the jet velocity. The procedure required the use of the measured pressures upstream of the nozzle and the compressibility data for the flowing fluid. In these experiments, the fluid used was a synthetic hydraulic fluid called Plexol 201 (Ref. 6).

The target in each experiment was a hardened steel flat plate. Two different stand-off distances were used; 16 nozzle diameters and 28 nozzle diameters.

6. COMPARISON OF NUMERICAL AND EXPERIMENTAL RESULTS

A comparison between the experimental and computed target forces is shown in Fig. 6 for the two stand-off distances considered; H/D = 16 and H/D = 28. Both the calculated and the experimental results indicate that there is no discernable difference in the target force when the stand-off distance is changed. However, there is a consistent disparity between the computed and experimental values of target force, with the computed values running approximately 20% higher.

An explanation for the disparity between the computed and experimental results can probably be found by considering the method used to determine nozzle exit velocity for the experiments. As mentioned above, the experimental nozzle velocities were really theoretical velocities derived by using a frictionless Bernoulli equation and the measured pressure just ahead of the nozzle.

The customary way to handle frictional losses in a nozzle is to introduce a velocity coefficient (C_V) as follows:

$$C_V = \frac{U \text{ actual}}{U \text{ theoretical}} \tag{20}$$

The magnitude of C_V is affected by the frictional losses in the nozzle. In the present case, the supply pipe losses will also tend to decrease the value of C_V. The values of C_V reported in the literature vary widely and are strongly affected by nozzle design. If a value of 0.9 is assumed for C_V and the experimental data replotted (Fig. 7) the agreement between computed and experimental results is much improved. This method can of course be used as a means to determine C_V for the nozzle, supply pipe combination.

7. DISCUSSION OF NUMERICAL RESULTS

In the present study, two cases of stand-off distance are considered; H/D = 16 and 28. The case of H/D = 28 was used to compute results for various nozzle conditions. The reason is that for the larger stand-off distance, the effect of the boundary condition at the upper surface is negligible.

It is discussed in the previous section that assuming the velocity coefficient at the nozzle exit to be 0.9, the agreement between computed and measured results is improved as shown in Fig. 7. In this figure the computed results with different numerical conditions are also compared for H/D = 28. Two different nozzle exit velocity profiles and two grid mesh (number of cells) are used. To check the effect of grid mesh on calculated results the number of cells was increased in the jet axial direction. For the same number of cells, the nozzle exit velocity profile affects the results by 12%. On the other hand, for the same nozzle exit velocity profile, the effect of increasing the number of cells is about 4.5%. For the assumed velocity coefficient (C_V) value of 0.9 and a realistic 5.5 power velocity profile, good agreement with the measured force values is obtained.

The calculated pressures and skin friction coefficients with different nozzle exit velocity profiles are shown in Figs. 8 and 9 for a nozzle Reynolds number of 2.16×10^4. From these results it can be seen that for the case of uniform velocity profile the computed pressures, plate forces and skin friction are all about 15% higher than those computed for a non-uniform velocity profile at the maximum deviation point. This is because the uniform velocity profile produces higher generation rates of turbulent kinetic energy at the outer edge of the jet due to high shearing action. This causes larger kinetic energy and consequently larger shear stress and pressure at the point of impingement on the plate.

In Fig. 10 the pressure distributions are compared for different turbulence intensities in the nozzle. Here a uniform velocity profile is assumed. In Fig. 11 and Fig. 12 the target forces are plotted as a function of nozzle exit velocity. Comparing these two figures it can be seen that the influence of different nozzle conditions on the force is much smaller in Fig. 12 than in Fig. 11. This means the turbulence intensity produced in the nozzle will not affect the total force but only the stagnation pressure. The influence of nozzle velocity profile is much more significant on the force. From this discussion we may conclude that the consideration of the inlet conditions is much more important than the choice of turbulence models.

8. CONCLUSIONS

The main conclusions to emerge from this study may be summarized as follows:

1. The influence of the number of cells in the computational domain is relatively small.

2. The influence of the nozzle exit velocity is much larger than that of the turbulence intensity level in the nozzle. This suggests that the consideration of the inlet conditions is more significant than the choice of turbulence model.

3 Finally, it appears that the present numerical method has significant potential for predicting the flow properties for not only impinging jets on a flat plate but also jet flow in an irregularly shaped cavity with geometric refinement in the computer program.

9. REFERENCES

1) Jones, W.P. and Launder, B.E., "The Prediction of Laminarization with a 2-Equation Model of Turbulence". Int. J. Heat Mass Transfer, 15, 1972, p. 301.

2) Gosman, A.D., Pun, W.M., Runchal, A.K., Spalding, D.B., and Wolfshtein, M., "Heat and Mass Transfer in Recirculating Flows". Academic Press, 1968, p. 229.

3) Patankar, S.V. and Spalding, D.B., "A Calculation Procedure for heat, mass and momentum transfer in three-dimensional parabolic flows." Int. J. Heat Mass Transfer, 15, 1972, p. 1787.

4) Chieng, C.C. and Launder, B.E., "On the Calculation of Turbulent Heat Transport Downstream from an Abrupt Pipe Expansion". Numerical Heat Transfer, 3, 1980, pp. 189-207.

5) Amano, R.S., "Numerical Study of a Turbulent Jet Impinging on a Flat Plate and Flowing into an Axisymmetric Cavity". Ph.D. thesis, 1980, Univ. California, Davis.

6) Bayer, J.L. and Neusen, K.F.: "Specific Energies for Jet Cutting of Polymeric Materials" In: Third International Symposium on Jet Cutting Technology (Chicago, 1976), BHRA, England, pp. C2-15.

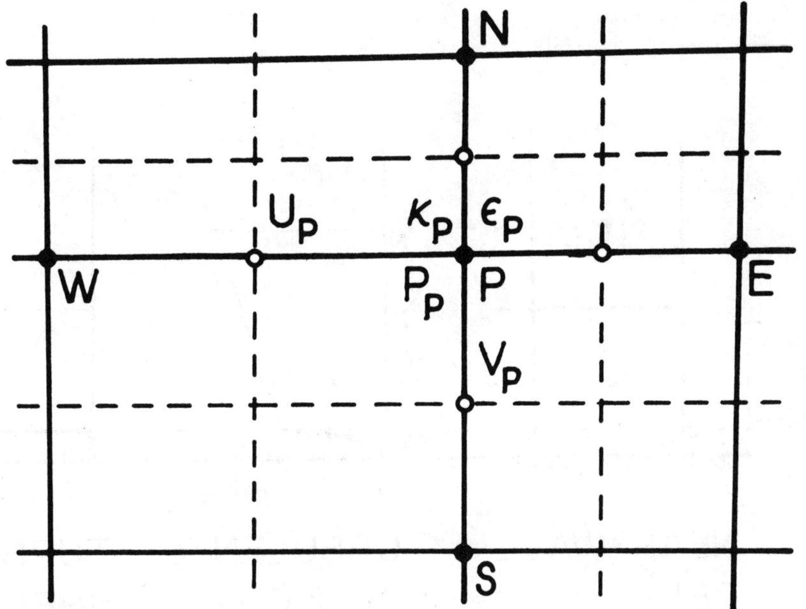

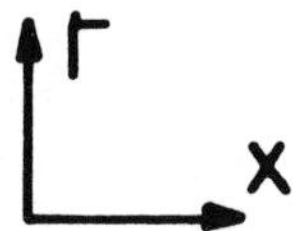

Fig. 1. Juxtaposition of Grid Nodes

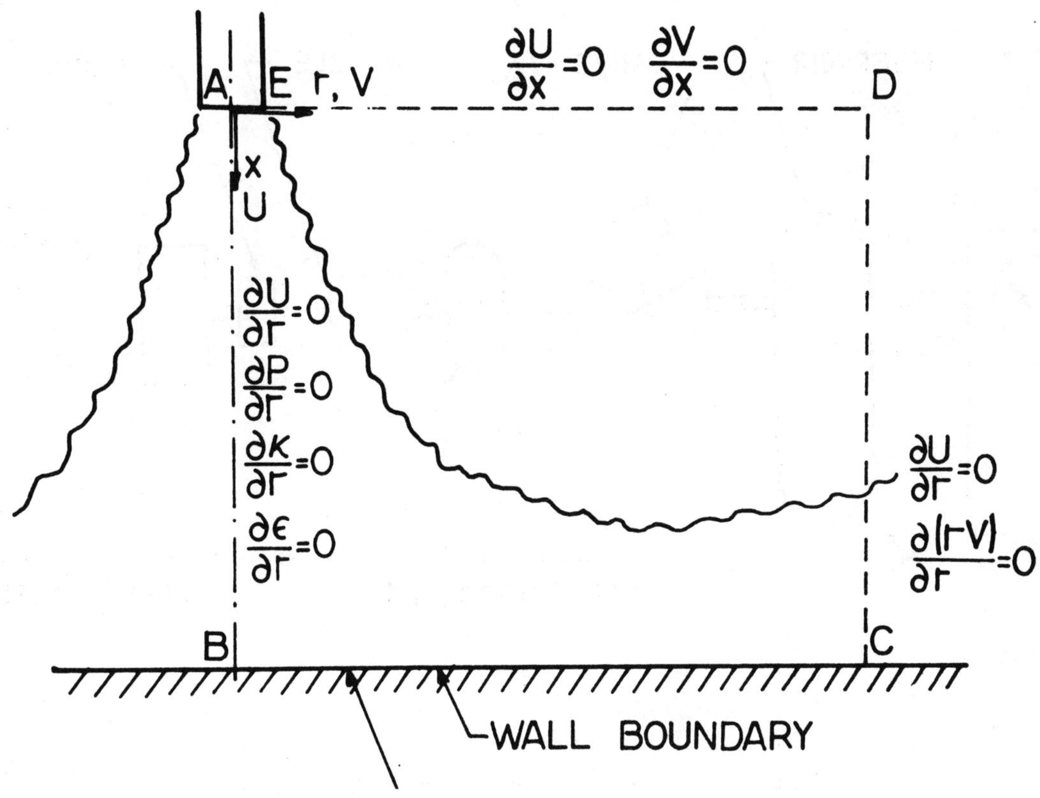

Fig. 2. Computational Domain and Boundary Conditions

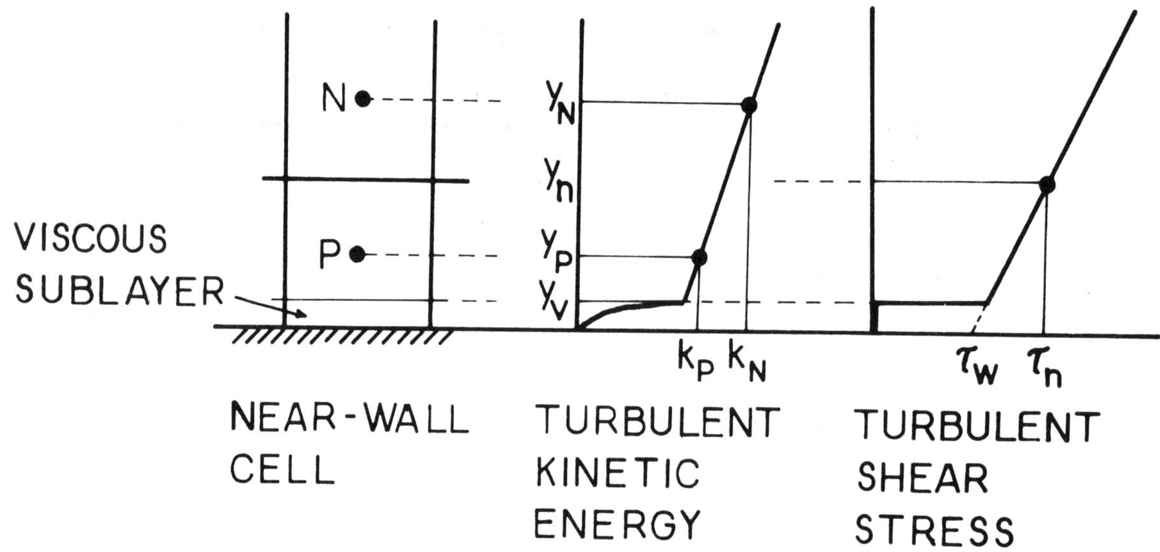

Fig. 3. Near Wall Physical Model

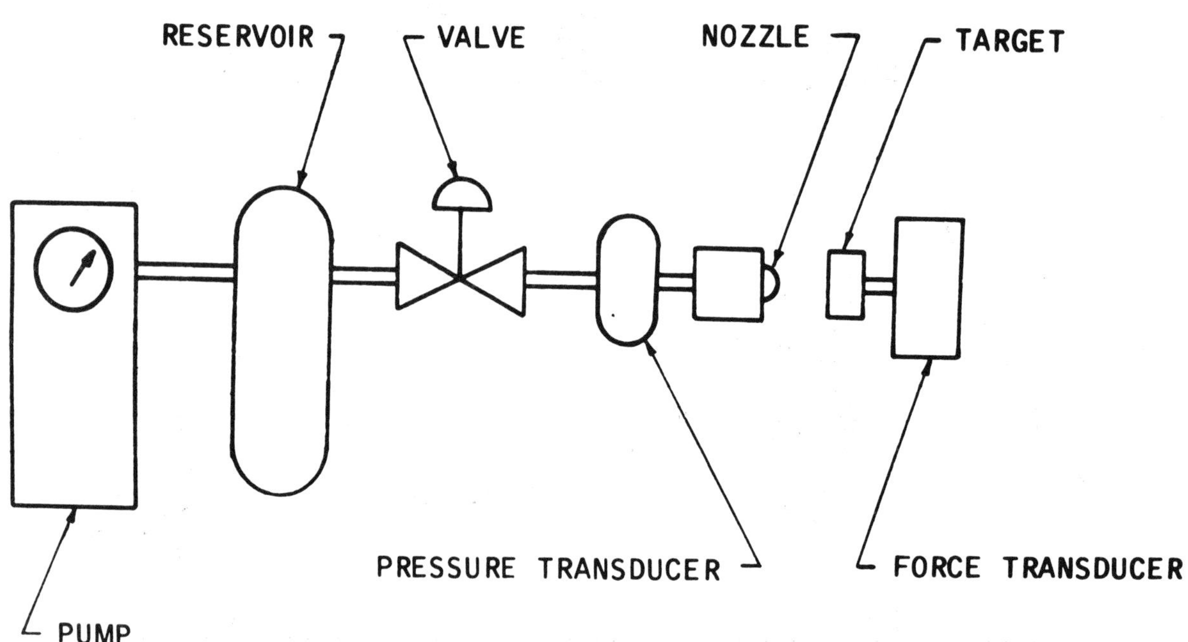

Fig. 4. Experimental System Schematic

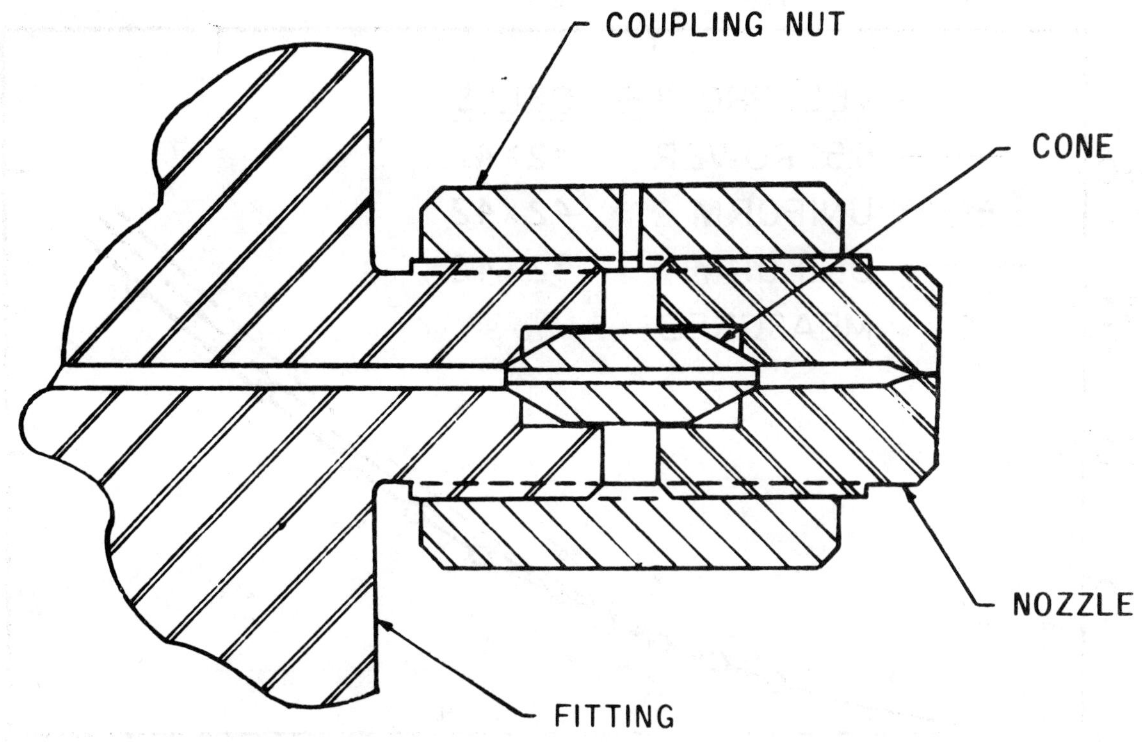

Fig. 5. Nozzle to Tube Attachment

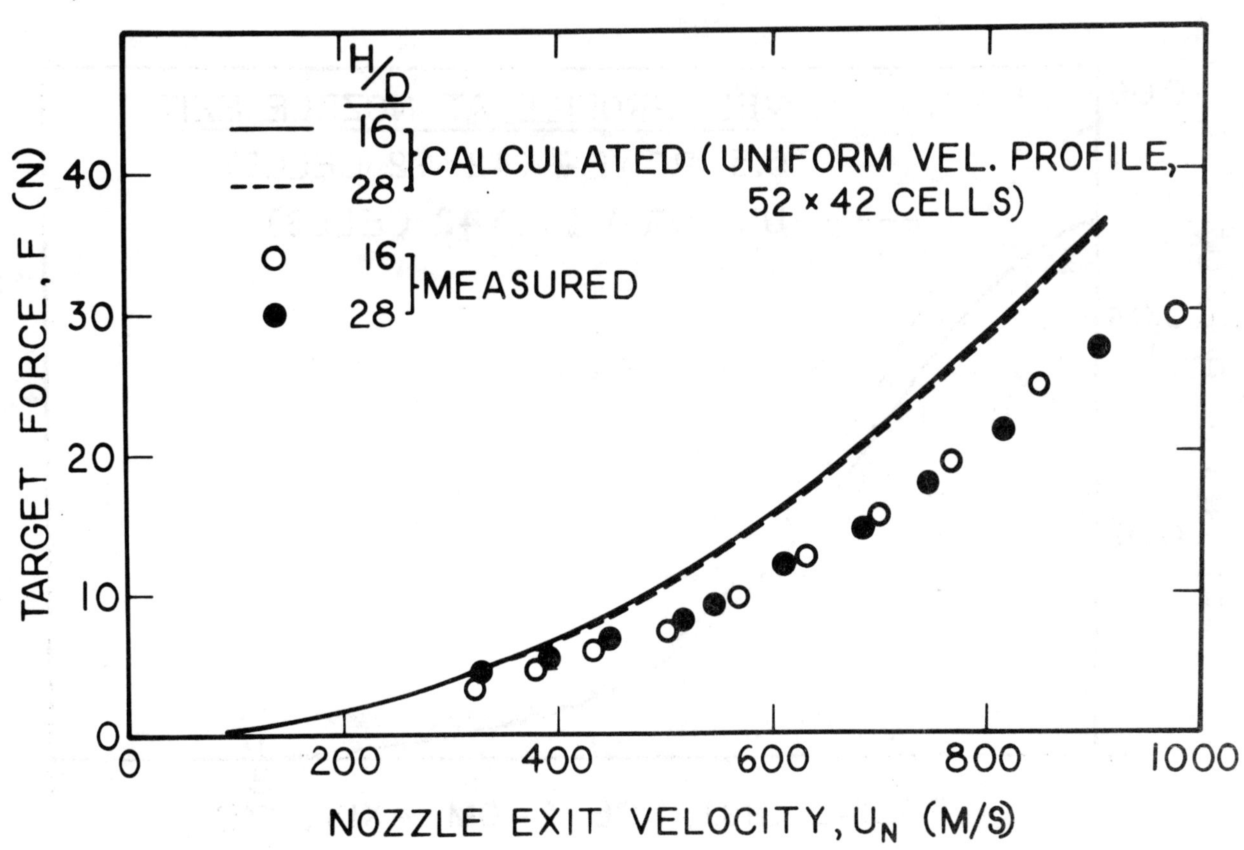

Fig. 6. Calculated and Measured Target Forces as a Function of Nozzle Exit Velocity

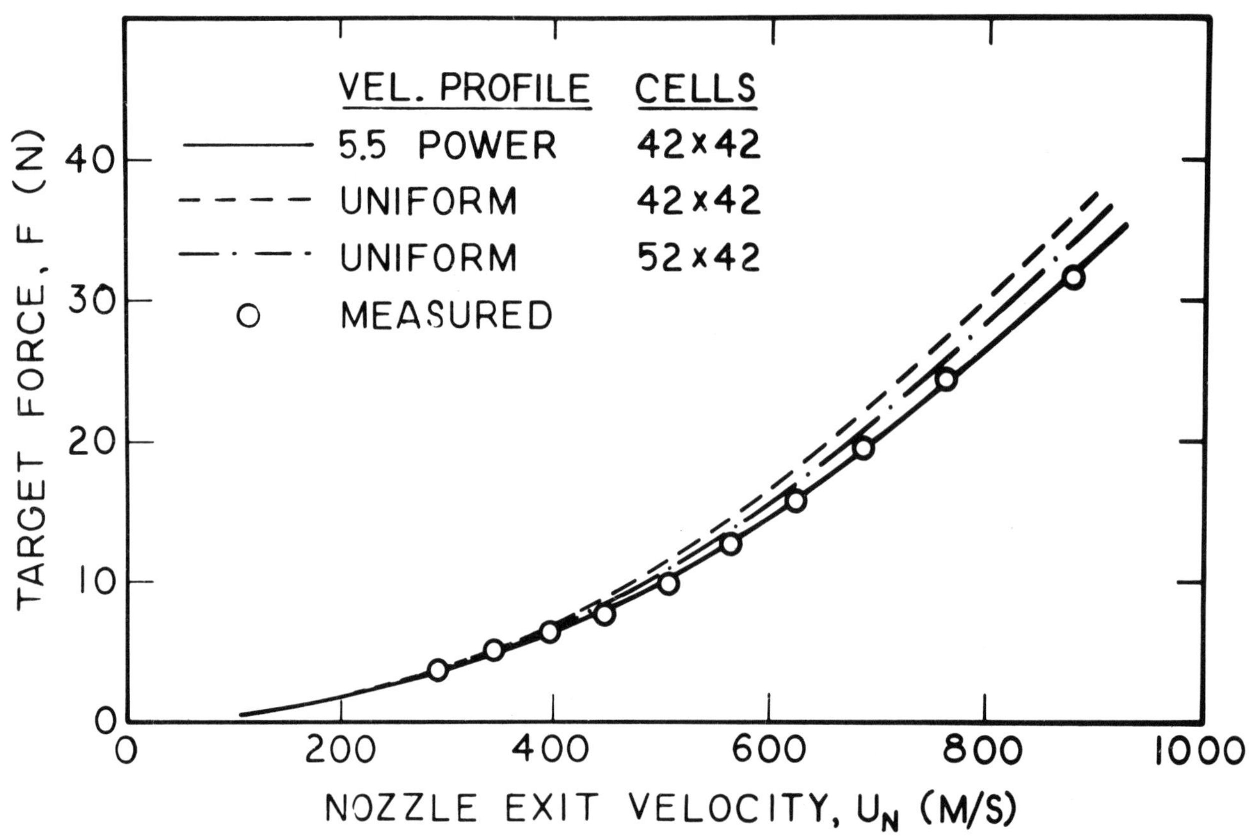

Fig. 7. Calculated and Measured Target Forces as a Function of Nozzle Exit Velocity (H/D = 28)

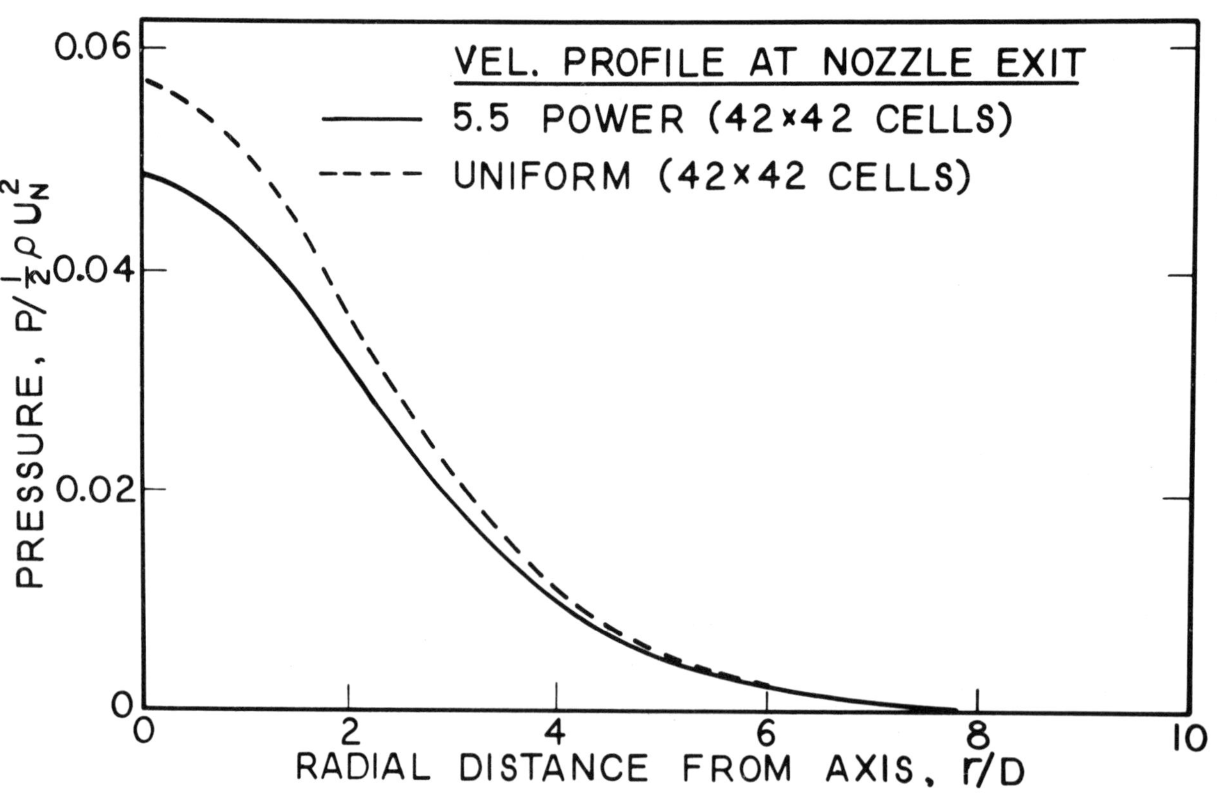

Fig. 8. Calculated Pressure Along the Plate for Different Nozzle Exit Velocity Profiles (Re = 2.16 x 10 , H/D = 28)

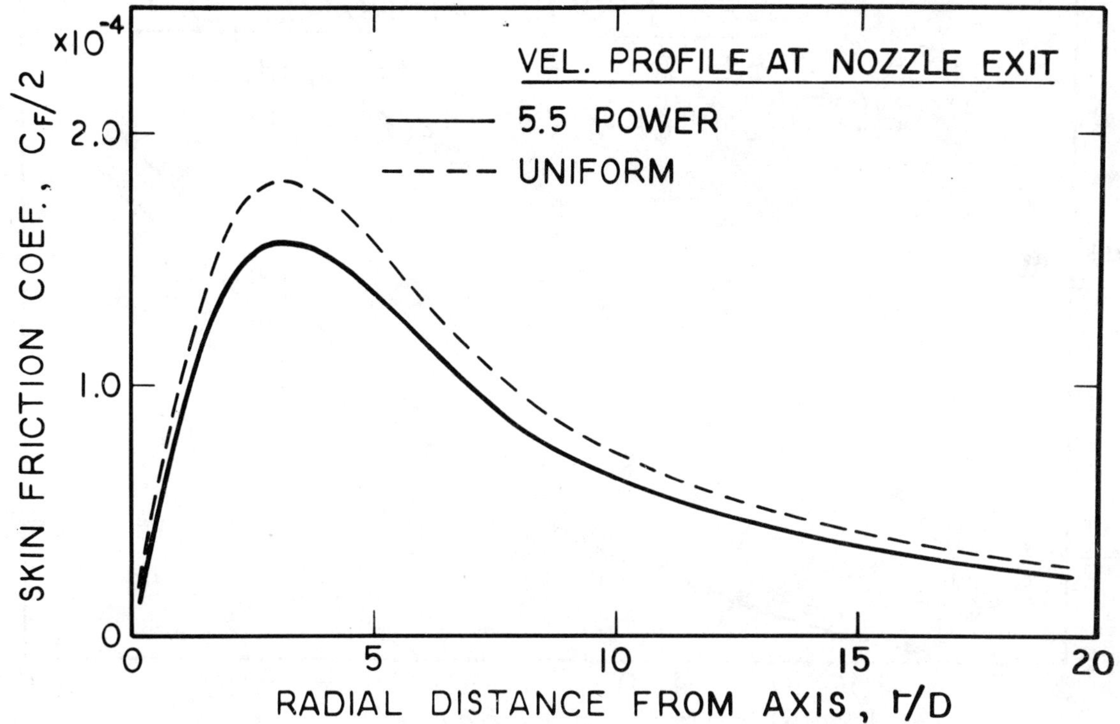

Fig. 9. Calculated Skin Friction Coefficient Along the Plate for Different Nozzle Exit Velocity Profiles (Re = 2.16 x 10 , H/D = 28)

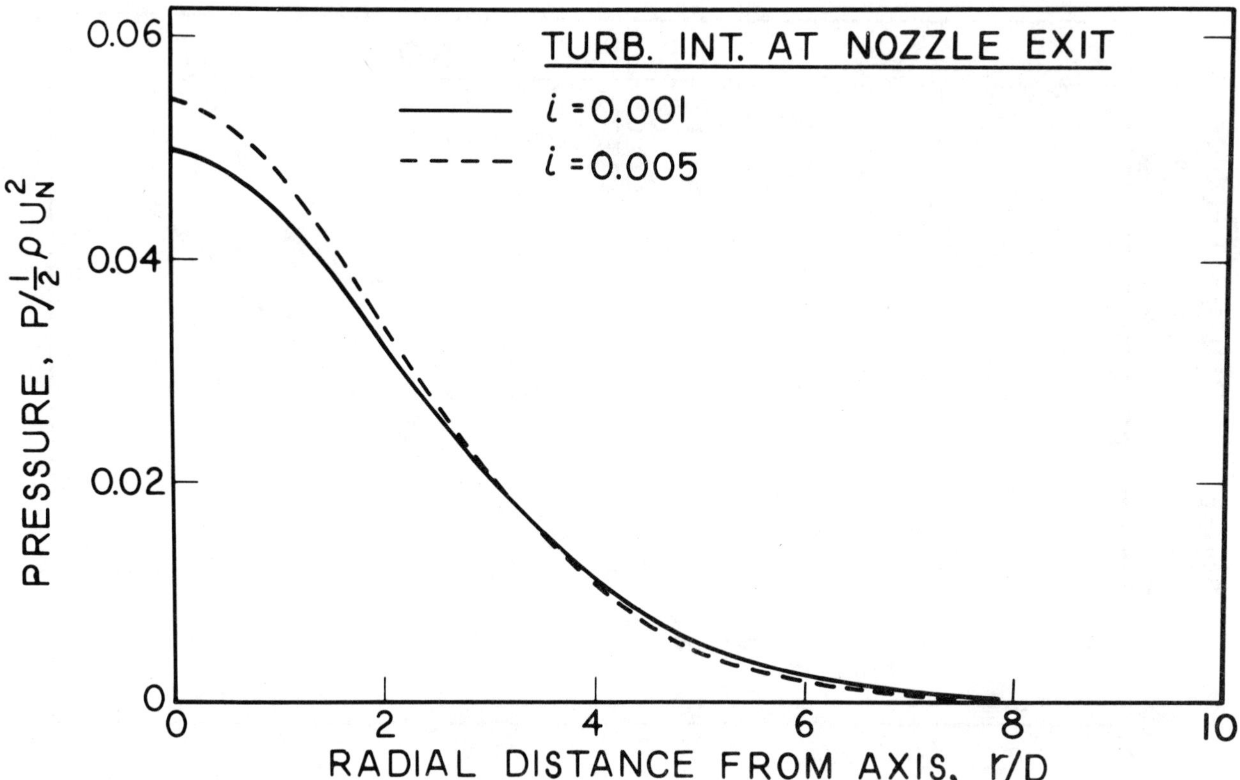

Fig. 10. Calculated Pressure Along the Plate for Different Turbulence Intensities at the Nozzle Exit (Re = 2.16 x 10 , H/D = 28)

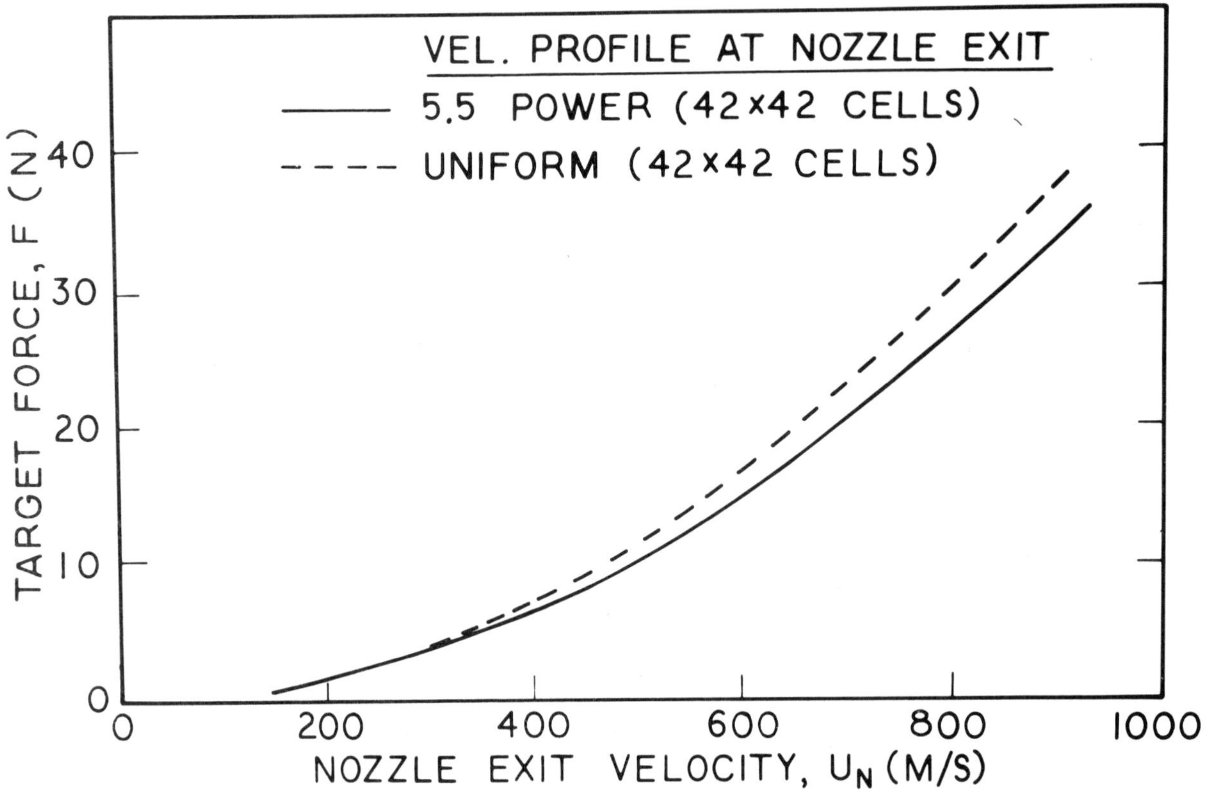

Fig. 11. Calculated Target Forces on the Plate for Different Nozzle Exit Velocity Profiles as a Function of Nozzle Exit Velocity (H/D = 28)

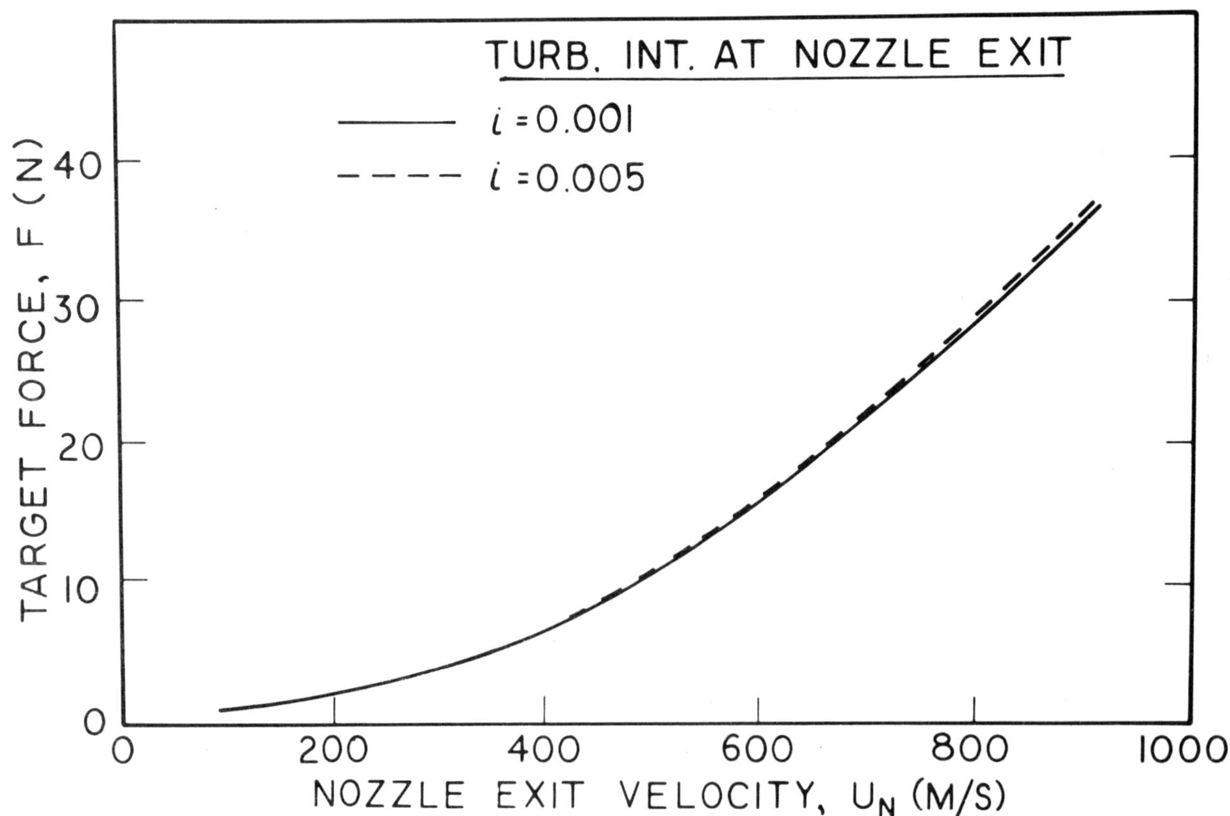

Fig. 12. Calculated Target Forces on the Plate for Different Turbulence Intensities as a Function of Nozzle Exit Velocity (H/D = 28)

6th International Symposium on
Jet Cutting Technology
6-8, April, 1982

PAPER C4

THE COHERENCE OF IMPULSIVE WATER JETS

D.G. Edwards, R.M. Smith and G. Farmer
University of Surrey, U.K.

Summary

Laboratory experiments have been carried out to assess the progressive loss of potential cutting effectiveness of impulsive water jets with increasing stand-off distance from nozzle to target, both in air and in a vacuum.

A mechanism has been described whereby the droplets separating from the jet boundary in air may be dispersed rapidly from the central core in a manner detrimental to cutting effectiveness.

Held at the University of Surrey, U.K.
Symposium organised and sponsored by
BHRA Fluid Engineering

NOTATION

U	jet velocity
ρ	water density
ρ_{air}	air density
c	acoustic velocity in water
ν	kinematic viscosity of water
μ	viscosity of water
x	axial distance along the jet
y	distance normal to nozzle surface
p	pressure
t	time
Γ	circulation
s	length dimension of liquid droplet
L	lift force

1. INTRODUCTION

The main aim of the present work has been to investigate the structure of impulsive liquid jets produced by an impact-cumulation water cannon. The accepted view of the formulation of such jets, in air, is that they consist essentially of a central rod-like core surrounded by a cloud of droplets. The core is thought to be the useful element of the jet in terms of cutting performance while the spray cloud is parasitic in that it represents a loss of momentum which would otherwise be concentrated in the core thereby producing a smaller target area.

The partition of momentum flux into the two zones has an important bearing on the cutting potentialities of the jet and an experimental examination of this aspect of the behaviour of impulsive jets has been carried out in the present work.

A family of target plates of different diameters, each attached to a force transducer, has been mounted so as to intercept the discharge from the impact-cumulation water cannon, at various stand-off distances from the nozzle exit.

In this way an impression has been gained of the progressive loss of cutting effectiveness with increasing stand-off distance, due partly to retardation of the jet caused by air-drag and partly to radial spreading of the jet.

Repetition of these force measurements in a vacuum chamber has indicated a considerable increase in jet coherence and cutting potential as compared with the coreresponding properties in air.

Spark-schlieren photography of the jets in air carried out in the present work and similar work by Edney[1] have provided both a qualitative insight into the character of the flow, and a quantative estimate of the forces required on typical spray droplets to give the observed cross-range movement during the total travel-time of the jet-head.

A mechanism is suggested whereby such forces may be generated through the combined effects of boundary layer development in the nozzle and the subsequent relative motion of rotating droplets through the atmosphere.

2. EXPERIMENTAL WORK

Laboratory experiments were carried out using the University-of-Surrey free-piston water cannon described in a previous paper by Edwards and Welsh[2] and illustrated schematically in Fig. 1.

2.1 Schlieren Photography

Qualitative information regarding the development of jets through various phases was obtained from single-spark schlieren photographs taken at different time intervals after the emergence of the jet-head. A selection of these is shown in Plates 1 and 2.

2.2 Jet-Head Velocity Measurement

The tests were carried out at a nominal initial jet-head velocity of 800 to 1000 m/s for initial piston velocities of 40 to 60 m/s. Measurements of the jet-head velocity over a distance of up to 85 mm from the nozzle exit were made using the system shown in Fig. 2. The main feature of the system is a parallel beam of light stopped down by a mask to form five thin filaments which cross the axis of the jet and pass on via a concave mirror to a photo-multiplier. Successive interruption of each light filament by the advancing jet produces a staircase output from the photo-multiplier from which the trajectory of the jet head may be determined.

The results obtained in this way were used to augment a previous calibration of the water cannon[3] shown in Fig. 3 in comparison with corresponding computer predictions. The latter are based on a "leaky Lagrangian" finite-difference scheme which has been used by Trulio and Trigger[4], Glenn[5], and Edwards and Welsh[6] to model the behaviour of impact-cumulation water cannon.

2.3 Jet-Force Measurement in Air

The impact-force produced by the jet was measured at a series of stand-off distances up to 85 mm (13 diameters) from the nozzle unit. Circular flat plates of various diameters from 9 mm to 80 mm were attached to a block of steel via a proprietary load washer* (Fig. 2). The vibration-frequency response of the system was checked initially by means of a separate impact test using the 80 mm diameter force-plate so as to give the lowest natural frequency of any of the plate/washer combinations. A steel rod of 16 mm diameter, 1.5 m long with its axis vertical was allowed to fall through a short distance onto the sensitive surface of the load washer, the latter being end-mounted on a second vertical steel road identical with the former. The result is shown in Fig. 4.

The measured amplitude is in good accord with the prediction of one-dimensional bar impact theory and the accompanying oscillation is considered to be of acceptably small amplitude for the present purposes.

* Kistler Type 9001

2.4 Jet Force Measurement in Vacuo

The techniques described in Section 2.3 in relation to jet-force measurements in the atmosphere was also used to obtain corresponding results for a jet travelling in a vacuum.

A special vacuum chamber was manufactured by means of which runs could be carried out with either the nozzle or both the nozzle and the space immediately downstream evacuated. The vacuum system gave steady pressures in the vicinity of 100 Pa abs.

3. EXPERIMENTAL RESULTS

3.1 Schlieren Photographs

The sequence of photographs shown in Plate 1 illustrates the various phases in the development of an impulsive water jet in atmospheric air. The first three pictures show the blast wave produced by the expulsion of the air initially contained in the nozzle, as a highly turbulent supersonic air-jet. The liquid jet then emerges from the nozzle and overtakes the air jet displacing it radially outwards. A lack of coherence is evident in the liquid jet shown in each of the photographs.

Field and Lesser[7] have examined the break-up in the atmosphere of supersonic liquid jets at pressures greater than ambient. Their explanation centres on the interaction of transverse oppositely-moving expansion waves which over-relax the initial pressure excess and produce cavitation on the central core. This explanation is not applicable to the present jets which travel at velocities below the speed of sound in water and so remain in pressure-equilibrium with the outside air.

The present schlieren photographs represent the jet region as an uniformly dark shadow, but provide no indication of the density of the spray cloud relative to that of the central core. Measurements of the radial and axial distribution of momentum flux are necessary in order to resolve this question and to provide an indication of the seriousness of the break-up of impulsive jets in air in the context of cutting potentiality.

3.2 Jet Forces in Air

A typical record of jet force vs time is shown in Fig. 5, accompanied by corresponding pressure histories in the water packet and the nozzle. The variation of maximum jet force with stand-off distance for each of the series of four force-plates used is shown in Fig. 6 with the force values expressed in non-dimensional form as a proportion of the ideal force based on nozzle exit area and a maximum jet stagnation pressure, $\frac{1}{2}\rho U^2$ rather than the impact-generated pressure rise ρcU which is of very short duration ($\simeq 5 \mu s$). The over-riding feature is the marked loss of maximum available cutting force as the stand-off distance increases especially at stand-off distances less than $\simeq$ 2-5 diameters.

The 80 mm plate was known to be sufficiently large to intercept the entire discharge over the range of stand-off distances covered in the present tests but even this showed a considerable short-fall in maximum force compared with that predicted from the known jet-head velocity.

The deduction from these results was that whilst the jets undoubtedly possess a central rod-like core, a very considerable fraction of the total momentum flux is diverted into the spray cloud which reduces the plate force partly by virtue of appreciably increased aerodynamic drag on a given mass of spray droplets compared with the same mass of coherent jet, and partly through the lateral diversion of momentum. Only the cosine component of this momentum contributes to the axial plate force.

The rapidity of the reduction in maximum force with stand-off distance in the immediate vicinity of the nozzle exit indicates a loss of coherence either on, or almost immediately after, the emergence of the jet into the atmosphere. This accords with the jet divergence downstream of the nozzle exit illustrated in Plate 2. Corresponding results showing the variation of force with plate diameter at various fixed stand-off distances appear in Fig. 7.

For each stand-off distance, following an initial fairly steep increase in plate force with diameter the curves flatten off appreciably. The early increase is associated with the impingement of progressively greater proportions of the spray cloud as the plate size increases while the levelling off indicates the complete capture of the discharge. It is notable however that the plate of 9 mm diameter even when placed a stand-off distance of only 1½ exit diameters produces less than half the force experienced by the 80 mm plate. This reinforces the impression that the jet has become diffuse at a very short distance from the nozzle exit, a factor which militates strongly against a good cutting performance.

A further deduction to be made from the results shown in Fig. 7 is that changing from a plate diameter of 40 mm to one of 80 mm produces only a small ($\simeq$ 10%) increase in jet force despite the approximate quadrupling of plate area on which droplets impinge. The implication is that the outer regions of the spray cloud are either less densely packed with droplets than those nearer the centre or are populated by particles of a smaller size than those closer to the centre, which as a result suffer a greater deceleration due to air drag before impinging on the plate.

3.3 Jet Forces in a Vacuum

The variation of jet force with plate stand-off distance has been measured under the three different sets of environmental conditions:

a) Atmosphere air in nozzle and outside

b) Vacuum in nozzle atmospheric air outside

c) Vacuum in nozzle and outside.

The results are shown in Fig. 8, for a plate diameter of 36 mm. A notable feature in all cases is that the magnitude of the maximum force coefficient invariably falls short of its ideal value of unity even at very small stand-off distance in a vacuum. Since the jet velocity and stagnation pressure used in the evaluation of the coefficient are those obtained from actual measurements, thereby eliminating discrepancy on this score, the only remaining factor, the effective jet impact area must be less than that of the nozzle exit, to account for the observed oss in force. Indeed the results presented by Field and Lesser[7] provide supporting evidence for this explanation in the form of photographs showing a jet-head tapering away appreciably as it leaves the nozzle.

At stand-off distances of up to some 50 mm (7 nozzle diameters) the results follow the anticipated pattern with the force in air being less than that produced in conjunction with an evacuated nozzle, and with the forces generated when the whole process occurs in a vacuum being greater still.

However, as the stand-off distance increases, though the force for case c) shows a downward trend, possibly through evaporation of the jet, the result for case b) stages a surprising recovery. At 85 mm stand-off it gives a jet force almost as great as it provides at 10 mm stand-off and clearly in excess of the corresponding force in the completely-evacuated case. The results for case a) (atmospheric air throughout) also display a tendency towards recovery of plate force at the largest stand-off distances though the effect is less marked than that described above.

A possible explanation for the apparently beneficial effects arising from the presence of atmospheric air is that the effect of aerodynamic drag possibly linked to Rayleigh-Taylor instability at the jet-head[7] though reducing the jet coherence and velocity, produces a locally enlarged effective jet cross-sectional area. The accompanying accumulation of mass in a particular region is the reverse of the process occurring nearer the nozzle exit and referred to above. However the relatively large area over which the enhanced force is distributed implies that no real benefit in cutting performace is likely at these large stand-off distances.

Figure 9 shows the variation of jet force with plate diameter at 35 mm stand-off distance. The results illustrate vividly the beneficial effects produced by operation in a vacuum both inside and outside the nozzle.

A plate of only 9 mm diameter produces a maximum force within 5% of that generated against the 80 mm plate. This indicates very little lateral spreading of momentum flux and a high degree of jet coherence.

In contrast, the jet in air with air-filled nozzle produces a markedly lower overall force level especially at small plate diameters. It also shows only a gradual rise in force with plate diameter. Indeed the 80 mm plate is only just sufficiently large to capture the whole of the momentum.

This result is of great significance in terms of cutting potential, indicating as it does that the presence of atmospheric air leads to a highly diffuse jet which, even at small stand-off distances, is likely to be significantly inferior in cutting performance to a jet produced wholly in a vacuum. Unhappily, operation with only the nozzle evacuated while readily obtainable in the context of a practical cutting device produces results markedly inferior to those obtained in the case where the jet travels in a vacuum throughout.

3.4 Summary of the force-measurement experiments

The over-riding impression gained from the present force-plate measurements is that of a considerable degree of incoherence in the jets in air especially when atmospheric air is present in the nozzle before firing. Only the jets produced completely in vacuo appear to show the desired degree of concentration of momentum on small cutting area. In no case did the maximum cutting force exceed 60% of its ideal value based on jet exit area and known maximum jet velocity. The marked inferiority of the potential cutting performance in air compared with vacuum prompts a further examination of the nature of jet motion under two differing circumstances which is undertaken in Section 4.

4. DISPERSAL OF DROPLETS BY AERODYNAMIC LIFT

4.1 General Consideration

In addition to the qualitative insight they afford into the development of the impulsive jet, the pictures shown in Plate 2 provide quantitative data on the rate of lateral advance of the jet. Taking a nominal jet-head velocity of 10^3 m/s, an axial movement of say 0.1 m implies a time scale of 10^{-4} s. During this period, particles at the outer edge of the spray cloud are seen to have moved a distance in excess of .025 m normal to the axis of the jet. Even assuming a constant lateral acceleration, applied throughout the whole trajectory of the droplet, the minimum rate of change of velocity necessary to accomplish the movement observed is some 5×10^6 m/s^2.

Clearly some strong mechanism of force-generation is at work to produce this large acceleration and maintain it for the required time duration. Following the reasoning advanced to account for the upward projection of coal dust into the wake of the primary shock front in a coalmine explosion, it is assumed here that aerodynamic lift acts on the individual spray particles which break away from the outer edges of the jet.

The fluid at the outer edge of the jet is highly rotational, being in the boundary-layer of the impulsive nozzle flow. Each droplet from this region of the jet retains a very high angular velocity as it travels through the atmosphere, and is then subject to Magnus-effect lift directed away from the jet centre-line. When impulsive jets travel into vacuum[1], they display a much reduced tendency towards dispersal. This observation supports the present hypothesis in that the absence of air eliminates the major source of lift, the small remainder being due to water vapour surrounding the jet.

An order-of-magnitude estimate of the lift forces on typical spray droplets in air is presented in Section 4.3 based on an analysis of boundary layer development given in Section 4.2.

4.2 Nozzle Boundary-layer Development

4.2.1 Development of the Laminar Boundary layer on an Impulsively-Accelerated Flat Plate

The total time-duration of the nozzle flow process is quite short (≈ 2.0 m/s) and the period is which the jet head region attains its highest velocities occupies only a fraction of this small total.

The portion of the flow in which the velocity is highest must give rise to very large velocity gradients in the boundary layer with consequent concentration of the greatest vorticity in this portion of the boundary layer. In the present case the jet-head may accelerate from say 25% of its maximum velocity to 100% over the final 30% of the total nozzle length where the cross sectional area becomes increasingly uniform.

This rapid change in velocity implies that an approach to boundary layer calculations based on Rayleigh's solution[8] to the problem of the impulsively started flate plate may be appropriate. The validity of the treatment of a circular pipe as a rolled-up flat plate depends on the thickness of the boundary layer remaining small compared with the pipe-radius.

4.2.2 Axial Pressure Gradient

A substantial axial pressure gradient arises during the motion of the water packet, typical computer calculations being illustrated in Fig. 9. Again this parameter is subject to substantial temporal and spatial variations being adverse in the flow direction over a substantial portion of the water packet from the rear end forwards, and favourable over a short length near the front. The aim in the present analysis is to estimate the influence of the pressure gradient on boundary layer development and the simplifying assumption has been made that an adverse pressure gradient extends along the complete water packet. The value of1.82×10^8 Pa/m chosen for the pressure gradient is based on the computed conditions of discharge (Fig. 9) and is therefore an over-estimate of the effects of pressure gradient at earlier times.

4.2.3 Radial Velocity Variation

The momentum equation for the laminar flow in the nozzle boundary layer including the effect of pressure gradient becomes:

$$\frac{\partial u}{\partial t} = \nu \frac{\partial^2 u}{\partial y^2} - \frac{1}{\rho}\frac{\partial p}{\partial x} \qquad (4.1)$$

subject to the boundary condition: $u = U$ at $y = 0,\ t > 0$

where ν is the kinematic viscosity of the fluid, u the velocity at (x, t), p the static pressure, y distance normal to the wall, x distance along the wall.

Rayleigh's solution[8] for the flow induced by a flat plate without pressure gradient is:

$$u = U[1 - \mathrm{erf}(0.5\, y/(\nu t)^{\frac{1}{2}})] \tag{4.2}$$

where erf represents the error function defined as:

$$\mathrm{erf}\; x = \frac{2}{\sqrt{U}} \int_0^x e^{-z^2} dz \tag{4.3}$$

For the calculations based on equation (4.2) and its derivatives a value of 10 ^{6} m^2/s has been used for the kinematic viscosity of water, together with a density value of 10^3 Kg/m^3. Taking a mean velocity equal to 50% of the maximum over the last 100 mm of the nozzle flow, gives a time scale of 0.2 m/s for a jet velocity of 1000 m/s, equation (4.2) becomes:

$$u = 500(1 - \mathrm{erf}\; 3.535 \times 10^4 y) \tag{4.4}$$

If a more comprehensive solution is sought[9], including the effects of pressure gradient, equation (4.1) becomes:

$$u = [U - (t + \frac{y^2}{2\nu})(\frac{1}{\rho}\frac{\partial p}{\partial x})]\; \mathrm{erf}\; [0.5\, y/(\nu t)^{\frac{1}{2}}]$$

$$- \frac{y}{}\frac{\partial p}{\partial x} (\frac{\nu t}{\pi})^{\frac{1}{2}}\; e^{-0.25\, y^2/(\nu t)} + 0.5 \frac{y^2}{}\frac{\partial p}{\partial x} \tag{4.5}$$

With the present data inserted, this gives:

$$u = [500 -.0002 + 5 \times 10^5 y^2) \times 1.82 \times 10^8]\; \mathrm{erf}\; 3.536 \times 10^4 y$$

$$- 1.452 \times 10^6 y e^{-1.25 \times 10^9 y^2} + 9.1 \times 10^{10} y^2 \tag{4.6}$$

Equations 4.4 and 4.6 form the basis for the calculated velocity profiles at the nozzle exit featured in Fig. 10. The effect of including substantial pressure-gradient terms is shown to be very small. Furthermore the 99% thickness of the boundary layer is seen to be of a smaller order than the nozzle exit diameter.

4.3 Estimation of Lift Forces on Spray Droplets

Assuming that spray droplets having a variety of different sizes become detached from the main course of the jet as it emerges from the nozzle, Fig. 10 provides the means of estimating the total circulation around prescribed "blocks" of fluid in the boundary layer. Neglecting transverse velocity components, the circulation around a square element of fluid may be obtained simply by subtracting the axial velocity along the edge nearest the nozzle surface from that along the edge nearest the nozzle surface from that along the further edge and multiplying the result by the length of side of the square.

Further application of Rayleigh's impulsive flat-plate solution to the motion of the air surrounding a rotating droplet shows that a very rapid radial diffusion of vorticity occurs and that the air very soon comes to behave as though a free vortex were implanted within the spray droplet.

The tangential velocity at a radial distance y from the outer edge of a circular cylindrical droplet is given by:

$$u = U \text{ erf } 0.5 \, y/\sqrt{\nu t} \tag{4.7}$$

Here ν is the kinematic viscosity of air and U the peripheral speed of the rotating droplet. Figure 11 shows the velocity distribution induced around a droplet of 5.0μm diameter after 10^{-6}s and 2×10^{-6}s respectively by the rotation of the latter, in comparison with the ideal free-vortex velocity distribution. The similarity achieved during the very short time scales allowed indicates the rapidity of the rise in air circulation about the droplet.

This factor lends support to the use of the well-known Kutta-Joukowski expression for the lift force L on a droplet

$$L = \rho_{air} U \Gamma \tag{4.8}$$

where Γ is the induced circulation, obtained from Fig. 10. The above expression applies to a prescribed unit length of droplet and in calculating the resulting acceleration, the mass must be evaluated on the same basis.

For an elementary droplet of square cross-section, of side s the acceleration f is then given by:

$$f = \frac{\rho_{air} U \Gamma}{\rho \cdot s^2} \tag{4.9}$$

Figure 12 shows the variation of acceleration with droplet size for a range of droplets each having its outer edge coincident with the jet boundary.

4.4 Results of Droplet Force/Acceleration Calculations

The calculated velocity profile in the boundary layer at the nozzle exit (Fig. 10) at the instant of discharge indicates that the boundary layer thickness is small compared with the nozzle exit radius. This result confirms the validity of the approach used in the present analysis, based on the use of flat-plate boundary layer results.

The influence of axial pressure gradient on the form of the profile is shown to be very small. The lateral acceleration of droplets due to Magnus-effect lift estimated on the basis of the present analysis is found to be more than adequate to produce the observed trajectory of the droplets (Fig. 12).

The initial excess of acceleration however is likely to diminish rapidly as the droplet drag force increases.

Also the values calculated apply to only the outer .05 mm thickness of the jet whereas a greater proportion of the total jet cross-section is known to enter the spray cloud. Nevertheless, the rapid removal of the outer layer exposes liquid beneath which is immediately involved in a vorticity transfer process in air similar to that analysed above in the nozzle and the lift generation process may continue at a diminished rate along the entire path of the jet.

5. CONCLUSIONS

a) Impulsive high-speed water jets produced in, and travelling through atmospheric air lack coherence even at short distances from the nozzle exit.

b) Similar jets produced in, and traversing a vacuum show a much greater coherence even at relatively large stand-off distances and offer appreciably greater potential for jet cutting.

c) A mechanism for the dispersal of the spray-cloud surrounding a diffuse water jet, based on Magnus-effect aerodynamic lift has been proposed.

d) Estimates of spray-droplet acceleration based on this theoretical model have predicted adequate magnitudes to account for observed droplet trajectories.

References

1. Edney, B.E.: "Experimental studies of pulsed water jets". Proceedings of the 3rd International Symposium on Jet Cutting Technology, pp. B2-11, B2-26, May 1976.

2. Edwards, D.G. and Welsh, D.J.: "The influence of design and material properties on water cannon performance". Proceedings of the 5th International Symposium on Jet Cutting Technology, pp. G3-358 June 1980.

3. Welsh, D.J.: "A study of nozzle design for pulsed water jets". PhD Thesis, University of Surrey, November 1979.

4. Trulio J.G. and Trigger, K.R.: "Numerical solution of the one-dimensional hydro-dynamic equation in an arbitrary time-dependent co-ordinate system". UCRL Report 6522, July 1961.

5. Glenn, L.A.: "The mechanics of the impulsive water cannon". Computers and Fluids, 3, pp. 197-215, June 1975.

6. Edwards, D.G. and Welsh, D.J.: "A numerical study of nozzle design for pulsed water jets". Proceedings of the 4th International Symposium on Jet Cutting Technology, pp. B1-1, B1-12, April 1978.

7. Field, J.E. and Lesser, M.B.: "On the mechanics of high-speed liquid jets". Proceedings Royal Society, London, A357, pp. 143-162, 1977.

8. Batchelor, G.K.: "An Introduction to Fluid Dynamics". (p. 189) Cambridge University Press.

9. Carslaw, H.S. and Jaeger, J.C.: "Conduction of Heat in Solids". Clarendon Press, Oxford.

Acknowledgements

The authors are indebted to the Science Research Council for their support for this work. They wish to thank Mr M. Bonia of the Mechanical Engineering Department Technical Staff for assistance in the development of the apparatus.

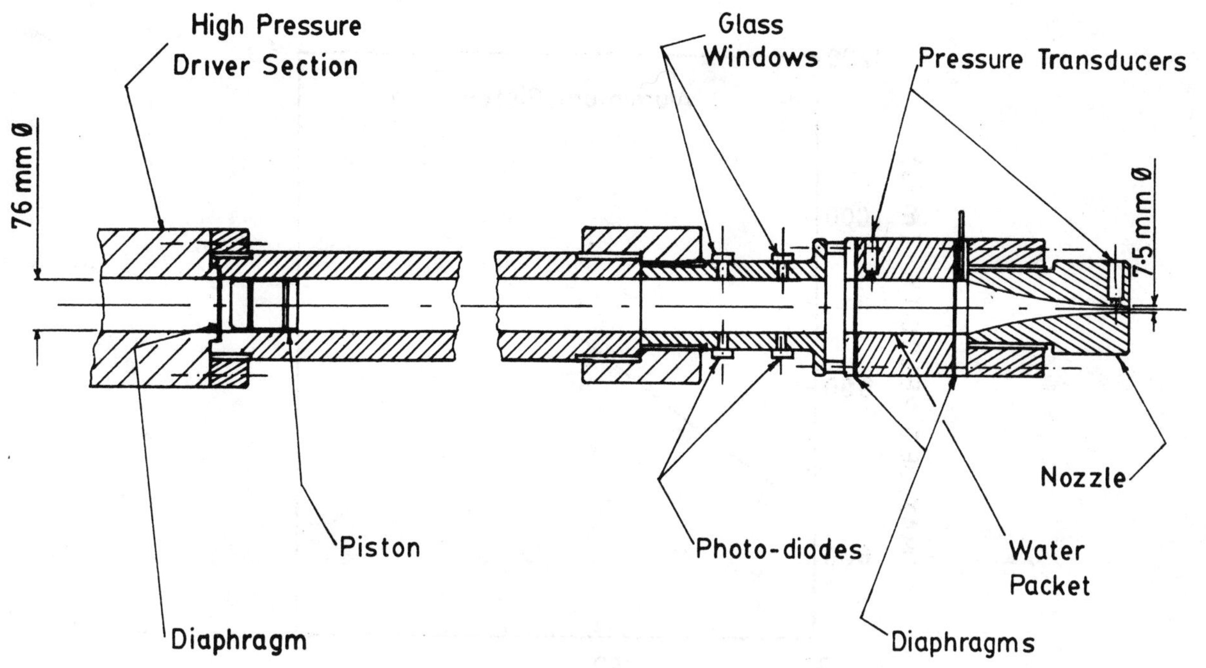

Fig. 1 Free-Piston Water Cannon

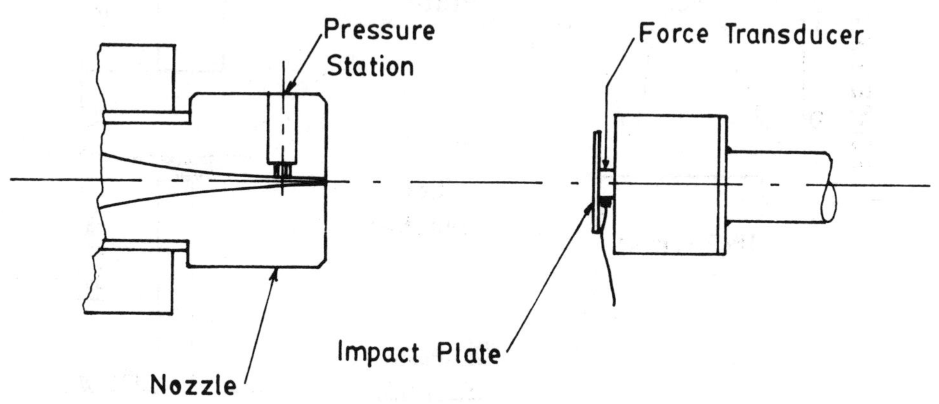

Fig. 2 Force-Measuring Assembly

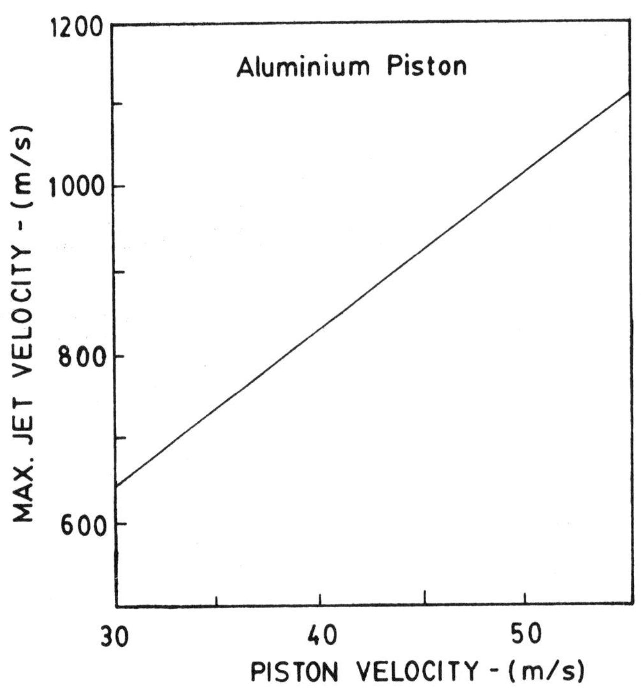

Fig. 3 Water Cannon Calibration Curve

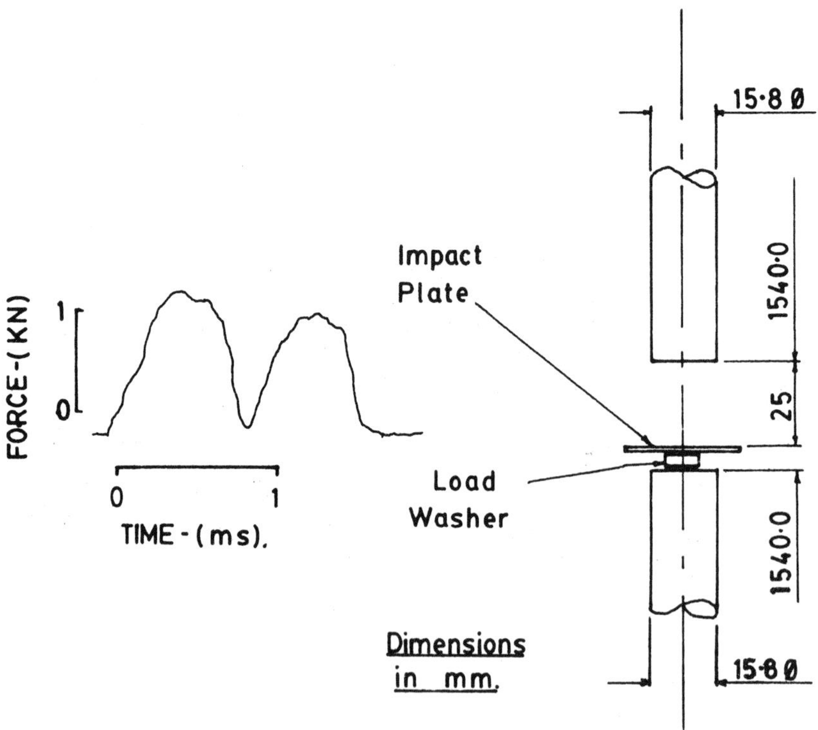

Fig. 4 Dynamic Calibration of Force Transducer

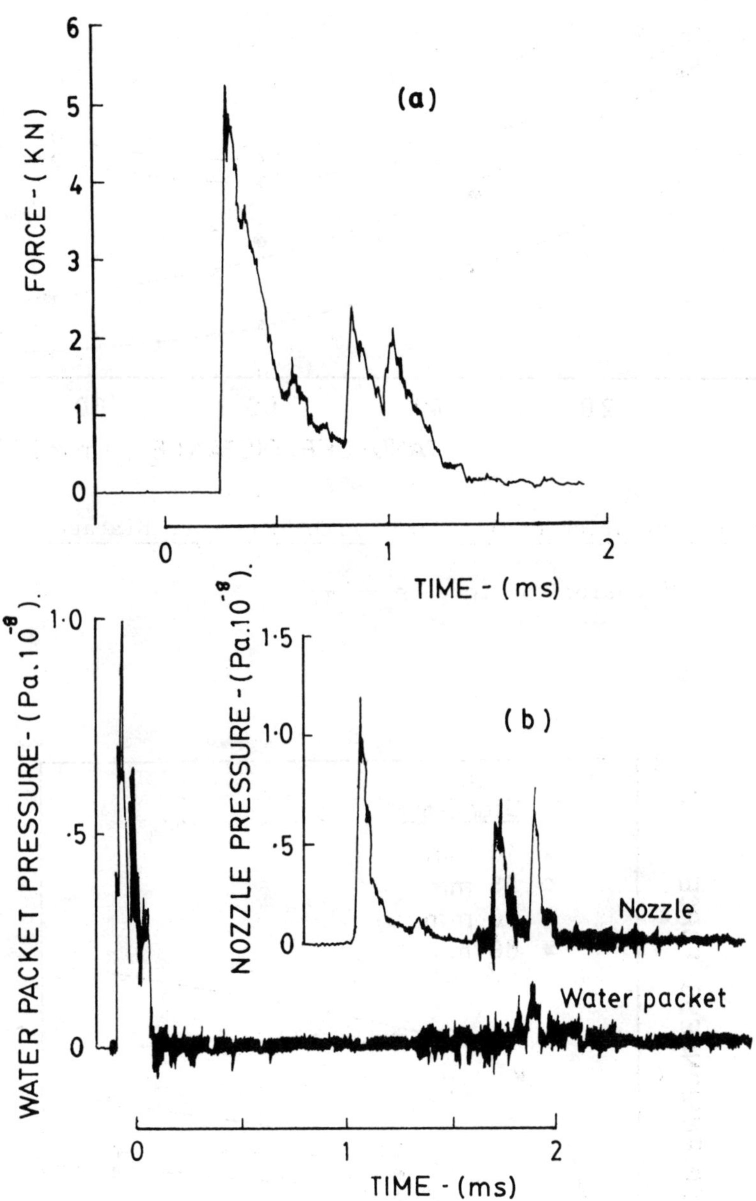

Fig. 5 Typical Jet Force (a) and Water Cannon Pressure Records (b)

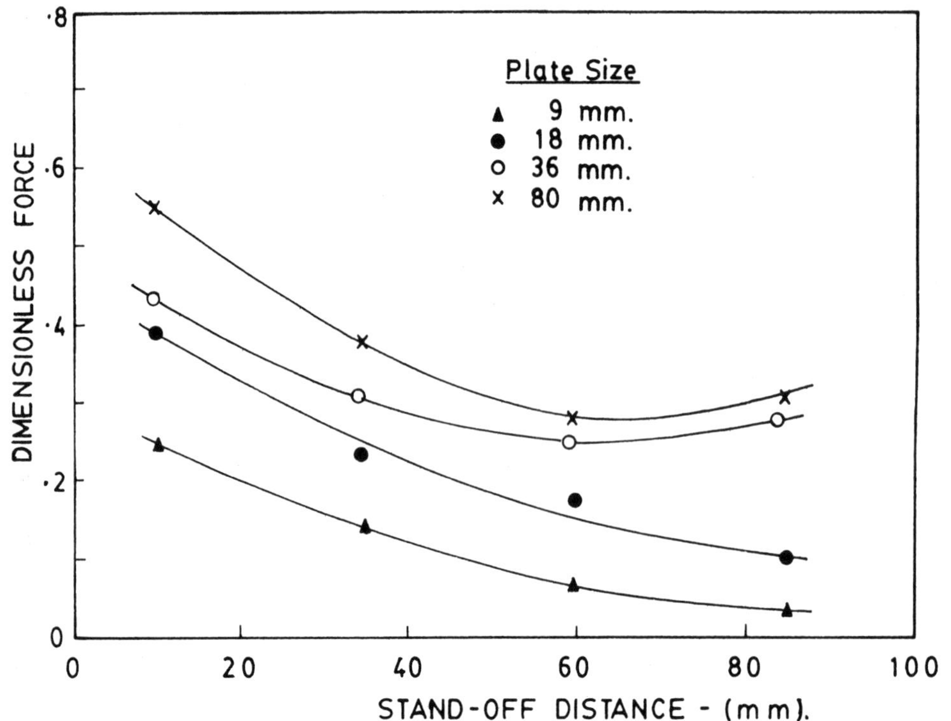

Fig. 6 Variation of Jet Force with Stand-off Distance

$$\text{Dimensionless force} = \frac{F}{\frac{1}{2}\rho U^2 A}$$

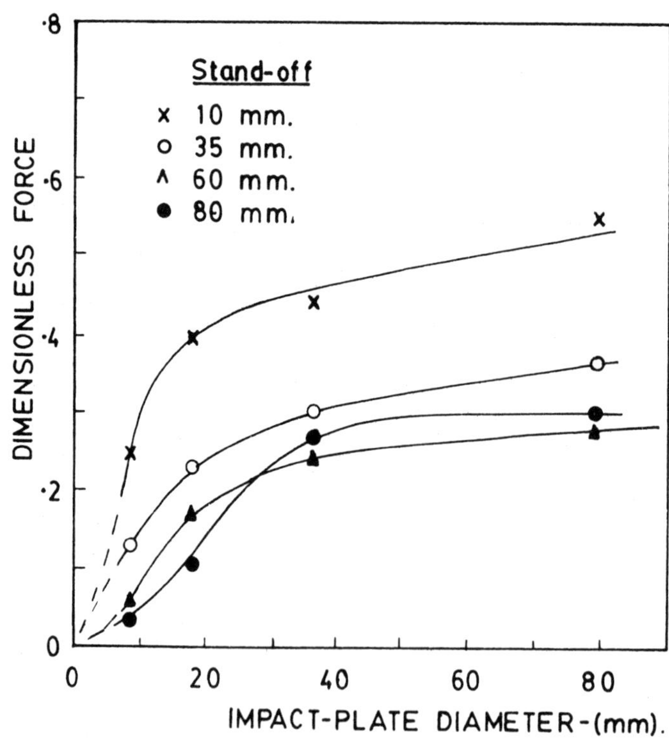

Fig. 7 Variation of Jet Force with Plate Diameter

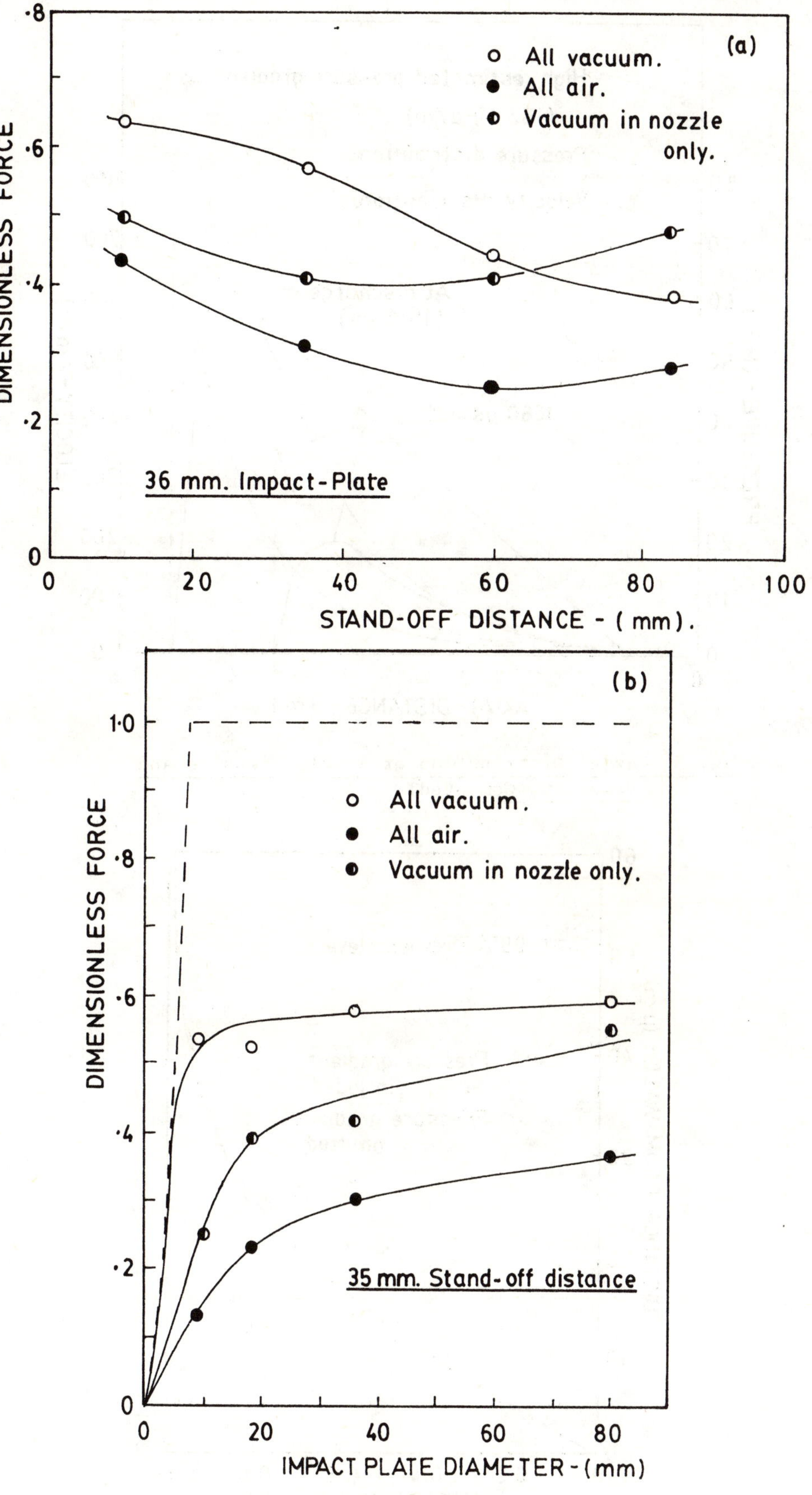

Fig. 8 Influence of Vacuum on Jet Force

(a) Force vs stand-off distance

(b) Force vs plate diameter

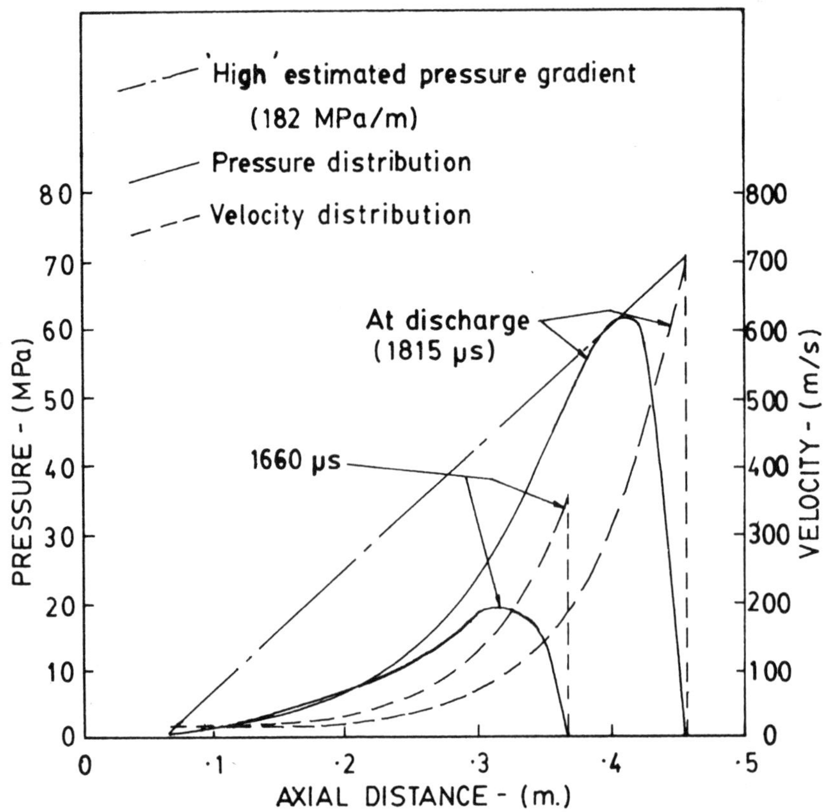

Fig. 9 Axial Distribution of Nozzle Pressure and Velocity, (Computed)

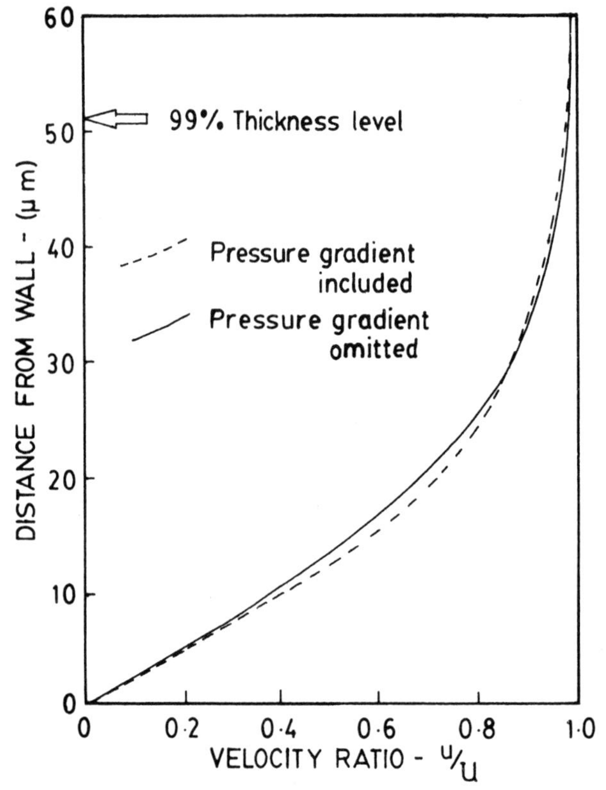

Fig. 10 Boundary layer Velocity Profile at Nozzle Exit

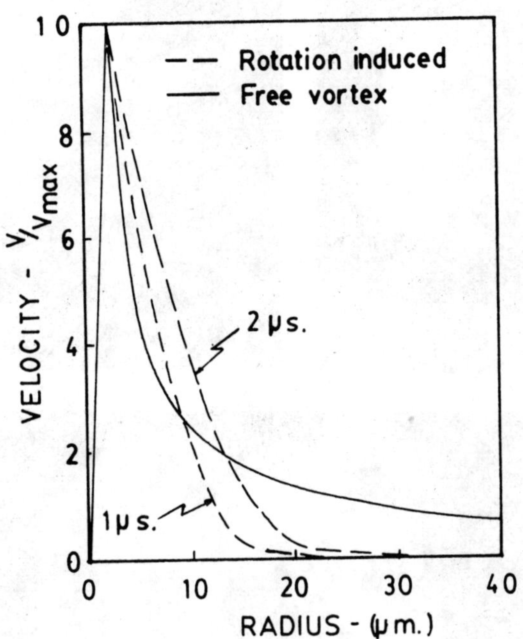

Fig. 11 Tangential Velocity Induced by Droplet Rotation

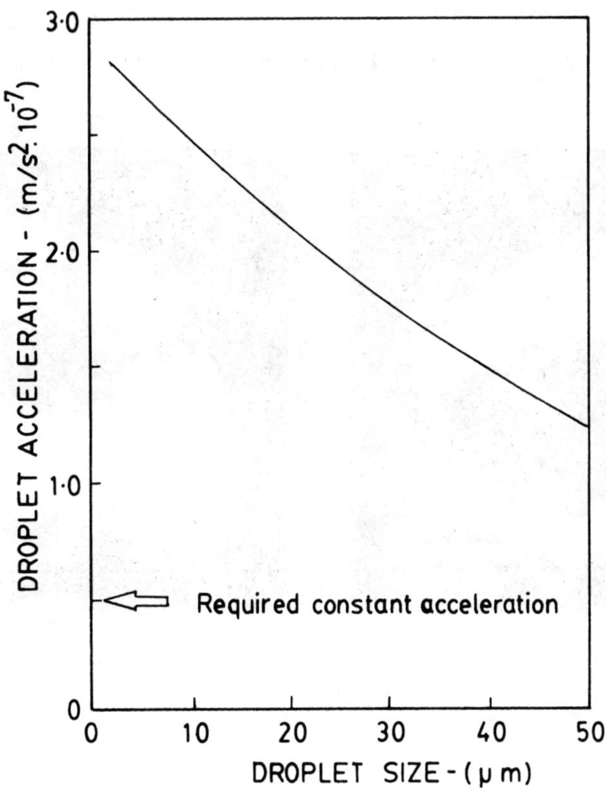

Fig. 12 Variation of Droplet Acceleration with Size

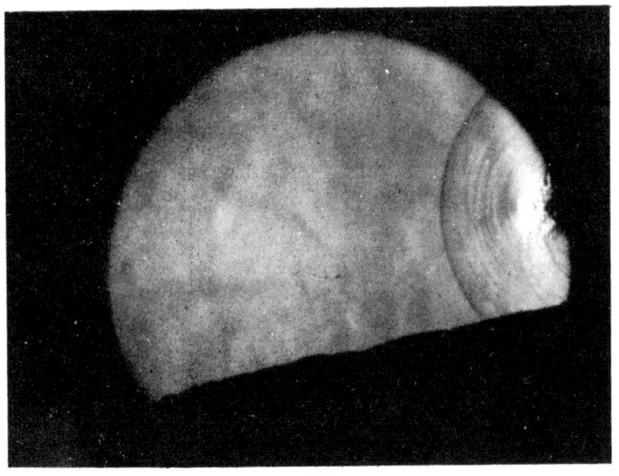

a) 150 μs

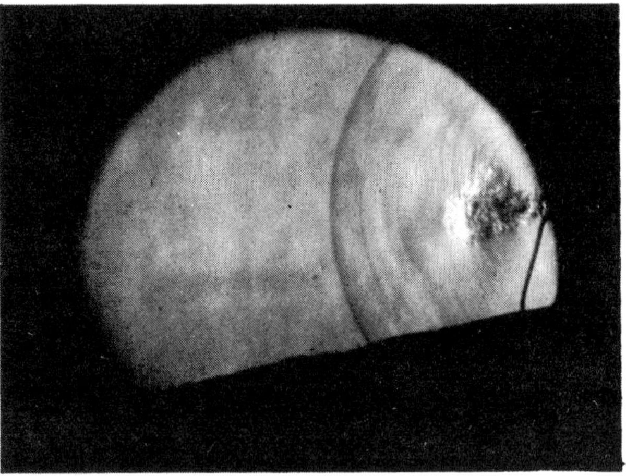

b) 250 μs

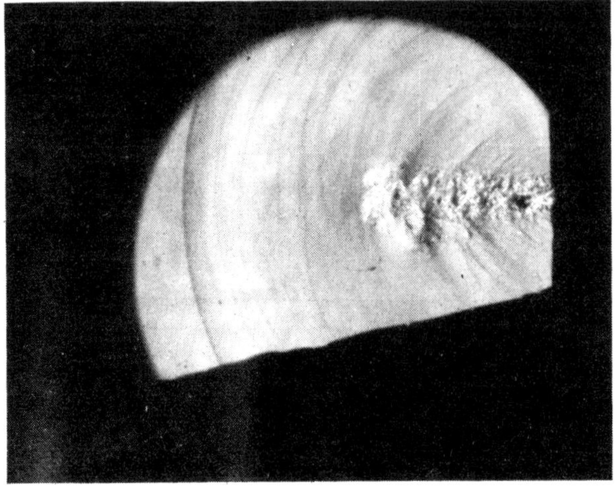

c) 340 μs

PLATE 1. Development of precursor air-jet.

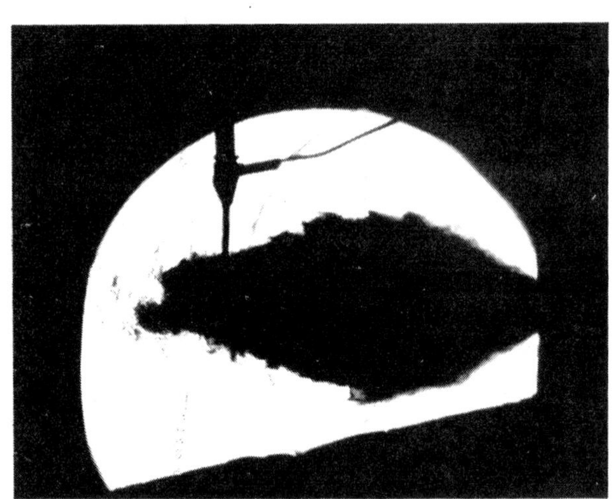

a) Atmospheric air in nozzle.

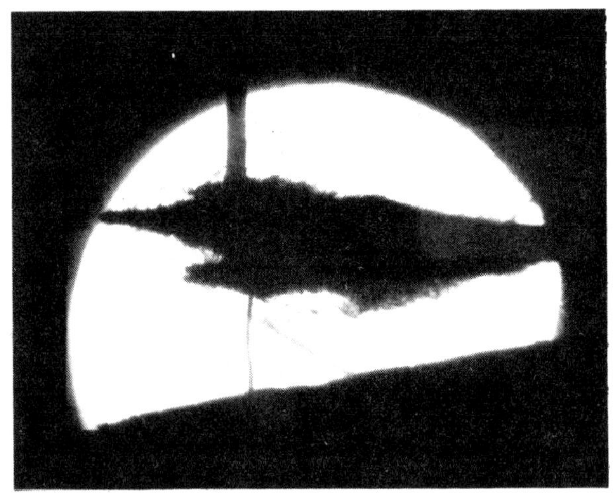

b) Evacuated nozzle.

PLATE 2. Effect of air in nozzle on development of liquid jet.

6th International Symposium on
Jet Cutting Technology
6-8, April, 1982

INTERIOR BALLISTICS OF IMPULSIVE WATER JET

G.A. Atanov

Donetsk State University, USSR

Summary

This paper treats the problem of the interior ballistic of one device for producing ultrajets - impulsive water jet. The mathematical model of the impulsive water jet was built, the similarity criteria were defined, the analysis of the influencing wave processes was carried out. It was found that considering compressibility is compulsory. The authenticity of the model was confirmed experimentally.

Held at the University of Surrey, U.K.
Symposium organised and sponsored by
BHRA Fluid Engineering

NOMENCLATURE

E	elastik modulus
F_*	input cross-sectional area of the narrowing nozzle or the critical cross-sectional area of Laval nozzle
F_1	piston cross-sectional area on the part of water
F_2	piston cross-sectional area on the part of gas
F_{ef}	effective aperture the area
I	total impulse of the get reaction force
I_1	unit impulse
L	barrel length
M_1	piston mass
M_{1r}	piston rod mass
M_2	casing mass
P	water pressure
P_g	gas pressure
P_e	pressure on the barrel cut-off
P_{cr}	critical pressure
a	sonic velocity
C_1, C_2, C_3	similarity criteria
d	barrel diameter
d_*	diameter of the nozzle outlet cross-section
h	clearance value between the piston and the barrel
k	adiabatic constant
l_1	barrel calculated length
l_2	nozzle length
l_p	optimum acceleration path of the piston
l_r	piston rod length
m	water mass
n	constant in Tait's equation of state
r_1	interhall radii of the barrel
r_3	outside radii of the barel
t_s	shot time
u	flow velocity
V_f	final gas volume
w	piston velocity
x	piston coordinate
α, β	similarity criteria
η	interhall efficiency
μ	poisson's ratio
ρ	water dencity

φ_e loss coefficient of gas energy
φ_v efflux coefficient

SUBSCRIPTS

o initial value
opt optimum value

1. INTRODUCTION

As we know it is possible to single out three types of devices for producing ultrajets: cumulative units, hydraulic cannon and impulsive water jet (Ref.1) . Cumulative units realize the effect of hydrodynamic cumulation (Ref.2), hydraulic cannon realizes the inertia principle of fluid acceleration (Ref.3,4,5), impulsive water jet realizes the extrusive effect (Ref.1,6). Fig.1 shows the simplest diagram of the impulsive water jet which functions on the following principle. A heavy piston 2, accelerating on a definite way from an extreme right position under the action of the compressed gas 1 in the receiver, closes the barrel 3 filled with water. As a result of accelerating the piston gets a large kinetic energy which then is transformed into water internal energy, the water pressure increases and water discharges through the nozzle 4. The whole process of the water compression and water effluxing (shot) lasts some milliseconds, in this case the essential role is played by the compressibility of water.

2. SYSTEM OF EQUATIONS OF INTERIOR BALLISTICS

When describing the shot of the impulsive water jet in exact arrangement it is necessary to integrate a complex system of differential equations. At the same time it is possible to build a suitable approximate model of the process, having accepted a series of the simplified assumptions (Ref.7) . The water is considered to be ideal in which case viscosity and heat conductivity are neglected; water parameters in the barrel are averaged over the whole volume and considered to be equal to the parameters of isentropic stagnated flow. The shot process is believed to be quasiundisturbed, i.e. the wave nature of the process is ignored, all parameters are considered to be dependent only on time and effluxing is calculated from the formula for undisturbed flow. For the case of Poisson's adiabat we apply Tait's equation (Ref.8) :

$$\frac{P+B}{\rho^{n}} = const, \qquad (2.1)$$

where B = 300 MPa and n = 7.15. The water flow in nozzle is described by quasi-one-dimentional flow equations (Ref.9) . Elastic deformation of the barrel as well as the piston and water flow can be neglected in clearance between the barrel and the piston.

Analysis shows that when describing the piston movement and the water jet casing it is possible to neglect both the reaction T of the jet and the reaction R of the shock absorber on which the water jet is mounted. Under these conditions the system of equations of the interior ballistics of the impulsive water jet takes the form (Ref.7)

$$\frac{dP}{dt} = n(P+B)\left(\frac{1}{m}\frac{dm}{dt} + \frac{1}{l_1 - x}\frac{dx}{dt}\right),$$

$$\frac{dm}{dt} = -F_* \rho_o a_o \sqrt{\frac{2}{n-1}\left[\left(\frac{P+B}{B}\right)^{\frac{n-1}{n}} - \left(\frac{P_e+B}{B}\right)^{\frac{n-1}{n}}\right]},$$

$$\frac{dw}{dt} = \frac{1}{M_1}\left(1+\frac{M_1}{M_2}\right)\left(F_2 P_g - FP\right),$$

$$\frac{dx}{dt} = w. \qquad (2.2)$$

The barrel calculated length l_1 ,determined by ratio of initial water volume of the barrel cross-sectional area. The case of rigidly fixed casing conforms to the value $M_2 = \infty$. Gas pressure on piston is assumed constant in the proces of shot(piston accelerating way is much greater than its way when shooting). For laval nozzle $P_e = 0$, for narrowing nozzle

$$P_e = \begin{cases} 0 \quad \text{when } P \leq P_{crit.}, \\ (P+B)\left(\frac{2}{n-1}\right)^{\frac{n}{n-1}} - B \quad \text{when } P > P_{crit.} \end{cases} \qquad (2.3)$$

where $P_{crit} = 1.25$ GPa - critical pressure when effluxing into air and only when exceeded is it possible to get the water to supersonic velocity. Practically it is always possible to make the nozzle in such a way that $P_e = 0$. From now on we will consider that this condition has been met.A thorough analysis of water effluxing is reviewed in reports (Ref.9,10) . Index "H" here and henceforth corresponds to water parameters at the initial moment of time.

The first equation of the system (2.2) is written in differential form of Tait's equation,the second - the efflux law of water from the barrel,the third and the fourth - equations of the piston and barrel relative motion.

The initial conditions for the system's integration (2.2) are determined as follows. Coordinate system is selected so that $x_0 = 0$. The initial velocity of the piston is determined by the process of its acceleration,more detail will be considered later on. The value m is made up of that part of water,parameters which can be considered isentropic stagnated. With an accuracy up to 15% on pressure this part of water located between the piston and section where $F_1/F_* = 2$, i.e. practically all the water being poured into the barrel. Initial pressure with the sufficient accuracy is determined by Zhukovsky's formula for hydraulic impact

$$P_0 = a_0 \rho_0 w_0 . \tag{2.4}$$

The efflux velocity is determined from formula

$$u_e = a^\circ \sqrt{\frac{2}{n-1}\left[\left(\frac{P+B}{B}\right)^{\frac{n-1}{n}} - 1\right]}, \tag{2.5}$$

where a°=1460 m/s .

Essentially the shot process is determined by water compressibility,in doing so the neglecting of water compressibility distorts not only the quantitative characteristics but also the qualitative shot pattern. The formulated above assumptions allow us to get the solution when neglecting the compressibility; the function $P = P(t)$ takes the form:

$$P = \begin{cases} P_c\left(\dfrac{\exp(cbt) - \left|\dfrac{\sqrt{P_0}-\sqrt{P_c}}{\sqrt{P_0}+\sqrt{P_c}}\right|}{\exp(cbt) + \left|\dfrac{\sqrt{P_0}-\sqrt{P_c}}{\sqrt{P_0}+\sqrt{P_c}}\right|}\right)^2 & \text{when } P_0 > P_c \\ P_c & \text{when } P_0 = P_c , \\ P_c\left(\dfrac{\exp(cbt) + \left|\dfrac{\sqrt{P_0}-\sqrt{P_c}}{\sqrt{P_0}+\sqrt{P_c}}\right|}{\exp(cbt) - \left|\dfrac{\sqrt{P_0}-\sqrt{P_c}}{\sqrt{P_0}+\sqrt{P_c}}\right|}\right)^2 & \text{when } P_0 < P_c , \end{cases} \tag{2.6}$$

where

$$b^2 = c^2 \frac{F_2}{F_1 P_g} , \quad c^2 = \frac{F_1 F_2 P_g \sqrt{2\rho}}{F_* M_1}\left(1 + \frac{M_1}{M_2}\right). \tag{2.7}$$

Initial pressure is not determined from formula (2.4) (there is no impact wave in incompressible fluid) but from the expression

$$P_0 = \frac{\rho F_1^2 w_0^2}{2F_*^2} , \tag{2.8}$$

pressure $P_c = F_1 P_g / F_2 = (c/b)^2$ is in accordance with the equation of forces acting in the piston on the part of water and gas.

Fig.2 shows the dependence $P = P(t)$ with allowance (curves 1 and 2) and without allowance (curve 3) for the compressibility of one of the versions of the water jet. The solution for incompressible fluid doesn't depend on initial mass of water, but taking the compressibility into account makes this dependence fairly strong (curve 1 is obtained when m_o = 0,35kg, curve 2 - when m_o = 0,2 kg) .

3. SIMILARITY CRITERIA

A greate many constructive parameters determining the shot hinders the analysis of the process and the culculation of the impulsive water jet parameters from meeting the specified requirements. A more advantageous form of writing the equations is the dimensionless one which permits determination of the device similarity criterion. Among these constructive parameters are the initial conditions for the integration of the system (2.2); at the same time it is advisable to have the universal initial conditions. That is why as the scales for piston velocity and water mass we will take the initial values of these parameters. Let us take as a scale l_1 for the piston movement and for time - magnitude l_1/w_0 . Numerical evaluations have shown weak dependence of the shot maximum pressure on the initial pressure, neglecting it even results in the decrease of the maximum pressure as low as some percentage. Therefore from the practical point of view it is convenient to take as initial pressure some average pressure value and as the scale - magnitude B .

The estimation of the influence on parameters of elastic deformation and water flow in the clearence between the piston and barrel was also carried out. The significance of these factors is small and their estimation can be done in the simplest way considering the tensile strain for the barrel as for the cylinder which is under the internal pressure and for the piston rod - the compressive strain; the flow through the clearance we accept also quasi-undisturbed, however the water viscosity should be taken into consideration.

The influence of the barrel deformation is taken into account by the complex equation

$$C_1 = \frac{2B}{E}\left(\frac{z_3^2 - z_1^2}{z_3^2 + z_1^2} + \mu\right), \qquad (3.1)$$

and the influence of the piston deformation- by the complex equation

$$C_2 = \frac{l_2}{l_1}\frac{B}{E}\left(1 - \frac{M_{12}}{M_2}\right). \qquad (3.2)$$

The analysis of the flow in clearance has shown that under big pressures the velocity loss due to viscosity amounts to not more than 10+15% . Thus one may consider that the water efflux takes place through the effective aperture the area F_{ef} of which is equal to the area sums of the nozzle and the clearance. To consider the imperfec - tion of the efflux process we introduce the efflux coefficient $\varphi_v=0.9$. The estimation shows that total effect of deformations and clearence decreases the maximum of the shot pressure by 15% + 20%.

Written in the dimensionless form the system of equations of the interior ballistics takes the form (Ref.11) :

$$\frac{d\bar{P}}{d\bar{t}} = \left[C_1 + 1/n(\bar{P}+1)\right]^{-1}\left(\frac{1}{\bar{m}}\frac{d\bar{m}}{d\bar{t}} + \frac{1}{1-\bar{x}}\frac{d\bar{x}}{d\bar{t}}\right),$$

$$\frac{d\bar{m}}{d\bar{t}} = -\alpha\sqrt{(\bar{P}+1)^{\frac{n-1}{n}} - 1},$$

$$\frac{d\bar{w}}{d\bar{t}} = \beta(C_3 - \bar{P}),$$

$$\frac{d\bar{x}}{d\bar{t}} = \left[C_1 + 1/n(\bar{P}+1) + C_2/(1-\bar{x})\right]^{-1}\left\{\bar{w}\left[C_1 - 1/n(\bar{P}+1)\right] - C_2\frac{d\bar{m}}{d\bar{t}}\right\}, \tag{3.3}$$

where similarity criteria

$$\alpha = \varphi_v\sqrt{\frac{2}{n-1}}\,\frac{a_o}{w_o}\,\frac{F_{ef}}{F_1}, \tag{3.4}$$

$$\beta = \frac{1}{n}\left(1+\frac{M_1}{M_2}\right)\frac{m_o}{M_1}\left(\frac{a_o}{w_o}\right)^2 = \frac{a_o^2 m_o}{2\varphi_e n E_g}, \tag{3.5}$$

$$C_3 = \frac{P_g F_2}{B F_1}, \tag{3.6}$$

loss coefficient of gas energy φ_e determined by the process imperfection of the piston acceleration. The dimensionless variables are designated by dashes.

The initial conditions are as follows:

$$\bar{P}_o = 0.1;\quad \bar{m}_o = 1;\quad \bar{w}_o = 1;\quad \bar{x}_o = 0. \tag{3.7}$$

The values C_1, C_2 and C_3 vary in sufficiently narrow ranges and because the effect of these parameters on the shot is small it is possible to accept them as constant values: $C_1 = C_2 = 0.005$; $C_3 \simeq 0.25$. Thus the shot process is characterized only by two similarity criteria α and β . The shot maximum pressure and the efflux velocity are important characteristics of the shot,which determine the dynamic pressure on the barrier. Along with this the destroying effect must be determined by total impulse of the jet reaction force during the shot which is determined by expression (Ref.12):

$$I = \varphi_v^2 a_o m_o \alpha \sqrt{\frac{2}{n-1}} \int_0^{\bar{t}_s} \left[(\bar{P}+1)^{\frac{n-1}{n}} - 1 \right] d\bar{t}. \tag{3.8}$$

The shot time has the following meaning. When shooting the piston velocity is severely decreased, and reaches its minimum at the instant t_s . At first sight, the most favourable from the constructive point of view, in the case when $w_{min} = 0$ because in this case there is no impact of the piston against the breaking gear. However in this case more stringent requirements are placed on the accuracy of the arrangement; if we assume the small positive value w_{min} then the distance allowance for the brake arrangement will substantially increase.

For specific unit consideration it is possible to introduce the specific or unit impulse showing which part of the compressed gas energy (Ref.11):

$$I_1 = I/E_g = \frac{2n\varphi_v^2 \varphi_e}{a_o} \alpha\beta \int_0^{\bar{t}_s} \left[(\bar{P}+1)^{\frac{n-1}{n}} - 1 \right] d\bar{t}. \tag{3.9}$$

Being the most significant of the shot effectiveness, I_1 at the same time depends only upon the similarity criteria and loss coefficients changing within fairly narrow limits; this makes the unit impulse a universal characteristic.

Another important characteristic of the shot quality is the internal efficiency showing which part of the compressed gas energy consumed on taking a shot is transformed into jet kinetic energy:

$$\eta = \frac{2n\varphi_v^3 \varphi_e}{(n+1)\left[1 + 2\varphi_e c_3 \beta \bar{x}(\bar{t}_s)\right]} \alpha\beta \int_0^{\bar{t}_s} \left[(\bar{P}+1)^{\frac{n-1}{n}} - 1 \right] d\bar{t}. \tag{3.10}$$

It is evident from the equation that $\eta = \eta(\alpha, \beta)$.

4. OPTIMUM IMPULSIVE WATER JET

For effective rock breakage the pressure being developed by the jet against a barrier as many investigations show must not be less than a definite magnitude. Therefore, as one of the characteristics when working out the impulsive water jet, the maximum shot pressure P_{max} must be preassigned. Until now the investigation of breakage which should have taken into account the force of the jet reaction has not been known . At the same time the impulse of the reaction force is determined by the action time of the jet against the barrier, and it is clear that considering it for the impulse jets is compulsory Thus the water jet must provide along with P_{max}. the preassigned total impulse of the jet reaction force. Among all the water jets the most effective (optimum) will be that which will fulfill the preassigned requirements with minimum consumption of energy, i.e. which will have the biggest magnitude of the unit impulse. Further we will show how to determine the similarity criteria for such a shot. The integration was performed numerically with the Runge-Kutta method.

Using the different values of α and β it is possible to get the given value of P_{max}. Fig. 3 shows the lines of equal maximum pressure near which the values of P_{max} are given in MPa. Fig.4 illustrated the solid lines showing the dependence of $I_1(\alpha)$ at constant values of P_{max}, indicated near the lines. As can be seen these functions have extremum which conforms to the optimum criterion of $\alpha : \alpha_{opt}$. As is seen from the Fig.2 the optimum values $\beta = \beta_{opt}$ are determined from P_{max} and α_{opt}. When $\alpha = \alpha_{opt}$ and $\beta = \beta_{opt}$ the function $\eta = \eta(\alpha)$ also has extremum.

The dash-and-dot lines, Fig.4 indicate the dependence of the piston minimum velocities, the dimensionless magnitudes of which are shown near the curves. The extremum of the function $I_1(\alpha)$ corresponds to the value $W_{min} = 0$. The obtained results are explained by the fact that when $W_{min} = 0$ all energy, given up by gas is transformed into the energy of water. When $W_{min} > 0$ the piston liberates not all the stored energy, when $W_{min} < 0$ part of energy given up to water by the piston returns to the piston.

After determining α_{opt}, β_{opt} and I_1 from formula (3.8) with a given value of I the value of water initial mass is found after which from formula (3.5) - the required gas energy. When given the piston mass (its tentative value is usually known) its velocity is determined as follows:

$$W_0 = \sqrt{\frac{2 \varphi_e E_g}{M_1}\left(1 + \frac{M_1}{M_2}\right)} . \tag{4.1}$$

Using gas parameters

$$W_0 = \sqrt{\frac{2 P_{g0} V_0 \left[1 - (V_0/V_f)^{k-1}\right]}{(k-1) M_1}\left(1 + \frac{M_1}{M_2}\right)} . \tag{4.2}$$

There are definite relationships between the parameters of the impulsive water jet, ensuring its minimum dimensions (Ref.13).

At a given initial gas pressure, the initial gas volume is calculated from the formula

$$\mathcal{V}_o = k \frac{E_g}{P_{go}} . \tag{4.3}$$

This value $\mathcal{V}_0$ gives the minimum gas volume (it is this volume which determines the water jet's dimensions)

$$\mathcal{V}_f = k^{\frac{k}{k-1}} \frac{E_g}{P_{go}} . \tag{4.4}$$

There also exists the optimum acceleration path of the piston

$$\ell_p = \left(k^{\frac{k}{k-1}} - k \right) \frac{E_g}{P_{go} F_z} \tag{4.5}$$

and the optimum ratio

$$\mathcal{V}_f / \mathcal{V}_o = k^{\frac{1}{k-1}} . \tag{4.6}$$

For diatomic gases,in particular,air ratio is equal to 2.32 .

The piston radius is found from the condition that the receiver must have a minimum weight and,for instance,for the receiver scheme Fig.1 where the receiver has the shape of a cylinder with radius equal to the piston radius

$$r_d = \sqrt[3]{k^{\frac{k}{k-1}} \frac{E_g}{\pi P_{go} (2\xi + 1)}} , \tag{4.7}$$

where ξ = 2.5 + 3 - constructive coefficient considering jamming of the length and the piston radius.

The diameter of the nozzle outlet cross-section is found from the formula

$$d_* = \sqrt{d_1 \left(\frac{a_o}{w_o} \frac{\alpha}{\varphi_v} d_1 \sqrt{\frac{2}{n-1}} - 2h \right)} . \tag{4.8}$$

5. WAVE - LIKE PROCESSES

The main and the strongest assumption when considering the shot model of the impulsive water jet is the assumption of a quasi - undisturbed process. To define the process features caused by its wave-like nature and fixing the limit of acceptability of the quasi-undisturbed approximation,it is necessary to carry out calculations as in the case of undisturbed process (Ref.14) . We confine ourselves to the consideration of quasi-one-dimensional flow without taking into account the deformation and flow in the clearance. In this case the motion is described by the following system of differential equations (let us write it in divergent form)

$$\frac{\partial \bar{u}}{\partial \bar{t}} + \frac{\partial}{\partial \bar{x}}\left(\frac{\bar{u}^2}{2} + \frac{\bar{\rho}^{n-1}}{n-1}\right) = 0, \tag{5.1}$$

$$\frac{\partial}{\partial \bar{t}} \bar{\rho} F + \frac{\partial}{\partial \bar{x}} \bar{\rho} \bar{u} F = 0.$$

The channel cross-sectional area F=F(x) . The variables are dimentionless, a_0, ρ_0, the nozzle length l_2, magnitude l_2/a_0 are taken as scales. The initial conditions are the following

$$\bar{u}(\bar{x},0) = 0; \ \bar{\rho}(\bar{x},0) = 1: \ -L/l_2 < \bar{x} \leq 1; \tag{5.2}$$

$$\bar{u}(-L/l_2, 0) = w_0/a_0.$$

The condition on the barrel cut-off

$$\bar{\rho}(1,\bar{t}) = 1, \tag{5.3}$$

on the piston

$$\frac{d\bar{u}(\bar{x}_p,\bar{t})}{d\bar{t}} = A\left[C_3 + 1 - \bar{\rho}(\bar{x}_p,\bar{t})\right], \tag{5.4}$$

$$A = \frac{1}{n} \frac{l_2}{l_1} \frac{m_0}{M_1} . \tag{5.5}$$

The magnitude A in this case is the similarity criterion. System (5.1) was integrated numerically by the Godunov's method of finite differences (Ref.15); some results of these calculations are shown in Fig.5. The solid lines show the changing of the water pressure on the piston and disturbed solution and the dash-and-dot lines-quasiundisturbed one. In the quasidisturbed solution the pressure has an oscillatory character which is caused by motion between the piston and the nozzle of the waves caused by the impact wave which arose during the impact of the piston against the water and reflacted from the nozzle by the compression wave. As is seen for version 1 the quasi-one-dimensional assumption yields fully reasonable accuracy. The condition of acceptability of quasiundisturbed assumption- the ratio of maximum shot pressure to the initial pressure (the pressure beyond the impact wave) has been found; the quasiundisturbed assumption is allowable when $P_{max}/P_0 = 8 \div 10$. Inasmuch as P_0 is determined by the piston initial velocity the said condition at the given maximum pressure means a limitation on the piston initial velocity. The dependance of the most allowable velocity of the piston of the maximum pressure estimated from the non-linear formulae is shown in the table

$\bar{P}_{max}$, GPa	0.2	0.4	0.6	0.8	1.0	1.2
w_o, m/s	14	25	37	49	62	76

The comparison of the solutions in the disturbed and quasi-undisturbed cases shows that their difference for the efflux velocities is much bigger than for the pressure on the piston especially at the initial stage of the process, in the case of disturbed solution the velocities are lower. In this manner the quasiundisturbed solution under the same pressure yields higher velocities. This property is explained by the water expansion (Ref.13,16). When water expands up to atmospheric pressure the following relation is true:

$$\frac{u_{undist}}{u_{dist}} = \sqrt{\frac{n-1}{2}\,\frac{a_o+1}{a_o-1}}, \qquad (5.6)$$

where U_{undist} - the velocity under undisturbed expansion(we recall that this is the same as in quasiundisturbed solution); U_{dist} - the velocity under disturbed expansion in one wave. Decreasing with the growth of a_0 , relation (5.6) tends to the magnitude 1.76. If the process is disturbed and the expansion occurs not only in one wave but in the system of waves relation (5.6) is reduced approaching 1 when increasing the number of interacting waves. The quasiundisturbed solution conforms to the infinite number of waves, the water efflux is culculated from the formula of undisturbed process. So, the impulsive water jet, the process of which is closely approximating to the quasi-undisturbed, under the same pressure in the barrel will have the biggest efflux velocity consequently, the shot of such a water jet will be the most efficient.

6. EXPERIMENTS AND COMPARISON WITH THEORY

Series of the impulsive water jet bench testings was carried out in order to examine the presented theory. When accelerating the piston, the loss of energy, water pressure in the shot process and the impulsive force of the reaction jet were determined (Ref.12,17). The coefficient of the energy loss was determined by way of comparison of the piston velocity estimated in the ideal case (when φ_c =1) and defined experimentally. The piston velocity was computed with electromagnetic transducer. This transducer consists of two coils wound on a pipe textolite frame in series in which the core, connected with the piston from annealed iron is freely moved. When feeding power (A.C.) to one of the coils an electromotive force, linearly dependent upon the movement of the piston is induced in the second one. Measurements revealed that in the interval of the piston velocities from 0 to 40 m/sec the coefficient of the energy loss is very close to a linear function, changing from 1.0 to 0.7.

For measuring the water pressure a probe with a sensitive element was used which is a flat spiral from manganine wire dia 0.07 mm (Ref.18) mounted in the steel case on an exposide compound. The probe was placed in the bridge arm, the signal from the measuring diagonal was amplified and fed to the oscillograph. Inasmuch as the probe recorded the voltage appearing in the solid body the frequency of its own fluctuations are not more than 200 kHZ. The frequency spectrum of the pressure impulse in quasiundisturbed solution, determined with the aid of the direct Fourier transformation to an accuracy of 1% in energy lies within the frequency limits from 0 to 3 kHZ. Considering that in actual fact the shot process has characteristic time equal to the time of disturbance propagation from the piston to the nozzle, the upper margin of the frequency spectrum is essential to increase to 40 + 50 kHZ. Thus the described pressure probe meets the frequency requirements for recording the impulse of the pressure of the impulsive water jet and assumes only the statistic calibrating which was carried out in the bomb of high pressure. The measurement error did not exceed 6%. Fig.6 shows a representative oscillogram of the pressure obtained from measurements for the water jet having the following similarity criteria $\alpha=0.67$ and $\beta= 4.04$. The dependence of the pressure upon time for this water jet calculated in quasi-one-dimensional assumption is shown by broken line. The maximum shot pressure is equal to 340 MP . The difference of the calculated and measured values of P_{max} is about 5%.

The measurments of the total impulse of the force of the jet reaction carried out with ballistic pendulum also showed reasonable accuracy of calculations (Ref.12).

7. CONCLUSION

Thus it is possible to draw the following conclusion from this work:

a) based on the hypothesis of the quasiundisturbedness, the model of the impulsive water jet shot was built taking into account water compressibility;

b) the similarity criteria were defined, the method of the optimum impulsive water jet calculation was developed;

c) the conditions of acceptability of the quasiundisturbed solution was defined. It was found that the most efficient water jet was one the shot of which was close to quasiundisturbed;

d) the suitable accuracy of the calculation was confirmed experimentally.

8. REFERENCES

1. Atanov,G.A.: "Methods of ultrajets obtained". (Методы получения ультраструй). Ref. "Mechanics", 4, 1975, ref. 4B696. (In Russian).

. Bowden, F.P. and McOnie, M.P.: "Formation of cavities and micro-jets in liquids and their role in initiation and growth of explosion". Phil. Trans. Roy. Soc. of London, A, 298, 38 (1967).

3. Atanov, G.A.: "Numerical investigation of water cannon". (Численное исследование сверхзвукового течения в гидропушке). Izv. Akad. Nauk, SSSR,"Mech. of Fluid and Gas", 1, 1973.(In Russian).

4. Atanov, G.A.: " Design of the splitting off flows in the water cannon". (Расчет течения с отколами в гидропушке). Izv. Vuzov SSSR, Energetics, 5, 1974. (In Russian).

5. Cooley, W.C. and Lucke, W.N.: "Development and testing of a water cannon for tunneling". In: Proc. 2nd International Symposium on Jet Cutting Technology, Cambridge, 1974.

6. Moodie, K. and Tailor, G.: "The fracturing of rocks by pulsed water jets". In: Proc. 2nd International Symposium on Jet Cutting Technology, Cambridge, 1974.

7. Atanov, G.A.: "Design of the shot parameters of the impulsive water jet". (Расчет параметров выстрела импульсного водомета). In col.: Hydromech., Kiev, 12, 1972. (In Russian).

8. Koul, L.: "Submarine explosions". (Подводные взрывы). Moscow, Publ. H. of For. Lit., 1950. (In Russian).

9. Atanov, G.A.& Povh, I.L.: "The influence of compressibility on

the water and other liquid flows". (Влияние сжимаемости на течение воды и других жидкостей). In col.: Hydromech. , Kiev, 15, 1969. (In Russian).

10. Atanov, G.A.,Semko, A.M.: "On the effluxing liquids in the underexpansion regime." (Об истечении жидкости на режиме недорасширения). In col.: Hydroaeromech. and theory of elasticity. Dnepropetrovsk,15, 1972. (In Russian).

11. Atanov, G.A., Chernikov, G.A.:"On the optimum impulsive water jet." (Об оптимальном импульсном водомете). Izv. Vuzov SSSR, Energetics,1, 1973. (In Russian).

12. Atanov, G.A. , Maksimenko, V.P.: "On the impulse of reaction force of the impulsive water jet". (Об импульсе силы реакции импульсного водомета). In col.: Theor. and appl. mech.,Kharkov, 5, 1974. (In Russian).

13. Atanov, G.A.:"The principles of the one-dimentional gasodynamics." (Основы нестационарной одномерной газодинамики). Kiev, Vishya shcola, 1979. (In Russian).

14. Atanov, G.A.:" Design of the impulsive water jet shot of calculation the wave processes". (Расчет выстрела импульсного водомета с учетом волновых процессов), Izv. Vuzov SSSR, Energetics, 3 , 1975. (In Russian).

15. Godunov, S.K.: "The difference method of the numerical calculation of the break decisions for the hydrodynamic equations". (Разностный метод численного расчета разрывных решений уравнений гидродинамики). In coll.: Mathematical coll., v.47, N 3,1959. (In Russian).

16. Atanov, G.A.: "On the unstationary expansion of liquid and the use of its pressure for running of the piston". (О нестационарном расширении жидкости и использовании её давления для разгона поршня). In coll.: Theor. and appl. mech. Kiev-Donetsk, Vishya shcola, 9 , 1978. (In Russian).

17. Atanov, G.A. , Ukrainskiy, .D.: "Experimental investigation of the impulsive water get". (Экспериментальное исследование импульсного водомета). Izv. Akad. Nauk,SSSR, "Mech. of Fluid and Gas", 1 , 1979. (In Russian).

18. Tsiklis, D.C. , Borodina, M.D.: "Uninertional manganine monometer". (Малоинерционный манганиновый манометр). Instr. and tech. of experiments. 2, 1965. (In Russian).

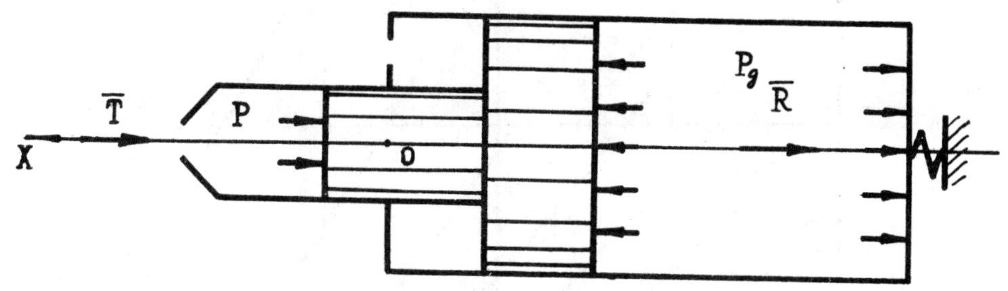

FIG.1. DIAGRAM OF THE IMPULSIVE WATER JET.

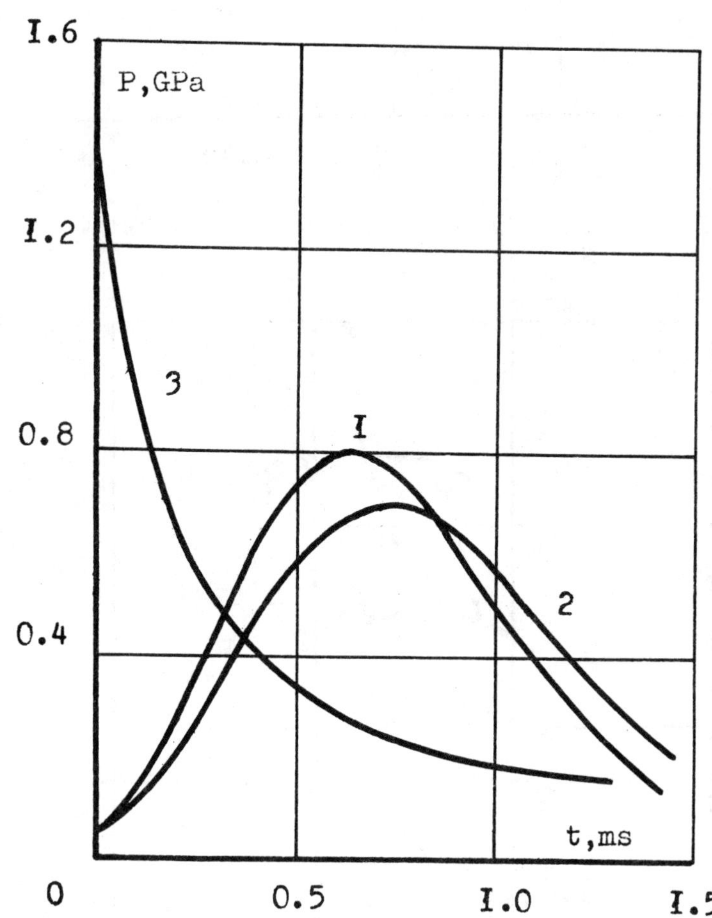

FIG.2. CHANGING OF THE WATER PRESSURE OF THE BARREL.

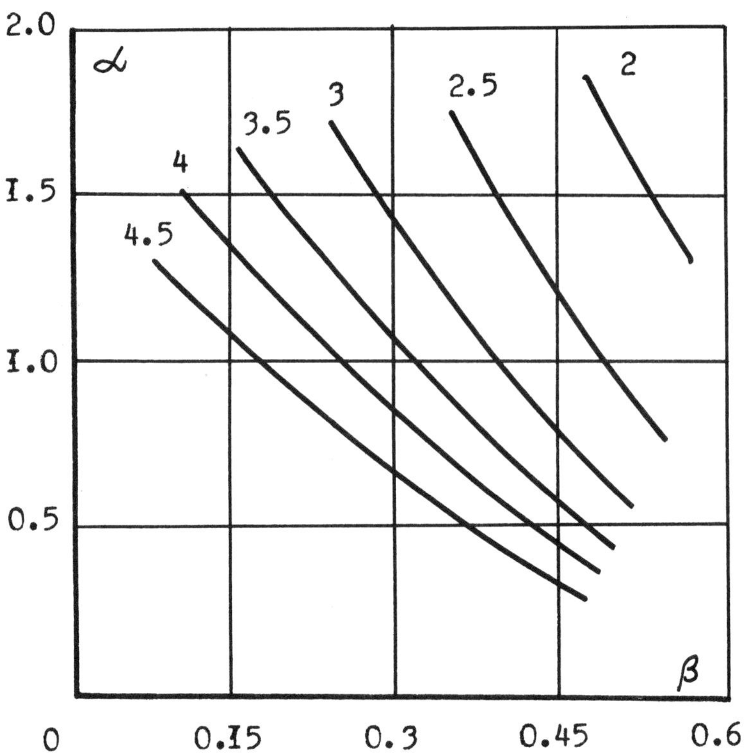

FIG.3. LINES OF EQUAL MAXIMUM PRESSURE.

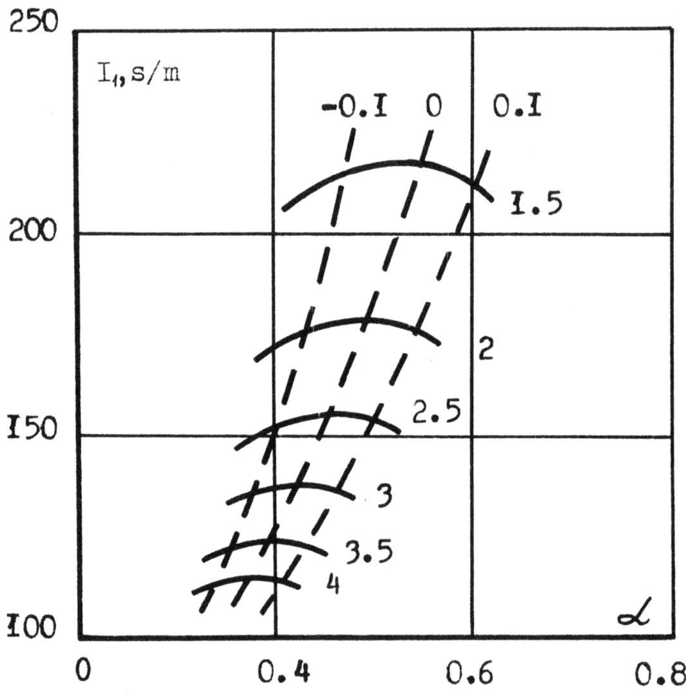

FIG.4. CHANGING OF THE UNIT IMPULSE.

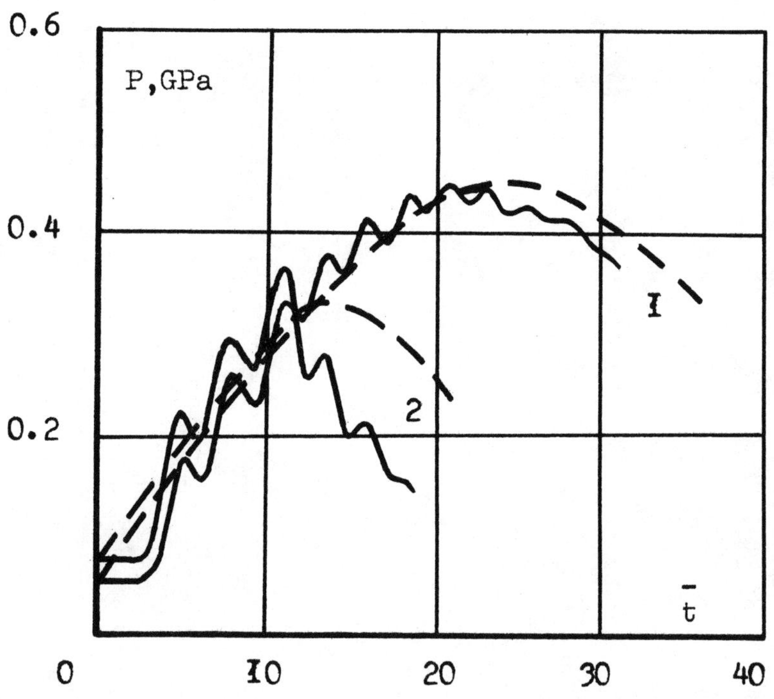

FIG.5. CHANGING OF THE WATER PRESSURE ON THE PISTON.

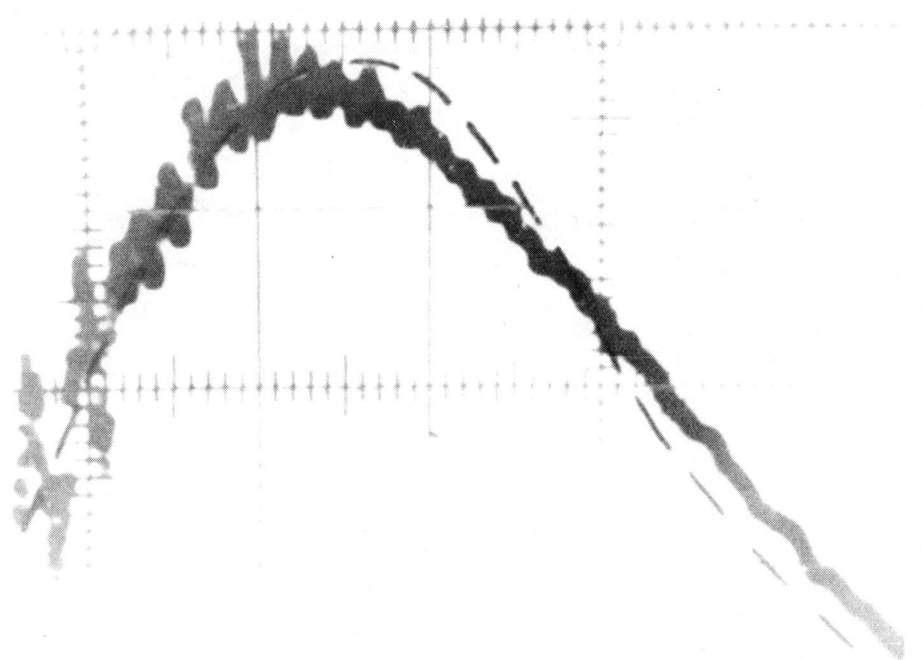

FIG.6. OSCILLOGRAM OF THE WATER PRESSURE OF THE BARREL.

6th International Symposium on
Jet Cutting Technology
6-8, April, 1982

JET NOISE MEASUREMENTS ON HAND HELD CLEANING EQUIPMENT

C.R. Barker and A. Cummings,
University of Missouri-Rolla, U.S.A.
M. Anderson
Partek Corporation, U.S.A.

Summary

In many cleaning applications, hand held water jet nozzles are used to produce the desired result. In these cases, the distance between the operator and the jet nozzle is relatively short and the operator will be exposed to any noise generated by the water jet. Therefore, it is important to determine the level and frequency content of the noise generated by typical nozzles before any attempts to silence the noise are proposed.

Results are reported here of noise measurements taken from a circular and a 15 deg fan jet when the jets were operating in air and when impacting normal to the surface of a steel sheet. Data were taken to characterize both the near and far field sound emissions of the free jets. In addition, a screen was used in conjunction with a tube housing the microphone to determine the manner in which the jet noise sources change with position along the jet.

Based on the data taken, three possible silencer designs are suggested, and these will be evaluated to determine their effectiveness in reducing the jet noise which reaches the operator.

Held at the University of Surrey, U.K.
Symposium organised and sponsored by
BHRA Fluid Engineering

1. INTRODUCTION

The number of applications in which hand held water jet cleaning equipment is being used has increased significantly in the last few years. The versatility of the water blasting technique accounts for its widespread use and the continuing expansion of the process to satisfy the needs of contractors in cleaning various surfaces. Recent papers (Refs. 1,2,3,4) illustrate the wide range of materials currently being cleaned by the use of an operator-held cleaning gun such as the Fail Safe Gun manufactured by Partek Corporation and illustrated in use in Fig. 1. Notice in this illustration that the distance from the operator's head to the water jet nozzle is approximately 2 m. The path for sound transmission is unimpeded, and the noise generated by the jet will in this case impinge on the operator with little or no attenuation (other than that caused by the "spreading" of the sound waves).

One possible method of reducing the exposure of the operator to the jet noise is to provide him with protective ear plugs or ear shields. However, the attitude of both labor unions and OSHA toward this approach is that personal protective equipment is a last resort (Ref. 5). The priorities listed in this reference are instead:

#1 engineering controls as the most desirable method of controlling noise;

#2 administrative controls, such as rotating workers through a noise area to reduce the period of the exposure, as the second most acceptable method;

#3 lastly, personal protective equipment, which might be advisable in combination with the first two until - ultimately - engineering controls solve the total problem.

The first approach, of using engineering controls in the case of jet noise, would involve the introduction of some sort of sound attenuating device between the operator and the noise source which would alter the sound path and energy transmission. This approach requires background data to define the noise characteristics of typical jets operating in air and impacting against a surface. Tests were therefore planned to gather as much data as required to define the noise source. All data were taken using a General Radio Portable Sound Level Meter, Model 1982.

2. NOISE MEASUREMENTS

2.1 Variation of jet noise with jet pressure

Various jet operating pressures are currently in use in water blasting. However, the bulk of current work is at pressures of 27.6 to 68.9 MNm^{-2} (4,000 to 10,000 psi). In order to determine the variation of the jet noise with operating pressure, a tungsten carbide nozzle with an exit diameter of 1.42 mm (0.056 in.) was mounted as shown in Fig. 2. With the microphone held at a fixed position as shown in Fig. 3, the jet pressure was adjusted to 27.6 MNm^{-2} (4,000 psi) by partially closing the pump bypass valve. The pressure was measured at the pump and would of course be lower at the nozzle. In the cleaning business, however, it is most common to use pump pressure rather than nozzle pressure as a reference.

Sound pressure level measurements were then taken in octave frequency bands in the range of 31.5 Hz to 16 kHz at this jet operating pressure. The jet pressure was then increased in increments of 6.89 MNm^{-2} (1,000 psi) and the data taken again at the new pressure. This procedure was followed until a pressure of 68.9 MNm^{-2} (10,000 psi) was reached, at which time the water flow through the nozzle was 0.448 ℓs^{-1} (7.1 gpm). Because the flow through the nozzle was not constant, the hydraulic power input at the highest pressure was about 4 times the power at the lowest pressure.

The results of these measurements are shown in Fig. 4, and the data are summarized in Table 1.

2.2 Near and far field noise measurements

Again using the 1.42 mm (0.056 in.) exit diameter tungsten carbide nozzle, the water pressure was set at 68.9 MNm^{-2} (10,000 psi) and the flow rate measured at

0.448 ℓs^{-1} (7.1 gpm). The microphone was then stationed at a fixed position along the arc of a semicircle as shown in Fig. 5. The angular position corresponding to the centerline of the jet was called 0 deg with the position along a line opposite to the direction of the jet being 180 deg. For the near field sound measurements the radius of the semicircle was 2.44 m (96 in.). The data taken at this distance are shown in Table 2 and the results are plotted in Fig. 6.

In the case of the far field measurements, the radius of the semicircle was 13.72 m (45 ft). These data are shown in Table 3 and are plotted in Fig. 7.

2.3 Source location

In order to determine the manner in which the jet noise changes with position along the axis of the jet, a large screen was placed between the jet and the microphone. In addition, the microphone was placed inside a pipe of nominal inside diameter 5 cm (2 in.) with the pipe attached to the screen as shown in Fig. 8; the pipe was partially packed with glass fiber to minimize acoustic resonances, and was closed off at the end remote from the jet. The principle of operation was that the sound waves entering the tube would, in general, originate in a region fairly close to the open end of the tube. Thus, at any given frequency, the position - along the jet - corresponding to a maximum microphone response should coincide with the region of greatest source strength. The location of the microphone within the pipe placed it at a distance of 54.6 cm (21.5 in.) from the centerline of the jet. Fig. 9 shows the opposite side of the screen with the opening leading into the pipe at the center of the photograph. The distance "X" from the nozzle to the center of the opening leading into the microphone chamber was varied from 0 to 152 cm (60 in.) while the round jet was operating at the pressure and flow rate of the previous test. Data were again taken at the same frequencies for each X value given in Table 4. These results are plotted in Fig. 10.

The procedure was repeated for the case of a 15 deg fan jet nozzle operating at a pressure of 68.9 MNm^{-2} (10,000 psi) and flow rate of 0.416 ℓs^{-1} (6.6 gpm). The results are given in Table 5 and Fig. 11.

2.4 Jet impingement tests

To complete the basic measurements which would characterize the noise source, the round and fan jets were directed to strike a steel plate normal to its surface as shown in Fig. 12. The microphone was positioned at the location where the operator's head would normally be located. Data were taken for the two nozzles operating at 68.9 MNm^{-2} (10,000 psi) for various standoff distances from 2.54 cm (1 in.) up to 38.1 cm (15 in.). The data are listed in Table 6 and are plotted in Figs. 13 and 14.

3. DISCUSSION OF NOISE MEASUREMENTS

3.1 Discussion of noise variation with pressure

Table 1 gives the complete set of data on the radiated noise spectrum (at a fixed location, relative to the jet nozzle) for a range of jet pressures. Fig. 4 shows (for clarity) only some of these spectra, although the full pressure range is covered. It can be observed that the spectra are almost parallel from about 500 Hz upward, and increasing pressure brings about increasing noise level. Even at the lower frequencies, this trend is still apparent, with the exception of a slight discrepancy - which may be taken to be unimportant - in the 27.6 MNm^{-2} (4,000 psi) curve at 63 Hz. A possible slight tendency for the peak in the spectrum - which is around 4-8 kHz - to shift to higher frequencies at the high pressure, may be detectable. The spectrum "turns up" in the 31.5 Hz octave band, and narrow-band analyses (not shown here) reveal that this is associated with noise generated by water pulsations from the pump, which give a spectral peak at about 26 Hz.

There had initially been some speculation concerning whether shock waves, caused by the supersonic* unsteady water jet flow, might be an important contributor to the

*That is, traveling faster than the adiabatic speed of sound in air.

radiated noise from the jet at the highest working pressure, where the nozzle exit velocity is in the region of 366 m/s (1,200 ft/s) and is therefore just supersonic.

No evidence of shock associated noise is present in Fig. 4 since the spectrum shape for the highest jet pressure is almost identical to the shapes of the spectra for the lower jet pressures. If shock noise had been important, then some change in spectrum shape or sharp change in level would be expected, but neither of these features is apparent. Presumably the water jet slows down fairly rapidly, on leaving the nozzle, to subsonic speeds, so that any shock noise would be concentrated very close to the nozzle; in any case, its effects - if at all present - are clearly negligble.

3.2 Discussion of near and far field data

Table 2 and Fig. 6 (which is a three-dimensional plot of noise spectra versus the angle to the jet axis of the microphone location) show the "near field" noise data. These would be reasonably representative of the sound field surrounding the free jet (operating without impingement) in the general vicinity of the jet, though farther away than the operator's head. Nearby personnel, at a distance of about 2.5 m from the jet nozzle, would experience approximately these noise levels.

The directional characteristics of the noise radiation in the near field are not directly indicative of the physical processes occurring in the jet, unless a series of contours of constant sound pressure level is plotted around the jet.

For example, at low frequencies, the sound level is considerably higher (20-30 dB) at 0 deg to the jet axis than it is at 180 deg to the axis. This could be either because of inherent directional characteristics of the radiated noise (and if so, would also be observed in the far field) or be simply because the microphone at an angle of 0 deg - was situated nearer to low frequency sources located down the jet at some distance from the nozzle (c.p. air jets) than it would be at an angle of 180 deg. Alternatively, a combination of both effects might explain the observations.

Pronounced "directional" characteristics are also evident in the mid and high frequency data. Again, they cannot be directly interpreted.

Table 3 and Fig. 7 show the "far field" data, measured at a distance from the jet nozzle of 13.72 m (45 ft), with the microphone at angles to the jet axis ranging from 0 deg to 180 deg. This distance should ensure that the sound level measurements were effectively in the acoustic far field, so that directional characteristics could reasonably be interpreted as "inherent" in the noise sources rather than being a feature of greater or less proximity of the microphone to the sources.

The low frequency results - up to about 125 Hz - are interesting in that pronounced directivity is apparent in the noise radiation, with the peak level occurring near to the axis. The feature which is unexpected is that a minimum in the radiation pattern occurs at 90 deg to the jet axis. If this directivity characteristic is associated with the nature of the the noise sources rather than reflections from the ground and nearby buildings, then it may indicate the presence of significant acoustic dipole components in the jet. A dipole with its axis aligned with the jet axis would typically display two maxima along the axis (in either direction) and a minimum level at 90 deg to the axis. The fluctuating (axial) force on the nozzle, caused by the pulsations in the water flow from the pump, could constitute a strong dipole component at low frequencies. Additionally, the fluctuating forces caused by the passage of water droplets through turbulent eddies of air could contribute, since these would also tend to be axial. The above speculation gives no quantitative explanation about the noise producing mechanisms, but merely suggests possible causes for the observed characteristics.

The directional patterns from 250 Hz to 16 kHz show generally less pronounced minima, although a "flattening off" of the curve is often noticeable, beginning at an angle of 75 deg to 90 deg to the axis and continuing up to 180 deg. In all of the curves, the peak level occurs close to the jet axis. At 16 kHz, an almost monotonic fall-off of level (discounting a small peak near the axis) is apparent. Convection effects, caused by the axial motion of the noise sources away from the jet nozzle, are probably responsible for the appearance of the noise peaks (at all frequencies) close to the jet axis.

One may say, in general, that the possibility of both quadrupole noise (from the Reynold's stresses in the turbulent eddies of air) and dipole noise (from the aforementioned fluctuating force) would appear to exist in water jets, although any definite quantification of these ideas must await more detailed measurements.

3.3 Discussion of source location measurements

Table 4 and Fig. 10 give the results of the "source location" measurements on the round jet. Fig. 10 is a three-dimensional plot of noise spectra versus distance along the jet. It can be seen that the low frequency radiation appears to originate predominantly far down the jet, whilst the high frequency radiation displays peaks much nearer to the nozzle. A broad "ridge" may be discerned in Fig. 10, running (very roughly) diagonally across the diagram such that the peak in the sound level versus distance curve at any frequency shifts nearer to the jet nozzle as the frequency increases.

It may well be that the low frequency results at fairly large distances from the jet are influenced by hydrodynamic (rather than acoustic) disturbances in the near field of the jet, and that these do not propagate as sound. Additionally, of course, any inherent directional characteristics of the radiation could influence the results at all frequencies, since the angle between the line drawn from the source to the open end of the microphone tube, and the jet axis, will vary according to the relative axial distance between the microphone tube and the source.

It is felt, however, that the general trend exhibited by the measurements is at least partially representative of reality, and that one can be fairly certain that the high frequency sources are generally closer to the jet nozzle than the low frequency sources.

Note also that even at high frequencies, relatively little noise appears to be generated very close to the jet nozzle; this has important ramifications when the noise from an "impinging" jet is considered.

Incidentally, the ridge in Fig. 10, running parallel to the distance axis, at 63 Hz, can be attributed to an axial, damped, resonance in the microphone tube.

Table 5 and Fig. 11 show the data measured on the fan jet. Very similar conclusions may be drawn, to those from the round jet, although the ridge in Fig. 11 is more pronounced than that in Fig. 10 and the levels generally appear to be somewhat higher.

3.4 Discussion of impingement data

Table 6 and Figs. 13 and 14 show the results of the noise measurements, made at the approximate location of a jet operator's head, with both round and fan jets impinging on a steel plate from distances varying between 2.5 cm and 38.1 cm (1 in. to 15 in.).

The levels, as one can see, are high. Typical dBA levels (which may readily be found from the octave band data) would be over 100 dBA and in some cases over 110 dBA.

There is a general tendency for the noise level to increase as the distance between the steel plate and the jet nozzle increases. If one speculates that the noise caused by impingement is relatively small, then the major effect of the plate will be to "cut off" the region of noise sources at the point at which the plate is situated. The structure of the jet between the nozzle and plate is probably affected relatively little by the presence of the plate. This speculation is, at least, consistent with the trend in the measurements.

Generally speaking, the peaks in the fan jet spectra are less sharp than those in the round jet spectra. The peak frequency in corresponding spectra is slightly lower (in the case of the fan jet) and the peak level lower, but there is generally more energy at low frequencies and less at high frequencies.

The fact that noise sources very close to the jet appear to be relatively weak is no doubt an important factor in determining the noise levels radiated by an impinging jet, when the separation of plate and nozzle is relatively small. The presence of the plate, acting as an acoustic reflector, no doubt exacerbates the situation.

4. PROPOSED DESIGNS FOR SILENCERS

4.1 General considerations

A silencer to reduce the noise reaching the operator must be designed in such a manner that it does not appreciably interfere with the use of the water jet in cleaning applications. This places rather severe restrictions on possible silencer designs. The overall size and weight will be important factors as well as the effect the silencer may have on reducing the operator's visibility of the surface he is attempting to clean.

Another factor is that devices to alter the path of sound transmission from the jet to the operator may involve surrounding the jet with some type of open ended enclosure for a given distance outward from the nozzle. This enclosure may interfere with the normal interaction between the jet and the ambient air, and result in premature jet breakup.

Rather than place too many restrictions on possible silencer designs in the prototype stage of development, it was decided first to investigate any possible enclosure which might result in the reduction of jet noise. Therefore three possible silencer configurations were proposed and are being tested to measure their effectiveness.

4.2 Silencer One

The simplest possible enclosure configuration is a tube surrounding the jet for some distance forward from the nozzle. This design is illustrated in Fig. 15. The tube will be fabricated from Plexiglass stock and attached to the nozzle supply pipe by a collar which can be locked to the supply pipe using set screws. In the event that the jet is disrupted by the tube, two air passages are drilled through the collar to allow a natural aspiration to occur. If the air passages are required, the sound path back through the air holes will be modified by using sintered bronze air mufflers commonly used on air driven tools.

During testing, the tube will be moved to a position enclosing a fixed length of the jet with the air passages sealed. Measurements will be taken to determine the potential sound reduction from this silencer. It is also planned to place a water-retardant liner of sound absorbing, synthetic fibrous material ("Kevlar", manufactured by DuPont) inside the tube during the tests to measure its effectiveness in reducing the noise.

4.3 Silencer Two

The design shown in Fig. 16 is a modification of the first silencer. A flange of Plexiglass sheet will be fitted around the tube as shown in the figure. Flanges of various diameters will be tested with and without a covering of Kevlar. The length of the jet contained within the tube can also be adjusted by this arrangement.

4.4 Silencer Three

The third design is shown in Fig. 17. In this case a cavity has been formed between the inner tube and a larger outside tube. The cavity will be lined with Kevlar as well as the face of the flange at the front of the outside tube. Again some adjustment can be made of the position of the silencer relative to the nozzle location.

5. CONCLUSIONS

Noise measurements were taken from a round and a 15 deg fan jet to define the basic characteristics of the noise generated by typical jets used in water blasting applications. These data serve as a basic guide in the design of possible noise silencers, and it will be used to measure the effectiveness of the three silencer designs proposed.

The nature of the noise sources in water jets is very much an open question at present. Some general ideas have been suggested here, but confirmation of these can only be forthcoming from a much more fundamental series of tests. At all events, the noise is clearly strongly directional, and is most intense in the direction of the jet flow.

It is almost certain that the particular geometry of the nozzle, which will determine the jet flow structure, will also be an influencing factor in the radiated noise. The difference between the round jet and fan jet data illustrate this point.

Despite this, the data presented here should give at least a rough idea of noise levels to be expected in typical cleaning applications.

6. ACKNOWLEDGEMENTS

The authors wish to acknowledge the support of Mr. Mike Ginn, President of Partek Corporation and Mr. Amos Pacht, Manager of Engineering, for suggesting this project and for the financial support which they provided. In addition, Mr. Robert Grower and Mr. Norman Wheeler assisted in conducting the noise measurements. Thanks should be given to Mr. Way King for his assistance in preparing the computer data plots, and Mrs. Mujde Erten for drawing the figures.

7. REFERENCES

1. Torpey, P.: "Some Experiences in the Manufacture and Application of High Pressure Water Cleaning Equipment." In: Proc. 1st International Symposium on Jet Cutting Technology (Coventry, U.K.: April 5-7, 1972) Cranfield, U.K., BHRA Fluid Engineering, 1972, Paper D1.

2. Pardey, P.H.: "Notes on Marine Applications of Jet Cutting." In: 1st International Symposium on Jet Cutting Technology (Coventry, U.K.: April 5-7, 1972) Cranfield, U.K., BHRA Fluid Engineering, 1972, Paper D2.

3. Gronauer, R.W.: "Cleaning and Descaling of Equipment with High Pressure Water." In: 1st International Symposium on Jet Cutting Technology (Coventry, U.K.: April 5-7, 1972) Cranfield, U.K., BHRA Fluid Engineering, 1972, Paper D3.

4. Crowe, A.R.: "Cleaning and Descaling Equipment with High Pressure Water - A Contracting Service to Industry." In: Proc. 2nd International Symposium on Jet Cutting Technology (Cambridge, U.K.: April 2-4, 1974) Cranfield, U.K., BHRA Fluid Engineering, 1974, Paper F3.

5. Liebich, R.E. and P.B. Ostergaard.: "Industrial Noise Pollution - The Nature and Extent of the Problem." Mechanical Engineering, 103, 7, July 1981, pp 34-36.

TABLE 1

VARIATION OF NOISE LEVEL WITH WATER PRESSURE

AT A FIXED LOCATION RELATIVE TO JET

(Sound level in dB)

Pressure-MNm^{-2}

Hz	27.6	34.5	41.3	48.2	55.1	62.0	68.9
31.5	75.0	75.0	78.0	82.0	83.0	84.0	85.5
63.0	76.0	75.0	76.0	79.0	80.0	82.0	82.0
125.0	76.0	77.5	78.5	80.0	82.5	83.0	83.5
250.0	78.0	80.5	82.0	84.5	86.5	87.5	88.5
500.0	81.0	84.0	85.5	88.0	90.0	91.0	92.5
1K	83.0	85.5	87.5	90.5	93.0	94.5	96.0
2K	86.0	88.5	89.5	93.0	95.0	97.0	99.0
4K	88.0	90.5	91.8	95.0	97.0	99.0	101.0
8K	87.0	89.5	91.5	94.5	97.0	98.5	100.5
16K	80.5	82.5	85.0	88.5	90.5	92.0	94.0

TABLE 2

NEAR FIELD SOUND MEASUREMENTS - DISTANCE 2.44m

(Sound level in dB)

Angular Position - Degrees

Hz	7.5	15	30	45	60	75	105	120	135	150	165	180
31.5	103.5	94.0	83.5	79.5	77.0	76.0	77.5	77.5	77.5	78.0	78.5	78.5
63.0	104.0	90.5	78.5	76.5	74.5	74.5	76.0	74.0	72.5	73.0	76.5	77.4
125.0	98.5	88.0	81.5	77.5	74.5	75.0	75.5	76.5	77.5	76.5	75.0	74.5
250.0	98.0	95.0	90.5	85.5	80.5	80.0	78.0	79.0	78.0	77.5	77.5	78.0
500.0	102.5	100.5	95.0	89.5	86.0	85.0	84.0	84.0	84.0	84.0	83.0	83.5
1K	107.5	106.0	100.0	93.5	91.5	90.5	89.5	89.5	89.0	88.5	88.0	86.5
2K	111.0	110.5	103.0	97.5	95.5	94.5	93.0	92.5	92.0	92.5	92.0	90.5
4K	114.0	113.0	105.0	100.5	98.5	96.5	95.5	95.5	95.5	95.5	95.0	93.5
8K	113.0	110.0	105.0	100.5	97.5	96.0	94.5	94.0	94.0	94.0	93.5	92.0
16K	102.5	100.0	99.0	93.0	89.0	87.0	85.0	85.0	84.5	84.5	84.0	82.5

TABLE 3. FAR FIELD SOUND MEASUREMENTS - DISTANCE 13.7 m

(Sound level in dB)

Hz	Angle - Degrees 0	7.5	15	22.5	30	45	60	75	90	105	120	135	150	165	180	Back-ground
31.5	87.0	84.0	77.5	78.0	75.5	73.5	69.5	67.5	65.0	67.0	69.0	70.0	70.0	70.0	71.0	63.0
63.0	75.0	75.0	69.0	70.0	68.0	68.5	67.0	66.0	64.5	67.0	72.0	69.0	68.0	67.0	67.0	65.0
125.0	80.0	77.0	77.5	77.0	76.5	77.0	74.5	71.5	68.0	69.0	71.0	72.0	71.0	71.0	70.5	65.0
250.0	82.0	82.5	83.0	82.5	81.5	80.0	76.0	72.5	71.0	71.5	71.0	71.0	70.5	71.0	72.0	57.0
500.0	87.0	86.5	87.0	88.0	86.5	82.5	79.5	76.5	76.5	76.5	76.5	77.0	76.5	76.5	76.0	53.0
1K	94.5	94.5	94.5	94.5	91.5	86.5	83.5	82.0	82.0	81.5	82.0	81.0	81.0	81.0	79.5	50.0
2K	99.0	101.5	102.0	101.0	97.5	92.5	89.0	87.5	85.5	85.0	84.5	85.5	86.5	86.5	84.0	45.0
4K	103.0	105.0	103.5	104.0	100.0	95.0	93.0	91.5	90.0	89.5	89.5	90.0	89.0	88.5	86.5	41.0
8K	100.0	102.0	101.0	101.0	96.5	93.0	92.0	90.5	88.5	87.5	87.5	86.5	86.5	86.5	84.0	37.0
16K	89.0	90.0	91.0	90.0	86.5	84.5	82.0	80.0	79.0	78.0	77.5	77.0	75.5	75.5	73.0	27.0

TABLE 4. VARIATION OF SOUND LEVEL WITH X DISTANCE - ROUND JET

(Sound level in dB)

Hz	"X" Distance - Inches 0	1	2	4	6	10	14	18	22	26	30	36	42	48	54
31.5	92.0	92.5	93.5	94.0	96.0	99.0	100.0	102.0	106.0	108.0	111.0	114.0	114.0	120.0	117.0
63.0	93.5	94.0	95.5	95.0	97.0	100.0	104.0	107.0	114.0	118.0	121.0	122.0	122.0	129.0	124.0
125.0	87.5	89.0	90.0	91.0	92.0	96.0	99.0	102.0	109.0	114.0	117.0	118.0	116.0	124.0	118.0
250.0	95.0	98.5	99.5	101.0	100.0	106.0	108.0	110.0	118.0	121.0	124.0	124.0	123.0	127.0	122.0
500.0	96.5	101.5	101.5	104.0	101.0	109.0	111.0	113.0	118.5	121.0	123.0	122.0	121.0	124.0	119.0
1K	101.5	107.5	108.0	113.0	106.0	115.0	117.5	117.5	123.0	125.0	125.5	123.5	122.5	123.0	118.0
2K	103.5	107.5	106.5	111.0	106.5	116.5	119.5	120.0	125.0	126.0	126.5	124.0	122.5	121.0	119.0
4K	99.5	98.5	98.5	103.0	104.5	113.5	117.5	118.0	122.0	123.0	122.5	119.5	117.5	116.5	113.5
8K	97.0	94.0	96.0	101.0	103.5	113.0	116.0	115.5	118.0	117.5	117.0	112.5	111.5	109.0	107.0
16K	93.5	91.0	92.5	98.5	101.0	108.0	110.0	110.0	111.5	112.0	111.0	106.5	105.5	103.5	100.0

TABLE 5

VARIATION OF SOUND LEVEL WITH X DISTANCE - FAN JET

(Sound level in dB)

"X" Distance - Inches

Hz	0	1	2	4	6	10	18	26	30	36
31.5	106.0	107.0	109.0	109.5	110.5	112.0	116.0	129.0	125.0	130.0
63.0	104.0	106.0	108.0	109.0	110.0	113.0	124.0	136.0	134.0	137.0
125.0	100.0	102.0	103.5	105.0	106.0	110.0	122.0	133.0	129.0	134.0
250.0	108.0	110.0	112.0	113.5	114.5	120.0	131.0	137.0	130.0	137.0
500.0	109.0	109.0	113.0	114.0	116.5	121.5	125.0	128.0	119.0	127.0
1K	111.5	112.0	115.5	117.5	120.0	123.0	120.0	119.0	115.5	117.0
2K	111.5	113.0	116.5	119.5	122.0	124.5	121.5	115.0	115.0	111.0
4K	109.0	111.0	115.0	118.5	121.0	123.5	119.0	108.5	109.0	105.5
8K	109.0	112.0	115.5	119.0	121.5	122.5	114.5	103.0	105.0	104.0
16K	106.0	110.0	113.5	118.0	120.5	120.5	109.5	101.0	101.5	98.0

TABLE 6

JET IMPINGEMENT DATA

(Sound level in dB)

	Round Jet "X" - Inches					Fan Jet "X" - Inches				
Hz	1	4	7	10	15	1	4	7	10	15
31.5	82.0	76.0	74.0	74.5	73.0	75.0	75.0	75.0	74.0	77.0
63.0	71.5	76.0	82.0	77.0	73.0	73.0	74.0	78.5	77.5	76.0
125.0	73.5	78.5	81.5	75.0	75.0	76.0	74.0	80.0	77.0	78.5
250.0	80.0	82.5	81.5	80.0	79.0	77.5	76.5	80.5	82.5	90.5
500.0	85.0	86.5	86.0	86.5	87.0	85.5	83.5	88.5	94.0	100.5
1K	91.0	89.5	91.0	94.0	96.0	86.0	94.0	100.0	104.0	103.5
2K	99.0	97.0	99.5	104.0	106.5	92.0	103.0	108.0	108.5	106.0
4K	106.0	107.0	109.0	112.5	114.0	96.5	109.5	111.0	109.0	106.0
8K	108.0	110.0	111.5	112.5	112.5	100.5	109.5	108.5	107.0	103.5
16K	104.0	104.0	102.0	101.0	103.0	98.0	103.5	102.0	100.5	97.0
dBA	110.7	111.9	113.6	116.0	117.2	102.6	113.3	114.9	114.1	111.7

Figure 1. A typical cleaning operation.

Figure 2. General arrangement of equipment.

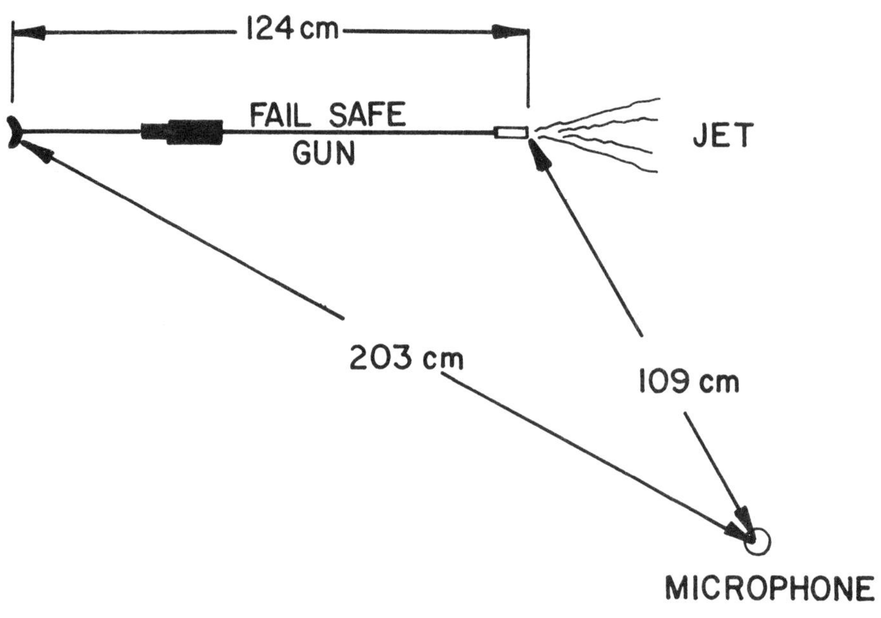

Figure 3. Location of the microphone for data in Table 1.

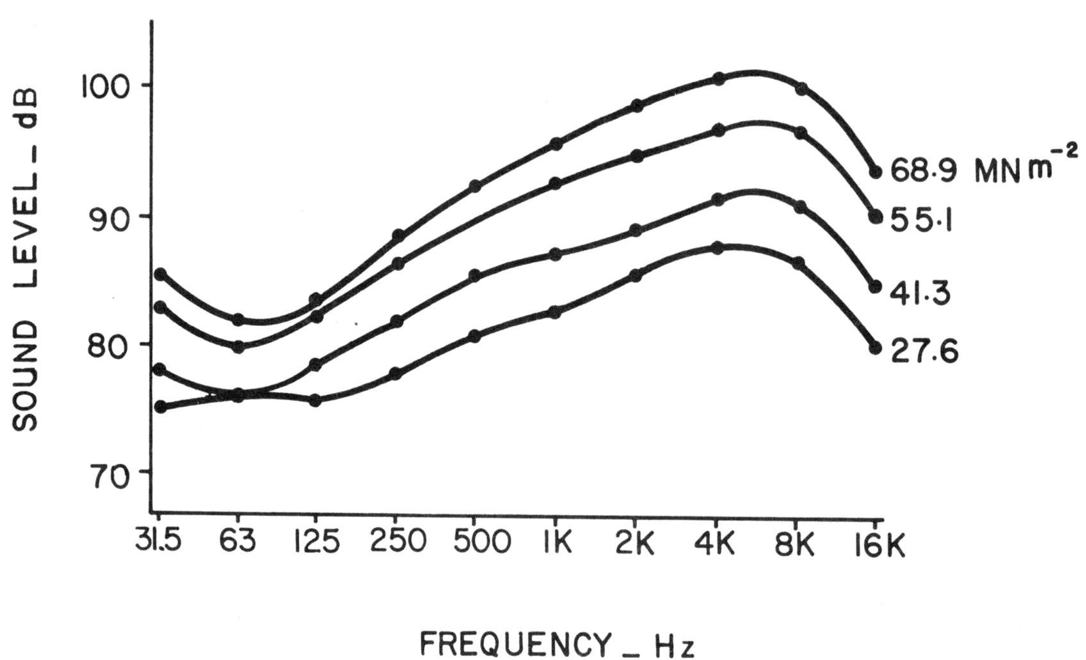

Figure 4. Variation of noise with jet pressure.

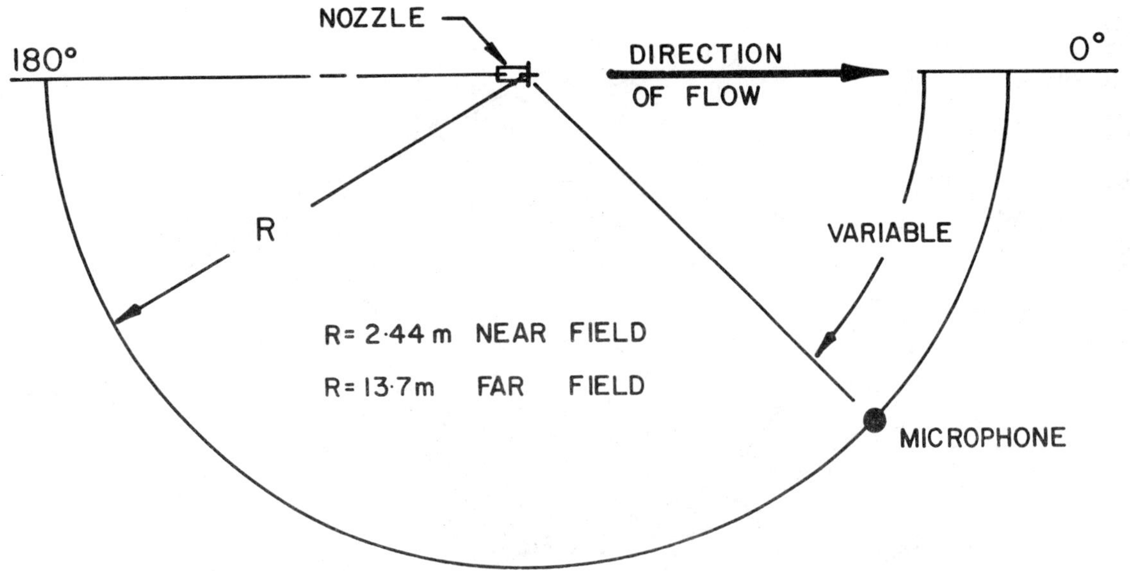

Figure 5. Microphone location for near and far field data.

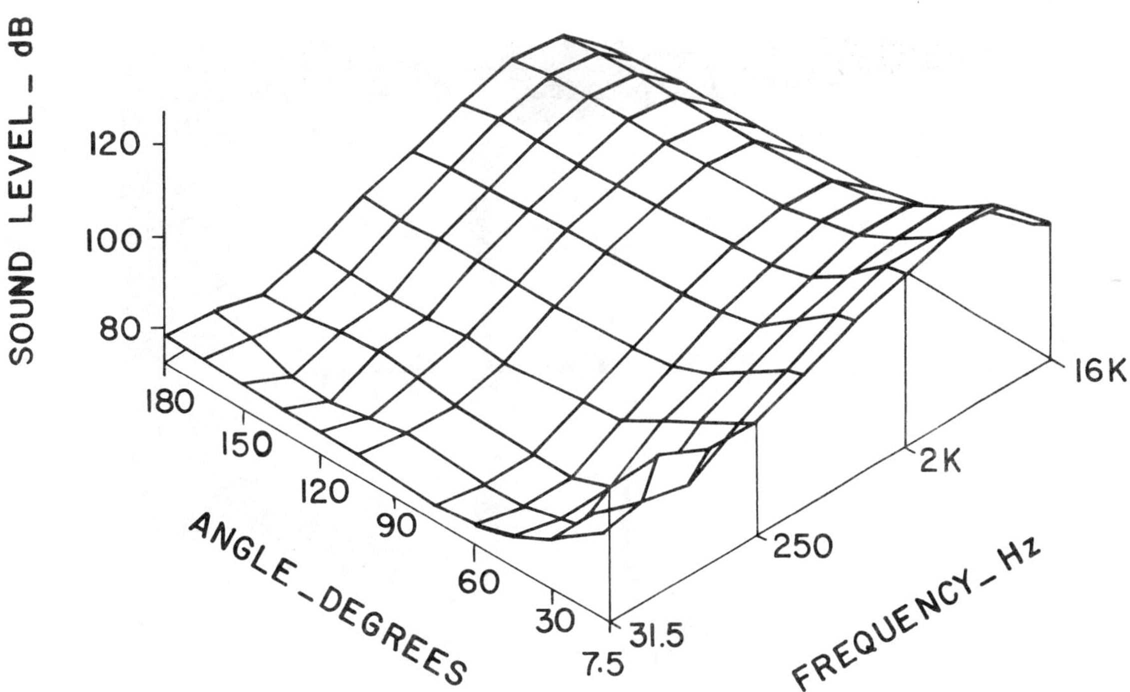

Figure 6. Plot of near field data in Table 2.

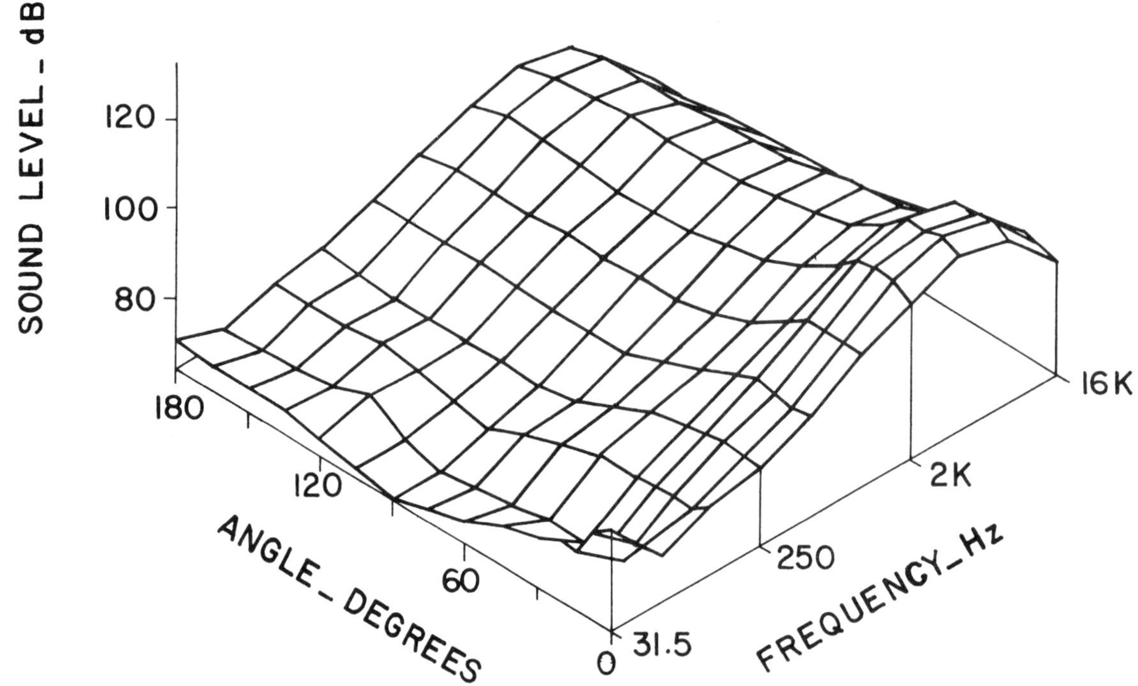

Figure 7. Plot of far field data in Table 3.

Figure 8. Pipe used to house microphone.

Figure 9. Opposite side of screen with opening into pipe.

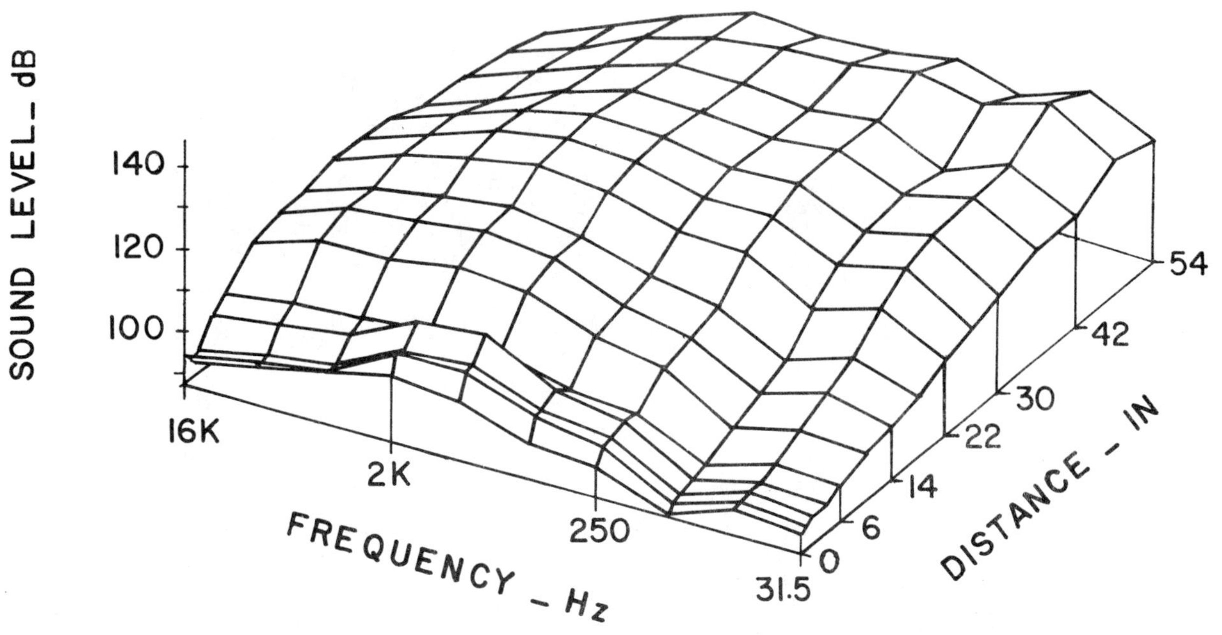

Figure 10. Plot of round jet data in Table 4.

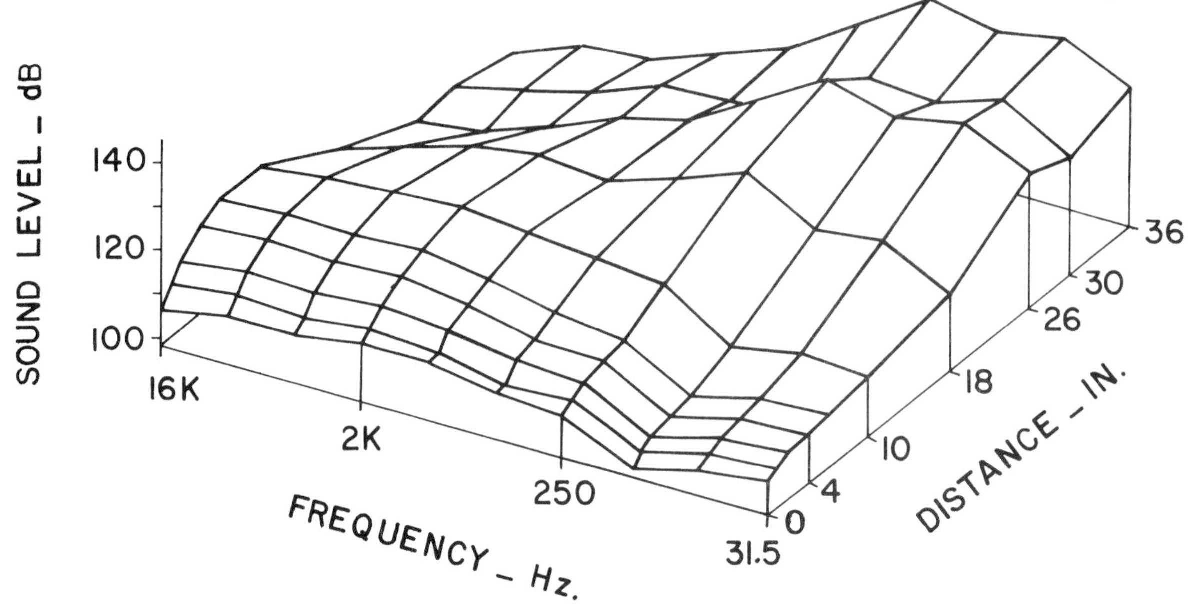

Figure 11. Plot of fan jet data in Table 5.

Figure 12. Arrangement for jet impingement tests.

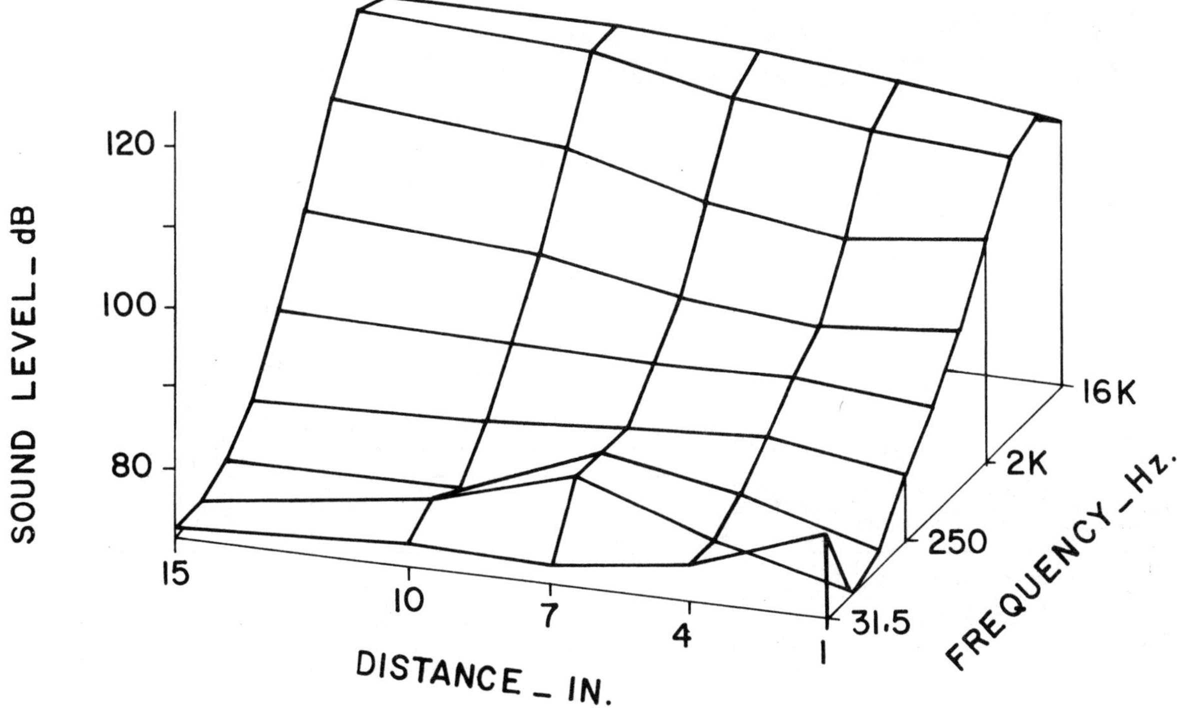

Figure 13. Plot of round jet impingement data in Table 6.

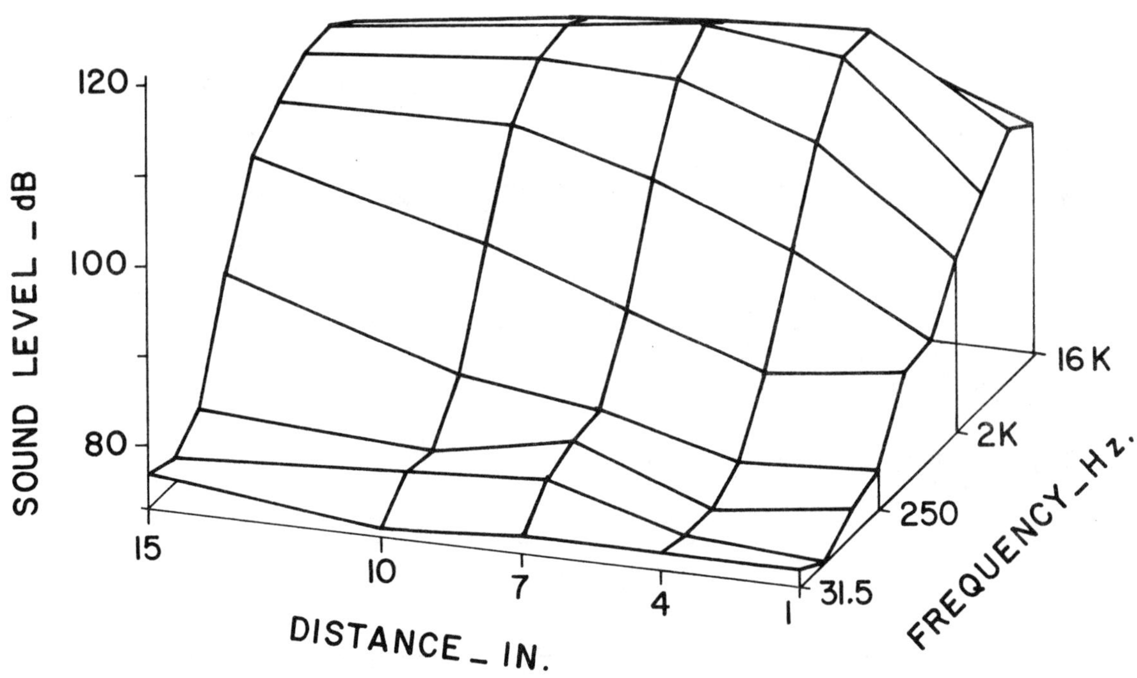

Figure 14. Plot of fan jet impingement data in Table 6.

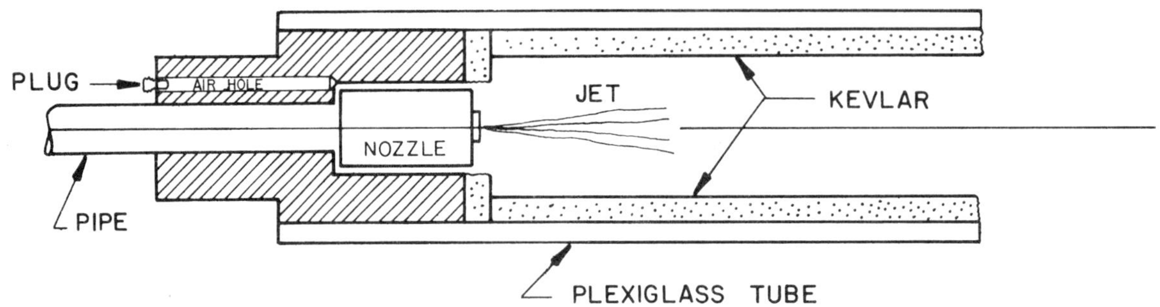

Figure 15. Silencer one design.

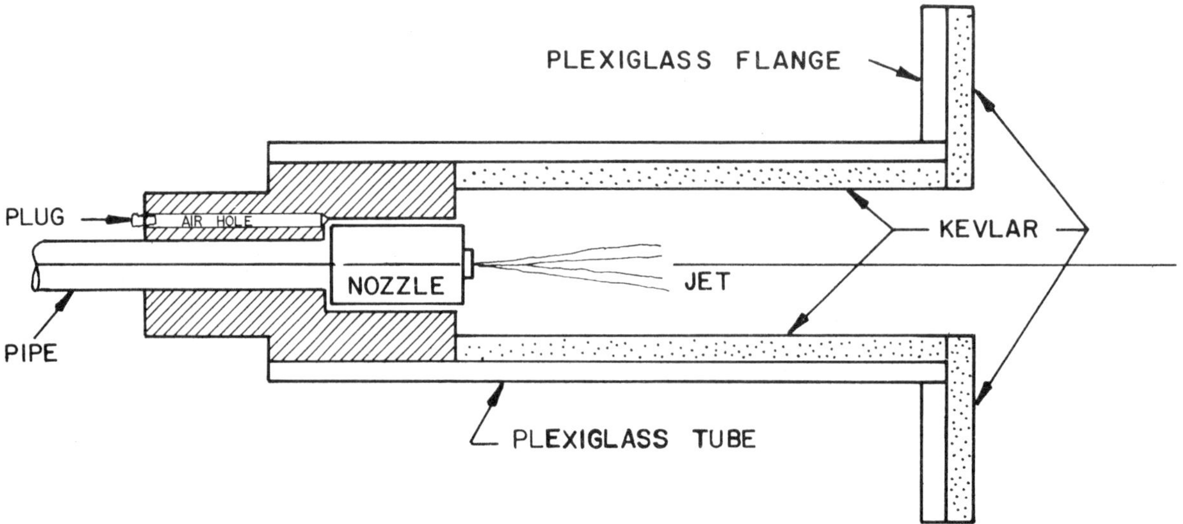

Figure 16. Silencer two design.

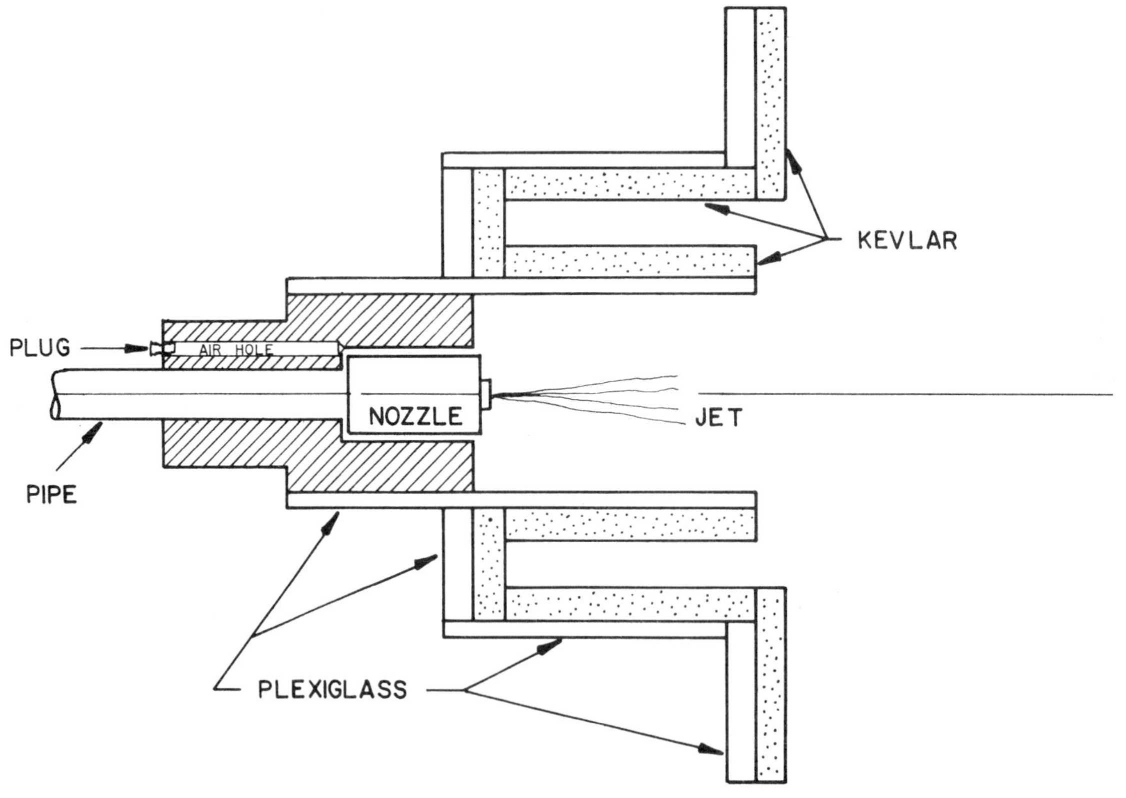

Figure 17. Silencer three design.

6th International Symposium on
Jet Cutting Technology
6-8, April, 1982

DRILLING OF ROCKS WITH ROTATING HIGH PRESSURE WATER JETS: INFLUENCE OF ROCK PROPERTIES

M.M. Vijay, W.H. Brierley and P.E. Grattan-Bellew

National Research Council, Canada

Summary

Drilling of rocks with high pressure rotating water jets is discussed. Presented data pertain to drilling of unconfined blocks of granite, limestone and marble with a series of dual orifice nozzles consisting of a central orifice (fixed at a constant angle of 10° to the axis of rotation) and a second orifice located at a radius of 5.8 mm (the angle of inclination varied over a range of 10° to 30°). The area of the outer orifice was varied systematically to allow 50 to 70 percent of the total available flow.

Tests were conducted at pressures up to 69 MPa and at three different rotational speeds (120, 240 and 420 RPM). Maximum power loading based on the nozzle body area was 20 kw/cm^2. It is shown that (i) optimum nozzle configuration depends on the type of rock material used, and the rock properties which are responsible for this dependence are not the compressive or tensile strengths, but the micro properties such as grain size, porosity and pre-existing micro fractures, (ii) the specific energy decreases considerably with increased pressure, over the pressure range tested and (iii) quite low rotational speeds ($\approx$ 200 RPM) were close to optimum.

Held at the University of Surrey, U.K.
Symposium organised and sponsored by
BHRA Fluid Engineering

NOMENCLATURE

a	Radial distance between orifices in the nozzle body, mm.
A_H	Mean cross sectional area of the drilled hole, cm^2
A_I	Cross sectional area of the inner orifice in the nozzle body, cm^2 (Fig. 1)
A_N	Cross sectional area of nozzle body, cm^2
A_O	Cross sectional area of outer orifice in the nozzle body, cm^2 (Fig 1)
A_R	$= A_H/A_N$
A_T	Total cross sectional area of orifices in the nozzle body, cm^2
C	Compression wave velocity in the rock, m/s
d	Penetration per revolution
D	Diameter of orifice, mm
D_H	Diameter of hole, cm
D_I	Diameter of inner orifice, mm (Fig. 1)
D_O	Diameter of outer orifice, mm (Fig. 1)
E	Specific energy, J/cm^3
E_R	Modulus of deformation of rock, MN/m^2
h	Depth of a slot, cm
H_P	Total hydraulic power of jets, kW
k,k'	constants
ℓ	Average grain size of a rock, mm
L	Length of cylindrical outlet of nozzle, mm
N	Rotational speed of drill, RPM
P	Pressure at nozzle inlet, MPa
R	Mean drilling rate (feed rate), cm/min
S	Standoff distance, cm
u	Traverse speed of a jet relative to the rock, m/s
W	Width of a slot, cm
θ_I	Angle of inclination of inner orifice, deg. (Fig. 1)
θ_O	Angle of inclination of outer orifice, deg. (Fig. 1)
α	Nozzle entry angle, deg.
β	Constant, Eq. 2
ρ_R	Dry bulk density of rock, kg/m^3
σ_c	Uniaxial compressive strength of rock MN/m^2
σ_t	Uniaxial tensile strength of rock, MN/m^2
κ	Permeability of a rock, darcy
μ	Viscosity of fluid, Ns/m^2

INTRODUCTION

The work presented in this paper is a continuation of the study of drilling of rocks with high pressure rotating water jets (Ref. 1). In Ref. 1, data pertaining to drilling of granite (Muskoka Pink Granite) with nozzle bodies of varying configurations (see Fig. 1 and table 1) were reported. The conclusions from this earlier study were:

1. A quite low rotational speed (≈200 RPM) was close to optimum.

2. An increase in the value of the parameter A_O/A_T (ratio of the cross-sectional area of the outer orifice to the total cross-sectional area of orifices in the nozzle body) from 0.50 to 0.60 had a beneficial effect on both the specific energy (E) and A_R (ratio of the mean cross-sectional area of the drilled hole to that of the nozzle body). Beyond 0.60, the effect was either insignificant or adverse.

3. Angle of inclination (θ_O) of the outer orifice had a profound influence on drilling performance. An increase in the value of θ_O from 10 to 20° reduced the measured specific energy by a factor of 12 and increased the value of A_R, on the average, by 25%.

4. The best nozzle body configuration tested, from the specific energy and hole area ratio (A_R) standpoint, for Muskoka Pink Granite was C2 (see table 1).

The paper recommended further work using:

1. Nozzle bodies for which angles of the outer orifice, θ_O, are greater than 20° (because the trend of the data indicated that an optimum value of θ_O was possible),

2. nozzle bodies incorporating nozzles with straight entry (i.e. $\alpha = 180°$) and

3. other materials to confirm the configurational trends observed on Muskoka Pink Granite.

Experimental data reported in the present paper were obtained in response to the recommendations made above. A set of new nozzle bodies (D2, D3, E2, E3, SC2-1 and SC2-2, table 1) was designed to investigate items (1) and (2) of the recommendations. These nozzle bodies were assessed, as before (Ref. 1) on the basis of holes drilled horizontally in unconfined blocks of Muskoka Pink Granite. With regard to item (3) of the recommendations, some tests were conducted on unconfined blocks of Amabel Marble (Dolomite) and Ottawa Limestone. These rocks had quite different mechanical properties and petrographic structure compared to granite (see table 2). As expected, each type of rock was observed to have a significant effect on the nozzle body performance. However, what was unexpected was that the type of chippings (debris) produced by the water jets, not the mechanical properties (such as compressive strength, hardness, etc.) appeared to control the drilling performance. A comparison of the microscopic structure of the chippings with that of the rock itself has shown that porosity (more importantly, permeability) and grain size are the two properties that control the penetrability of rocks. These findings appear to confirm the recent theoretical observations of Rehbinder (Refs. 2 & 3), formulated particularly for the linear slotting of rocks with single, non-rotating high pressure water jets (discussed more fully in the Appendix).

Other (very limited) experimental results included in this paper are:

1. effect of nozzle pressure on specific energy (marble and limestone),
2. effect of the direction of drilling with respect to the bedding planes of the rock (limestone).

EQUIPMENT AND PROCEDURE

Except for the six new nozzle bodies, the equipment employed was the same as described in Ref. 1. Nozzle bodies with which the data reported herein were acquired are shown schematically in Figs. 1A and 1B and their design specifications are listed in table 1. Total cross-sectional area (A_T) of the orifices = $0.023\ cm^2$.

Straight entry (that is, with $\alpha = 180°$) nozzle bodies SC2-1 and SC2-2 were designed with the same dimensions as C2 except for the L/D ratios. The reason for this was that nozzle body configuration C2 was the best for drilling Muskoka Pink Granite (Ref. 1). As a straight entry nozzle with L/D = 2 (nozzle body SB1 in Ref. 1) was found to produce a poor jet (the jet dispersed and broke down into droplets immediately after emerging from the nozzle), nozzle bodies SC2-1 and SC2-2 were designed with L/D = 5 and 10 respectively, with the expectation that this would improve the quality of the jet.

Experimental procedure and the method of analysis of data were the same as described in Ref. 1.

TEST MATERIAL

The rocks used in the present series of experiments were unconfined blocks of Muskoka Pink Granite (approximately 15 nozzle body diameters in length), Amabel Marble (Dolomite, 30.5 x 24 x 5.1 cm) and Ottawa Limestone (15 to 20 nozzle body diameters in length). Some of the properties of the rocks measured in the laboratory are listed in table 2. Petrographic description, which is important in understanding the mechanism of interaction between the rock and the high pressure water jet, of each rock is as follows:

Muskoka Pink Granite: A detailed description of this rock is already given in Ref. 1. Fig. 11A shows the typical grain size, shape and composition of this rock. The rock is dominantly composed of discrete crystals of quartz and feldspar mostly with irregular boundaries. Since the porosity is low ($\approx$1.2%), the crystal boundaries form the main planes of weakness within the rock. Also a very weak foliation is imparted to the rock by the alignment of mafic minerals and occasional elongate quartz grains. However, it seems unlikely that the foliation is sufficiently pronounced to give rise to discontinuities within the rock which would affect water jet drilling.

Amabel Marble (Dolomite): This is a buff coloured dolostone which is composed of 95% dolomite, with quartz, calcite and iron oxides as accessory minerals. The dolomite grains vary from irregular shapes to well formed rhombs, as shown in Fig. 12A. The mean grain size is about 0.150 mm. This is a moderately fine grained rock with coarse pores (porosity $\approx$ 8.7%). Coarse pores or voids are visible in the rock with the unaided eye.

Ottawa Limestone: This is a typical field stone obtained from a local quarry. The rock is variable in composition, texture, grain size and porosity. As shown in Fig. 13A, it varies from a coarse grained biosparite consisting of large crystals of calcite (up to 2 or 3 mm in grain size) mixed with finer grained fossil remains to a very fine grained lithographic limestone (omicrete, indicated in Fig. 13A by 'L'). SEM (scanning electron microscope) examination showed this region 'L' to consist dominantly of fine calcite crystals with an average size of about 0.003 mm (Fig. 13B). The porosity of this rock varies with texture and grain size. The coarsely crystalline material has a porosity of 2.90%, but the porosity of lithographic limestone was so low ($\approx$0%) that it could not be measured with the available porosimeter. In the coarse limestone, the grain boundaries of the calcite crystals form discontinuities along which fracturing might occur. Calcite also has a well developed rhombohedral cleavage (Fig. 13C) and the pressure applied by the water jet could easily cause the calcite to cleave. The fine lithographic limestone, on the other hand, has virtually no discontinuities, apart from occasional microfractures which are cemented by iron oxide. As shown later, this rock was found to be very resistant to drilling.

PRESENTATION OF DATA

The experimental data for all the rocks investigated are listed in table 3 and are presented graphically in Figs. 2 to 10 (refer to table 1 for description of nozzle bodies). For ease of comparison, Figs. 2 to 8 are plotted in the same way as reported in Ref. 1. Also, the data points shown for nozzle bodies A2, A3, B2, B3, C2 and C3 pertaining to granite were taken from Ref. 1. In Fig. 9 specific energy is plotted against nozzle pressure for dolomite and limestone, including the effect of direction of drilling with respect to bedding plane for limestone (only three points are plotted because while conducting these tests, the fluid-end of the high pressure pump failed and the work came to a stop). Fig. 10 compares the performance of straight entry nozzles (SC2-1 and SC2-2) with the corresponding conical entry nozzle C2. The microphotographs showing the petrographic structure of the rocks and of the chips collected during experimentation appear in Figs. 11 to 13.

It should be stated that the data points in the plots are connected by lines (dotted, dashed, etc.) only for the sake of clarity.

DISCUSSION

An examination of Fig. 2 (plot of rate of penetration against rotational speed) and Fig. 3 (plot of specific energy against rotational speed) brings out the following points:

1. Some degree of influence of the rotational speed (N) on the rate of penetration (R) and the specific energy (E) can be seen, but the trend is not well defined. The effect of rotational speed on these parameters depends to some extent on the nozzle body configuration and the type of rock. However, for all the rocks tested to date, rotational speeds in the neighbourhood of 200 RPM appear to be quite satisfactory from the standpoint of specific energy or penetration rate.

2. It can be readily observed that each type of rock requires a different nozzle body configuration for optimum performance (defined in terms of the minimum specific energy). The best configurations were C2 and D3 for Muskoka Pink Granite; D2 and D3 for Ottawa Limestone and C3 for Amabel Marble. An explanation for this observed difference in response of the rocks is given later.

3. The best (that is, the minimum) value of the specific energy obtained for limestone ($\approx 10^5$ J/cm^3) was higher, by a factor of 10, than the value achieved for granite ($\approx 1.2 \times 10^4$ J/cm^3) or marble (1.1×10^4). This was rather surprising because the compressive strengths of limestone and marble were respectively 44% and 58% of that of granite. The measured tensile strengths of these rocks were also lower than that of granite by about 30 percent (see table 2). If compressive and tensile strengths were the controlling factors in cutting or drilling rocks as suggested by many investigators (some of these investigators have been cited in Refs. 4 & 5; see also Refs. 6 & 7), results should have been quite contrary to those quoted above (that is, $E_{granite} > E_{marble} > E_{limestone}$). Therefore, it appears that rock properties other than compressive and tensile strengths are important in cutting or drilling of rocks. A more detailed explanation is given later.

4. Performance of the straight entry nozzle body configurations SC2-1 and SC2-2 compared to its counterpart the conical entry nozzle C2 is very poor. However, the significant reduction in the values of specific energy obtained with the increase in L/D ratio of the orifices in these nozzle bodies should be noted. This observation is further clarified in Fig. 10 where the specific energy ratio ($E_{straight}/E_{conical}$) is plotted against L/D ratio. The trend of the data, although limited, indicates that it might be possible to obtain the same level of specific energy as that achieved by C2 (1.2×10^4 J/cm^3 for granite) by increasing the L/D ratio beyond 10. As discussed in detail by McCarthy and Molloy (Ref. 8), the L/D ratio has a great effect on the development of velocity profile within the orifice which in turn influences the stability and the coherent length of the jets. A stable jet would be produced when the velocity profile becomes fully developed and this requires a certain L/D ratio. Because of the simplicity of design, further study of these nozzle bodies would be worthwhile.

It was stated above that a nozzle body configuration which is best for drilling one type of rock may not necessarily be the best for drilling another type of rock. Although this was anticipated, an attempt was made to seek an explanation for this difference in response of the rocks. The steps taken were (i) to collect the debris (chippings) during the drilling tests and (ii) to examine and compare the petrographic structures of the chips with those of the parent rocks. It should be stated that the recent experimental and theoretical work of Rehbinder (Refs. 2 & 3) on linear slot cutting of rocks with single non-rotating jets lends support to such an approach.

In Fig. 11, the petrographic structure of granite and its debris are depicted. Since the porosity of this rock is low (≈1.2%), the crystal boundaries form the main planes of weakness and it is possible that upon impact by the water jets, fracture would occur along the grain boundaries. Fig. 11B shows that the debris consist mostly of single crystals indicating that fracture has indeed taken place around grain boundaries. Occasionally, some multiple crystal fragments also occur (Fig. 11C) due to fracturing around blocks of crystals, possibly following pre-existing microfractures in the rock.

Although marble is a fine grained rock (Fig. 12A), it has an abundance of coarse pores or voids. The presence of these voids would allow penetration of the water jet into the rock and facilitate splitting off of blocks of dolomite crystals. Fig. 12B shows a SEM (scanning electron microscope) micrograph of a typical chip collected from this rock and it is clearly seen to consist of a mass of dolomite crystals. This may explain why nozzle body C3 performed better than C2, because the higher flow available through the outer orifice of C3 (see table 1) could more effectively remove these relatively large size chippings through the annular space between the nozzle body and the hole.

Of the three rocks investigated, limestone had the most interesting structure and composition (Figs. 13A and 13B). Its structure varied in the direction of drilling (perpendicular to the bedding plane), being coarse at some places and fine at others. This variation dictated the rate at which the rock could be penetrated. The coarse layers being moderately porous (≈3%) could be penetrated much faster than the fine layers (porosity ≈ 0%). As a matter of fact, at times it took almost 2 minutes to penetrate through a 2.5 cm thick layer of fine material. In addition to this problem, the rate of progress through the rock was also hindered by the relatively large size of the chippings produced by the jets (Figs. 13C and 13D). As shown in these figures, various sizes (1 - 10 mm) of the chips were produced depending on the mode of fracture of the rock. For the coarse crystalline material, fracture seems to occur along grain boundaries (Fig. 13D) and also due to cleavage (Fig. 13C). For the fine material, on the other hand, the mechanism is probably the reopening of micro fractures already present in the rock, causing large pieces to be removed (Fig. 13D). These factors account for the observed high value of the specific energy for this rock compared to granite or marble, although its measured values of compressive and tensile strengths were lower than the other two rocks. It is worth mentioning here that the so-called hybrid systems (mechanical cutters incorporating nozzles to produce high speed liquid jets) might eliminate some of these problems, thus making it possible to achieve reduced values of specific energy (Refs. 4 & 9).

Observations from Figs. 4 and 5 are as follows:

The effect on specific energy of the distribution of flow area between the inner and outer orifices (A_O/A_T) in the nozzle body is shown in Fig. 4. For granite and limestone, specific energy decreases by approximately 30 and 50% respectively when A_O/A_T is increased from 50 to 60%. Further increase of A_O/A_T to 70%, has no significant effect on E. For marble, on the other hand, E decreases by 56% when A_O/A_T is increased from 60 to 70%. These results again indicate that the optimum nozzle body configuration depends on the material.

Fig. 5 shows the influence of the angle of inclination (θ_O) of the outer jet on the specific energy for granite and limestone. The data indicate the optimum angle, that is, the angle at which E is minimum, is in the neighbourhood of 20° for granite and 25° for limestone. This is presumably due to the difference in structure of the rocks as discussed above.

Figs. 6, 7 and 8 show the plot of the ratio of the areas of hole and the nozzle body ($A_R = A_H/A_N$), which was the other parameter studied in the investigation, against N, A_O/A_T and θ_O respectively. As stated in Ref. 1, depending on the type of rock, the value of A_R should be considerably greater than one for uninterrupted penetration into the rock. The observations from these figures are:

Fig. 6: The data once again confirm that low rotational speeds (≈200 RPM) are satisfactory for drilling. The better performance of nozzle body C2 for granite, C3 for marble and D2 for limestone, compared to other nozzle bodies, should be noted.

Fig. 7: Generally, A_R increases significantly with increase of A_O/A_T from 50 to 60%. Beyond this range, the effect appears to depend on the type of rock.

Fig. 8: It is interesting to note that the optimum angle, that is, the angle at which A_R is maximum, falls in the neighbourhood of 20° for granite and 25° for limestone.

Fig. 9 shows the effect of nozzle pressure and to a very limited extent, the effect of direction of drilling with respect to the bedding plane on the specific energy for limestone and marble. Although the decrease of E with increase of the nozzle pressure (that is, jet power) is as anticipated, the rate of decrease (slope) seems to be a function of the type of rock. The petrographic structure of the rocks, as discussed elsewhere, is responsible for this behaviour. Also, the specific energy would not decrease indefinitely, and should be expected to level off, depending on the type of rock, at some value of the pressure (Refs. 4 & 5), greater than 70 MPa in the present investigation.

With respect to the orientation of the bedding plane, the figure shows drilling parallel to the bedding plane to be more efficient, for pressures up to 50 MPa. The present data are in contradiction to those reported by Bresee et. al. (Ref. 10) for Berea Sandstone. The reasons given below not only explain this observed difference in specific energies in the two directions, but also uphold the data presented here.

In limestone the bedding planes are defined by clay layers (shale). Due to compaction of the sediments, the clay platelets are aligned parallel to the layering. These platelets are held together, not by chemical bonding, but largely by interlocking with one another. Cleavage occurs readily parallel to the layering but not normal to it. When a water jet impinges on the clay layers parallel to bedding, the shale readily cleaves allowing the jets to progress rapidly through the rock. In a direction normal to the layering, there are no major discontinuities on which the jets can act, instead they have to work their way slowly into the material, following grain boundaries, pores or microcracks. In more massive rocks where 3-dimensional interatomic forces come into play or when the jet pressures are high (>60 MPa), as shown by the trend of the present data and those of Bresee et. al. (Ref. 10), these differences should disappear.

CONCLUSIONS

The conclusions from the experimental data presented in this paper and also in Ref. 1 are:

1. A quite low rotational speed (≈200 RPM) was close to optimum.

2. The performance of nozzle bodies with straight entry orifices was poor compared to the performance of similar nozzle bodies with conical entry orifices. However, the L/D ratio of straight entry orifices was found to have a significant effect on specific energy and therefore deserves further study.

3. The influence of the flow area ratio (A_O/A_T) on the specific energy and the hole area ratio (A_R) was found to depend on the type of rock. For example, for marble, the specific energy decreased by 56% when A_O/A_T increased from 60 to 70%, while over the same range, for granite and limestone, the effect on specific energy was insignificant.

4. From both the specific energy and the hole area ratio (A_R) standpoint, the optimum angle of inclination of the outer jet (θ_O) was observed to be close to 20° for granite and 25° for limestone.

5. The best nozzle body configurations, from the specific energy (also A_R) standpoint, were C2 and D3 for granite, D2 and D3 for limestone and C3 for marble.

6. The dependence of the best nozzle body configuration on the type of rock is shown to be due to the microstructure of the rocks. The experimental data seem to confirm that porosity (more importantly, permeability), grain size and possibly the existence of micro fractures are the factors, not the compressive or tensile strengths, that control the rate at which the water jets penetrate rocks.

7. For an increase in nozzle pressure from 35 to 67 MPa, the specific energy decreased by factors of 3 and 2 respectively for marble and limestone. The orientation of the bedding plane with respect to the direction of drilling was found to have some influence on specific energy. However, further work is required to fully resolve this effect.

ACKNOWLEDGMENTS

The authors are thankful to Mr. R. A. Tyler, Head, Gas Dynamics Laboratory of N.R.C., for showing continuing interest on this project. Thanks are also due to Mr. P. J. Lefebvre, of the Division of Building Research, N.R.C., for rendering assistance in measuring the properties of rocks and to Mrs. J. Robinson for typing the manuscript of the paper.

REFERENCES

1. Vijay, M.M. and Brierley, W.H.: "Drilling of Rocks with Rotating High Pressure Water Jets: An Assessment of Nozzles." In: Proc. 5th Int. Symp. on Jet Cutting Technology (Hanover, Germany, June 2-4, 1980), BHRA Fluid Engineering, Cranfield, U.K., 1980, Paper G1, pp. 327-338.

2. Rehbinder, G.: "Some Aspects of the Mechanisms of Erosion of Rock with a High Speed Water Jet." In: Proc. 3rd Int. Symp. on Jet Cutting Technology (Chicago, U.S.A.: May 11-13, 1976), BHRA Fluid Engineering, Cranfield, U.K., 1976, Paper E1, pp. 1-19.

3. Rehbinder, G.: "A Theory about Cutting Rock with a Water Jet." Rock Mechanics, 12, 1980, pp. 247-257.

4. Maurer, W.C.: "Advanced Drilling Techniques." U.S.A., Petroleum Publishing Co., P.O. Box 1260, Tulsa, OK 74101, 1980, chp. 12, pp. 229-301.

5. Vijay, M.M. and Brierley, W.H.: "Drilling of Rocks by High Pressure Liquid Jets: A Review." ASME Paper No. 80-Pet-94, American Society of Mechanical Engineers, New York, U.S.A., 1980, pp. 1-11.

6. Singh, M.M. and Huck, P.J.: "Correlation of Rock Properties to Damage Effected by Water Jet." In: Dynamic Rock Mechanics, Ed., G. B. Clark, The University of Missouri Rolla, Missouri, U.S.A., Nov. 16-18, 1970, chp. 35, pp. 681-695.

7. Cooley, W.C.: "Correlation of Data on Erosion and Breakage of Rock by High Pressure Water Jets." Ibid, chp. 33, pp. 653-665.

8. McCarthy, M.J. and Molloy, N.A.: "Review of Stability of Liquid Jets and the Influence of Nozzle Design." The Chemical Engineering Journal, 7, 1974, pp. 1-20.

9. Bonge, N.J. and Wang, F.D.: "Advanced Application of Water Jets to Small Hole Drilling." In: Proc. 1st U.S. Water Jet Symposium (Golden, Colorado, U.S.A., April 6-9, 1981), Colorado School of Mines, Golden, Colorado, U.S.A., 1981, pp. 1-14.

10. Bresee, J.C., Cristy, G.A. and McClain, W.C.: "Research Results Show Interesting Potential of Hydraulic Tunneling." E/MJ (Engineering and Mining Journal), July 1970, pp. 75-80.

APPENDIX

Rehbinder's theory (Refs. 2 & 3) on linear slot cutting of rocks with nonrotating jets is based on the premise that all rocks are permeable. Water penetrates the voids between the grains and thus exposes the grains to hydraulic forces which make them come off. There is a condition that the net hydraulic force on a grain must overcome the cohesive forces between the grains. The rate at which a crater is formed depends on the rate at which water can penetrate the pores. This suggests that parameters controlling the flow of liquids through porous media are relevant to the problem. Therefore, starting with Darcy's equations of fluid flow through porous media, the author derives the following equation for the depth of a slot when a water jet is traversed over the surface of a rock.

$$\frac{h}{W} = \frac{1}{\beta}\left(1 - e^{-\frac{\beta D \kappa P}{\mu \ell W u}}\right) \quad \text{(for } h/W \leqslant 10\text{)} \tag{1}$$

where

h = depth of the slot
W = width of the slot
D = diameter of the nozzle
κ = permeability of the rock
P = nozzle pressure
μ = viscosity of water
ℓ = average grain size of the rock
u = relative velocity (traverse speed) of the jet with respect to the rock

β is a constant that arises due to the assumption that pressure is a function of the depth of the slot. That is,

$$P(h) = Pe^{-\beta h/W} \tag{2}$$

Further, for fine grained rocks ($\ell/D < 1$), the author shows that $W/D \approx 4$ and $\beta = 0.025$. Also, if $\beta D\kappa P/\mu\ell W u \ll 1$, then Eq. 1 reduces to:

$$\frac{h}{D} = \frac{\kappa}{\mu\ell} \cdot \frac{P}{u} \tag{3}$$

Differentiation of Eq. 3 yields:

$$\frac{1}{D}\frac{\partial h}{\partial P} = \frac{\kappa}{\mu\ell} \cdot \frac{1}{u} \tag{4}$$

and

$$\frac{1}{D}\frac{\partial h}{\partial u} = -\frac{\kappa}{\mu\ell} \cdot \frac{P}{u^2} \tag{5}$$

Eqs. 4 and 5 show that the parameter $\kappa/\mu\ell$, which is a measure of the erosion resistance of the rock, can be determined from the experimental data, by plotting h against P or 1/u.

Expressions for the rate of penetration (R) and the specific energy (E) obtained with a rotating nozzle body can now be derived in terms of Eq. (1), or the much simpler Eq. 3 as follows:

Referring to Fig. 14, the traverse speeds (peripheral speeds) of the jets relative to the rock are:

$$u_o = \text{speed of outer jet} = 2\pi N(a + S \tan\theta_o) \tag{6}$$

and

$$u_i = \text{speed of inner jet} = 2\pi N S \tan\theta_i \tag{7}$$

where

N = rotational speed
S = standoff distance

Since u_i is $< u_o$, the inner jet penetrates deeper into the rock than the outer jet. Consequently, the rate at which the nozzle body can penetrate through the rock is dictated by the outer jet. As shown in the figure, let 'd' be equal to the penetration per revolution. Then,

$$R = Nd = Nh_o \cos\theta_o \tag{8}$$

From Eq. 3 (for the sake of simplicity),

$$h_o = \frac{\kappa}{\mu\ell} \cdot D_o \cdot \frac{P}{u_o}$$

$$= \frac{\kappa}{\mu\ell} \cdot D_o \cdot \frac{P}{2\pi N(a+S\ \tan\theta_o)} \tag{9}$$

Therefore,

$$R = \frac{\kappa}{\mu\ell} \cdot \frac{D_o}{2\pi} \cdot \frac{P\cos\theta_0}{(a+S\ \tan\theta_o)} \tag{10}$$

Now, the diameter of the hole (D_H) depends on the widths W_o and W_i, cut by the jets (Fig. 14). By geometrical considerations, it can be shown that:

$$D_H = 2b(D_o \sec\theta_o + D_i \sec\theta_i) \tag{11}$$

where it is assumed that $W = bD$ and b is a constant (≈2 to 5).

The rate of volume removal is given by:

$$\dot{V} = \frac{\pi}{4} D_H^2 R \tag{12}$$

Substituting for R and D_H from Eqs. (10) and (11), one obtains:

$$\dot{V} = \frac{\kappa}{\mu\ell} \cdot \frac{b^2}{2} \cdot \frac{D_o P(D_o\sec\theta_o + D_i\sec\theta_i)^2\cos\theta_o}{(a + S\ \tan\theta_o)} \tag{13}$$

Next, the total hydraulic power of the jets is obtained as follows:

$$H_p = \text{Total hydraulic power} = (H_p)_o + (H_p)_i$$

$$= kD_o^2P^{3/2} + kD_i^2\ P^{3/2}$$

$$= kP^{3/2}(D_o^2 + D_i^2) \tag{14}$$

where k = constant

Hence, E = specific energy = $H_p/\dot{V}$

$$= k' \frac{\mu\ell}{\kappa} \cdot \frac{(a + S\ \tan\theta_o)(D_o^2 + D_i^2)}{D_o P^{1/2}\cos\theta_o(D_o\sec\theta_o + D_i\sec\theta_i)^2} \tag{15}$$

where k' = constant

This expression for E, although derived using the simple equation for h (Eq. 3), brings out some important points. It shows that E is directly proportional to ℓ and inversely proportional to κ, in accord with the experimental results. Also, for small values of θ_o and θ_i, $S \approx 0$ and $D_o \approx D_i$, Eq. 15 reduces to:

$$E = \kappa'' \frac{\mu\ell}{\kappa} \cdot \frac{a}{D_o} \cdot \frac{1}{P^{1/2}} \tag{16}$$

It is clear from this equation that E can be reduced by reducing 'a', other factors remaining constant. This result is useful for designing nozzle bodies to drill small holes (≈1 cm) or for cutting deep narrow slots in rocks. Work in this direction is in progress in the laboratory.

Table 1. Design Specification of Nozzle Bodies (Fig. 1)

Diameter of Nozzle Body = 19.05 mm

$A_T/A_N = 0.008$

Nozzle Body Identification	D_I (mm)	D_O (mm)	θ_O (deg)	Remarks
A2	1.090	1.320	10	$\theta_I = 10°$ for all
A3	0.914	1.400	10	nozzle bodies
B2	1.090	1.320	15	investigated.
B3	0.914	1.400	15	
C1	1.180	1.180	20	$\alpha = 20°$ for all
C2	1.090	1.320	20	nozzles except
C3	0.914	1.400	20	where noted.
D2	1.090	1.320	25	
D3	0.914	1.400	25	
E2	1.090	1.320	30	
E3	0.914	1.400	30	
SC2-1	1.090	1.320	20	L/D = 5 $\alpha = 180°$
SC2-2	1.090	1.320	20	L/D = 10 $\alpha = 180°$

Table 2. Measured Properties of Rocks

1. Muskoka Pink Granite

Property	Mean Value	Range	Coefficient of Variation (%)*
ρ_R	2627	2603-2649	0.5
q_t	8.0	5.3-10.0	16.6
q_c	165.5	139-175	7.2
C	4115	3767-4456	4.8
E_R	44540	37230-52400	9.9
Grain size,mm	1.98	0.2-4.4	14.9
Porosity,%	1.235	0.61-2.65	56.9

* Standard deviation expressed as a percentage of the mean value.

2. Ottawa Limestone

Property	Mean Value	Range	Coefficient of Variation (%)
ρ_R	2635	2592-2675	1.4
q_t	5.3	2.3-9.2	35.5
q_c	72.2	56.9-89.0	17.6
C	4760	4480-5150	5.7
E_R	59800	53340-69000	11.1
Grain size,mm	1.5	0.003-3.0	-
Porosity, %	2.9	0.00-3.10	-

3. Amabel Marble (Dolomite)

Property	Mean Value	Range	Coefficient of Variation (%)
σ_t	5.90	2.90-7.60	23.5
σ_c	96.50	72.5-118.2	12.8
Grain size, mm	0.10	0.05-0.30	-
Porosity, %	8.72	-	-

Table 3. Mean Drilling Data

1. Muskoka Pink Granite P = 69.0 ∓ 3.4 MPa

Nozzle Body	N (RPM)	R (cm/min)	D_H (cm)	A_R	E x 10^{-4} (J/cm^3)
D2	120	26.7	2.48	1.689	2.69
D2	240	46.8	2.46	1.674	1.63
D2	390	41.5	2.58	1.830	1.78
D3	120	24.2	2.87	2.269	2.16
D3	240	40.1	2.64	1.935	1.52
D3	390	30.9	2.63	1.907	2.06
E2	120	28.2	2.52	1.766	2.77
E2	240	31.4	2.54	1.780	2.25
E2	408	34.6	2.52	1.745	2.05
E3	120	19.4	2.87	2.274	2.70
E3	240	36.7	2.58	1.821	1.76
E3	390	27.1	2.63	1.913	2.28
SC2-1	120	2.1	2.82	2.209	26.53
SC2-2	120	10.1	2.71	2.026	6.52
SC2-2	240	10.7	2.65	1.933	6.10
SC2-2	390	16.3	2.66	1.955	4.17

2. Ottawa Limestone

Nozzle Body	P (MPa)	N (RPM)	R (cm/min)	D_H (cm)	A_R	E x 10^{-4} (J/cm^3)
B2	64.9	120	2.69	2.73	2.049	20.60
B2	64.2	240	3.42	2.37	1.546	21.10
B2	64.2	420	2.05	2.27	1.420	38.30
C1	69.0	120	3.03	2.42	1.617	24.10
C1	68.3	240	2.16	2.41	1.594	33.70
C1	67.6	420	1.54	2.31	1.473	50.60
C2	64.9	120	4.01	2.80	2.155	13.10
C2	64.9	240	3.76	2.65	1.937	15.60
C2	64.9	420	3.27	2.35	1.517	22.80

Ottawa Limestone (continued)

Nozzle Body	P (MPa)	N (RPM)	R (cm/min)	D_H (cm)	A_R	E x 10^{-4} (J/cm^3)
C3	64.2	120	3.81	2.99	2.469	11.20
C3	64.2	240	2.96	2.39	1.578	22.60
C3	64.9	420	2.90	2.41	1.604	23.10
D2	69.0	120	7.31	2.73	2.055	8.27
D2	69.0	240	3.72	2.80	2.161	15.50
D2	69.0	420	4.78	2.51	1.741	15.00
D2	34.5	240	0.71	2.43	1.633	38.10
D2	44.8	240	1.47	2.42	1.609	27.50
D2	55.2	240	2.16	2.55	1.787	23.10
D2	67.6	240	3.96	2.29	1.448	21.00
D2*	34.5	240	1.11	2.17	1.294	30.60
D2*	44.8	240	2.00	2.28	1.438	22.60
D2*	55.2	240	2.37	2.30	1.460	25.70
D3	69.0	120	6.67	2.67	1.964	9.00
D3	69.0	240	4.90	2.61	1.883	12.80
D3	69.0	420	4.63	2.41	1.596	16.00
E2	69.0	120	6.21	2.66	1.946	10.30
E2	69.0	240	6.01	2.36	1.531	13.50
E2	69.0	420	9.07	2.33	1.491	9.19
E3	69.0	120	6.44	2.45	1.657	11.10
E3	69.0	240	6.59	2.35	1.516	11.80
E3	69.0	420	3.57	2.40	1.586	20.80

*These tests were conducted parallel to the bedding planes of the rock.

3. Amable Marble (Dolomite)

Nozzle Body	P (MPa)	N (RPM)	R (cm/min)	D_H (cm)	A_R	E x 10^{-4} (J/cm^3)
C2	41.4	420	3.60	2.38	1.559	10.30
C2	55.2	420	10.12	2.48	1.689	5.20
C2	66.9	420	24.91	2.66	1.949	2.45
C2	66.9	120	16.19	2.83	2.209	3.32
C3	41.4	420	9.81	2.66	1.949	2.87
C3	55.2	420	23.13	2.83	2.209	1.65
C3	66.2	420	46.26	2.83	2.209	1.09
C3	66.2	120	26.99	2.91	2.339	1.76

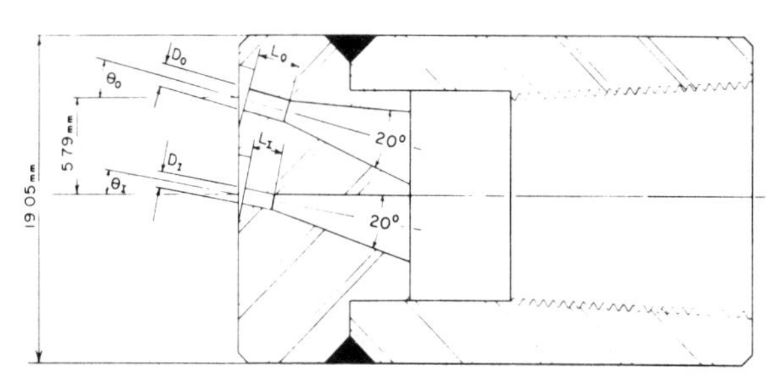

FIG. 1A NOZZLE BODY INCORPORATING CONICAL ENTRY ORIFICES (SEE TABLE 1)

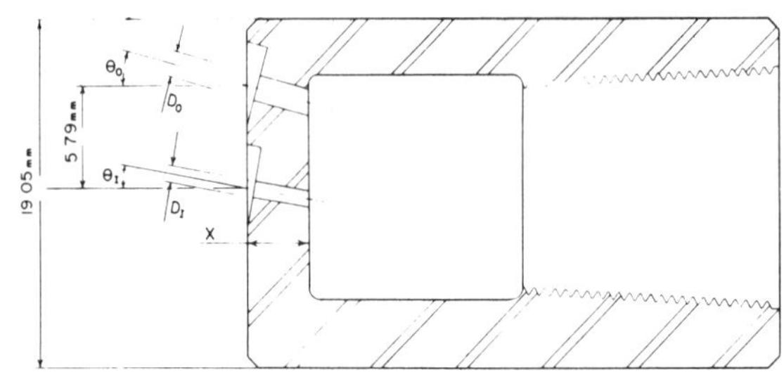

FIG. 1B NOZZLE BODY INCORPORATING STRAIGHT ENTRY ORIFICES (SEE TABLE 1)

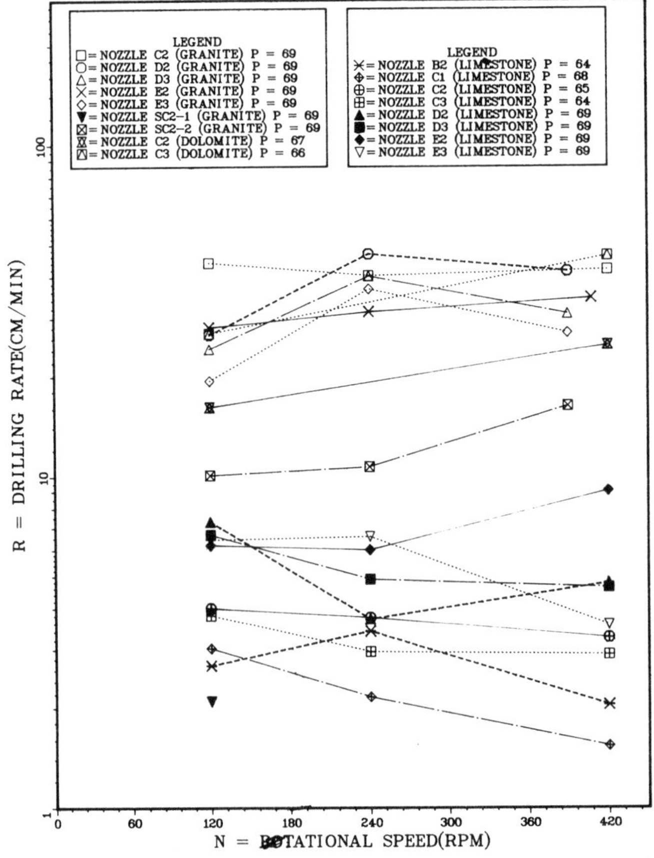

FIG.2 PLOT OF DRILLING RATE(R) AGAINST ROTATIONAL SPEED(N)

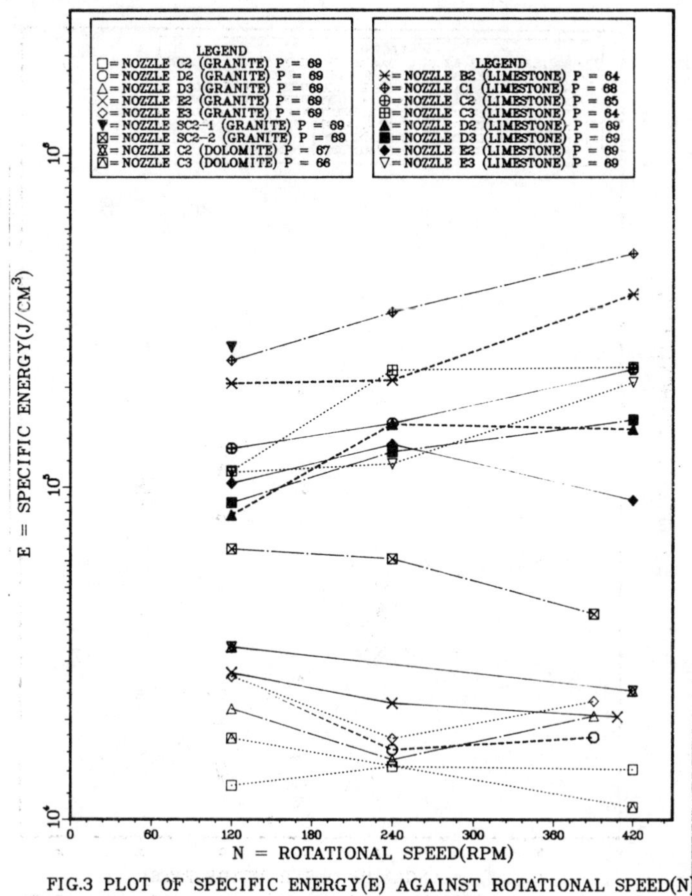

FIG.3 PLOT OF SPECIFIC ENERGY(E) AGAINST ROTATIONAL SPEED(N)

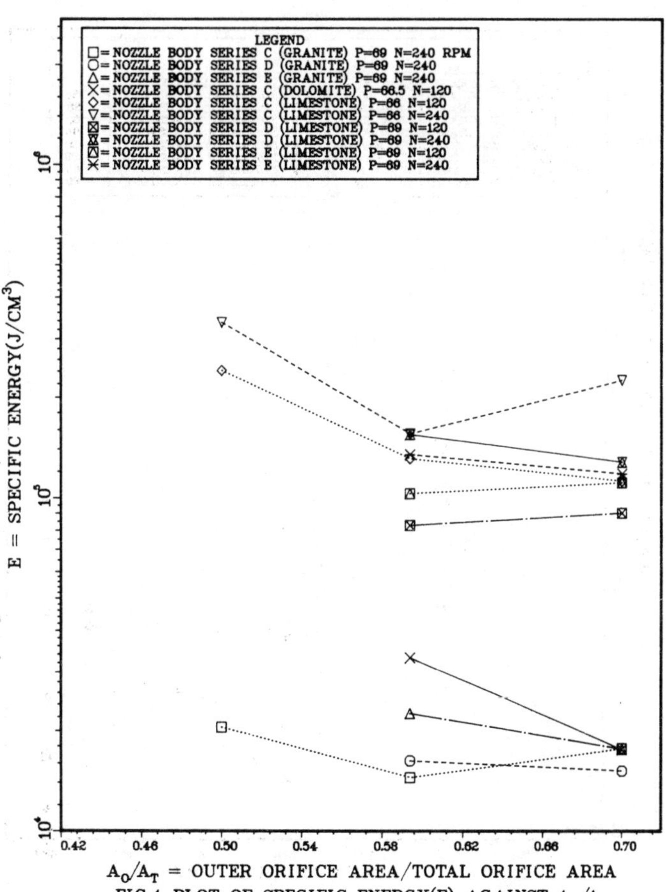

A_O/A_T = OUTER ORIFICE AREA/TOTAL ORIFICE AREA

FIG.4 PLOT OF SPECIFIC ENERGY(E) AGAINST A_O/A_T

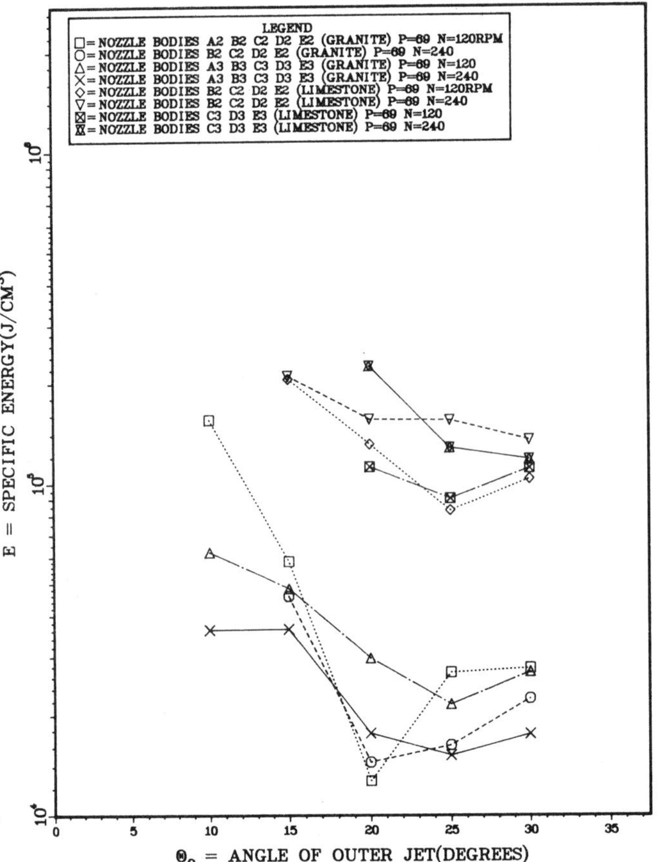

FIG.5 PLOT OF SPECIFIC ENERGY(E) AGAINST OUTER JET ANGLE(Θ_0)

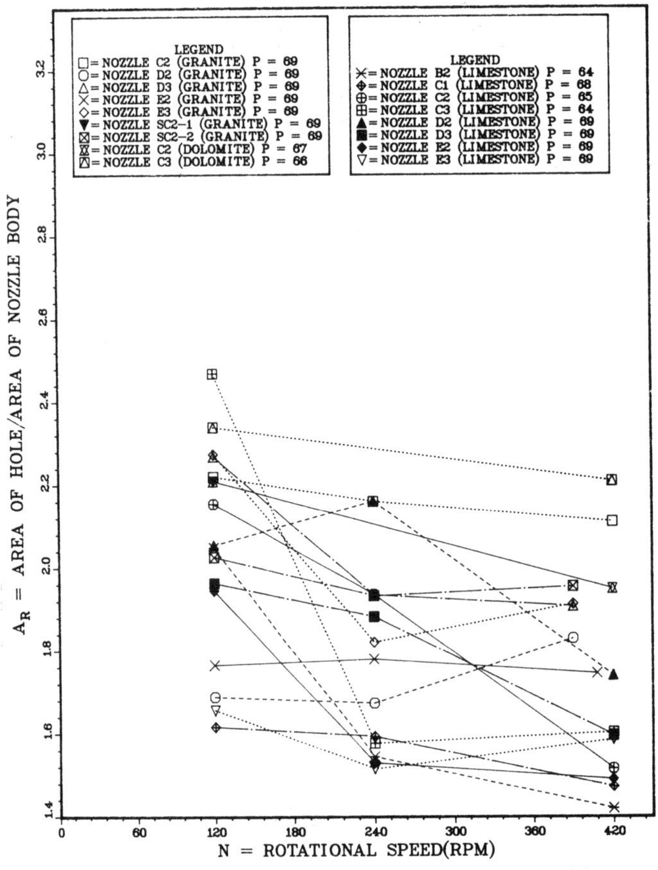

FIG.6 PLOT OF A_R AGAINST ROTATIONAL SPEED(N)

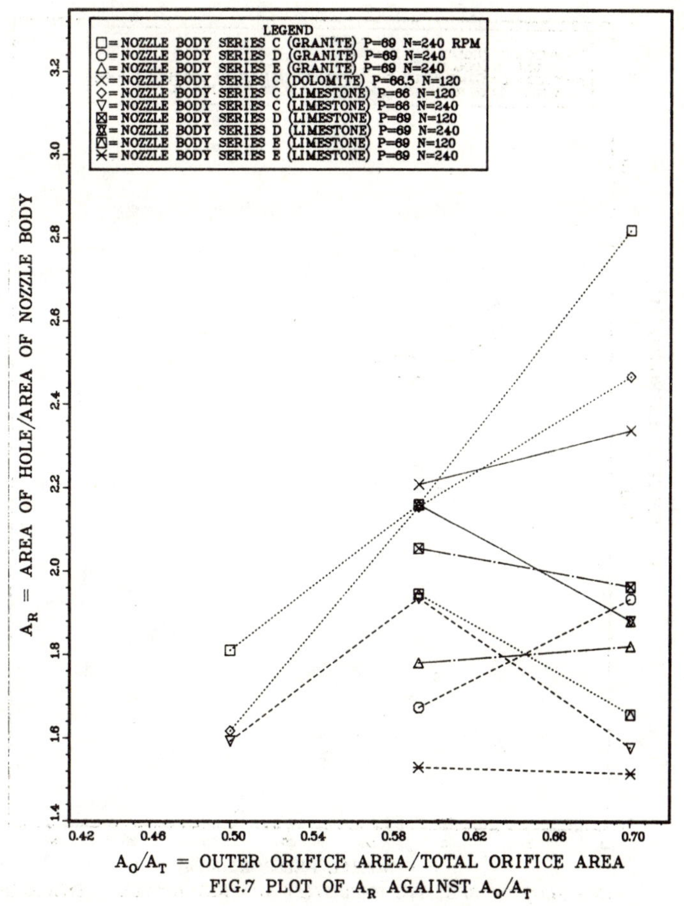

FIG.7 PLOT OF A_R AGAINST A_O/A_T

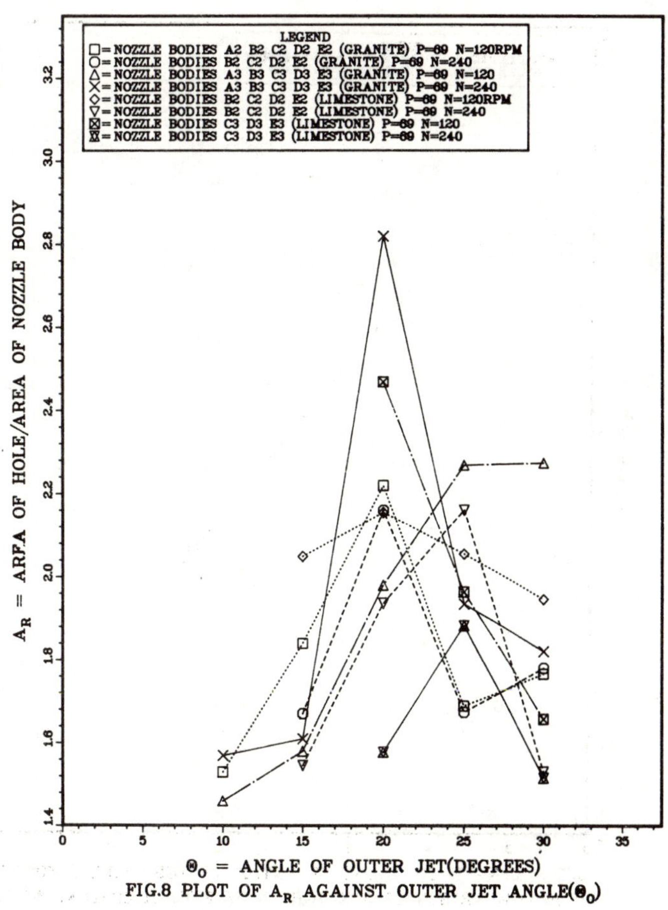

FIG.8 PLOT OF A_R AGAINST OUTER JET ANGLE(Θ_0)

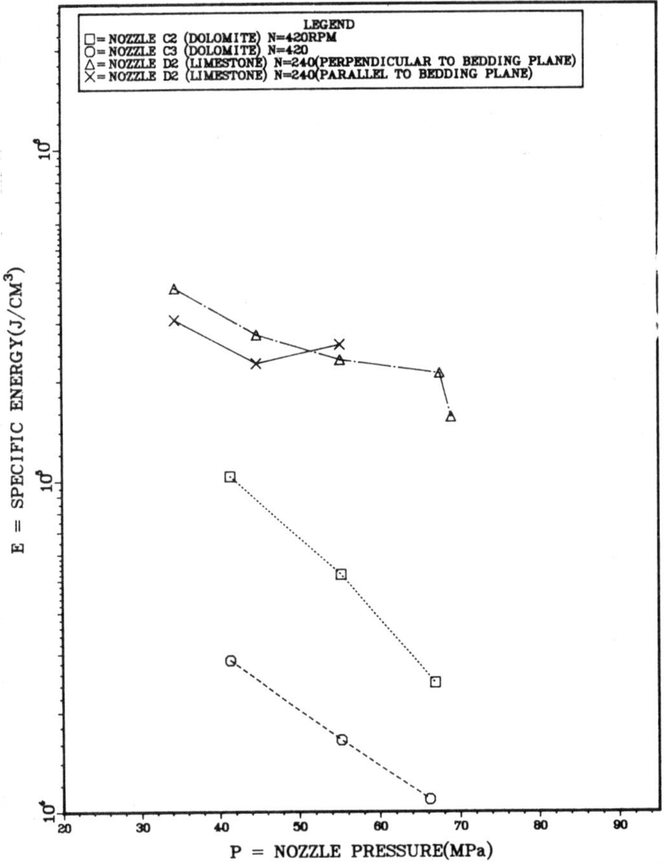

FIG.9 PLOT OF SPECIFIC ENERGY(E) AGAINST NOZZLE PRESSURE(P)

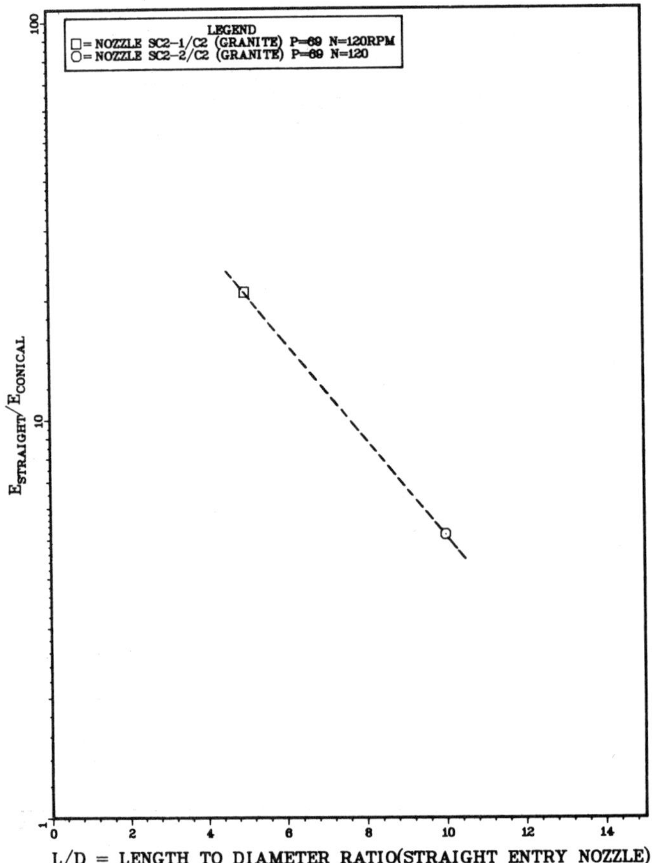

FIG.10 COMPARISON OF STRAIGHT NOZZLE WITH CONICAL NOZZLE

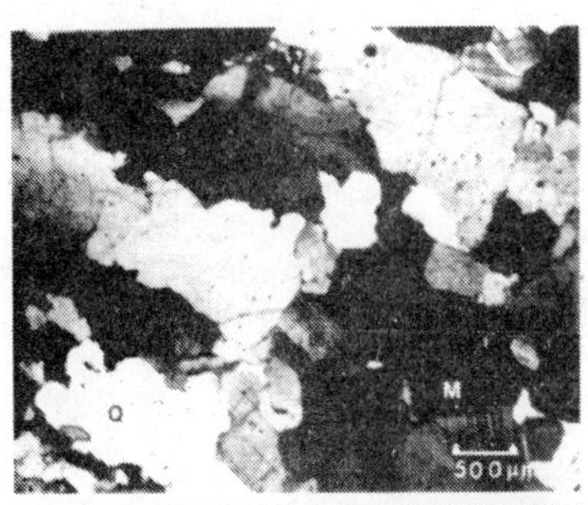

FIG. 11 A

THIN SECTION OF GRANITE OBSERVED IN POLARIZED LIGHT SHOWING IRREGULAR GRAIN BOUNDARIES.

Q: QUARTZ CRYSTAL

M: MICROCLINE FELDSPAR

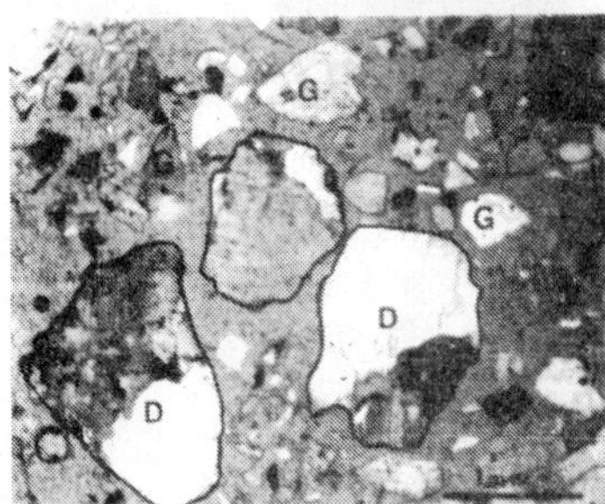

FIG. 11 B

TYPICAL CHIPS OF GRANITE OBTAINED IN DRILLING TESTS. EXAMINATION OF THIN SECTIONS IN POLARIZED LIGHT SHOWS THE PRESENCE OF MANY SINGLE (G) AND DOUBLE (D) CRYSTALS OF QUARTZ AND FELDSPAR

FIG. 11 C

SAME AS IN 11 B. THE CHIP CONSISTS OF MULTI CRYSTALS OF QUARTZ AND FELDSPAR

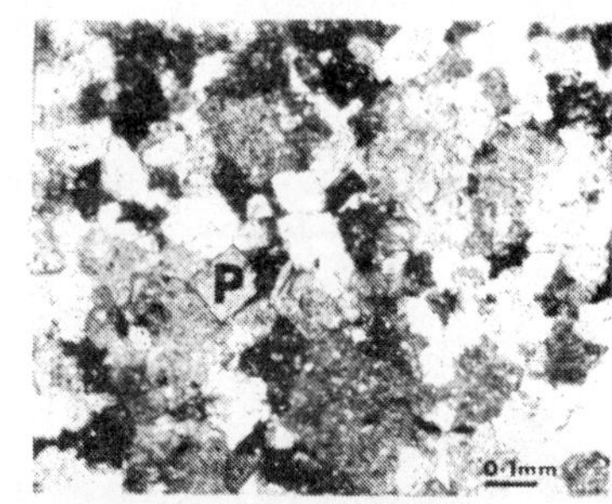

FIG. 12 A

THIN SECTION OF DOLOSTONE (DOLOMITE) OBSERVED IN POLARIZED LIGHT, SHOWING THE GRAIN SIZE AND SHAPE OF DOLOMITE CRYSTALS. A TYPICAL DOLOMITE RHOMB IS SEEN AT P.

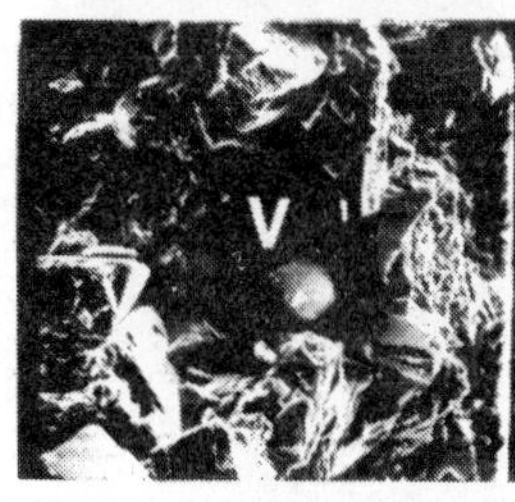

FIG. 12 B

A TYPICAL S.E.M. (SCANNING ELECTRON MICROSCOPE) MICROGRAPH OF A CHIP OF DOLOMITE.
THE CHIP CONSISTS OF A CLUSTER OF RHOMB SHAPED DOMONITE CRYSTALS.
A LARGE VOID (V) IS CLEALY VISIBLE.

FIG. 13 A

THIN SECTION OF FINE AND COARSE LIMESTONE OBSERVED IN POLARIZED LIGHT.
THE COARSE LIMESTONE CONSISTS OF LARGE CALCITE CRYSTALS (C) AND FOSSILIFEROUS MATERIAL (F).
THE FINE LITHOGRAPHIC LIMESTONE IS SEEN ON THE RIGHT (L).
IT CONSISTS OF A MASS OF SPHERICAL BODIES (OOLITES) IN A FINE MATRIX.

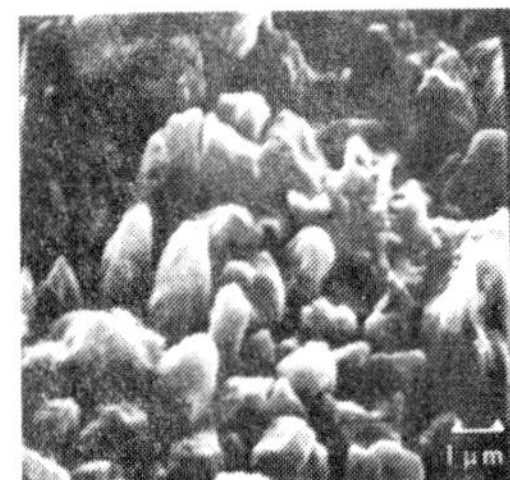

FIG.13B

A SEM MICROGRAPH OF ETCHED FINE GRAINED LIMESTONE.
THE ROCK CONSISTS OF SMALL (0·003mm) CALCITE CRYSTALS.

FIG.13C

THIN SECTION OF CHIPPINGS OF COARSE LIMESTONE SHOWING ONE SINGLE CRYSTAL OF CALCITE.
FRACTURING FOLLOWED ALONG THE CRYSTAL BOUNDARIES AND THEN CUT ACROSS THE GRAIN BY CLEAVING IT ALONG THE RHOMBOIDAL CLEAVAGES (CL).

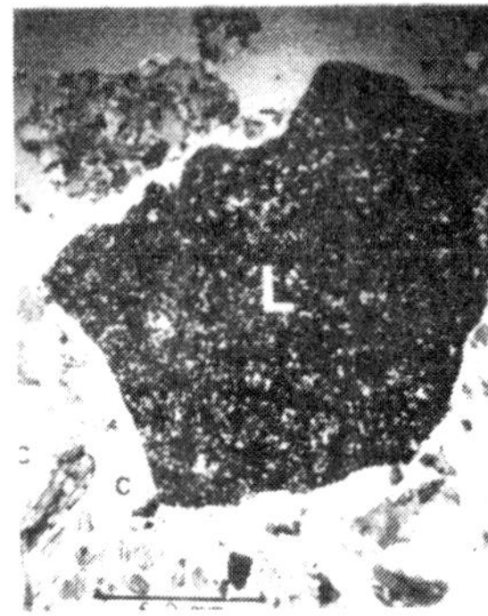

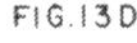

FIG.13D

THIN SECTION OF CHIPS OBSERVED IN POLARIZED LIGHT SHOWING A LARGE GRAIN OF FINE LIMESTONE (L) SURROUNDED BY SMALL GRAINS OF EITHER SINGLE OR MULTIPLE CRYSTALS OF COARSE LIMESTONE (C).

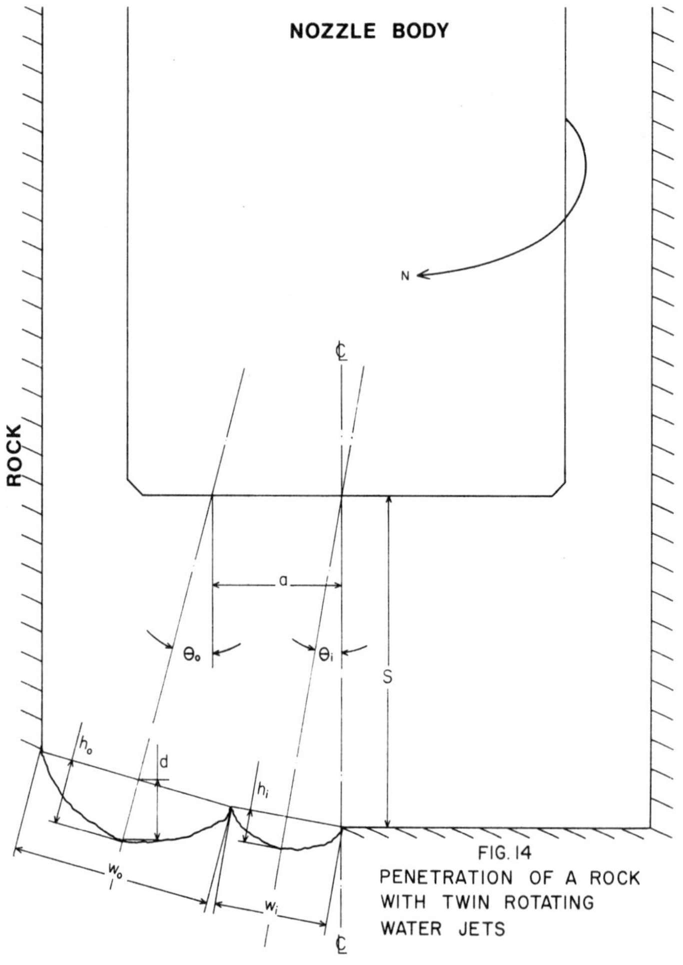

FIG.14
PENETRATION OF A ROCK WITH TWIN ROTATING WATER JETS

6th International Symposium on
Jet Cutting Technology
6-8, April, 1982

PAPER E2

HYDRAULIC INTRUSION BY A TRAVERSING WATER JET

J. L. Evers, D. L. Eddingfield and W. S. Man
Southern Illinois University at Carbondale, U. S. A.

Summary

A theoretical model has been developed and experimentally verified that describes the flow of a fluid column, produced by a traversing water jet, into an air filled capillary tube. The study shows the potential for hydraulically exploiting certain internal spaces in minerals that can be modeled by a dead-end capillary tube of equivalent hydraulic radius. The effect of cavity size, jet traverse rate and impingement pressure on the pressure pulse produced in the trapped gas is analyzed. Pressure pulses that are orders of magnitude higher than the jet impact pressure have been predicted for flow spaces with relatively large diameters and short lengths.

Held at the University of Surrey, U.K.
Symposium organised and sponsored by
BHRA Fluid Engineering

NOMENCLATURE

A = area of the tube.

D = hydraulic diameter of the flow space.

f = Darcy-Weisbach friction factor.

g = acceleration of gravity.

H_1 = pressure at the capillary inlet.

H_2 = pressure in the trapped gas.

H_o = centerline pressure of the jet at the impact surface.

H_{atm} = atmospheric pressure.

k = constant of Eq. 7 for the pressure profile.

L = length of the tube.

m = constant of Eq. 7 for the pressure profile.

n = polytropic exponent.

P_d = driving pressure.

P_o = centerline pressure at the impact surface.

P_{max} = maximum pressure during the hydraulic transient.

P_p = pump pressure.

P_r = perimeter of the liquid column.

r = distance from jet centerline.

r_o = radius of jet at the nozzle.

t = time.

S_o = nozzle standoff distance.

V = velocity of fluid-air interface.

V_o = jet traverse velocity

x = distance of liquid-air interface from tube inlet.

z_o = initial position of sweeping jet.

ρ = fluid density.

τ = fluid shearing stress

1. INTRODUCTION

The hydraulic transient produced by a water jet as it pressurizes a pore space, a crack or a bedding plane has been judged to be the dominant mechanism of fragmentation in certain minerals, especially those formed by a sedimentary process (Ref. 1). Even minerals that do not exhibit such large scale fractures have been shown to erode at a rate that is related to the permeability of the material (Ref. 2, 3). However, the relative importance of this mechanism of fragmentation compared to water hammer effect, stagnation pressure damage, or the erosional drag effect has not been determined. Therefore, a clearer understanding of the internal flows through pores, fractures and bedding planes produced by a traversing water jet is expected to reveal parameter optimizations that will increase the efficiency of water jet cutting.

In this study a theoretical model has been developed to describe the pressurization of a crack or cleat fracture that is small enough to be modeled by a dead-end capillary tube of equivalent hydraulic radius. Even though the pore spaces modeled in this study are only representative of a very limited occurrence in the broad spectrum of natural minerals, they are much more typical of spaces found in materials such as coal that were formed by a sedimentary process and then cracked by earth shocks and strains as well as thermal cycling. In addition, the fact that high speed steady flows in porous materials have been modeled with bundles of capillary tubes equipped with internal orifice plates (Ref. 4), leads one to speculate that transient capillary flows might serve as a basic element in describing hydraulic intrusion into more general porous spaces.

Although this study is an experimental assessment of an analytical model that includes only a limited number of features of a general porous matrix, the results can help to provide insight into the mechanism of internal pressurization produced by a sweeping water jet. There is also an indication of the relative importance of various parameters such as jet traverse rate, jet size, standoff distance, the pressure profile shape, the hydraulic radius and the length of the space in maximizing the internal pressure rise. More importantly, the success of this initial model can provide a starting point for subsequent studies that sequentially add features to the model such as gas release, tortuosity and interconnectivity.

2. THEORETICAL MODEL

The fluid column illustrated in Fig. 1 is driven into the gas filled capillary by the pressure imposed at the inlet, H_1. The motion is retarded by the fluid viscous forces and the pressure in the trapped gas, H_2. The net force on the fluid slug is equal to the rate of change of momentum of the fluid element. The resulting equation becomes

$$\rho g\ H_1 A - \rho g\ H_2 A - \tau\ P_r\ X = \frac{d}{dt}\left(\rho\ A\ X\ \frac{dx}{dt}\right) \qquad (1)$$

where ρ is the density of the liquid, g is the acceleration of gravity, x is the length of the column, A is the area, P_r is the perimeter and τ is the shearing stress. The shearing stress can be expressed in terms of the liquid velocity v, the Darcy-Weisbach resistance coefficient f and the fluid density as follows:

$$\tau = f\ \frac{\rho\ v|v|}{8} \qquad (2)$$

The absolute value sign is required to provide a sign change in the shear stress as the velocity changes direction. For laminar flow the friction factor is known to be inversely proportional to the Reynolds number

$$f = \frac{64}{Re} \qquad (3)$$

In the fully turbulent flow regime, the Moody diagram has normally been used to determine the friction factor. However, Chen (Ref. 5) has developed the following explicit expression for the friction factor that is particularly useful for computer solutions.

$$f = \left\{-2.0 \log_{10}\left\{\frac{\varepsilon}{3.7065D} - \frac{5.0452}{Re} \log_{10}\left[\frac{1}{2.8257}\left(\frac{\varepsilon}{D}\right)^{1.1098} + \frac{5.8506}{Re^{0.8981}}\right]\right\}\right\}^{-2} \qquad (4)$$

For the transition region, the friction factor was calculated from a linear interpolation between the laminar flow and the fully turbulent flow values.

The pressure in the trapped gas is determined from the polytropic compression and expansion relationship

$$H_2 = H_{atm} \left(\frac{L}{L-x}\right)^n \qquad (5)$$

where L is the length of the capillary, H_{atm} is the initial pressure inside, and n is the polytropic exponent. When a more accurate prediction of the pressure transient is required, heat transfer to the compressed gas should be included to provide a more realistic model (Ref. 6). However, the simpler polytropic assumption with the exponent reduced from its isentropic value to compensate for the heat transfer has been successfully employed by Parmakian and Evans (Ref. 7, 8) in analyzing air chamber performance on pump lines.

Combining Eqs. 1, 2 and 5 yields

$$\frac{dx^2}{dt^2} + \left(\frac{f}{2D} + \frac{1}{x}\right)\frac{dx}{dt}\left|\frac{dx}{dt}\right| + \frac{g}{x}\left[H_{atm}\left(\frac{L}{L-x}\right)^n - H_1 - H_{atm}\right] = 0 \qquad (6)$$

where D is the hydraulic diameter of the flow space.

The upstream pressure, H_1 in this investigation was either supplied by a large accumulator or a traversing water jet. The accumulator which provided a nearly constant pressure was used to simplify the upstream boundary condition while the transient internal flow model was evaluated. Whenever the upstream pressure was produced by a sweeping water jet, a pressure profile, determined for each pump pressure and standoff distance was expressed by a curve suggested by Rehbinder and Shavlosky (Ref. 2, 9)

$$H_1 = H_o e^{-\frac{1}{k}\left(\frac{r}{r_o}\right)^m} \qquad (7)$$

where r is the distance from the jet centerline to the point where the pressure is H_1, H_o is the centerline pressure of the jet at the impact surface, r_o is the jet radius, and the constants k and m were chosen for the best curve fit. For a moving jet the distance r from the jet centerline to the capillary is a function of the jet traverse velocity V_o and the elapsed time, t. The expression for upstream pressure becomes

$$H_1 = H_o e^{-\frac{1}{k}\left(\frac{-z_o + v_o t}{r_o}\right)^m} \qquad (8)$$

where z_o is the initial position of the jet. This expression was substituted into Eq. 6 and solved numerically using the fourth order Runge-Kutta method.

This analysis assumed a totally trapped gas, a uniform velocity profile, a stable gas-liquid interface, a friction factor that was a function of Reynolds number and relative roughness, a liquid viscosity that was a function of temperature and pressure, an average polytropic exponent between 1.1 and 1.2 (Ref. 10, 11), and an incompressible liquid phase.

3. EXPERIMENTAL APPARATUS, PROCEDURE AND RESULTS

In assessing the validity of the analytical model the experimental study was divided into two types of tests. In the first experiments the capillary was pressurized with a constant pressure source to simplify the upstream boundary condition. In the second group of experiments a sweeping jet supplied the driving pressure.

3.1 Constant driving pressure tests.

In order to evaluate the internal transient flow model without the uncertainties associated with describing the impact pressure of a traversing jet, the flow circuit shown in Fig. 2 was devised. The air filled tube was pressurized by the sudden opening of the ball valve, allowing high pressure liquid to flow from the accumulator into the tube. The time history of the pressure in the trapped gas and in the fluid just downstream of the ball valve were measured by pressure transducers located at each end of the flow circuit and recorded on two channels of a cathode ray tube (CRT) strip chart recorder.

The pressure in the gas end of the accumulator was provided by a tank of nitrogen equipped with a pressure regulator. The fluid end of the accumulator was either filled manually or with a hydraulic hand pump.

The ball valve was manually opened producing a pressure rise to a steady value in about 20 milliseconds. This pressure rise was described in the analytical model by fitting a polynomial through the measured pressure points. Therefore the upstream boundary condition was provided by the polynomial which described the pressure rise at the tube inlet for the first 20 milliseconds followed by a constant pressure for the remaining time of the calculation.

3.2 Results-constant driving pressure tests.

The pressure magnification, which is the pressure in the trapped gas divided by the driving pressure in the accumulator, is shown for 100 milliseconds after the valve is opened in Fig. 3. For this curve a 2.4 mm inside diameter tube, 1.14 m long was pressurized with an accumulator pressure of 4.14 MN/m^2. The fluid started from rest and accelerated into the tube compressing the gas trapped at the end. Analogous to a weight being dropped on a spring, the moving liquid column compressed the gas above the equilibrium pressure by an amount governed by the mass and velocity of the liquid column and the stiffness of the equivalent spring. In the case shown, the maximum pressure was five times the equilibrium pressure. The fluid column momentarily came to rest at the maximum pressure and then oscillated about the equilibrium position with decreasing pressure peaks due to the fluids viscous resistance.

Fig. 3 also illustrates the agreement between the pressures predicted by the analytical model and those measured by the transducer at the end of the tube. The measured periods of oscillation were for the most part lower than those predicted by the theory and the difference became larger with each oscillation. There are at least two effects that would contribute to this difference. First, the temperature of the compressed gas is elevated causing heat transfer to occur, continually reducing the effective polytropic exponent as time elapses. Second, liquid particles are torn from the leading edge of the fluid column and mixed with air during the period when the valve opening is much smaller than the tube. Leading edge particles of liquid are also torn away by the high speed motion of the interface. This air-liquid mixture will distort the analysis in proportion to the extent of the mixing due to the change of mass of the liquid column and the change of the characteristics of the trapped gas. In addition, as the column changes its direction the air-liquid mixture can separate from the column reducing its effective length. However, since the first pressure spike is of primary interest in this study and only order of magnitude indications are used to judge the importance of the effect, there was not sufficient motivation to attempt to refine the analysis to include these factors.

The effect of driving pressure on the magnification is illustrated in Fig. 4. Numerical solutions using polytropic exponents of 1.1 and 1.3 are both included on the figure. The experimental results lie within that range.

By reducing the length of the tube from 830 mm to 300 mm the magnifications were doubled as illustrated in Fig. 5. With the experimental method used in this study there was a practical limit to the reduction of tube length for each tube diameter. Whenever the pressure peak occurred before the valve had completely opened, the inaccuracies associated with measuring the low pressures at the beginning of the pressure rise became more significant. This made the agreement between theory and experiments more difficult than when the large constant pressure dominated the upstream boundary condition.

The period of oscillation of the fluid column can be seen in Fig. 6 to decrease with increasing driving pressure, leveling off at 4 MN/m^2. The experimental periods of oscillation were usually lower than those predicted by the theory. It is speculated that most of the gas-liquid mixture created at the leading edge of the liquid column would be separated from the column as it receded after each pressure pulse. This reduction in the portion of the column that moves would reduce the period of oscillation.

The data presented in these figures are typical of the agreements between theory and experiments for tube diameters ranging from 3 mm to 0.75 mm and tube lengths from 1.2 m to .15 m. It would have been desirable to test smaller tubes that would model micro-cracks. However, extensive alterations will be required to reduce the pressure transducer volumes so that they are negligible compared to the tube volumes before smaller tubes can be studied.

3.3 Traversing jet tests.

The hydraulic pressurization of a dead-end pore space by a moving water jet was experimentally studied using the equipment shown in Fig. 7. Capillary tubes of various lengths and diameters were placed in a vertical impingement plate with an open end exposed to the jet. A pressure transducer was placed at the closed end of the tube to measure the pressure in the trapped gas. The transducer voltage was recorded by a cathode ray tube chart recorder with chart speeds up to 4 m/sec. The 3.6 mm diameter Leach and Walker 3D nozzle was placed in a positioner that was mounted on a swivel plate that could be rotated in a horizontal plane. Pressures up to 34.4 MN/m^2 were supplied by a triplex pump with a flow rate of 2000 liters per minute.

At the outset of these tests, the pressure profiles at the impact surface produced by a stationary jet were compared to those produced by a traversing jet. This comparison was intended to determine if existing empirical descriptions for stationary jets could be used to represent sweeping jets. The stationary jet profile was obtained by positioning the nozzle so that the center of the jet coincided with center of the pressure transducer which was located in the center of the impingement plate. The nozzle positioner was then indexed by a known distance and the pressure recorded.

For the traversing jet impact pressure profile, the compound positioner was rotated about the swivel plate at various rotational speeds. The pressure produced at the impingement surface was recorded by a quartz crystal pressure transducer mounted in the center of the plate. Considerable care was taken to minimize the water filled volume between the impact surface and the transducer diaphragm. The jet traverse velocity was determined by measuring the angle of rotation of the swivel with the voltage output from a potentiometer that was connected by a small wheel and friction contact surface to the circumference of the swivel. The potentiometer voltage indicating the swivel position was recorded on one channel of the CRT strip chart recorder which moved at a known chart speed. The angular velocity and the distance from the center of the swivel to the impact surface were used to calculate the traverse rate of the jet. The traversing jet pressure profile was developed from the pressure versus time plot and the measured traverse velocity as the jet swept past the transducer.

During a hydraulic intrusion measurement the capillary being tested was cleared of liquid and the water jet was positioned to rotate in a plane containing the tube inlet. The jet was directed so as not to strike the plate and the pump was started. The CRT recorder was turned on and the nozzle was rotated so that the jet swept past the tube inlet. The transducer pressure at the end of the capillary and the swivel position were recorded on the same strip chart for each run. These measurements were made for a range of traverse rates and a variety of tube sizes and lengths.

3.4 Results-traversing jet tests.

A comparison between sweeping jet and stationary jet pressure profiles indicated significant differences in profile sizes as well as average centerline pressures. Therefore, it was concluded that the empirical profile shapes presented in the literature (Ref. 9) could not be used for sweeping jet studies. Average surface pressure profiles were determined for sweeping jets for each standoff distance, sweep rate and pump pressure. These profiles were used as the upstream boundary conditions in the analytical prediction of pressure magnification in the trapped gas.

A typical sweeping jet pressure profile is shown in Fig. 8. This profile was obtained for a standoff distance of 0.9 m, a pump pressure of 14 MN/m^2 and a range of traverse rates from 2.4 to 4.2 m/sec. The points presented in this figure were taken from six different traverse rates as indicated by the different symbols.

The range of traverse rates used in this study did not reveal a discernable trend in profile size, shape or centerline pressure. Therefore, an average shape and centerline pressure was chosen for each standoff distance and pump pressure. The curve represented by Eq. 7 was used to describe the average shape by determining the constants k and m. The exponent m was chosen to be 2.75 for all profiles and k was determined for the best curve fit.

The variation in pressure profiles exhibited in Fig. 8 is believed to be due to fundamental variations in the jet structure as a function of time as well as fluctuations in flow conditions produced by the triplex pump. For example, a cycle of pressure at the pump manifold illustrated in Fig. 9 shows that the pressure variation due to flow fluctuations was 4 MN/m^2. The pressure spike produced when each discharge valve opens was twice as large. A highly snubbed measurement of centerline stagnation pressure at a 0.9 m standoff distance and 27.6 MN/m^2 pump pressure indicated that a pressure fluctuation of 2 MN/m^2 was persisting through the 15 m flexible hose, the standoff distance and the snubbing tube. The variation in the centerline pressures for the range of traverse rates used in this study were of the same order. The value for k for the data in Fig. 8 varied from 3.7 to 50. The value of k for the best curve fit was determined to be 8.8.

Pressure magnifications produced in a 2.4 mm diameter tube, 152 mm long by a sweeping water jet at 0.9 m standoff distance and pump pressure of 13.8 MN/m^2 are presented in Fig. 10 as a function of jet traverse rate. The pressure magnifications were calculated by dividing the maximum pressure spikes in the trapped gas by the average centerline pressure at the impact surface, which in this case was 2.76 MN/m^2. A jet sweep rate of 1.5 m/sec produced a maximum pressure magnification of 28. A doubling of the traverse rate reduced the magnification by an order of magnitude. The theoretically predicted values of magnification are presented as the solid line curve in Fig. 10. These values agree satisfactorily with the averages of the experimental points.

The scatter in the experimental points is also predictable by the analytical model. In Fig. 11, the dotted line curve is the same as the theoretical curve of Fig. 10. The upper curve assumes a centerline pressure of 3.8 MN/m^2 and the lower curve a pressure of 1.7 MN/m^2 which was the maximum centerline pressure variation at these conditions. The calculated maximum pressure magnifications varied from 10 to 65. When the experimental data was non-dimensionalized by these extremes of centerline pressures, the magnifications varied from 6 to 45. In all cases, the optimum traverse rate agreed very closely with the predicted values.

Fig. 12 illustrates the effect of reducing the pump pressure by a half. The optimum traverse rate was reduced to less than 1 m/sec. At this flowrate, the centerline pressure variation was a larger percentage of the average value and the increase of scatter in the experimental results is evident. Again, the theoretical model predicts the optimum traverse rate and the average pressure magnifications within the expected tolerances.

Additional theoretical calculations were made to illustrate the effect of various parameters on the pressure magnification and optimum traverse rate. The dotted line curve in Fig. 13 is the same as the theoretical curve of Fig. 10. The solid line curve was calculated for the same conditions except that the tube length was half as long. For the third curve the length of the tube was halved again. The maximum pressure magnification almost doubled for each reduction in length. However, this effect is reduced as the diameter of the tube is reduced as will be illustrated below.

Fig. 14 illustrates the effect of tube diameter while the tube length was held at 152 mm. By reducing the diameter to one third the original value the maximum pressure magnification was reduced by a factor of 6. When the diameter was reduced to one tenth the original size, the maximum pressure magnification was reduced by a factor of 15.

Fig. 15 shows the effect of jet centerline pressure on the pressure magnifications in small diameter tubes (.2 mm) with lengths of 152 mm. Doubling the centerline pressure only increased the maximum pressure magnification from 1.8 to 2.2. Doubling the pressure again increased the magnification to 2.8.

The effect of tube length was also calculated for the smallest tube diameter (.2 mm) of this study. Fig. 16 illustrates that by reducing the length by a factor of 4, only doubled the pressure magnification. These calculations were made for a centerline pressure of 2.76 MN/m^2. Higher pressures will produce correspondingly higher magnifications. However, the current range of pressures used in hydraulic mining would not be expected to produce magnifications in excess of 10 for tubes in this size range.

4. CONCLUSIONS

A theoretical model has been developed that describes the pressurization of a dead-end capillary by a sweeping water jet. The model agrees with the experimental results within predictable tolerances over the range of parameters utilized in this investigation.

Experimental results and the theory have been used to show that pressure magnifications can be maximized by finding the proper traverse rate for a particular capillary diameter and length. As the tube gets smaller and longer the potential for large magnifications decreases. For example, the smallest tube theoretically modeled in this investigation (diameter = 0.2 mm, length = 38 mm) indicated a maximum pressure magnification of only 3.5 for a centerline pressure of 2.76 MN/m^2 and a traverse rate of 6 m/sec. For cavity dimensions in the micro-fracture range, this mechanism of producing large internal hydraulic pressure pulses does not appear to be very promising. However, for the large macro-fracture in which gas can be trapped, pressure pulse can be produced by a sweeping jet that exceed the impact pressure by orders of magnitude.

It is anticipated that this model can be used for future studies in which additional features of a porous matrix such as gas release, tortuosity and interconnectivity are added.

5. REFERENCES

1. DuPlessis, M. P. and Hashish, M.: "Experimental and Theoretical Investigation of Hydraulic Cutting of a Western Canadian Coal." Fifth International Symposium on Jet Cutting Technology, BHRA Fluid Engineering, June 1980, Paper E3, pp. 269-286.

2. Rehbinder, G.: "Some Aspects on the Mechanism of Erosion of Rock with a High Speed Water Jet." Third International Symposium on Jet Cutting Technology, BHRA Fluid Engineering, May 1967, Paper E1.

3. Crow, S. C., Lade, P. V. and Hurlburt, G. H.: "The Mechanics of Hydraulic Rock Cutting." Second International Symposium on Jet Cutting Technology, BHRA Fluid Engineering, April 1974, Paper B1.

4. Blick, E. F.: "Capillary-Orifice Model for High Speed Flow through Porous Media." I and EC Process Design and Development, Vol. 5, No. 1, January 1966, pp. 90-94.

5. Chen, N. H.: "An Explicit Equation for Friction Factors in Pipes." Ind. Eng. Chem. Fundamentals, Vol. 18, No. 3, 1979, pp. 296-297.

6. Graze, H. R.: "The Importance of Temperature in Air Chamber Operations." First International Conference on Pressure Surges, BHRA Fluid Engineering, September 1972, Paper F2.

7. Parmakian, J.: Waterhammer Analysis, Dover, 1955.

8. Evans, W. E. and Crawford, C. C.: "Design Charts for Air Chambers on Pump Lines." Trans. ASCE, Vol. 119, 1954, pp. 1025-1045.

9. Shavlovsky, D. S.: "Hydrodynamics of High Pressure Fine Continuous Jets." First International Symposium on Jet Cutting Technology, BHRA Fluid Engineering, April 1972, Paper A6.

10. Martin, C. S.: "Entrapped Air in Pipelines." Second International Conference on Pressure Surges, BHRA Fluid Engineering, September 1976, Paper F2.

11. Graze, H. R.: "A Rational Thermodynamic Equation for Air Chamber Design." Third Australian Conference on Hydraulics and Fluid Mechanics, The Institution of Engineers, Australia, November 1968, pp. 57-61.

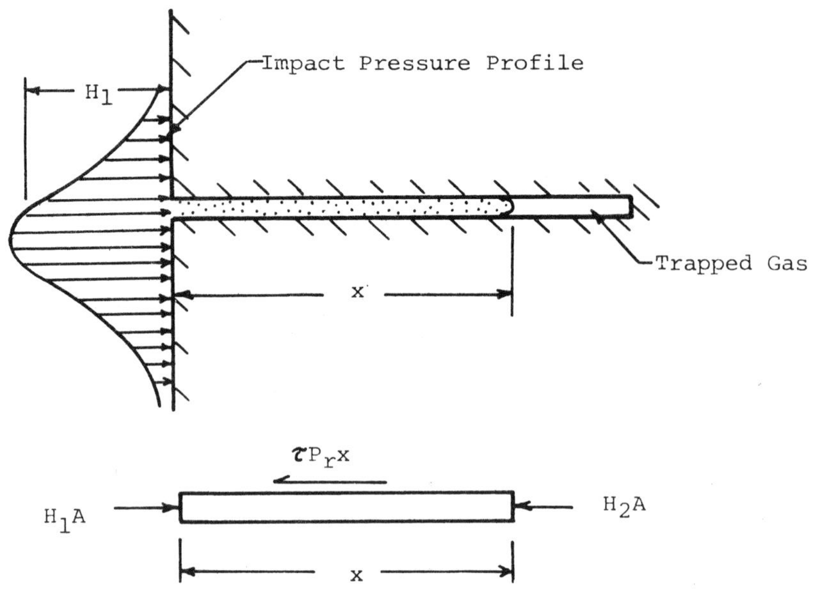

Fig. 1. Forces on Liquid Column

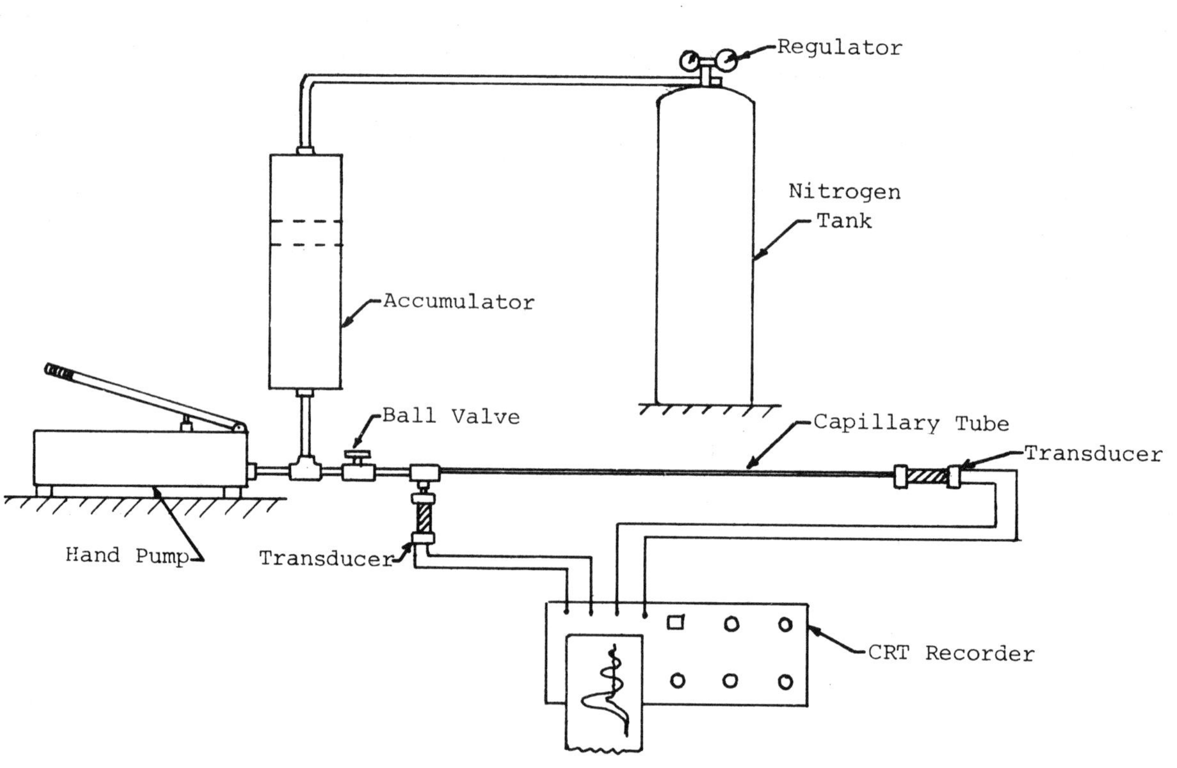

Fig. 2. Constant Driving Pressure Test Apparatus

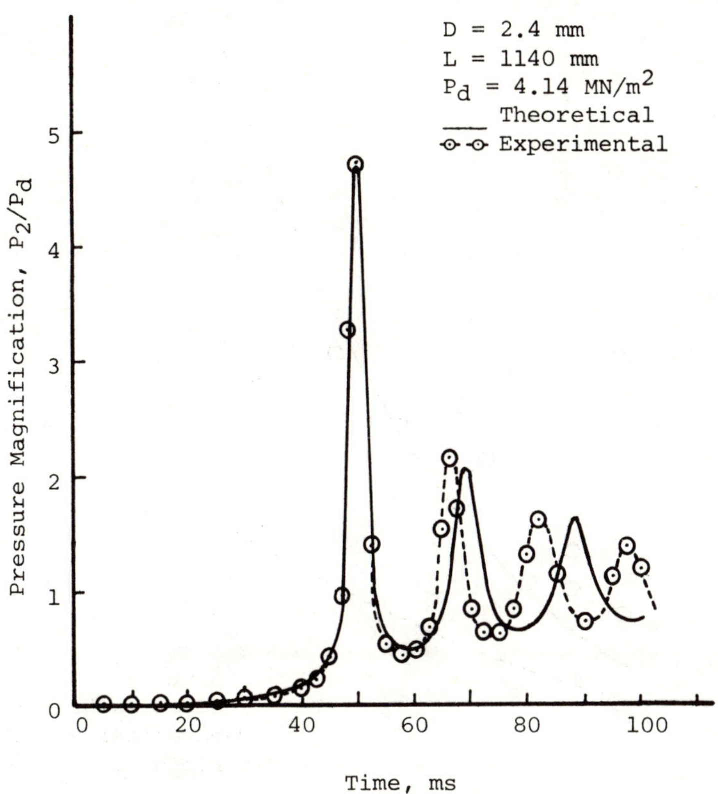

Fig. 3. Pressure Pulse-Constant Driving Pressure

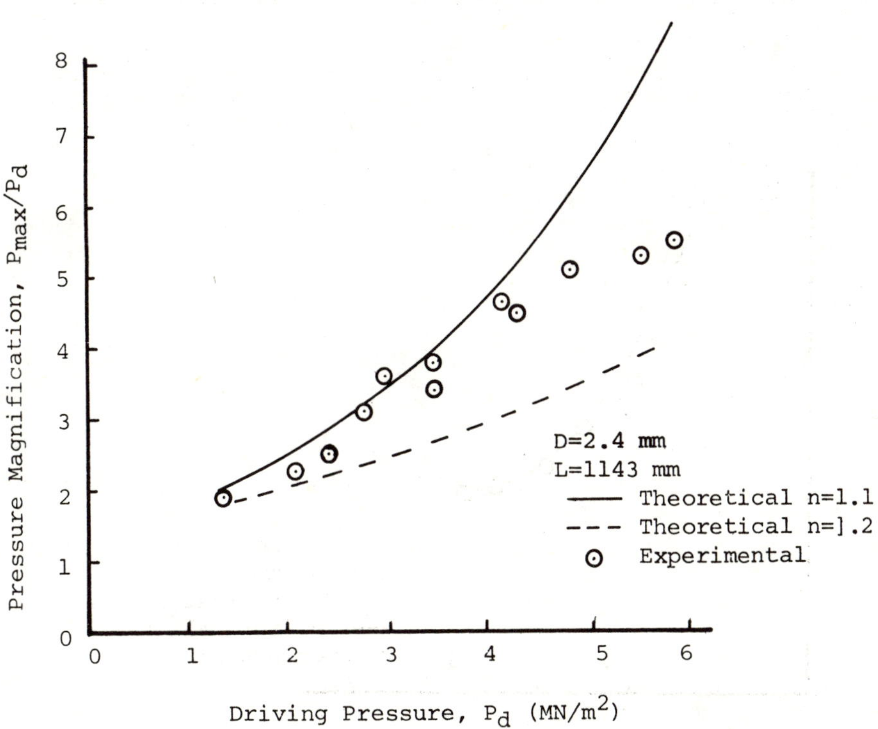

Fig. 4. Pressure Magnification versus Driving Pressure

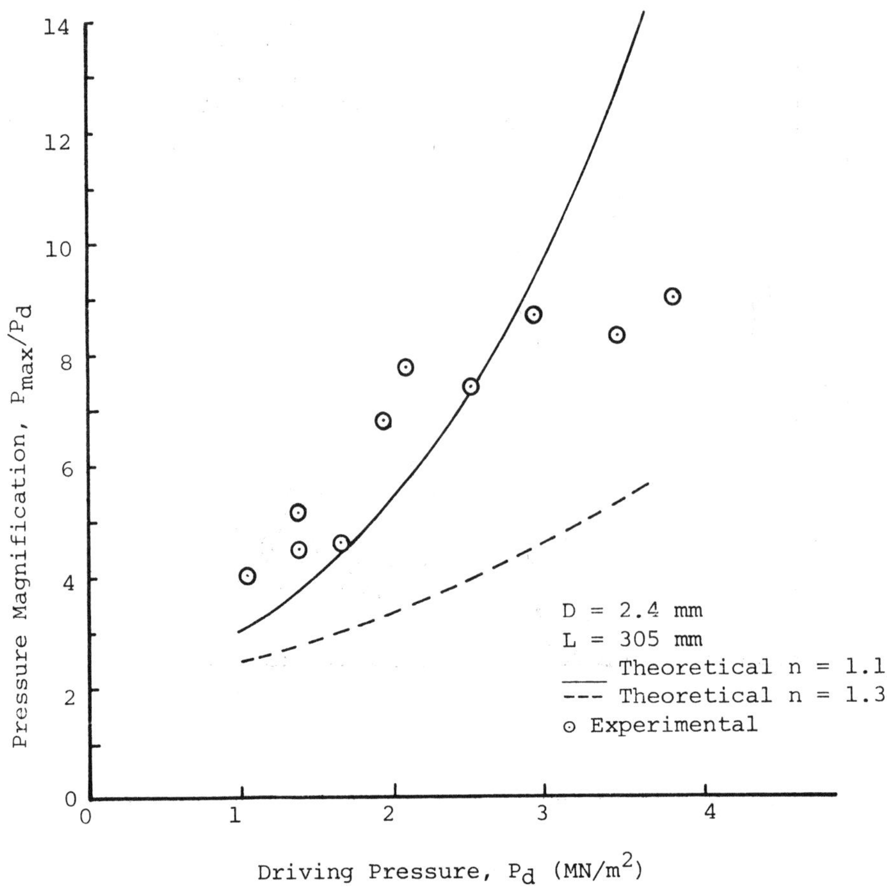

Fig. 5. Pressure Magnification versus Driving Pressure

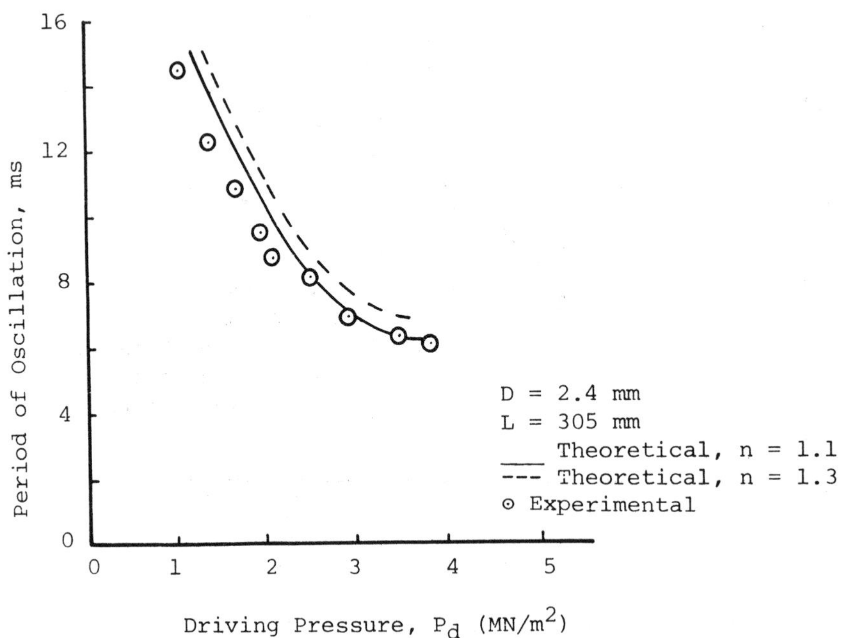

Fig. 6. Period of Oscillation versus Driving Pressure

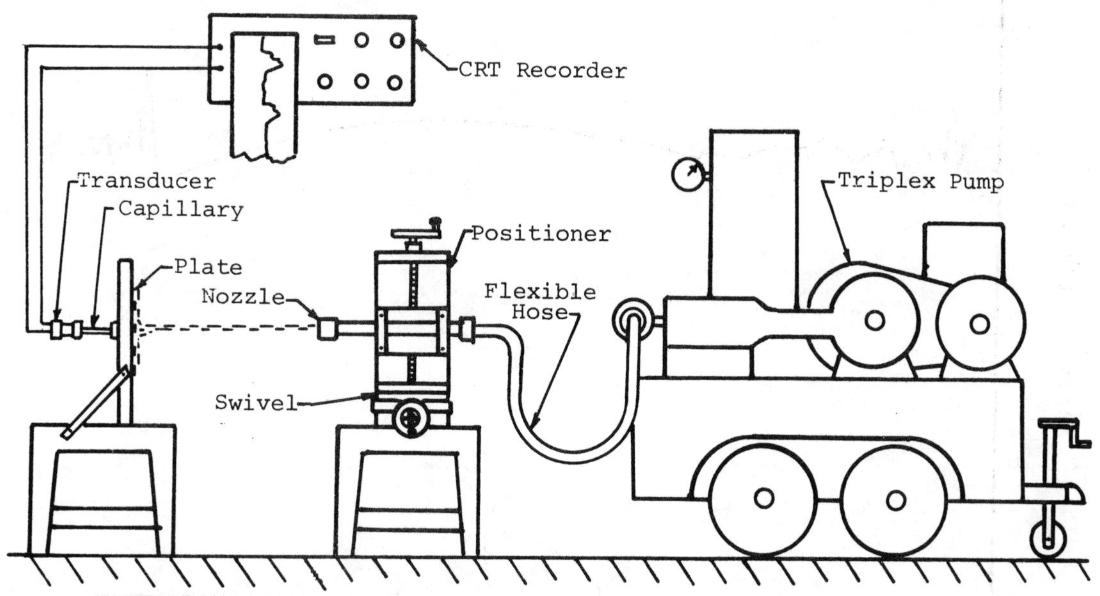

Fig. 7. Sweeping Jet Test Apparatus

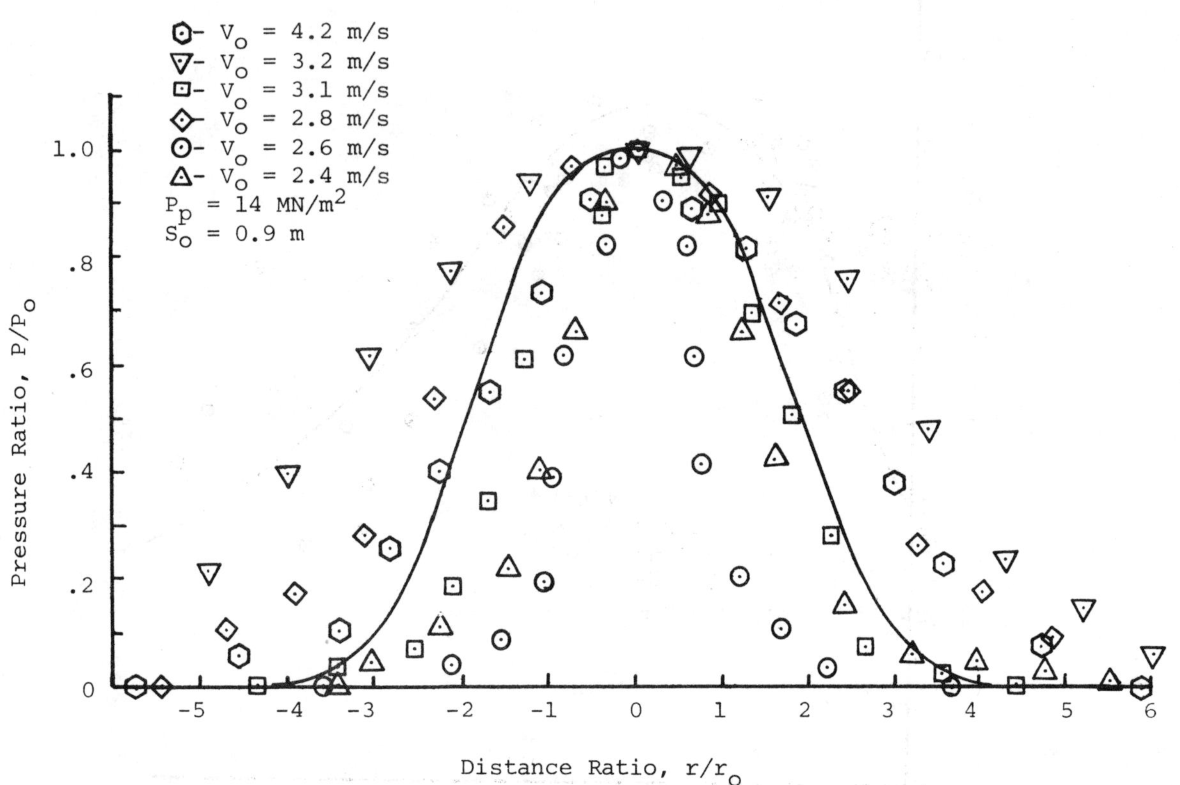

Fig. 8. Pressure Profile for a Sweeping Jet

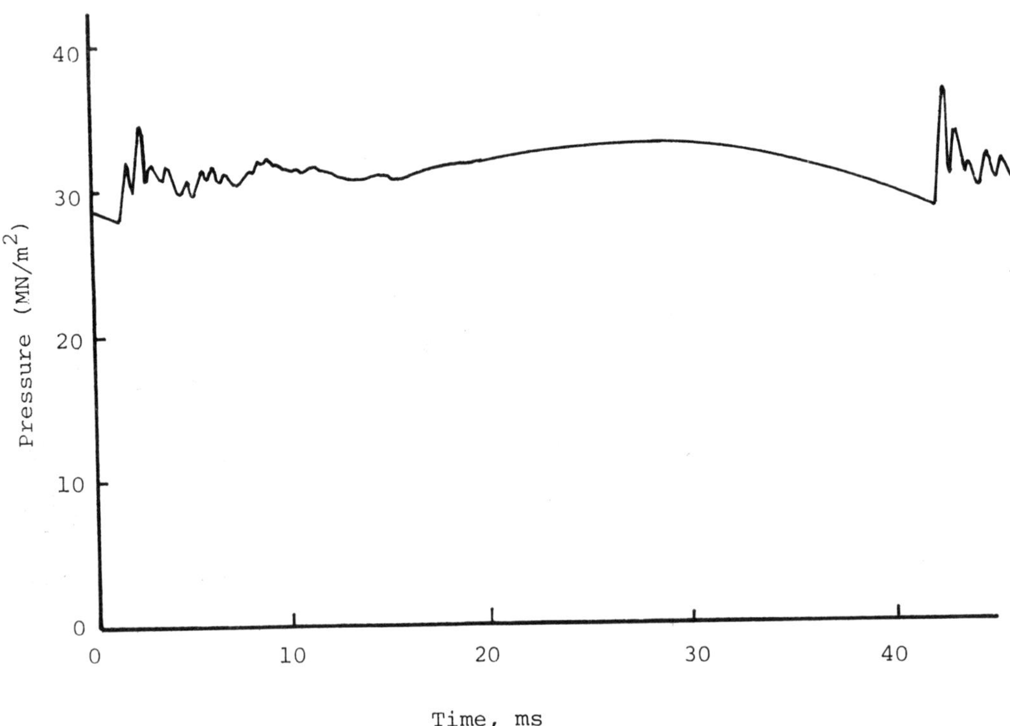

Fig. 9. One Cycle of Pump Manifold Pressure

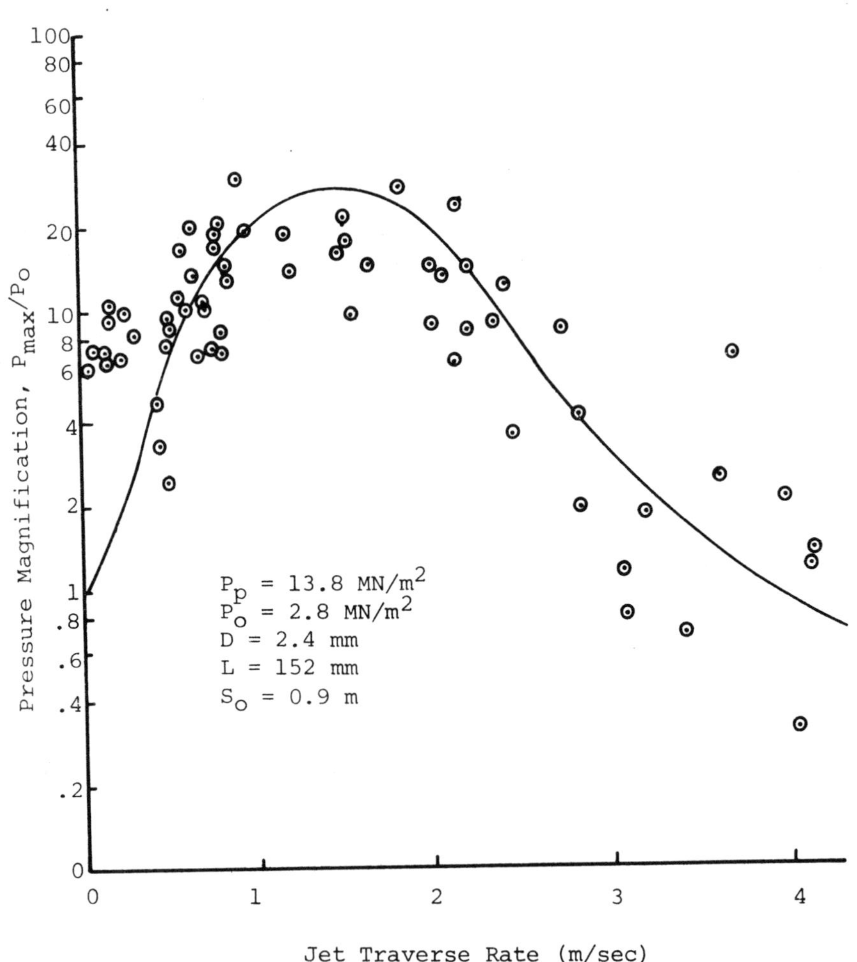

Fig. 10. Magnification versus Traverse Rate

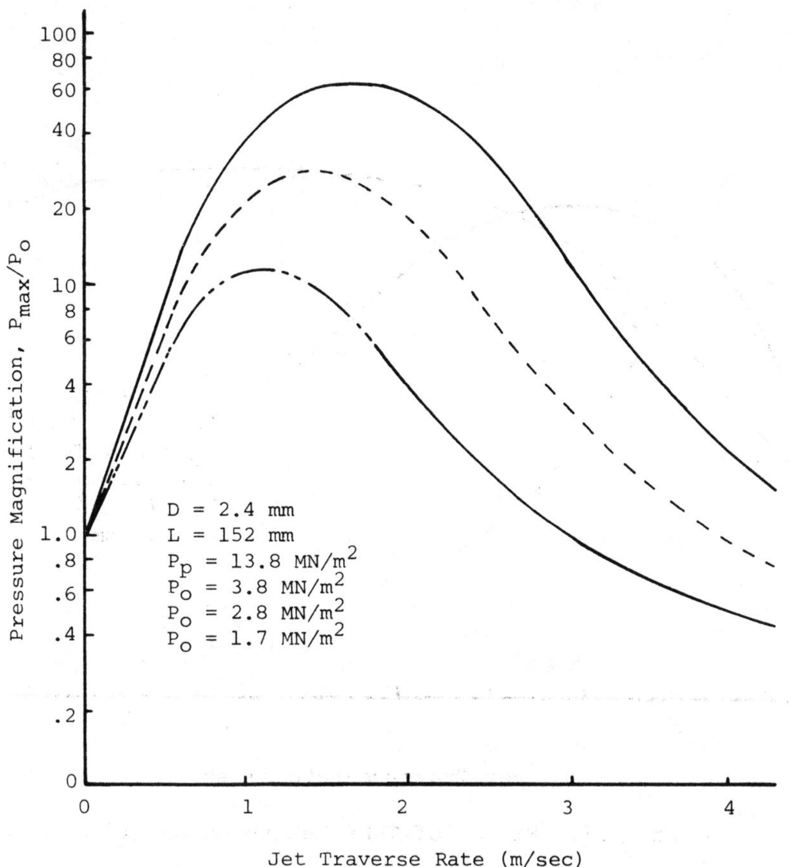

Fig. 11. Effect of Jet Centerline Impact Surface Pressure

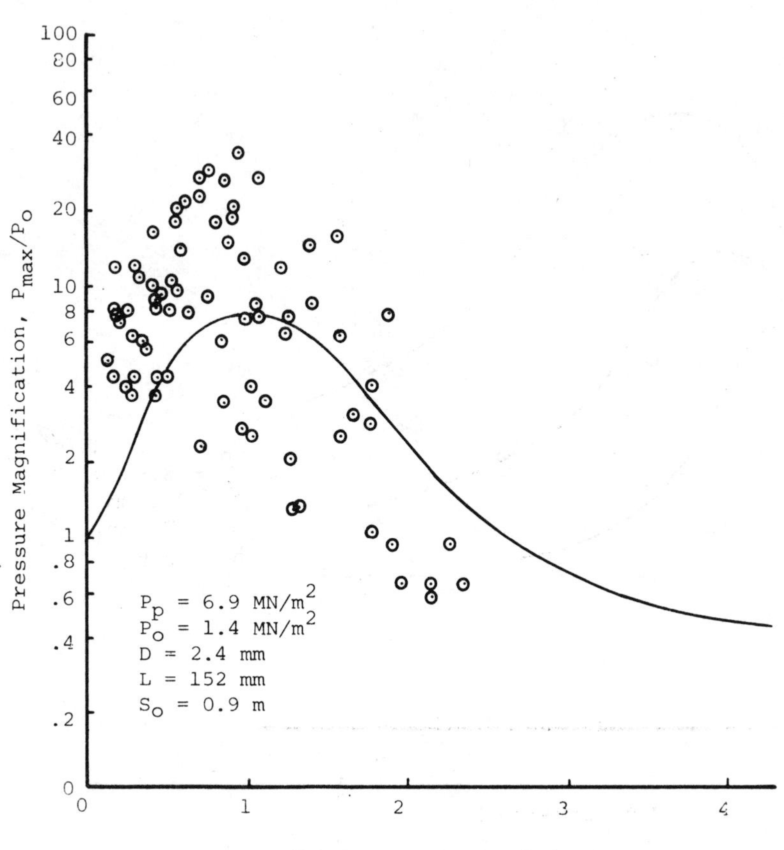

Fig. 12. Magnification versus Traverse Rate

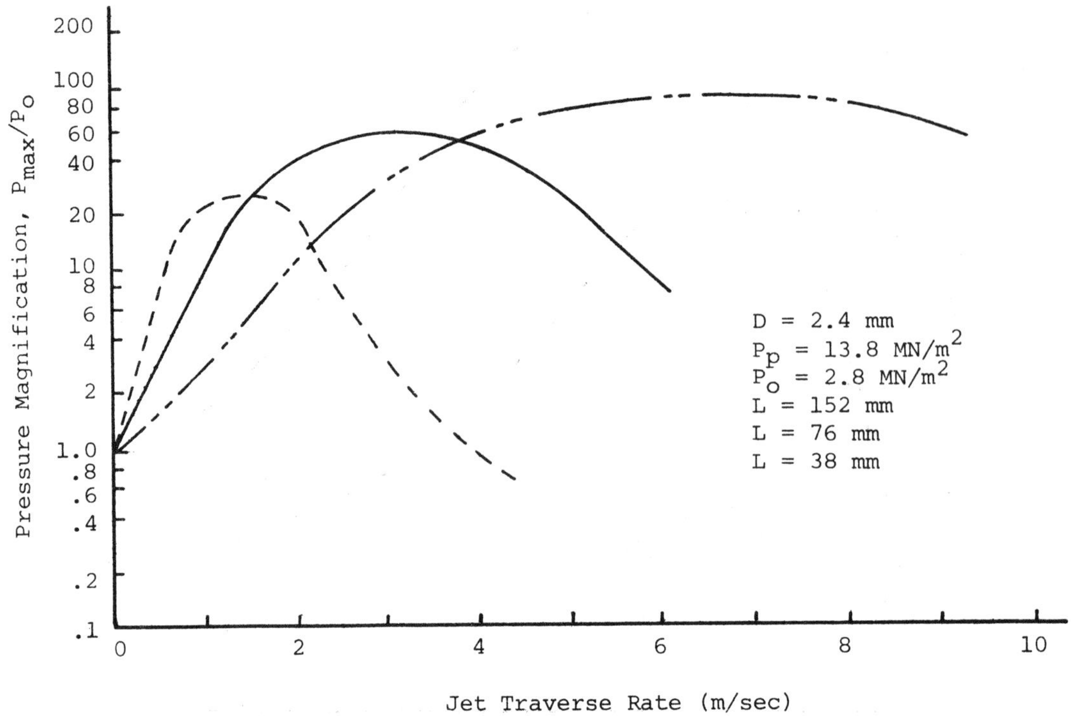

Fig. 13. Effect of Tube Length on Magnification

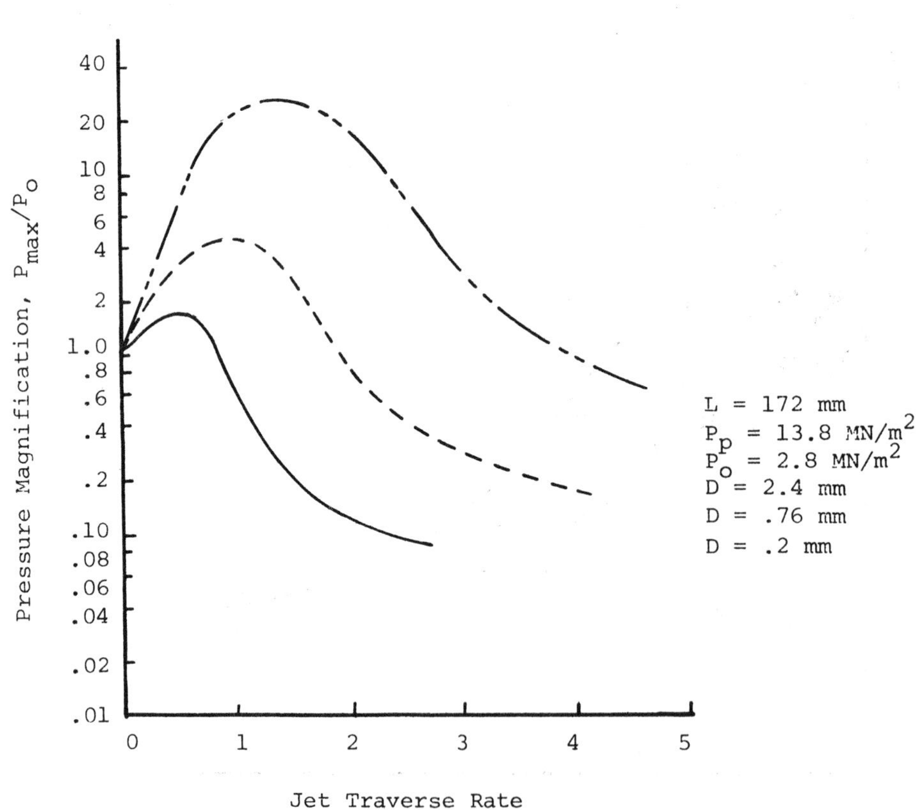

Fig. 14. Effect of Tube Diameter on Magnification

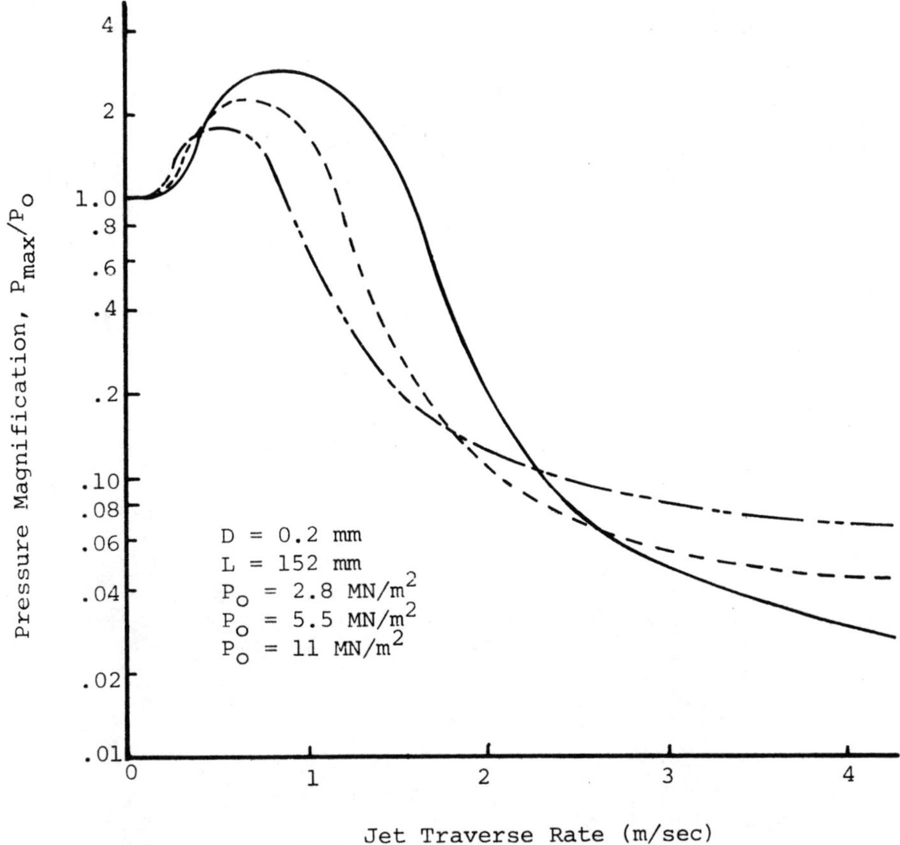

Fig. 15. Effect of Jet Impact Pressure for Small Tubes

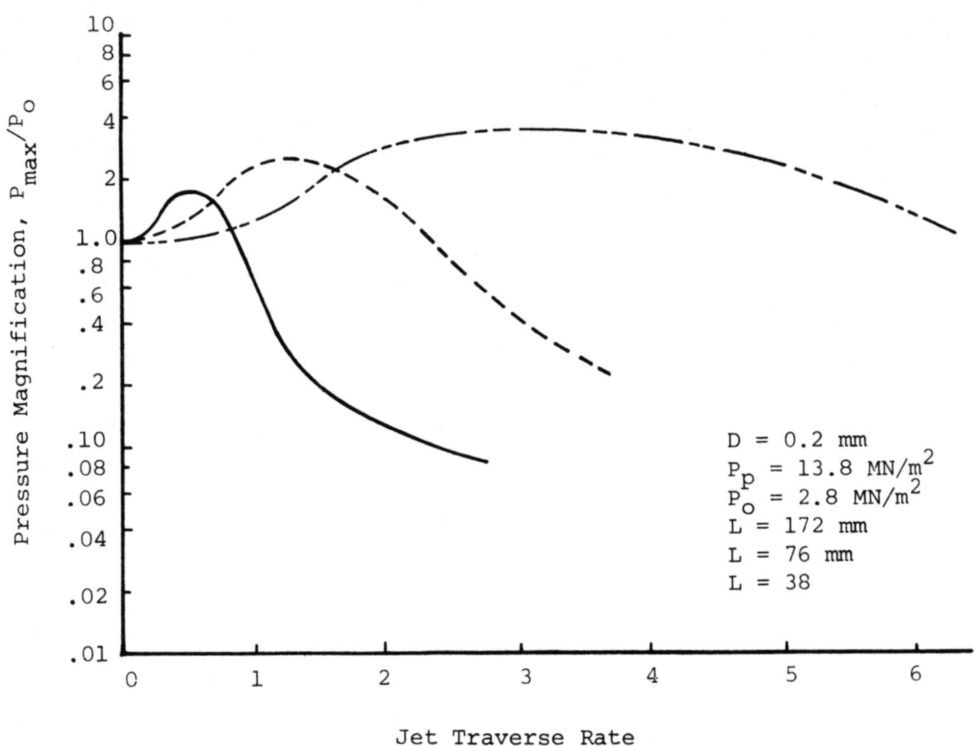

Fig. 16. Effect of Tube Length for Small Tubes

6th International Symposium on
Jet Cutting Technology
6-8, April, 1982

FUNDAMENTAL ASPECTS IN CLEANING WITH HIGH-SPEED WATER JETS

H. Louis and W. Schikorr

Universität Hannover, F.R. Germany

Summary

The use of high-speed water jets for cleaning applications is increasing. Surface of e.g. steel or concrete constructions are cleaned from different types of layers.

These layers may consist of e.g. rust, oil, paint, rubber, organic layers like mussels on undersea constructions or combinations of these. Up to the present the parameters for jet cleaning have been optimized empirically.

To optimize the cleaning conditions, it is necessary to classify the different types of layers and to investigate the interaction between jet, layer and substrata. One the one hand one can differentiate between brittle, ductile and elastic layers as well as between firm and infirm adhesivness to the substrata.

On the other hand the character of a jet varies with the change of pressure, distance and type of nozzle. The loading time or the traverse speed of the jet and the number of traverses are further parameters. The main demand on every jet cleaning application is the most efficient removal of the layer without damage to the substrata.

Held at the University of Surrey, U.K.
Symposium organised and sponsored by
BHRA Fluid Engineering

I. Introduction

The application of high-speed water jets for cleaning purposes was already mentioned in the first third of this century (1). The Co. Lanz in Mannheim used pressure water jets to clean cast machine parts from casting sand. The maximal pressure was 75 bar.
With modern pump constructions and better materials the pressure plants became more durable and within the last 15 years the use of high-speed water jets for many purposes spread enormously. The jets generated by pressures up to 4000 bar and more are used as cutting tools for many materials like asbestos, reinforced plastics, leather, rubber etc. Other applications are in mining e.g. drilling of bore-holes or for supporting mechanical cutting and impacting.
But in most cases high-speed water jets up to 1000 bar are applied for cleaning purposes. Often it is more economic to use water jets instead of mechanical tools like brushes or scrapers for cleaning periodically, especially if it has to be done during maintenance- or reparing-works, as described e.g. in (2). Moreover, the use of water jet blaster for cleaning is perceived to be less displeasing than using manual tools. But inspite of the wide spread use of the water jet cleaning in most cases cleaning parameters are found empirically. Often it is left to the experience of the man serving the high pressure pistol to get satisfactory cleaning results.

II. What is jet-cleaning and where it is used today?

To give a definition of jet-cleaning one can say: jet-cleaning seperates one or more materials from the surface of other materials using high-speed water jets (from 200 m/s up to 450 m/s).
The cleaning mechanisms depend, as later described, on the material behaviour of layer and substratum and on the interaction of both with the jet.
The possibilities of occurance of different layers and substratums are very frequent.
Several authors (Ref.) have described the using of jet cleaning for many different purposes, e.g.:
marine applications (4), (11), applications in chemical industry (13) or other industrial applications of high-speed cleaning jets.
The following shows some of the applications of water jets for cleaning purposes.

1) Removal of layers composed of grease, oil and dirt
2) Cutting-off the bark of tree-trunks in saw-mills
3) Jet-cleaning of paint-layers e.g. on the parts of a varnishing-plant
4) Cleaning of concrete-surfaces (rubber-layers on runways, damaged or worn building walls)
5) Removal of rust from steel-constructions
6) Jet-cleaning of the surfaces of hot rolled steel during the rolling process (breaking the oxides)
7) Cleaning of surfaces from growth (e.g. mussels, barnacles or sea-weed)
8) Cleaning of reaction vessels or tubes in chemical or other plants
9) Stripping down the rubber layer from rubber-roller
10) Removing the photosensitive layer from the cylinder of copy-apparatus
11) Preparation of concrete or asphalt road-surfaces
12) Stripping the insulating layers from conductor-profiles before soldering in the manufacturing of electric-engines
13) Removal of adherend forming-material from cast parts
14) Deburring and cleaning of aluminium castings in the automobile-industry
15) Removal of high-contaminated concrete-layers during the pulling down of nuclear plants.

The following list shows possible substratums and layers:

Substratum	Layer	Remarks
wood	bark paint grease-oil-mud growth	saw-mills marine buildings, ships
asphalt	asphalt grease-oil-mud plastics	regeneration of road-surfaces route-indicators, road-marks
plastics	grease-oil-mud plastics	
rocks, stones	grease-oil-mud weathering products growth	 cleaning of buildings
concrete	concrete plastics grease-oil-mud paints rubber growth ice	regeneration of road surfaces worn layers on the outer walls of buildings rubber abrasions on run-ways
metals	grease-oil-mud growth rust oxides paints metals	 marine steel buildings roller-mill deburring

III. Description of characteristics of substratums-layers-adhesion

a) Substratums:

1. Metals show from middle up to very high tensile and compression strength. As a result of their structure metals are liable to be deformed. Most metals suffer a high rate of deformation but there are also some very brittle metals like zinc.
   The undamaged metal does not show cleavages. Under stress of alternating and dynamic loads metals show the so called "Wöhler-behaviour". The movement and the accumulation of vacancies and dislocations cause an increase of inner damages and brittleness and lead to fracture far below the static tensile strength. A further characteristic of metals is corrosion. In connection with an electrolyte like water, electrochemical processes take place and metal-ions are set free. Often the corrosion process leads to a layer that may stop further corrosion if the layer prevents the contact between electrolyte and metal. But if the layer is porous, further corrosion may continue, e.g. the corrosion of iron with growing of rust. On their own metals are not able to absorb water.

2. Rocks and concrete are materials with similar properties. They show a good compression strength but a bad tensile strength. The crystall-structure of rocks does not allow any plastic deformation. They are brittle. Concrete is an artificial compound-material from splitted rocks in cement matrix. Normally rocks are very inhomogenous. They show many micro- and macro-cleavages. The degree of absorption of water is dependent on the kind of rocks. Rocks may wear by weathering.

3. Other substratums are e.g. wood, asphalt and plastic materials. Normally they show a less tensile- and compression strength than rock or metal materials. They are often easy to deform but in some cases they are brittle. There is no tendency to corrosion. but often a strong dependency of properties on temperature.

b) Layers:

1. Anorganic layers like rust, oxides, scale, carbon, ceramics, enamel are normally hard and brittle. Sometimes they consist of more than one layer. Because their brittleness tensions may not be reduced by plastic deformation. If there is a different coefficient of temperature-expansion to the substratum, a change of temperature may cause a lot of tears in the layer. The anorganic layers may not be soluble in water but often absorb water as a result of the layer's porosity.

2. Organic layers are divided into natural grown layers and artificial organic layers.

2.1 Natural grown layer-properties range from soft (e.g. moss, algal growth) to middle (e.g. bark) and to very hard (e.g. mussels, barnacles).
Because these layers partially consist of living organisms with anorganic components (e.g. mussel-chalk) the properties change rapidly with different water-percentages. Anybody who has cleaned the hule of a boat or a ship would confirm that this has to be done as soon as the vessel is out of the water. Otherwise, the layer will dry out and adhere much faster.

2.2 Artificial organic layers are produced with the most widespread mechanical properties. According to the usage of these layers the mechanical properties range from viscous to ductile, from rubber-elastic to brittle. Because these layers are developed from plastics or elastomer-materials, there is often a water absorption of some percentage. Often these layers are filled with anorganic materials for deminishing costs and improving the mechanical properties. With rising temperature the hardness of these layers decreases and the deformation increases with some materials in a wide range.

c) Mechanisms of adhesion (18):

1. mechanical adhesion
The layer is mechanically fixed in kerfs, pores and roughnesses of the substratum. The rougher the effective surface is, the higher is the degree of adhesion of the layer. The possible adhesion strength reached is relatively low.

2. specific adhesion (Van der Vaals-forces, side group-reactions)
The specific adhesion is caused by effects of polarisation and by electrostatic forces between substratum and layer-material. The adhesion may be of middle strength (200 N/mm^2).

3. Chemosorption (dissociation-forces)
   Adhesion by chemosorption happens e.g. with epoxi-resin coated metals. To get a chemosorption it seems necessary to activate the metal-surface to cause an emission of free electrons just before and a chemical reaction of the coating material (cross-linking) during the coating process. The reached adhesion strengths are very high (theoretical 5000 N/mm^2).

   In normal cases of jet-cleaning adhesion mechanisms are mainly the mechanical and the specific adhesions, seldom the chemosorption. Chemosorption may be found at wanted layers with high adhesion strength (e.g. layers of epoxide-resin, enamel, ceramics).

IVa. Description of free water jets

To generate cleaning water jets, generally two kinds of nozzles are used:

a) round-jet nozzles
b) flat-jet nozzles.

They are used for different purposes. The jet out of a round-jet nozzle is rotation-symmetrical. It has been tried to minimize the divergence of the round-jet to reach a good efficiency at a greater distance from the nozzle. Contrary to this, the flat-jet diameter enlarges only in one direction, depending on the nozzles geometry. The diverence orthogonal to this direction is less. Flat-jets are used in smaller distances from the nozzle to load large areas. Different authors have published papers on the behaviour of free-jets (20), (21).
The transition between laminar and turbulent flow is found at Reynold-numbers of 2300. The Reynolds-number at the nozzle's mouth is very high, e.g. the water-jet out of a round-jet nozzle is of 1.5 mm diameter and has an outlet velocity of 300 m/s, characterized by a Reynold-number of 450000.
It means, that the free-jet is fully turbulent.
The very high rate of turbulence causes statistic components of velocity orthogonal to the jet's direction of some percentage of the jet's velocity. This will lead to a spreading of the jet (also if the jet moves in a vacuum).
Additionally the friction to the ambient medium will have an intensifying influence on the jet's divergence.
The turbulent free-jet coheres only in a short distance from the nozzle.
The jet disintegrates into departed volumes of water and with greater distances from the nozzle into even smaller drops. The relation between surface to the volume of the fluid increases, which means, that the retarding effect by the ambient medium also increases.
The special geometry of the flat free-jet leads to a more rapid disintegration into small fluid particles than the round-jet. The diminishing of jet-velocity is more rapid, which means, that the effiency is limited to smaller nozzle-distances.

IVb. Loading of solid surfaces by impinging jets

a) If a continous jet flows orthogonal onto a plane surface it will be deflected and flows away from the center of impinging parallel to the surface.
   The loaded area has to bear a force which may be calculated by:

$$F = A \cdot \rho \cdot v^2$$

   with: A - loaded area
   ρ - density of jet-fluid
   v - jet-velocity.

Additional to the normal force, the surface is loaded by a shearing force in the direction of the fluid stream parallel to the surface.

If the material has been deformed by the jet or the jet flows into an existing kerf, it is deflected along the walls of the kerf. In the extreme case, the direction of flow turns at an angle of 180 degree.
The momentum to the surface has doubled and so has the force to the loaded area.
In the kerf the moving direction of the water changes and there will be a point where a great part of the kinetic energy of the jet will be changed into pressure.
This pressure causes a tensile-strength in the ground of the kerf. If it is high enough a crack will open. Additionally the walls of the kerf are loaded by shearing stress by the high-velocity flow.

b) The continuous impinging jet is an idealized case. Normally the jet has been disintegrated in more or less small fluid particles. If a fluid particle impacts a solid surface, one may calculate an impact pressure lasting for a very short time, using the following approximation:

$$p_{impact} = \rho \cdot c \cdot v$$

with ρ - density of the jet-fluid
c - sound velocity in the jet-fluid
v - velocity of the jet.

These impact pressure peaks may be several times higher than the steady-state pressure which is given by:

$$p_{cont} = \frac{1}{2} \rho \cdot v^2.$$

Materials that are not damageable by a continuous jet may be eroded by a stream of impacting drops. The fig. 1 shows the loading of a surface by different kinds of impingements. Another possibility for producing discontinuities in a jet and thus getting impact pressure peaks are gas-bubbles or cavitation in the jet (11).

c) Influence of temperature
The mechanical properties of plastic materials and other layers depend on the temperature. Fig. 2 shows the changing of the shear-modulus versus temperature. Layers consisting of oil or grease are easy to clean if the jet-temperature is high enough. Often the jet fluid is heated up before it leaves the nozzle. Combined water and steam jets are described in (7). A rough estimate of the jet's warming up by the power-loss in a normal nozzle gives a temperature increase of far below 10 K and may be neglected. But at the moment of the jet-impinging on the surface, a great part of the jet-power is changing into heat. Only a part of the jet's energy is used for the cleaning process and the removal of separated particles. In (19) is shown that there is an important temperature increase of the target material during jet-cutting.

V. Characterization of some typical layers

For a systematic investigation of the cleaning behaviour of substratum-layer compounds it is significant to classify all occuring types into three groups:

1. - hard, brittle layers
2. - ductile layers
3. - viscoelastic to viscous layers.

To simplify the classification the substratum may always be a material with mechanical properties much higher than the layer's itself, e.g. steel.

A further point of differentiation is the adherence to the substratum. It may range from very high to very low strengths.

a) To clean s substratum from a hard and/or brittle layer (anorganic layers, scale, ceramic, enamel) it is necessary to produce cracks in the layer. When two or more cracks are crossing a particle will break out.
In brittle materials, cracks are running with very few plastic deformation; i.e. a started crack uses a relatively small amount of energy for crackgrowth.
If there is only a static load, brittle materials are very resistent. There has to be a certain amount of dynamic (impact) in the loading to open the first crack.
To get a homogenous erosion, it is convenient to produce many short cracks in the layer and so break out many small particles.
To get a good efficiency, the number of cracks has to be small, the length of cracks must be long and so only few but bigger particles are broken out for the same volume of eroded layer.
If the adherence is optimal, a crack is stopped when reaching the substratum. The length of cracks is limited by the thickness of the layer.
If the adherence between layer and substratum is not optimal, i.e. the adherence is less than the adherence of the layer material itself, the crack may spread between layer and substratum. The layer is peeled off.

b) The second group consists of ductile layers (most plastic materials, kinds of rubber and metals).
Metals are found to be very brittle to very ductile. Ductile materials are characterized by plastic deformability. With increasing load, an elastic elongation is followed by a plastic elongation.
The growth of cracks in ductile materials is joined to a strong plastic deformation in the tip of the crack. These plastic deformations need a lot of energy. In ductile materials the opening of a crack and it's propagation needs similar loadings. While in a brittle material with the same tensile strength the load for crack-opening is the same, the crack propagation in a ductile material needs more energy than in a brittle material.
One may define a coefficient of crack-growth Z as the quotient of energy for crack-opening divided by energy necessary for crack-propagation ($Z = E_{opening} / E_{propagation}$).
For very brittle materials where an opened crack runs almost alone, Z is very high. The more ductile a material behaves, the smaller values Z will accept.

Influence of loading-velocity:
The material behaviour of ductile materials (metals and plastic materials) depends on the velocity of loading. Generally, with higher velocities a ductile material will get more brittle. The bearable strength will rise, but the elongation until fracture will decrease rapidly. If a notched bar impact test is carried out with different impact-speeds, a greater part of brittle fracture will be found at higher speeds. It means that the coefficient Z may change to higher values with higher loading-speeds.
Therefore the needed energy for destruction of a ductile material may be diminished with higher loading-speeds.

c) The third group consists of viscous and viscoelastic layers (bitumen, chlorine-caoutchouc, oil-grease).
These layers behave as very viscous fluids. A pure elastic range does not exist (except at very low temperatures). Already a slight loading causes an irreversible deformation. A quasi-static load is enough to percuss the layer to the substratum. Shear stresses applied on the surface displace the layers. The cleaning process is aided if the wetting of the substratum by the layers is bad. The mechanical properties of these layers in most cases depend very strongly on the temperature. With an increase in the temperature of the jet an increase in cleaning results is expected.

VI. Optimization of the cleaning process

Principally, for almost every cleaning purpose it is assumed that the layer has to be removed without damaging the substratum. In many cases the substratums are metals. Metals show under load of high-speed water jets (dynamic load) typical incubation periods. The macroscopic perceptable damaging starts after a loading time that depends on pressure, nozzle, distance and on the material itself.
The reason for this is the "Wöhler-behaviour" of metals for dynamic load. The higher the dynamic load (below breaking load) the less is the number of bearable stresses. Layers often show no "Wöhler-behaviour" when loaded by water jets. The load is too high, i.e. the incubation-period until damaging is very short. These significant differences in incubation times are used for every jet-cleaning purpose.
The first step in an optimization is the setting up of boundary values and the second step is the optimization under the given conditions.
To demonstrate the possibility of optimization in jet-cleaning, an example of disisolation of copper work-pieces from a plastic-material by water-jets is given below.
The first step was the separate loading of both materials. The principal mass-loss versus time curves at a special pressure, nozzle and distance are shown in fig. 3.
For further experiments a loading time has to be chosen where copper is not yet being damaged, but a satisfactory removal of plastic-material is expected.
The next step is the test of pressure-dependence of mass-loss as shown in fig. 4.
It is significant that the first mass-loss of copper takes place at a pressure, where the plastic material already shows a strong damaging. The loading time for this test was longer than the incubation period of copper in the mass-loss versus time graph. The loading time should be orientated to the expected later cycle-time in the production line.

The later nozzle-pressure has to be lower than the found "incubation-pressure" of copper.
A further parameter for the cleaning process is the nozzle's distance from the surface being cleaned.
The mass-loss versus nozzle-distance has a maximum (dependent on pressure and nozzle geometry) and so has the plastic material.

In the following graph such mass-loss versus distance curves with the pressure as a parameter are schematically drawn up for copper and the coating plastic-material in fig. 5.
To have a good cleaning result there have to be maximal differences in mass-loss. The graph is parted into five regions.

Region I is not suitable for this cleaning purpose. There is no mass-loss of copper, but the coating-material too shows only very few mass-loss.

Region II: In this region a cleaning process may take place. The copper has not yet been damaged but there is a satisfactory mass-loss of the plastic-material;
jet: quasi-coherent, high velocity but less dynamic

Region III: This is a forbidden region. The mass-loss of copper has its maximum.

Region IV: Similar to region II a cleaning process may take place. The jet has spread enough to damage only the plastic material but not the copper;
jet: diverged, high dynamic, but less velocity.

Region V is uninteresting. The jet has not enough power anymore to damage the coating.

The considerations made on jet cleaning optimization are interesting for layers that are very tight and fast e.g. layers of epoxide-resin, enamel, ceramic or metal-layers or if the substratum must not be damaged.

If a copper conductor has been damaged at its surface when being jet-disisolated, the kerfs in the surface may cause a fatigue fracture after a soldering in an electric engine and this may have fatal consequences.

If there is an automatic cleaning-plant with always the same charging then for achieving a good efficiency to optimize the cleaning process it is reasonable to proceed in the described form.

VII. Conclusion

Some fundamental considerations about cleaning processes by high-speed water jets were made.

Because there are a lot of different cleaning problems, a systematic optimization is difficult.

In practice, the parameters for a special jet-cleaning use are found out with only a few experiments. It is not sure, that there parameters are optimal.

With the example of stripping conductor profiles from isolation-material it is demonstrated how an optimization may work.

An optimization is significant in view of two aspects:

1. To minimize the requirement of energy resp. time e.g. in automatic jet-cleaning plants with always the same cleaning problem.

2. To expand the using of jet-cleaning into nowadays still uneconomic regions (very solid and/or very good adherent layers).

It is known, that different couples of materials need different loading by a jet for optimal cleaning.

Principally jet-cleaning is possible with layers of bad adherence to the substratum and brittle layers.

If it is possible to adapt the loading of the jet to the material and so change the ductile material's behaviour to brittleness, then it becomes possible to clean satisfactorily in these difficult cases too.

References:

(1) Lohse, U.
"Versuche an einer Naßputzanlage"
Die Gießerei 49 (1929); pp 1137

(2) Schuler, W.
"Hoher Druck kontra Schmutzkrusten"
Produktion 44 (1981)

(3) Torpey, B.
"Some experiences in the manufacture and application of high pressure water cleaning equipment"
Paper D 1 in the First International Symposium on Jet Cutting Technology, BHRA, 1972

(4) Pardey, H.P.
"Notes on marine applications of jet cutting"
1. ISJCT, BHRA, 1972, Paper D 2

(5) Gronauer, R.W.
"Cleaning and descaling of equipment with high pressure water"
1. ISJCT, BHRA, 1972, Paper D 3

(6) Ward, G.M.
"Safty considerations arising from operational experience with high pressure jet cleaning"
1. ISJCT, BHRA, 1972, Paper F 1

(7) Bury, L.; Housden, R.K.; Stephensen, R.; Thompson, R.C.
"Steam-boosted water jet cleaning at 300 bar"
2. ISJCT, BHRA, 1974, Paper F 1

(8) Crowe, A.R.
"Cleaning and descaling equipment with high pressure water"
2. ISJCT, BHRA, 1974, Paper F 3

(9) Erdmann-Jesnitzer, F.; Hassan, A.M.; Louis, H.
"A study of the oscillation's effects on the cleaning and cutting efficiency of high speed water jet"
3. ISJCT, BHRA, 1976, Paper C 3

(10) Calkins, D.; Mekor, M.
"Investigations of water jets for lock wall deicing"
3. ISJCT, BHRA, 1976, Paper G 2

(11) Conn, A.F.; Rudy, L.S.; Metha, G.D.
"Development of a cavijet™ -system for removing marine fouling and rust"
3. ISJCT, BHRA, 1976, Paper G 4

(12) Houlston, R.; Vickers, G.W.
"Surface cleaning using water-jet cavitation and droplet erosion"
4. ISJCT, BHRA, 1978, Paper H 1

(13) Barrick, I.D.
"An approach to remote controlled jets cleaning of chemical reaction vessels to gain the maximum benefit from the technique"
4. ISJCT, BHRA, 1978, Paper H 7

(14) Erdmann-Jesnitzer, F.; Louis, H.; Schikorr, W.
"Cleaning, drilling and cutting by interrupted jets"
5. ISJCT, BHRA, 1980, Paper B 1

(15) Veltrup, E.M.
"Experience of the industrial application of HP-water technology for stripping down rubber layers from rubber rollers"
5. ISJCT, BHRA, 1980, Paper H 2

(16) Dombrowski, H.
"Semi- and fully automatic handling and cleaning plant with high-pressure water"
5. ISJCT, BHRA, 1980, Paper H 3

(17) Engemann, B.K.
"Reinigen und Schneiden mit Hochdruckwasserstrahlen"
Werkstatt und Betrieb 113 (1980)

(18) Haferkamp, H.; Buhmann, K.-P.
"Beitrag zur Theorie, Technologie und Prüfung des Verbundes Kunststoff/Metall"
Konstruktion 26 (1974) pp. 129

(19) Neusen, K.F.; Schramm, S.W.
"Jet induced target material temperature increases during jet cutting"
4. ISJCT, BHRA, 1978, Paper E 4

(20) Zimm, W.
"Über die Strömungsvorgänge im freien Luftstrahl"
Forschungsarbeiten auf dem Gebiete des Ingenieurwesens"
Heft 234, Berlin (1921)

(21) Yanaida, K.; Ohashi, A.
"Flow characteristics of water jets in air"
4. ISJCT, BHRA, 1978, Papaer A 3

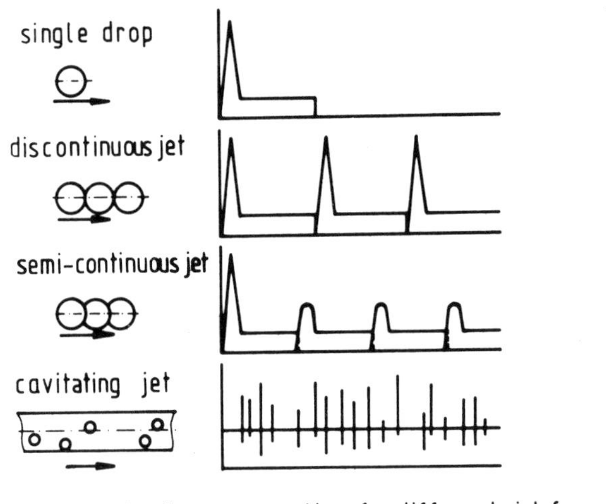

surface loading versus time for different jet-forms

fig. 1

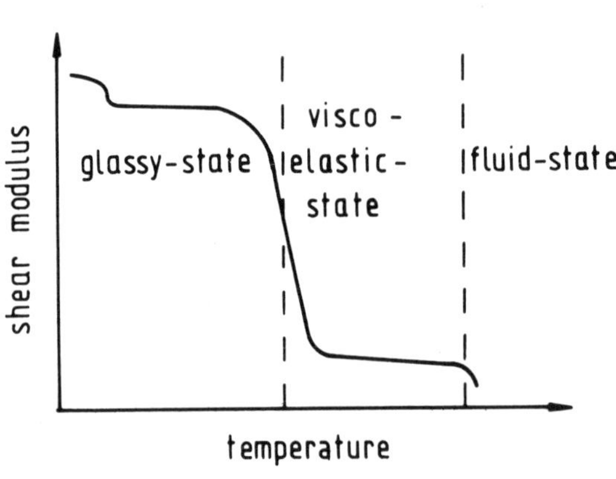

changing of mechanical properties versus temperature (plastic-material)

fig. 2

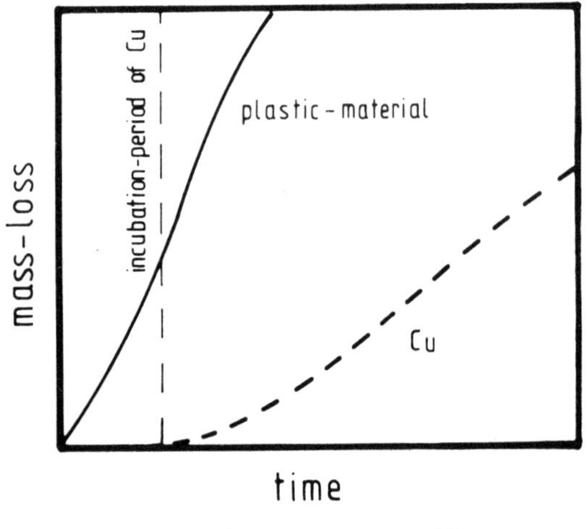

mass-loss versus time

fig. 3

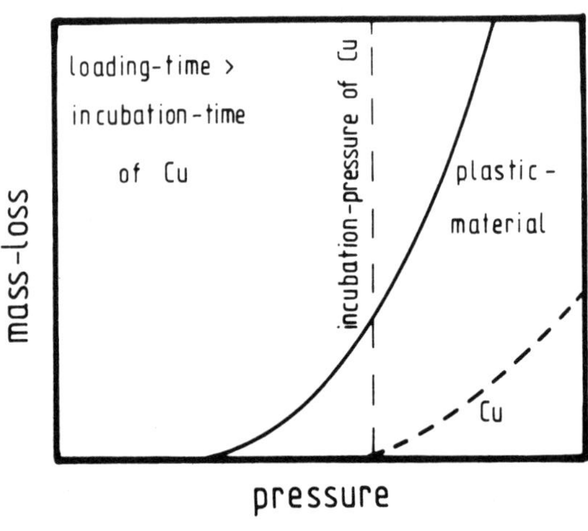

mass-loss versus pressure

fig. 4

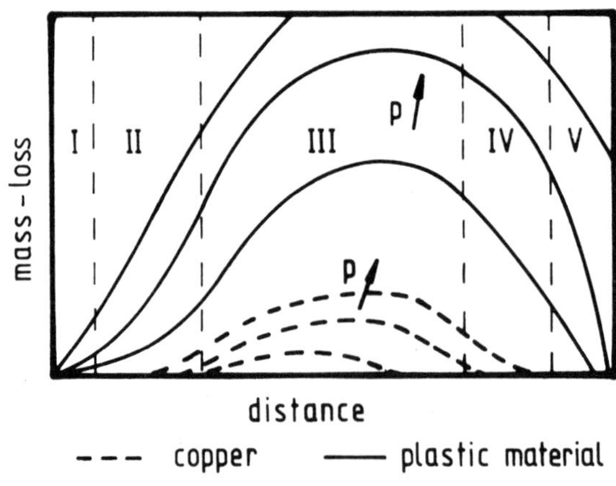

mass-loss versus nozzle-distance

fig. 5

6th International Symposium on

Jet Cutting Technology

6-8, April, 1982

A COMPARISON OF PULSED JETS VERSUS MECHANICAL BREAKERS

T.J. Labus

SCIRE Corporation, U.S.A.

Summary

A study of the performance characteristics of pulsed jets versus mechanical devices was conducted for concrete and rock breakage applications. The mechanical breakers were characterized in terms of size, breakage rate, power requirements and overall system parameters. A pulsed jet breaker model was developed, and published data used to construct a breakage rate-ouput characteristic relationship. The model was compared against data in the literature, and a parameter study undertaken. Results indicate that pulsed jets are competitive with mechanical breakers even at low average power levels, as long as the output pulse energy is maintained at a minimum level.

Held at the University of Surrey, U.K.
Symposium organised and sponsored by
BHRA Fluid Engineering

NOMENCLATURE

B_{rm} - breakage rate for mechanical devices

B_{rh} - breakage rate for hydraulic devices

h - depth of penetration

d - nozzle diameter

p - jet pressure

s - standoff distance

t - pulse duration

S_r - stroke ratio

σ - material tensile strength

V_j - jet velocity

V - broken material volume

V_1 - pulse fluid volume

E - elastic modulus

G - fracture energy

c - crack length

K_o, K_1, α, α_S - constants

η - machine efficiency

E_s - specific energy, mechanical breakage

E_{sh} - specific energy, hydraulic breakage

1.0 INTRODUCTION

The use of pulsed jets as substitutes for mechanical impact devices has been the subject of numerous investigations (1-7). The jet devices investigated have been relatively large machines, and direct comparisons with mechanical breakers have not been reported. The purpose of the current investigation was to characterize the mechanical breakers, concentrating mainly on portable hand tools, and compare them against pulsed jets. The study which follows is an analytical effort based on laboratory and field data.

2.0 MECHANICAL BREAKER CHARACTERIZATION

To characterize current mechanical breakers a survey of the commercial manufacturers was undertaken. The demolition tools were divided into four distinct classes:

1. The 13 kg hammer is a light weight unit designed for surface breaking of asphalt and concrete in the medium strength range.
2. The 27 kg hammer is used in utility construction, street maintenance and water and gas line repair.
3. The 41 kg hammer is a heavy duty unit used for general demolition service in mills, mines and road construction.
4. The XL hammer is a large capacity unit which is usually mounted on the arm of a backhoe. These units are large energy per blow devices, capable of breaking concrete up to 23 cm. thick in a single blow.

In each of the classes, the following information was collected:

1. Tool weight -kg
2. Energy per blow - N-m/blow
3. Blow rate - blows/minute
4. Power output - kW
5. Power input - kW
6. Efficiency - %
7. Tool specific weight - power/unit weight - kW/kg
8. Unit size -diameter and length - m
9. Unit cost - $

A total of 50 devices were surveyed and Table 1.0 shows the average characteristics in each class of demolition tool. This data includes hydraulically, pneumatically and electrically powered devices.

With the hammer parameters established the next step was to determine the breakage capabilities of each device. Fig. 1 shows a plot of specific energy versus pulse energy from ref. 9. This plot is a compilation of data from various types of devices (i.e. rock drills, drop hammers, impactors, projectiles, etc.). A best fit curve to the data has the following fora.

$$E_s = K_o \left(E_b \right)^{\alpha_s} \quad \text{where} \quad K_o = 2.874 \times 10^3$$

and E_s - Specific energy -N-m/cm^3 $\quad \alpha_s = -0.4764$

E_b - Blow energy - N-m/blow

Now, the breakage rate (B_r) is related to the specific energy by the following relationship.

$$B_r = E_b \times BPM/E_s$$

where BPM - blow rate

Combining the two relationships the breakage rate can be written as.

$$B_r = 2.087 \times 10^{-8} BPM \times E_b^{1.47} \; (m^3/hr)$$

Thus, for mechanical breakers the breakage rate is directly proportional to the blow rate, but more sensitive to the energy per blow. The data in Table 1.0 indicates that mechanical breakers are characterized by high blow rates and relatively modest blow energies. This illustrates the limitation that mechanical stress places on conventional breakers. To have effective breakage rates, the impact energy must be kept at a level which provides adequate component life, hence the high blow rates. These high blow rates require large flow volumes and the associated flow lossses, as indicated by the low efficiencies. Fig. 2 shows a plot of input power versus blow energy for various blow rates. The power input requirements are determined based on the average breaker characteristics from Table 1.0. The dotted line on the 600 BPM curve is a projection at a constant efficiency to a blow energy of 1219 N-m. The solid line is the actual relationship of existing devices. Note, that there is a decline in overall efficiency for the larger units, although there is a distinct advantage in specific energy by operating at higher blow energies, as shown in fig. 1. Thus, devices which could operate at higher blow energies and greater overall efficiency should provide superior performance. A pulsed jet breaker has some of these traits.

3.0 PULSED JET PERFORMANCE CHARACTERIZATION

Theoretical approaches to predicting material breakage due to pulsed jet impact, are limited due to the complex material response and fluid dynamics involved. Hence, test results have been used to develop empirical relationships, but these are also limited due to exclusion of important variables, and limited test data. Examination of the literature (ref. 1-8) produced a data base that is plotted in fig. 3. This plot includes the mechanical breakage curve from fig. 1 for reference. For the hydraulic breakage curve the specific energy is given as:

$$E_{sh} = 2.0 x 10^{6} \quad (E_b)^{-1.074} \qquad r^2 = .942$$

The data shown in fig. 3 was analized using ref. 7 and 10 to provide values for machine efficiency. This curve is for single shots only. Hydraulic breakage provides better specific energy values than mechanical breakage for pulse energies greater than 58,265 N-m, as shown in fig. 3. This high value of pulse energy suggests that hydraulic breakage would be effective for large units, but this does not take into account the overall efficiency of the breakage device, or the high energy density nature of pulsed jet units.

The following breakage rate equations are based on the previous specific energy relationships, and include machine efficiency effects for both mechanical and hydraulic breakage.

$$B_{rm} = 2.08\, 10^{-6} (P_{in} \eta_{om}/BPM)^{1.476} \quad x\ BPM$$

$$B_{rh} = 3.0\, 10^{-11} (P_{in} \eta_{oh}/BPM)^{2.074} \quad x\ BPM$$

Both of the above relationships are dependent on the overall efficiency of the breakage device, but the hydraulic breakage is more sensitive due to the larger exponent. Fig. 4 shows a plot of unit efficiency versus pulse energy for two types of hydraulic breakers and mechanical breakers. The mechanical curve is derived from the data in Table 1.0. The IP (impact pulse) hydraulic breaker curve is typical of water cannon type of units. The data for the curve was obtained from refr. 7 and 10. The DAI (pressure extrusion) breakage curve was generated using the PULSEJET computer code of ref. 4. These results correlate well with those reported in ref. 2 for actual field testing. The key factor to note is that hydraulic breakers have overall efficiencies greater than mechanical units. In general, there is a substantial difference in these efficiencies which reflect in improved breakage rates as shown in fig. 5. This plot shows the ratio of mechanical to hydraulic breakage rates verus pulse energy. Note, that even at low power levels hydraulic breakage can provide a significant benefit if the pulse energy is kept high. Increasing input power at a fixed pulse energy results in a less favorable ratio because pulse rate is increased, and this change has more effect for mechanical breakers than hydraulic units. Thus, pulsed jet breakers should have high energy outputs, at a given power input, and work at relatively low pulse rates to obtain maximum performance. This high energy output does not require a physically large unit as shown in fig. 6. This unit is a gas powered DAI type of pulsed jet generator, having a pulse energy of 7046 N-m, and a physical envelope of 12.7 cm diameter by 46 cm. long, and weighing 17.2 kg. A commercial boulder splitting unit based on pulse jet impact is also characterized by a high energy output in a compact package (ref. 11).

One additional point is worthy of consideration, and this is the difference between the DAI and IP pulse generators. Returning to fig. 4, the curves indicate that at high pulse energies the DAI and IP generators would have equivalent machine efficiencies, but there is an additional consideration to be made. Breakage of materials by pulsed jet impact is a fracture dominated process. As a results, crack propagation plays an important role, and to have efficient breakage sufficient energy must be maintained at the crack tip to insure continued growth. The pressurized jet fluid and static pressure in the borehole are the medias that distribute this energy to the crack tip. Also, since cracks grow in tensile stress fields, and brittle materials such as rock are weakest in tension, the failure mode is the most efficient. Thus, broken volume and specific energy should be sensitive functions of the flow rate and volume of high pressure fluid in the pulse. Fig. 7 shows some data for concrete that exhibits these trends (ref. 2). The broken volume increases while the specific energy decreases as the flow rate increases. Note, the flow rate and the total fluid volume are both important. The flow rate must be kept high enough to maintain the pressure in the borehole and crack by overcomming porosity, crack growth losses and fluid ejected up the borehole. The total volume of fluid controls the maximum distance that cracks can grow from the point of impact.

Flow characteristics contrast the difference between DAI and IP generators. The DAI generator delivers all of the fluid in the high pressure barrel in a controlled fashion, at a sustained and relatively constant flow rate. The IP generator does not deliver all of the water in the high pressure barrel, and the flow rate is uncontrolled. This uncontrolled flow rate is characterised by an initially high flow, followed by a low flow (ref. 10). These flow properties may not produce maximum breakage since the flow rate and total volume are cut short during the drilling and pressurization phases.

Also, in fig. 7 the pressures required to produce effective breakage do not have to excessively high. Better breakage was obtained for jet pressure of 2965 to 3586 bars versus 4206 bars. This again illustrates the flow sensitivity of the breakage process, since the lower pressures coincide with larger nozzle diameters and increased flow rates. This data is for a DAI type of pulse generator.

4.0 EMPIRICAL CORRELATION

Although test data is available on pulsed jet performance a scaling relationship would be a valuable design tool. Fig. 8 shows a correlation between the depth of penetration in concrete by a pulsed jet versus a group of jet parameters and material properties. This correlation take the form of:

$$h/d = K_o \left[(S_r (p/\sigma)(d/s)(V_j t/d)^{0.5} \right]^{\alpha} \qquad \text{for concrete } K_0 = 5.958 \quad \alpha = .4161$$

The data for this plot was taken from ref. 3, and although there is considerable scatter a trend is clearly evident. Based on the discussion in the previous sections the correlation should be modified to include pulse volume effects and fracture properties of the target material. Following this reasoning the correlation could be modified as follows:

$$V/V_1 = K_1 \left[(d/s)(V_j t/d)^{0.5} \ (p/\sqrt{EG/c}) \right]^{\alpha}$$

An evaluation of this correlation will have to await the development of a test base including these variables.

5.0 CONCLUSIONS

From the foregoing discussion the following observations are pertinent:

1. For low pulse energies (less than 813 N-m) mechanical breakers are more efficient, because of the high pulse energy required to initiate fracture hydraulically.
2. At lower power levels, pulsed jets can fracture material more effectively than mechanical breakers if the pulse energy is kept high.
3. The overall efficiency of hydraulic breakers is substantially higher than mechanical units. This high efficiency has a substantial impact on the range of application of a pulse jet.
4. Pulse jet generators are high energy output devices, but have physical dimensions comparable to mechanical breakers and overall provide better breakage rates per power input than mechanical units of equivalent power input.

REFERENCES

1. Yie, G.G., Burns, D.J. and Mohaupt, U.H., "Performance of a High-Pressure Pulsed Water-Jet Device for Fracturing Concrete Pavement", Proc. of Fourth Int. Sym. on Jet Cutting Tech., Paper No. H6, BHRA Fluid Engineering, Canterbury, England, April, 1978
2. Labus, T.J., Hilaris, J.A., and Yie, G.G., "A Study of High Pressure Water Jets for Highway Surface Maintenance", IITRI Report No. D6124, under NSF Grant No. ISP76-12230, Feb., 1978
3. Burns,D.J., Kalbfeisch,J.G., Mohaupt, U.H., and Vallve', F.X., "Relationship Between Jet Penetration and Design Parameters of a Pulsed Water Jet Machine", Proc. Fifth Int. Sym. on Jet Cutting Tech., Paper No. D2, BHRA Fluid Engineering, Hanover, FRG, June, 1980
4. Labus, T.J., "A Feasibllity Study of a Pulsed Jet Concrete Breaker", SCIRE Corp. report no. 80001, Aug., 1980
5. Cooley, W.C., "Fabrication and Testing of a Water Cannon for Rock Tunneling Experiments", Dept. of Transportation Report no. FRA-ORD &D 74-38, Feb., 1974
6. Clipp,L.L. and Cooley, W.C., "Development, Test and Evaluation of and Advanced Design Experimental Pneumatic Powered Water Cannon", under USA MERDC contract no. DAAK02-68-C-0444, March, 1969
7. Petrakov, A.I. and Krivortoko,OD., "Some Experiences in Developing Minning Roadways Using the Experimental Heading Machine with Pulsed Water Jets", Proc. of Fourth Int. Sym. on Jet Cutting Tech., Paper No. J3, BHRA Fluid Engineering, Canterbury, England, April, 1978
8. Gnirk, P.F. and Grams, W.H., "Rock Drilling with Pulsed High-Velocity Water Jets", Proc. of 14th Sym. on Rock Mechanics, Penn State Univ., University Park, Pa., June, 1972
9. Grantmyre,I. and Hawkes, I., "High Energy Impact Rock Breaking", Canadian Mining and Metallurgical Bulletin, 68, P63-70, Aug., 1975
10. Glenn, LA., "The Mechanics of the Impulsive Water Cannon", Computers and Fluids, Vol. 3, pp. 197-215, Pergamom Press, 1975
11. "Water Cannon Splits Boulders", Mining Magazine, June, 1980

Table 1.0

Average Class Characteristics of Demolition Hammers

Class	Weight (kg)	N-m/Blow	BPM	N-m/min	Power Out kW	Power In kW	Eff. %	N-m/min-kg	Length (m)	Cost $
13 kg	14.5	25.3	2043	51687	0.86	3.64	23.7	3564	0.56	725
27 kg	28.5	78.6	1556	122301	2.04	6.33	32.2	4291	0.67	887
41 kg	43.6	128.8	1301	167568	2.82	7.21	39.1	3843	0.70	1138
XL	409.9	1219.0	590	719210	12.0	39.49	30.4	1754	-	-

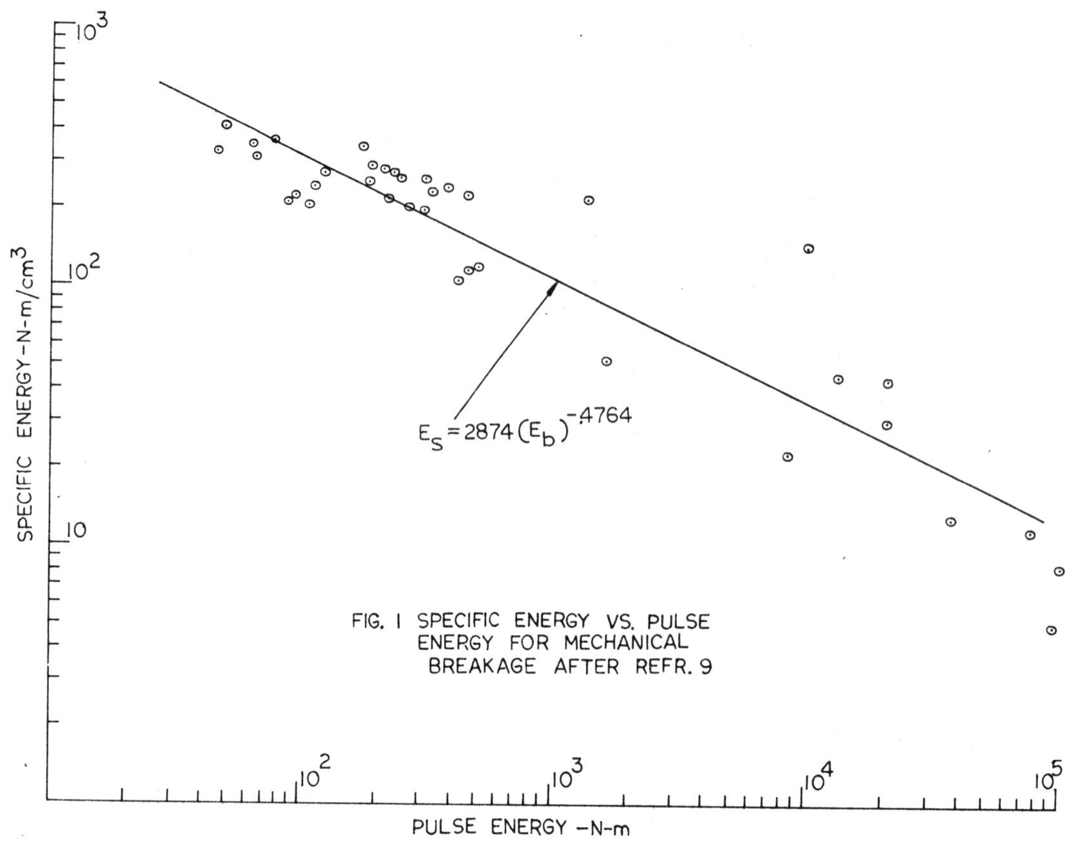

FIG. I SPECIFIC ENERGY VS. PULSE ENERGY FOR MECHANICAL BREAKAGE AFTER REFR. 9

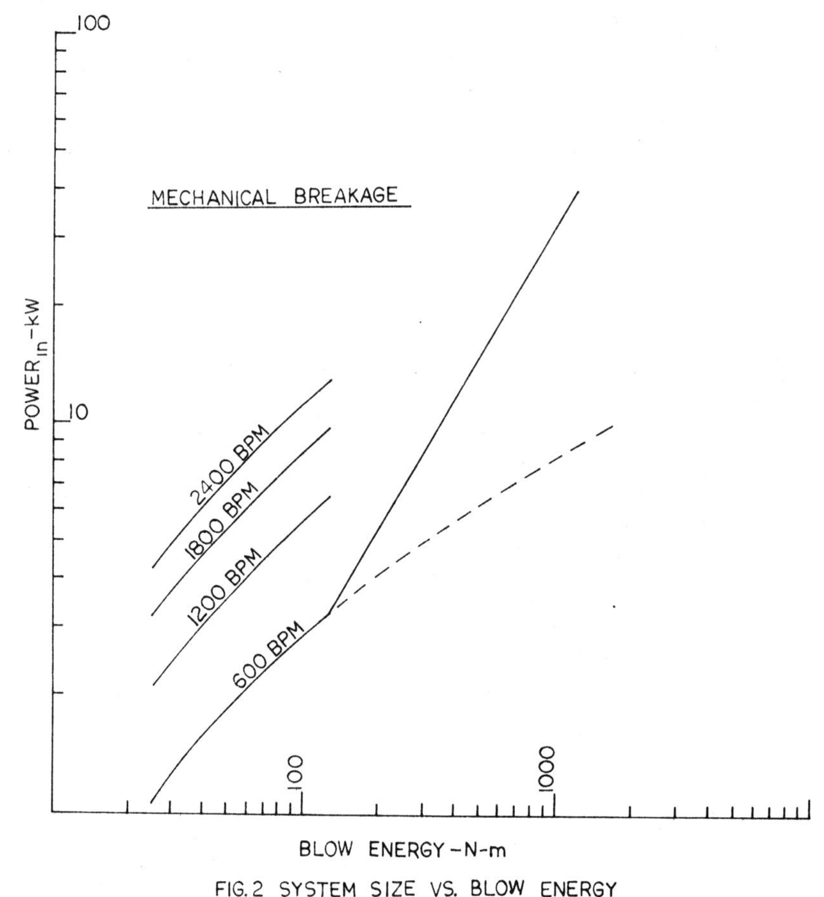

FIG. 2 SYSTEM SIZE VS. BLOW ENERGY

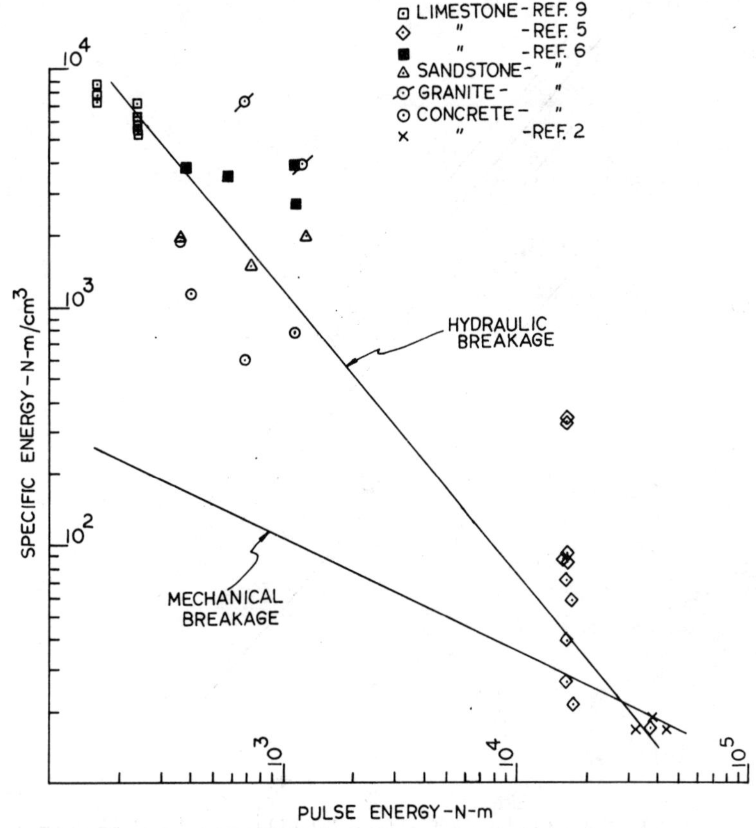

FIG. 3 SPECIFIC ENERGY VS. PULSE ENERGY

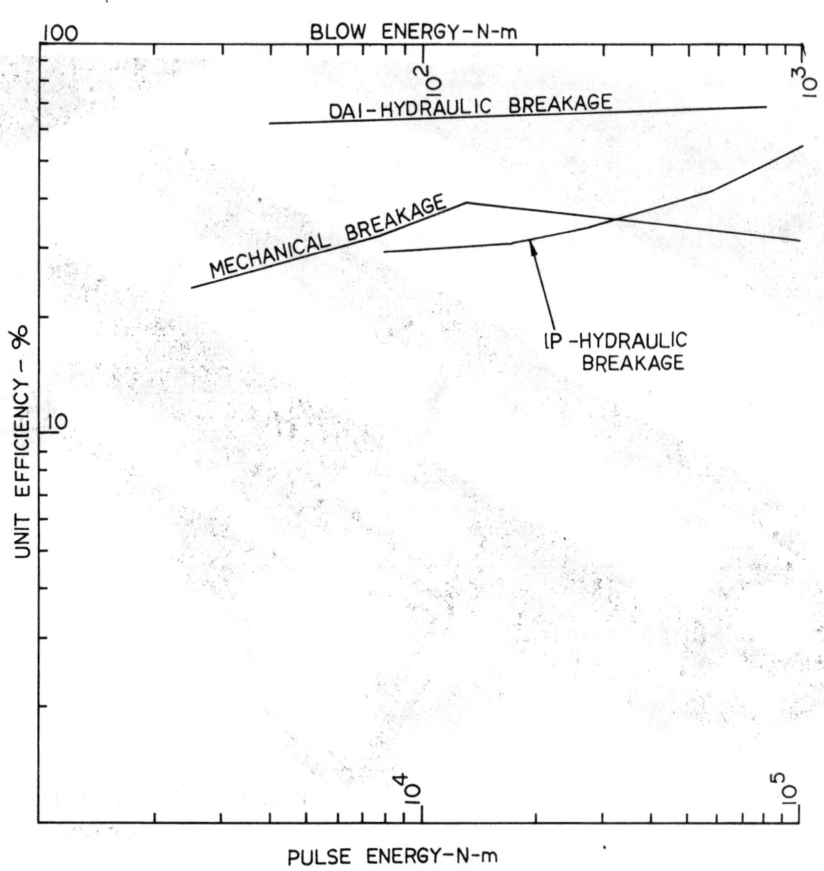

FIG. 4 EFFICIENCY VS. PULSE/BLOW ENERGY

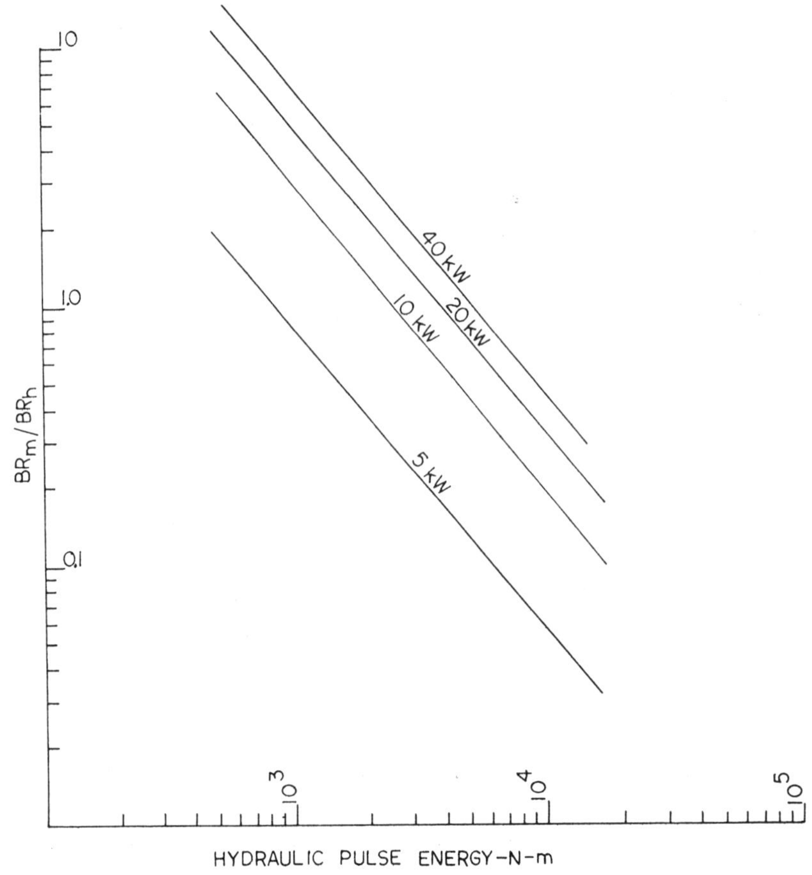

FIG. 5 BREAKAGE RATE RATIO VS. PULSE ENERGY

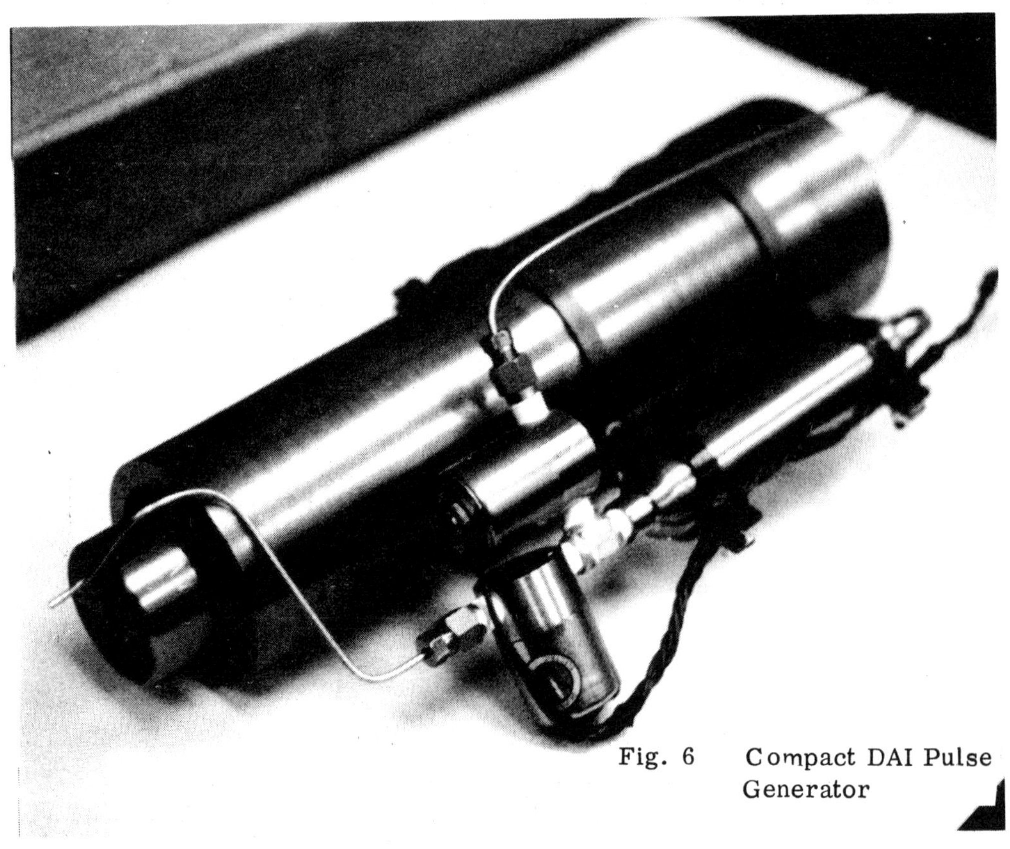

Fig. 6 Compact DAI Pulse Generator

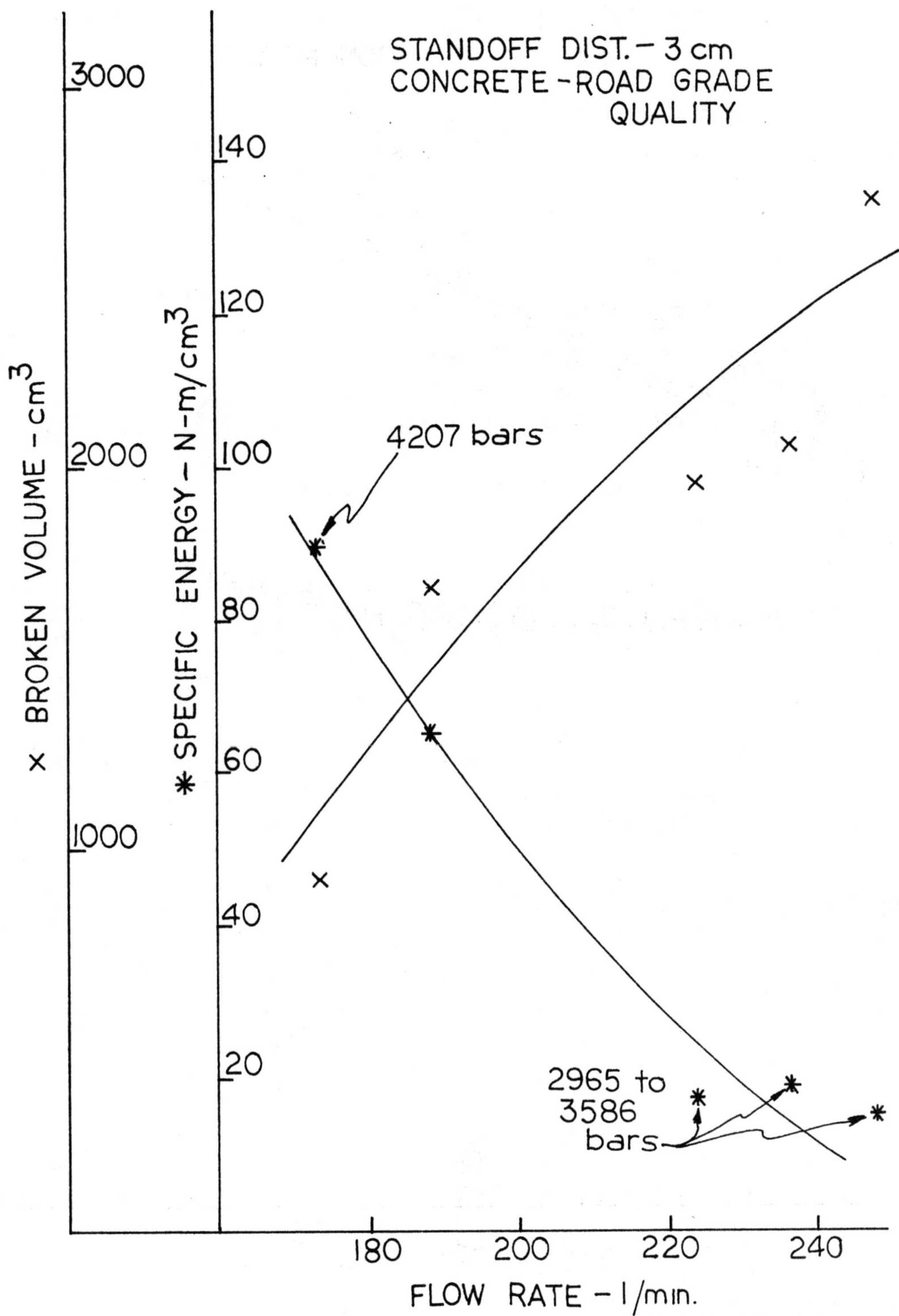

FIG. 7 FLOW SENSITIVITY OF HYDRAULIC BREAKAGE

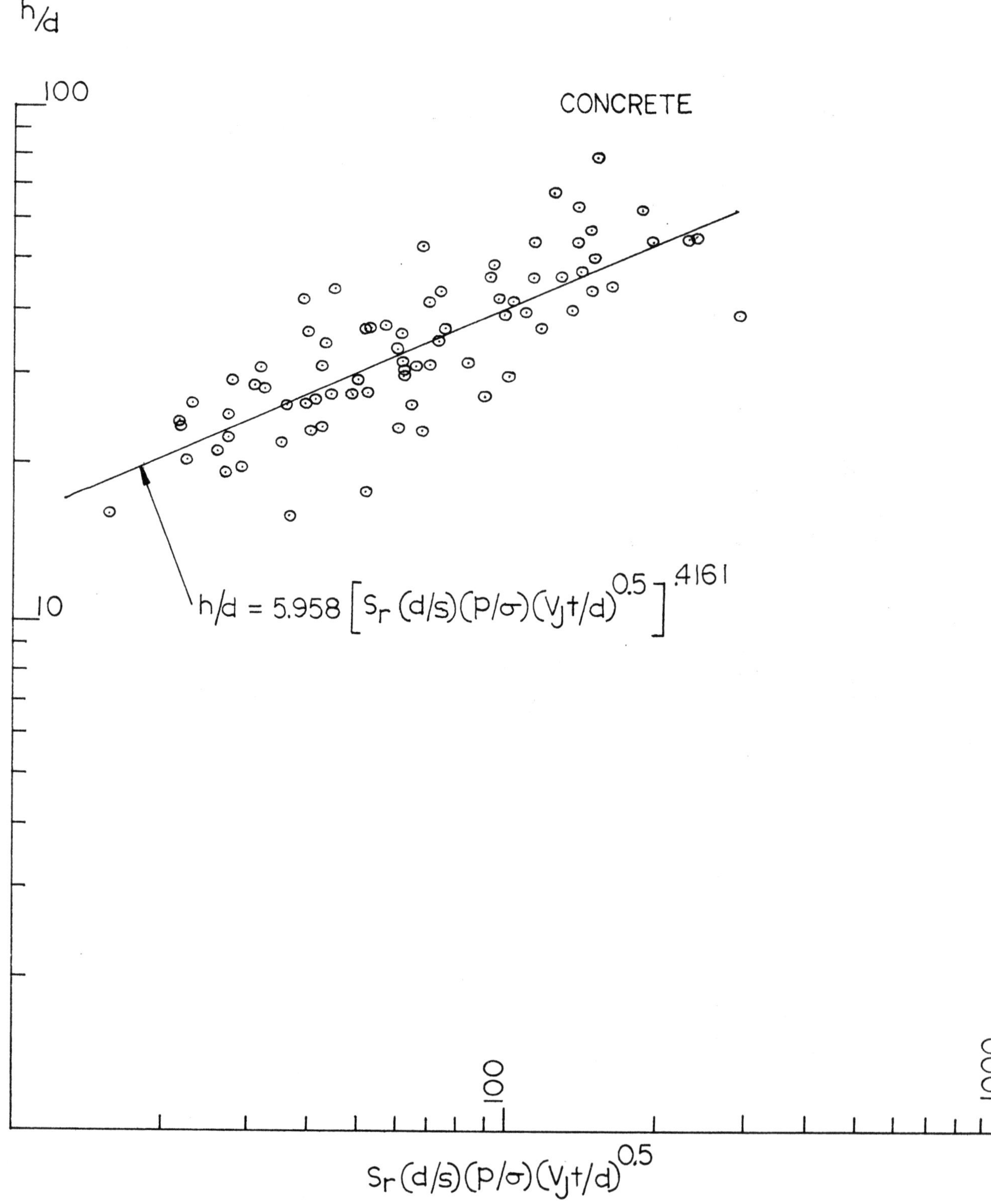

FIG. 8 EMPIRICAL CORRELATION

6th International Symposium on
Jet Cutting Technology
6-8, April, 1982

THE EFFECT OF CONCRETE PREHEATING ON WATER JET BREAKING OF CONCRETE

H. Kiyohashi, M. Kyo and W. Ishihama

Tohoku University, Japan

T. Yahiro and H. Yoshida

Kajima Institute of Construction Technology, Japan

Summary

Influences of preheating temperature, θ, on water jet breaking of a preheated concrete (exactly speaking, cement mortar) have been experimentally studied as a function of standoff distance, L. θ was varied from room temperature (no preheating in this case) to 600°C. Water jet speed, U_o, was kept constant at 210 m/s (water pressure, P_o, = 22.1 MPa) at nozzle exit (nozzle diameter, D_o, = 0.5 mm I.D.). Temperature of the jet water was normal. Dimensionless standoff distance, L/D_o, was taken as 20, 28.6, 100, 200, 400 and 600. Jet operation time was fixed at 10 s.

The experimental results indicated that the breaking pattern and efficiency varied characteristically with θ and L/D_o. Hole-shaped cavities, hole surrounded with eroded area-shaped cavities were macroscopically observed in the surface of the specimen struck by the water jet. Breaking augmentation ratios, $\ell_\theta / \ell_{\theta = 15°C}$ and $v_\theta / v_{\theta = 15°C}$ reached about 28 at L/D_o = 28.6 and about 100 at L/D_o = 600 at θ = 600°C, where ℓ and v are the depth and the volume of the cavity.

Held at the University of Surrey, U.K.
Symposium organised and sponsored by
BHRA Fluid Engineering

NOMENCLATURE

D_o = nozzle diameter (mm)

L = standoff distance (mm)

ℓ = depth of cavity yielded by breaking (breaking efficiency index) (mm)

P_o = static pressure at the nozzle exit (MPa)

t_o = temperature of water jet at the nozzle exit (^oC)

U_o = speed of water jet at the nozzle exit (m/s)

v = volume of cavity yielded by breaking (breaking efficiency index) (cm^3)

v_w = volume of jet water in jet operation (cm^3)

x_b = breakup length of water jet (mm)

x_c = length of initial region of water jet (mm)

x_p = length of potential core of water jet (mm)

θ = specimen temperature (^oC)

τ = jet operation time (s)

σ_c, σ_t = compressive and tensile strength (MPa)

Dimensionless number

C,Cn = dimensionless factors (n=1~4)

h,h' = dimensionless exponents

i,i' = dimensionless exponents

j,j' = dimensionless exponents

k,k' = dimensionless exponents

L/D_o = dimensionless standoff distance

ℓ/D_o = dimensionless depth of cavity

$\ell_\theta/\ell_{\theta=15^oC}$ = breaking depth augmentation ratio

v/v_w = dimensionless volume of cavity

$v_\theta/v_{\theta=15^oC}$ = breaking volume augmentation ratio

Subscripts

h = pertaining to the hydraulic power

max = pertaining to the maximum

min = pertaining to the minimum

t = pertaining to thermal cooling power

INTRODUCTION

Recently, new methods having no public nuisance of demolishing of old reinforeced concrete buildings and useless constructions in order to redevelop old urban areas have been sought (Refs. 1 and 2). Also, safe and effective methods for demolition of concrete shieldings of expired nuclear reactors will be needed in the near future. The application of high speed water jets for the demolition of concrete has been proposed (Refs. 1 and 2). The use of water jets to cut concrete and asphalts has been considered by various investigators (Refs. 2, 3, 4, 5, 6, 7, and 8). However, this method has not found to be either effective or economical.

The authors thought that the efficiency of concrete demolition by water jets would be augmented if the concrete were preheated before the water jets were used. This was thought to be true. Because, the preheating of concrete changes it's strength from hard to brittle (Ref. 1) and because the rapid cooling of the hot concrete by the impinging of the water jet induces thermal stresses. This is in addition to the hydraulic jet erosion effect (Refs. 9, 10, 11, 12 and 13). The authors believe that the Joule heating by applying an electric current to reinforcements (Ref. 1) and the dielectric heating by microwave (Ref. 14) are applicable for the practical preheating of reinforced and unreinforced concretes, respectively.

This paper deals with an experimental study, which was carried out, on a cement mortar specimen as a first step experiment of our studies on the demoliton of concretes. The influence of preheating temperature, θ, on water jet breaking of the preheated cement mortar have been studied as a function of standoff distance, L, using a relatively low speed water jet of U_O=210 m/s from a nozzle of diameter D_O=0.5 mm.

Experimental results indicated that the breaking patterns and efficiency varied characteristically with the specimen temperature, θ, and the standoff distance, L. Maximum breaking augmentation ratios of the depth and the volume of the cavity broken by the water jet reached about 28 and about 100, respectively, at θ=600°C in the tested ranges of preheating temperature, θ, from 100 to 600°C and dimensionless standoff distance, L/D_O, from 20 to 600. These results were used to predict breaking efficiencies in some assumed cases of preheated concretes from normal jet breaking data by practical high pressure water jets.

2. EXPERIMENTAL MATERIAL

The cement mortar specimen (75 mm in diameter and 150 mm in length) was made of special strong portland cement, river sand from 20 to 200 mesh sizes and water. Their mixing ratio was 1.80:2.3:1.0 cement:sand and water, respectively. Then they were mixed and poured into cylindrical molds. After solidifying in the molds for 24 hr., the solidified mixture, that is, the cement mortar specimen was kept in water for 3 days and dried in the shade for 30 days or more.

Table 1 shows the values of some physical properties of the specimens at different preheating temperatures. Each physical property except thermal conductivity was obtained as the test results for the specimens cooled gradually from test preheating temperature, θ, to room temperature. As for their compressive strength, it is assumed that the compressive strength in the hot state is almost similar to that in the cold state (Ref. 10). Over the preheating temperature range of 100 to 600°C the cement mortars show that their strengths rapidly decrease with increasing temperature.

3. EXPERIMENTAL APPARATUS AND METHOD

The main components of the experimental equipment consisted of a water jet generator, test chamber, measuring apparatus and cement mortar preheating furnace. The outline of the first three parts is shown schematically in Fig. 1. The water jet generator was set up from air compressor A, air cleaner and cooler B, pressure regulator C, intensifier type high pressure pump E driven by compressed air, water feed valve F, accumulator H, jet on-off valve I and nozzle J. A detailed drawing of the nozzle J in Fig. 1 is shown in Fig. 2.

Inner dimensions of the test chamber K in Fig. 1 are 820 mm x 830 mm x 1140 mm.

The specimen M was fixed on the test stand P equipped inside of K through the medium of refractory bricks N. Detailed descriptions of the test stand P in Fig. 1 have been reported elsewhere (Refs. 9 and 10). The jet water shutter L hung in front of M was pulled when the water jet was impinged on the M.

The measuring apparatus consisted of pressure transducer S, trigger switch T, microphone U, strain meter V, electromagnetic oscillograph W, tape recorder X and digital voltmeter Z. The temperature of jet water was measured through thermocouple Y. Preheating of the specimens was carried out in an electric furnace (15 kw) equipped with a programmed temperature controller.

The experimental procedure was as follows: (1) Manually established a tested standoff distance, L. (2) Maintained the air source pressure at 971 kPa (rated pressure) with the air compressor A. (3) Fed the jet water into the pump E through value F and opened valve I. (4) Established a jet water pressure tested, P_O, by controlling the valve C and by obtaining the reading of P_O with the EM-osillograph W. (5) Took out a specimen preheated to a certain test temperature in the electric furnace by tongs, and moved and fixed it quickly to the test stand P. (6) Operated the tape recorder X and the EM oscillograph W. Then, pulled up the jet water shutter L, allowing water jet to impinge against the specimen M. (7) After the water jet struck the specimen for a certain test period, the valve I was closed, and one test run was accomplished.

Experimental variables and conditions were as follows. The temperature of the specimens, θ, were normal temperature, 100, 200, 300, 400, 500 and 600^oC. The water jet speed at nozzle exit, U_O, was kept constant at 210 m/s which corresponded to static pressure of water at nozzle exit, P_O, of 22.1 MPa. Temperature of the jet water, t_O, was at 15.5 ± 5^oC.

Nozzle diameter, D_O, was fixed at 0.5 mm. Standoff distance, L was varied from 10 mm to 300 mm. Dimensionless standoff distance tested, L/D_O, were taken as 20, 28.6, 100, 200, 400, and 600. Jet operation time, τ, was fixed at 10 s.

Other conditions were taken as follows. When the specimen was preheated, the heating rate of the furnace containing the specimens to desired test temperature, θ, was set at $100^oC/h$. After the temperature was reached, it took subsequently one hour or over to establish an almost steady state on the specimen temperature.

Two kinds of breaking test by the water jet were undertaken in this experiment. The one is a hot breaking test and another a cold breaking test. The former and the latter are tests for the hot state specimens having a test preheating temperature, θ, higher than 100^oC and for the cold state specimens cooled gradually from the preheating temperature, θ, to room temperature, respectively. The number of specimens used for both breaking tests were five for each different condition (each combination of a specimen temperature, θ, and a dimensionless standoff distance, L/D_O). When the tested specimen was moved from the furnace to the test stand, it took 25-40 s until the water jet impinged on the specimen.

As indices of the breaking efficiency, the maximum depth and the space volume of the cavities drilled in the specimens by the impinging water jet within a certain jet operation time were adopted in this experiment. The depth of cavities of the drilled specimens, judged from their original surface before drilling, was measured as the length of the inserted part of the thin wire (0.3 mm in diameter) into the cavities with slide callipers. The volume of the cavities of the drilled specimens was filled up with dried carborundum (150 meshes in size), whose weight was measured afterwards. Hence, the volume of the cavities could be obtained by the known weight and the apparent specific gravity of the carborundum.

4. EXPERIMENTAL RESULTS AND DISCUSSIONS

4.1 Characteristics of the Water Jet

The general structure of a turbulent free water jet in air consists of several parts, namely, the initial region with a constant jet axial dynamic pressure, the main region with a constant jet axial velocity and the final region characterized by diffused droplet flow according to a jet structure model proposed by Yanaida and Ohashi (Refs. 15 and 16). In the initial region, there are the potential core region and the transition region. On the other hand, viewed differently, the jet is divided into the continuous flow region and the discontinuous flow region. The latter consists of the droplet and the diffused flow regions. Analytical and experimental results on the water jet structure within the range of the nozzle exit velocity, U_o, from 50 to 250 m/s given by Yanaida et al. (Refs. 15 and 16) show that the length of the potential core, x_p, is equal to about 19 D_o, that the length of initial region, x_c, which varies with nozzle configurations and Reynolds number of the jet, is equal to (50-120) D_o, and that the breakup length, x_b, namely, the length from the nozzle exit to the boundary of the continuous and the discontinuous flow regions, is equal to 3.55 x_c.

Figure 3 shows a photograph of the configuration of the water jet used in this experiment. The photograph was taken under a reflective light illumination. Considering Yanaida's results, a photographic study of the water jet used and a geometrical study of the cavities broken by the water jet as described later led to the following conclusions on the flow characteristics of the water jet at the dimensionless standoff distance, L/D_o, tested: the flow region at L/D_o=20 might be at the end of the potential core region; L/D_o=28.6 is just outside of the potential core region and in the transitional one; L/D_o=100 in the transitional region; L/D_o=200, 400 and 600 in the main region with continuous or droplet flow.

4.2 Observational Descriptions of the Drilled Specimens

Figure 4-1 shows front views of typical fracture patterns on the surface of the specimens broken by the water jet at the different dimensionless standoff distances, L/D_o, and specimen temperatures, θ, in the hot breaking test. Figure 4-2 shows front views of the typical fracture patterns on the surface of the specimen, which were preheated to test temperature, θ, and gradually cooled from θ to room temperature, broken by the water jet at the different dimensionless standoff distances, L/D_o.

Figure 5-1 shows typical cross sections of the cavity of the specimen broken by the impinging water jet at the different L/D_o tested at θ=200, 300, 400, 500 and 600°C. These figures show the results of the hot breaking test. Also, Fig. 5-2 shows typical cross sections of the cavity of the gradually cooled specimen broken by the impinging water jet at the different L/D_o tested at the same temperatures as the previously preheated specimens, i.e. θ=300, 400, 500 and 600°C. These cross sections of the cavity were obtained by its replicas made of synthetic resin. The following are drawn from the above figures.

1) When the strength of the specimen weakened by preheating becomes smaller than the power of the water jet, the specimen is damaged by the water jet. On the surface of the specimen struck by the water jet, a hole-shaped cavity, a hole surrounded with eroded area-shaped cavity or a crater-shaped cavity was observed.

2) There were no remarkable differences between the fracture patterns of the specimens in the hot and cold breaking tests except in a few cases of θ higher than about 300°C, as described later.

3) A hole-shaped cavity was observed in cases of L/D_o=20 and 28.6 in all the tested θ ranges, and in cases of L/D_o=100 at θ higher than about 500°C.

4) A crater-shaped cavity was observed in cases of L/D_o=100 and θ lower than about 400°C, and in cases of L/D_o larger than about 200 in all the tested θ range.

5) An eroded area surrounding the hole rarely appeared in cases of the smaller L/D_o, and many more appeared in the hot breaking test than in the cold one.

6) With a larger L/D_o than 100, the eroded area surrounding the hole or the crater was almost created and somewhat wider in the hot breaking test than in the cold one.

7) In very rare cases of θ=300, 400 and 500°C, and relatively many cases at θ=600°C, concentric irregular large circular cracks having zero to four lines starting from the jet stagnation point or irregular small honeycombed cracks having small fissures at center area and large ones at outer area of the impinged surface of the specimen appeared.

4.3 Relations between Breaking Efficiencies and Temperature of Preheated Specimens

The depth and the volume of the cavity in the specimen broken by the water jet were measured to evaluate the breaking efficiency. Figures 6-1 (a) to 6-2 (f) show the relationship between the depth of the cavity from the initial surface of the specimens ℓ (mm) and the specimen temperature θ(°C) at each dimensionless standoff distance, L/D_o. In these figures, open circles and open triangles designate the experimental points for each depth, namely, ℓ and ℓ_h, of the specimens by the hot and the cold breaking tests, respectively. Each closed symbol represents the arithmetical mean values of each five experimental value for the former and for the latter, respectively. Solid and chain lines were drawn as smooth curves connecting the closed symbols for each of the hot and the cold breaking tests, respectively. As shown in these figures the depth, ℓ, increased very rapidly with increasing θ within the tested temperature range in all cases of the tested standoff distances. The depth of the cavity drilled in the cold breaking test, ℓ_h, also increased rapidly with the increase of θ in all cases of the tested standoff distances. This tendency between ℓ_h and θ seems to depend directly on the temperature dependency of the physical properties, especially, porosity, compressive strength and dynamic Young's modulus of the specimen shown in TAble 1. That is to say, there are a fairly distinct positive relation for porosity and negative relations for compressive strength and dynamic Young's modulus between ℓ_h and magnitudes of their physical properties. Broken lines in the above figures show relations between a depth contributed by the thermal cooling power of the impinging water jet, ℓ_t, and the specimen temperature, θ. Where, ℓ_t was obtained by subtracting ℓ_h from ℓ. An explanation for it will be described later in detail.

Figures 7-1(a) to 7-2(f) illustrate the relation between the space volume of the cavity drilled by the water jet and the specimen temperature, θ, at the tested different dimensionless standoff distance, L/D_o. Open circles and open triangles also show experimental points for cavity space volumes, v and v_h, obtained in the hot and the cold breaking tests, respectively. Each closed symbol represents each mean experimental value as previously stated. In these figures, solid and chain lines drawn as smooth curves connecting the closed symbols for the hot and the cold breaking tests, respectively, lack smoothness in comparison with the corresponding curves, namely, the ℓ-θ curves and the ℓ_h-θ curve shown in Figs. 6-1 and 6-2. This seems that accidental errors for v and v_h become three times in contrast with ones for ℓ and ℓ_h, because v and v_h have three dimensions, and ℓ and ℓ_h one dimension, respectively.

The following matters are inferred from the above figures. The volume of the cavity of the specimen drilled in the hot breaking test, v, was larger than the volume of that in the cold breaking one, v_h, in either case. In cases of L/D_o=20, 28.6 and 100, v increased very rapidly with increasing θ, reaching the maximum at θ equal from 400 to 500°C, and then decreased with θ. v_{max}, each peak value of v on the v-θ curves, reached equally about 0.1 cm^3 in the above cases. In the cases of L/D_o=200, 400 and 600, v had no peak value and increased progressively with increasing θ. The volume of the cavity of the specimen drilled in the cold breaking test, v_h increased rather rapidly with increasing θ in all cases of the tested standoff distances, L/D_o, except L/D_o=20. In the case of L/D_o=20, the v_h-θ curve had a maximum at θ=300°C and a clear minimum at θ=400°C. These relations between v_h and θ seem to depend on the temperature dependency of the physical properties as well as the relations between ℓ_h and θ, as previously described. Broken lines in the above figures show the relation between the volume contributed by the thermal cooling power of the jet, v_t, obtained by subtracting v_h from v, and θ.

4.4 Relations between Breaking Efficiencies and Standoff Distance

Dimensionless depths of cavity, ℓ/D_o, ℓ_h/D_o and ℓ_t/D_o for ℓ, ℓ_h, and ℓ_t, and dimensionless volumes of cavity, v/v_w, v_h/v_w and v_t/v_w for v, v_h and v_t, respectively, were used as dimensionless indices of the breaking efficiencies. Where, D_o is the nozzle diameter and v_w the volume of jet water in the jet operation time 10 s.

Figures 8-1 (a) to 8-2 (f) show both relations between ℓ/D_o and L/D_o, and between ℓ_h/D_o and L/D_o at every tested temperature, θ, of the preheated specimens excepting θ=100°C. Also, Figs. 9-1 (a) to 9-2 (f) show both relations between v/v_w and L/D_o, and between v_h/v_w and L/D_o at various specimen temperature, θ, except θ=100°C. Figures on the ℓ/D_o-L/D_o, the ℓ_h/D_o-L/D_o, the v/v_w-L/D_o and the v_h/v_w-L/D_o curves in the case of θ=100°C are omitted because of minor changes of ℓ/D_o, ℓ_h/D_o, v/v_w and v_h/v_w, respectively between θ=15°C and θ=100°C. In these figures, each closed circle and each triangle represent each mean experimental value for the hot and the cold breaking tests, respectively, and correspond to each closed circle and each closed triangle explained in Fig. 6 and Fig.7.

The relation between ℓ_h/D_o and L/D_o shown in Fig. 8-1 (a) and the relation between v_h/v_w and L/D_o shown in Fig. 9-1 (a) illustrate that at 15 C there are maximum breaking efficiencies at an optimum standoff distance between L/D_o=200 and 400. Profiles of the ℓ_h/D_o-L/D_o and the v_h/v_w-L/D_o curves are popular in cases of the water jet drilling and cutting of normal temperature specimens in which such profiles have been shown by many researchers (Refs. 17, 18, 19, 20, 21, 22, 23 and 24). According to these researchers, nozzle configuration (Refs. 21 and 23), surface finish in the nozzle interior (Ref. 20), nozzle diameter (Refs. 19, 20 and 23), nozzle pressure or jet velocity (Refs. 17, 18, 19, 20 and 24), fluid additives to jet liquid (Ref. 18) , kind of materials (Refs. 17, 19 and 22) and jet surrounding (in air or in water) (Refs. 21 and 22) etc. affect the maximum breaking efficiencies and the optimum standoff distance to give them.

The followings may be suggested from the figures shown in Figs. 8-1 and 8-2 obtained in this study. ℓ_h/D_o increased slightly with increasing L/D_o, reached a muximum at an optimum standoff distance between L/D_o=200 and 400, and decreased slowly in cases of θ=15 and 200°C. In cases of higher temperature than θ=300°C, each value of ℓ_h/D_o at smaller L/D_o than L/D_o=28.6 was fairly large, and increased with increasing θ. So, each ℓ_h/D_o in these temperature range decreased with increasing L/D_o at smaller L/D_o than L/D_o=28.6, reached each minimum at about L/D_o=30, further increased with increase of L/D_o slowly, reached each maximum at about L/D_o=400, and again diminished steeply with L/D_o. This peculiar phenomena which appeared at L/D_o smaller than L/D_o=28.6 when θ is higher than 300°C seems to be caused by the stronger wedge effect of the water jet at the smaller L/D_o and the more fissures of the specimen preheated to the higher temperatures. Further the maximums of ℓ_h/D_o as well as the other values of ℓ_h/D_o at different L/D_o tested increased with increasing θ. This obviously shows that the mechanical resistivity of the cement mortar specimens against the water jet became weaker by preheating of the specimens.

It is no wonder that the above discussions on the ℓ_h/D_o-L/D_o relations are also applied to the v/v_w-L/D_o relations shown in Figs. 9-1 (a) to 9-2 (f) for the same reasons.

Next, let us now consider the results of the hot breaking tests briefly in order. These results are shown by the closed circles linked with solid lines as the ℓ/D_o-L/D_o relations in Figs. 8-1 and 8-2 and as the v/v_w-L/D_o relations in Figs. 9-1 and 9-2. The ℓ/D_o-L/D_o relations are divided roughly into two categories by the values of the specimen temperature, θ, in this study. In cases of θ=200, 300 and 400°C, the ℓ/D_o at the smaller L/D_o and at the higher θ was the largest. The ℓ/D_o which decreased rather quickly with increasing L/D_o, reached each minimum at about L/D_o=100, and further increased moderatly with increase of L/D_o. Furthermore, the ℓ/D_o passed through each maximum at a L/D_o between 200 and 400 with increasing θ and then decreased steeply. Another pattern appeared in the cases of θ=500 and 600°C. In these cases, the ℓ/D_o-L/D_o relation curve had two peaks (one of them is large and at about 30 of L/D_o, the another small and at L/D_o=200) and a valley at L/D_o=100.

The profiles of the v/v_w-L/D_o relation curve shown in Figs. 9-1 and 9-2 were somewhat analogous to those of the ℓ/D_o-L/D_o curve described above. Each curve in both groups had something common with a valley or ℓ/D_o at L/D_o=100 and a peak/hill at L/D_o=28.6 in cases of having a peak/hill at L/D_o smaller than 100. Changes of the profiles of the v/v_w-L/D_o curve by the specimen temperature, θ, were complicated and could be divided into following four types: (a) one valley-one hill type at θ=200°C. (b) one valley-two hills type at θ=300°C, (c) one peak-one valley-one hill type at θ=400 and 500°C, and (d) one hill-one valley type at θ=600°C.

As described above, the relations between ℓ/D_o and L/D_o as well as between v/v_w and L/D_o are very complex, because they may be affected by jet structures, impinging water jet heat transfer accompanied by boiling on the hot surfaces and properties of target materials. Furthermore, the first and the second are also functions of the dimensionless standoff distance, L/D_o, and the second and the third are functions of the specimen temperature, θ. This fact complicates further an analytical explanation on the breaking efficiencies of high temperature target materials by impinging water jets.

Somewhat detailed descriptions about pure thermal effects on the breaking efficiency, the cracking mechanism for hot target materials with impinging water jets at relatively short standoff distances and on thermal spallability indexes have been reported elsewhere (Refs. 9,10, 11, 12 and 13). However, the analytical solution of the more complex problem which appeared in this study will be postponed to a future time.

4.5 Breaking Efficiency Augmentation Due to Preheating.

It was found that preheating concrete was a very effective method to break concrete by the water jet as previously stated. In order to show the effect of concrete preheating on the water jet breaking of concrete practically, the next two breaking efficiency augmention ratios were defined. One of them is a breaking depth augmentation ratio, $\ell_\theta/\ell_{\theta=15^oC}$ and the another is a breaking volume augmentation ratio, $v_\theta/v_{\theta=15^oC}$. These ratios, namely, $\ell_\theta/\ell_{\theta=15^oC}$ and $v_\theta/v_{\theta=15^oC}$, are defined as the ratios of the breaking depth and the volume of the cavities yielded by impinging of water jets in the cases of specimens preheated to each set temperature, θ, against those of unpreheated specimens (namely, θ=15°C), respectively.

Figures 10 and 11 represent $\ell_\theta/\ell_{\theta=15^oC}$ and $v_\theta/v_{\theta=15^oC}$ against dimensionless distance, L/D_o, with θ as a parameter, respectively. In Fig. 11, each $v_\theta/v_{\theta=15^oC}$-$L/D_o$ curve at smaller L/D_o than about 90 is drawn as a broken line. This means that the reliability of each curve decreases a little at this L/D_o range, because the size of measurement errors becomes larger with decreasing L/D_o for the unpreheated specimen because of the small volume of the cavity produced by the water jet. The followings are inferred from the above figures.

For the ratio, $\ell_\theta/\ell_{\theta=15^oC}$: (1)The ratio increases with increasing θ. (2) In most cases, the ratios has each maximum at $L/D_o \doteqdot 30$ and has each minimum at a certain L/D_o between 200 and 400. (3) The largest maximum ratio in the tested θ and L/D_o ranges was obtained at $L/D_o \doteqdot 30$ in the case of θ =600°C and was about 28.

For the ratio, $v_\theta/v_{\theta=15^oC}$: (1) The ratio increases with increasing θ. (2) The ratio is the larger at the smaller L/D_o and at the larger L/D_o than 400. (3) The ratio at the different tested θ has each minimum at a certain L/D_o between 200 and 400. (4) The largest maximum ratio in the tested θ and L/D_o ranges was obtained at L/D_o =600 and θ =600°C and was about 100.

Now, let us try to predict the breaking efficiencies for some preheated concretes in assumed cases using normal jet drilling data for unpreheated concretes by practical high pressure water jets.

Two of the authors carried out an experiment on water jet drillings and cuttings for unpreheated concretes by using practical high pressure water jets (Ref. 25). Main experimental conditions on them are shown in the left four columns in Table 2. A few parts of the data obtained are shown in Fig. 12 and Fig. 13 using open symbols. Figure 12 shows the relations between the depth of the cavity, ℓ, broken by the impinging water jets and the compressive strength of the unheated concretes tested, σ_c. Also, Fig. 13 shows the relations between the volume of the cavity, v, and σ_c. In these figures, open circles, open squares and open triangles represent the data for each tested water jets having nozzle pressure P_o=68.6, 294.2 and 441.3 MPa, respectively.

If, preheating temperature, θ, for the concretes is assumed at 300 or 600°C, the breaking depth augmentation ratio, $\ell_{\theta=300^oC}/\ell_{\theta=15^oC}$ or $\ell_{\theta=600^oC}/\ell_{\theta=15^oC}$ for each tested nozzle exit pressure, P_o, is obtained from Fig. 10 and the conditions given in Table 2 as shown in the right fourth or third columns of the same Table, respectively. In the same manner, breaking volume augmentation ratios $v_{\theta=300^oC}/v_{\theta=15^oC}$ and $v_{\theta=600^oC}/v_{\theta=15^oC}$ for each P_o are given from Fig. 11 as shown in the right second and first columns of Table 2, respectively.

Predicted values of the depth of the cavities for the preheated concretes on a hypothesis can be calculated for each P_o as each product of ℓ for the unpreheated concretes (real data) and each value of $\ell_\theta/\ell_{\theta=15^oC}$ estimated above in Table 2. These results are shown as plots of modified symbols for θ=300°C and solid ones for θ=600°C in Fig. 12. Equally, results for the volume of the cavities are illustrated in Fig. 13.

These predicting methods are based on a following analogy. l_θ, the depth of cavities yielded by the hot breaking test for the specimen preheated to temperature θ, must depend on nozzle diameter, D_o, static pressure at the nozzle exit, P_o, strengths of the target material, for example, compressive strength, σ_c, and jet operation time, τ. This relation may be expressed as follows:

$$\ell_\theta = C \cdot D_o^{\,h} \cdot P_o^{\,i} \cdot \left(\frac{1}{\sigma_c}\right)^{j} \cdot \tau^{k} \qquad (1)$$

where C is a dimensionless factor and h, i, j and k are dimensionless exponents. Also, C must be a function of dimensionless standoff distance, L/D_o, and specimen temperature, θ, as we can infer from the experimental results. Thus,

$$C = C_\ell\left(\frac{L}{D_o}, \theta\right)$$

Substituting this into Eq.(1) gives:

$$\ell_\theta = C_\ell\left(\frac{L}{D_o}, \theta\right) \cdot D_o^{\,h} \cdot P_o^{\,i} \cdot \left(\frac{\ell}{\sigma_c}\right)^{j} \cdot \tau^{k} \qquad (2)$$

Similarly, for $\ell_{\theta=15^\circ C}$, the depth of cavities yielded by the breaking test for the normal temperature target material,

$$\ell_{\theta=15^\circ C} = C_2\left(\frac{L}{D_o}\right) \cdot D_o^{\,h'} \cdot P_o^{\,i'} \cdot \left(\frac{1}{\sigma_c}\right)^{j'} \cdot \tau^{k'} \qquad (3)$$

where $C_2(L/D_o)$ is a dimensionless factor and h', i', j' and k' are dimensionless exponents, also.

If it shall be assumed that relations of h'= h, i'= i, j'= j and k'= k exist among the exponents between Eq.(2) and Eq.(3), substituting Eq.(3) into Eq.(2) yields

$$\ell_\theta = C_\ell\left(\frac{L}{D_o}, \theta\right) \cdot \frac{\ell_{\theta=15^\circ C}}{C_2\left(\frac{L}{D_o}\right)}$$

$$= C_3\left(\frac{L}{D_o}, \theta\right) \cdot \ell_{\theta=15^\circ C} \qquad (4)$$

where $C_3(L/D_o; \theta)$ is a modified dimensionless factor and equal to $C_\ell(L/D_o, \theta)/C_2(L/D_o)$.

Equally, for v_θ, the volume of cavities yielded by the hot breaking test for the specimen preheated to θ, a following resultant equation is given.

$$v_\theta = C_4\left(\frac{L}{D_o}, \theta\right) \cdot v_{\theta=15^\circ C} \qquad (5)$$

At the above equations, factors C_3 in Eq.(4) and C_4 in Eq.(5) must be obtained by experiments and are shown in Fig. 10 and Fig.11, respectively.

These results show that the demolishing efficiencies of the concretes by the water jets are obviously augmented if the concrete samples are preheated before the water jets impinge on them. However, the validity of these predicting methods will require confirmation by many real data on the preheating concrete breaking by practically high pressure water jet impingements which is going to be conducted in the near future.

5. CONCLUSION

A new application of high speed water jets to demolish concrete buildings and constructions effectively has been basically established by using concrete preheating methods.

The experimental results carried out, as a first step of the study, using cement mortar specimens preheated from 100 to 600°C and unpreheated and using a water jet having a constant speed, U_o, equal to 210 m/s at nozzle exit (D_o=0.5 mm I.D.) were as follows.

1) In both the hot breaking and the cold breaking tests, a hole-shaped cavity, a hole surrounded with eroded area-shaped cavity or a crater-shaped cavity was macroscopically observed in the surface of the specimen struck by the water jet for a constant period of 10 s. The pattern of the hole shape varied characteristically with the preheating temperature, θ, and the dimensionless standoff distance, L/D_o as shown in Figs. 4-1 and 4-2, and Figs. 5-1 and 5-2 for the tested L/D_o range from 20 to 600 and for the tested θ range from 15 to 600°C.

2) The depth of the cavity of the specimen drilled in the hot breaking test, ℓ, was larger than the depth of that in the cold breaking one, ℓ_h, in each case. increased very rapidly with increasing θ within the tested θ in all cases of the tested L/D_o. ℓ_h also increased rapidly with the increase of θ in all cases of the tested L/D_o. This tendency between ℓ_h and θ seems to depend directly on the temperature dependency of the physical properties, especially, porosity, compressive strength and dynamic Young's modulus of the specimen.

3) The volume of the cavity of the specimen broken in the hot breaking test, v, was larger than the volume of that in the cold breaking one, v_h, in every case. In cases of L/D_o=20, 28.6 and 100, v increased very rapidly with increasing θ, reached the maximum at a certain θ equal from 400 to 500°C, and decreased with θ. v_{max}, each peak value of v on the v-θ curves reached equally about 0.1 cm^3 in the above cases. In cases of L/D_o=200, 400 and 600, v had no v_{max} and increased progressively with increasing θ. The volume of the cavity of the specimen broken in the cold breaking test, v_h, increased rather rapidly with increasing θ in all cases of the tested L/D_o except L/D_o=20. In the case of L/D_o=20, the v_h-θ curve had a maximum at θ=300°C and a minimum at θ=400°C.

4) In order to show the effect of concrete preheating on the water jet breaking of concrete practically, the breaking depth augmentation ratio, $\ell_\theta/\ell_{\theta=15°C}$, and the breaking volume augmentation ratio, $v_\theta/v_{\theta=15°C}$ were defined.

For the ratio, $\ell_\theta/\ell_{\theta=15°C}$: (1) The ratio increases with increasing θ. (2) In most cases, the ratio has each maximum at L/D_o=30 and has each minimum at a certain L/D_o between 200 and 400. (3) The largest maximum ratio in the tested θ and L/D_o ranges was obtained at L/D_o=30 in the case of θ=600°C and was about 28.

For the ratio, $v_\theta/v_{\theta=15°C}$: (1) The ratio increases with increasing θ. (2) The ratio is larger at the smaller L/D_o and at the larger L/D_o than 400. (3) The ratio at the different tested θ has each minimum at a certain L/D_o between 200 and 400 . (4) The largest maximum ratio in the tested θ and L/D_o ranges was obtained at L/D_o =600 and θ=600°C and was about 100.

ACKNOWLEDEMENTS

The authors wish to express their thanks to Dr. Enomoto, H., Mr. Uchimi, E. and Mr. Kanno, A. of Tohoku University for their help. The authors also acknowledge the assistance of Mr. Kawakita, Y. in conducting some of the experiments as a student in Tohoku University then.

REFERENCES

1. Kakizaki, M., Harada, M. and Nishikawa, I. : "Demolition method of building.". (Biru kaitai koho). Tokyo. Kajima-shuppan-Kai., 1973, (In Japanese).

2. Kasai, Y. (ed.): "Demolition methos of concreate constraction.". (Konkuriito Kouzoubutsu no kaitai koho). Tokyo. Nikkan-kogyo-shinbunsha Co., 1973, (In Japanese).

3. Frank, J.N. and Chester, J.W. : "Fragmentation of concrete with hydraulic jets". U.S. Bureau of Mines, RI-7572. (PB 205 499). 1971.

4. McCurrich, L.H. and Browne, R.D. : "Application of water jet cutting technology to cement grouts and concrete". In: Proc. 1st Int. Sym. Jet Cutting Technology (Coventry, U.K. : Apr. 5-7, 1972), Cranfield, U.K., BHRA Fluid Engineering, 1972, Paper G7, 23 pp.

5. Hilaris, J.A. and Labus, T.J. : "Highway maintenance application of jet cutting technology.": In: Proc. 4th Int. Sym. Jet Cutting Technology (Canterbury, U.K. : Apr. 12-14, 1978), Cranfield, U.K., BHRA Fluid Engineering, 1978, Paper Gl, 8 pp.

6. Yie, G.G., Burns, D.J. and Mohaupt, U.H. : "Performance of a high-pressure pulsed water-jet device for fracturing concrete pavement". In: Proc. 4th Int. Sym. Jet Cutting Technology (Canterbury, U.K. : Apr. 12-14, 1978), Cranfield, U.K., BHRA Fluid Engineering. 1978, Paper H6, 19 pp.

7. Reichman, J.M., Kirby, M.J. and Rodenbaugh, T.J. : "The development of a water jet cutting system for trenching in concrete". In: Proc. 5th Int. Sym. Jet Cutting Technology (Hanover, E.R.G. : June 2-4, 1980), Cranfield, U.K., BHRA Fluid Engineering, 1980, Paper Dl, 13 pp.

8. Vallve, F.X., Mohaupt, U.H., Kalbfleish, J.G. and Burns, D.J.: "Relationship between jet penetration in concrete and design parameters of a pulsed water-jet machine". In: 5th Int. Sym. Jet Cutting Technology (Hanover, F.R.G. : June 2-4, 1980), Cranfield, U.K., BHRA Fluid Engineering, 1980, Paper D2, 13 pp.

9. Kiyohashi, H, Kyo, M. and Ishihama, W. : "Effect of rock temperature on the thermal fracturing of an imitation rock specimen-Studies on the fracturing of hot dry rocks by high speed water jets". (Mogi-ganseki no netsu hasai ni oyobosu ganseki ondo no eikyo- Kousoku suifunryu niyoru kouon gantai no hasai ni kansuru kenkyu). J. Mining and Metal. Inst. Japan, 94, 1086, Aug. 1978, pp. 515-521. (In Japanese).

10. Kiyohashi, H., Kyo, M. and Ishihama, W. : "Thermal fracturing patterns and effects of an imitation hot dry rock by impinging of water jets". "Alternative Energy Sources (ed. Veziroglu, T.N.)", Vol. 6. Washington, Hemisphere Pub. Corp., 1978, pp. 2727-2745.

11. Kiyohashi, M., Kyo, M. and Ishihama, W. : "Water jet breaking of imitation hot dry rock". In: Proc. 4th Int. Sym. Jet Cutting Technology (Canterbury, U.K.: Apr. 12-14, 1978), Cranfield, U.K., BHRA Fluid Engineering, 1978, Paper C2, 21 pp .

12. Kiyohashi, H., Kyo, M. and Ishihama, W. : "Effects of rock temperature and standoff distance on water jet drilling performance for an imitation hot dry rock ". Technology Reports, Tohoku Univ., 45, 2, Dec. 1980, pp. 137-168.

13. Kiyohashi, H., Kyo, M. and Ishihama, W. : "The influence of standoff distance on fracturing patterns and effects of an imitation hot dry rock by impinging of water jet." "Alternative Energy Sources (ed. Veziroglu, T.N.)", Vol. 5. Washington, Hemisphere Pub. Corp. 1981, pp. 1965-1990.

14. Maurer, W.C. : "Thermal spalling experimental drills use heat, electricity, microwaves, chemicals". E/MJ, 169, 6, June 1968, pp. 101-106.

15. Yanaida, K. and Ohashi, A. : "Studies on the flow characteristics of the high speed water jets in air-The flow characteristics of continuous water jets (1st Report) -".. (Kichu kousoku suifunryu tokusei ni kansuru kenkyu-Renzokuryu ryoueki ni tsuite (dai ittsupou). J. Mining and Metal. Inst. Japan, 93, 1072, June, 1977, pp, 423-428. (In Japanese).

16. Yanaida, K. and Ohashi, A. : "Studies on the flow characteristics of the high speed water jets in air-The breakup characteristics of the water jet atomizer

(2nd Report)-".. (Kichu Kousoku suifunryu tokusei ni kansuru kenkyu-Muka ekitekiryu ryoueki ni tsuite (dai nihou). J. Mining and Metal. Inst. Japan, 93, 1073, July, 1977, pp. 423-428. (In Japanese).

17. Franz.N.C. : "The influence of standoff distance on jet cutting with high velocity fluid jets". In: Proc. 2nd Int. Sym. Jet Cutting Technology (Cambridge , U.K. April 2-4, 1974), Cranfield, U.K., BHRA Fluid Engineering, 1974, Paper B3, 10 pp.

18. Franz, N.C. : "The interaction of fluid additives and standoff distance in fluid jet cutting". In: Proc. 3rd Int. Sym. Jet Cutting Technology (Chicago, U.S.A. : May 11-13, 1976), Cranfield, U.K., BHRA Fluid Engineering, 1976, Paper A5, 9 pp.

19. Cheung,J.B. and Hurlburt, M.S. : "Submerged water-jet cutting of concrete and granite". In : Proc. 3rd. Int. Sym. Jet Cutting Technology (Chicago, U.S.A. : May 11-13, 1976), Cranfield, U.K., BHRA Fluid Engineering, 1976, Paper E5, 14 pp.

20. Barker, C.R. and Selberg, B.P. : "Water jet nozzle performance tests." In: Proc. 4th Int. Sym. Jet Cutting Technology (Canterbury, U.K. : April 12-14, 1978), Cranfield, U.K., BHRA Fluid Engineering, 1978, Paper A1, 20 pp.

21. Erdmann-Jesnitzer, F., Hassan , A.M. and Louis, H. : "A study of the effect of nozzle configuration on the performance of submerged water jets". In: Proc. 4th Int. Sym. Jet Cutting Technology (Canterbury, U.K. : April 12-14, 1978), Cranfield, U.K., BHRA Fluid Engineering, 1978, Paper A2, 18 pp.

22. Erdmann-Jesnitzer, F., Louis, H. and Wiedemeier, J. : "Material behaviour, material stressing, principle aspects in the application of high speed water jets ". In: Proc. 4th Int. Sym. Jet Cutting Technology (Canterbury, U.K. : April 12-14, 1978), Cranfield, UK., BHRA Fluid Engineering, 1978, Paper E3, 16 pp.

23. Houlston, R. and Vickers, G.W. : "Surface cleaning using water-jet cavitation and droplet erosion". In: Proc. 4th Int. Sym. Jet Cutting Technology (Canterbury, U.K. : April 12-14, 1978), Cranfield, U.K., BHRA Fluid Engineering, 1978, Paper H1, 18 pp.

24. Erdmann-Jesnitzer,F., Louis, H. and Wiedemeier, J. : "Rock excavation with high speed water jets-a view on drilling and cutting results of rock materials in relation to their fracture mechanical behaviour". In: Proc. 5th Int. Sym. Jet Cutting Technology (Hanover, F.R.G. : June 2-4, 1980), Cranfield, U.K., BHRA Fluid Engineering, 1980, Paper C3, 14 pp.

25. Yahiro, T. and Yoshida, H. : It will be contributed elsewhere.

Table 1 Physical properties of the specimen.

Properties	Temperature [°C]						
	15	100	200	300	400	500	600
Apparent specific gravity [-]	2.01	1.91	1.86	1.82	1.81	1.79	1.75
True specific gravity [-]	2.70	2.71	2.71	2.73	2.79	2.86	2.93
Porosity [%]	25.6	29.5	31.4	33.3	35.1	37.4	40.3
Compressive strength [MPa]	56.4	45.4	45.3	39.4	33.4	22.1	12.8
Tensile strength [MPa]	2.57	3.41	4.15	3.23	2.16	1.65	0.801
Elastic wave velocity [km/s]	3.8	3.4	3.4	2.9	2.8	2.2	1.1
Static Young's modulus [GPa]	9.71	8.32	8.72	5.37	4.09	2.02	1.04
Dynamic Young's modulus [GPa]	28.8	22.8	21.8	15.5	14.6	8.63	2.35
Brittleness index [-]	22.0	13.3	10.9	12.2	15.5	13.4	16.0
Thermal conductivity [W/(m·K)]	1.38	2.00	0.896	0.628	0.709	-	-

Table 2 Breaking augmentation ratios evaluated on the other different conditions.

Nozzle exit pressure P_o [MPa]	Standoff distance L [mm]	Nozzle diameter D_o [mm]	L/D_o [mm]	depth aug. ratio $l_\theta / l_{\theta=15°C}$ [-]		Volume aug. ratio $V_\theta / V_{\theta=15°C}$ [-]	
				$\theta = 300°C$	$\theta = 600°C$	$\theta = 300°C$	$\theta = 600°C$
68.6	50	1.2	41.7	12	26	(50)	(400)
294.2	50	0.3	167	4	7.5	9	32
441.3	50	0.2	250	4	7	8	28

N.B : ( ) means extrapolated value.

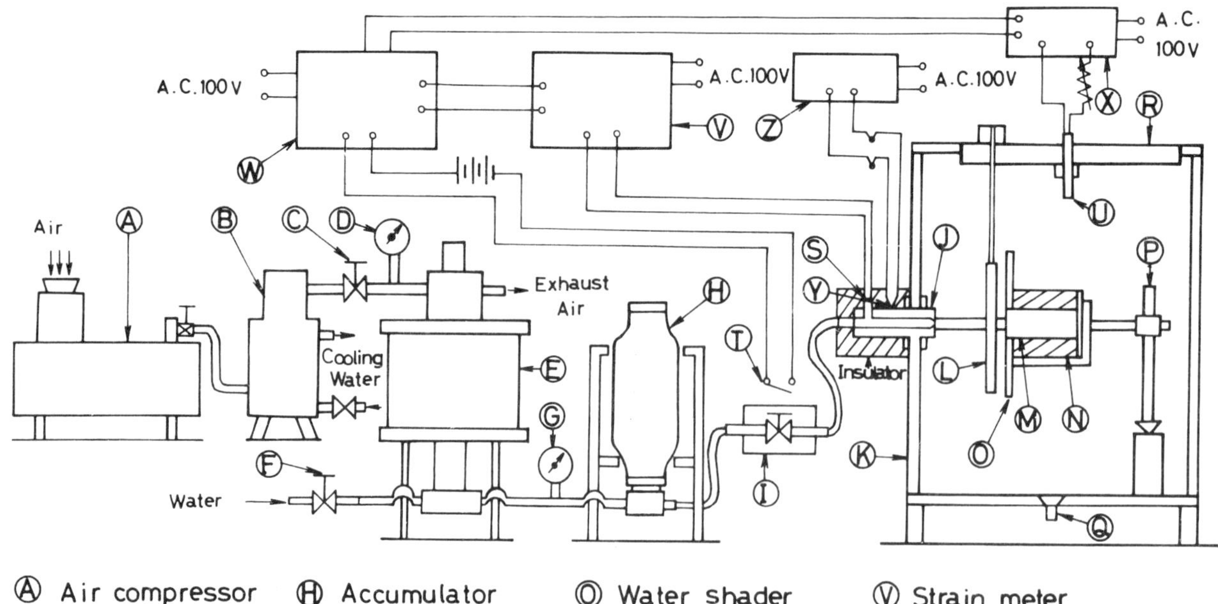

Fig.1 Schematic diagram of the experimental equipment.

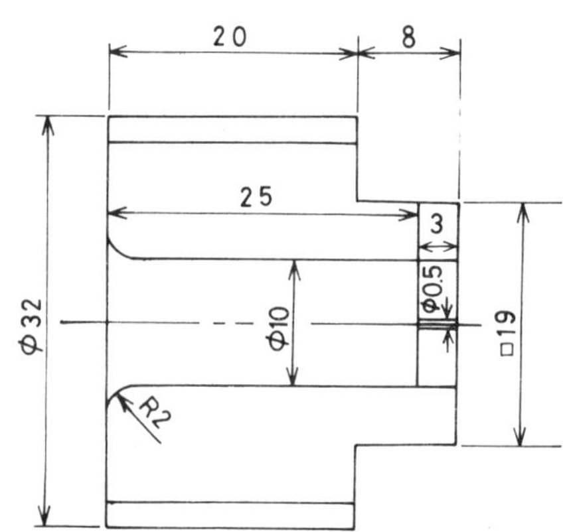

Fig.2 Detail of nozzle design.

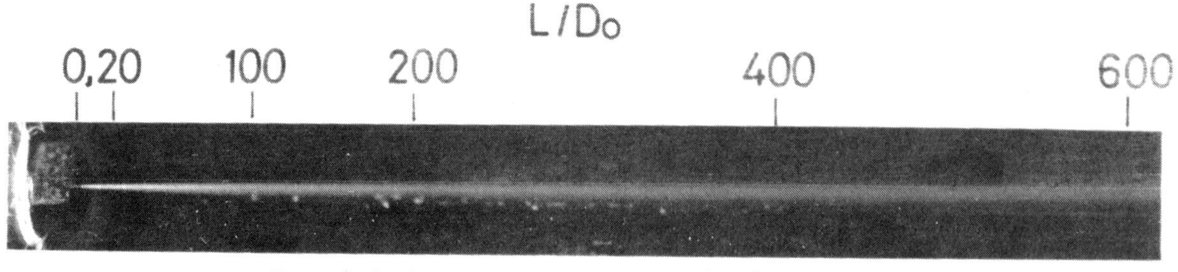

Fig.3 Configuration of the water jet used in this experiment.

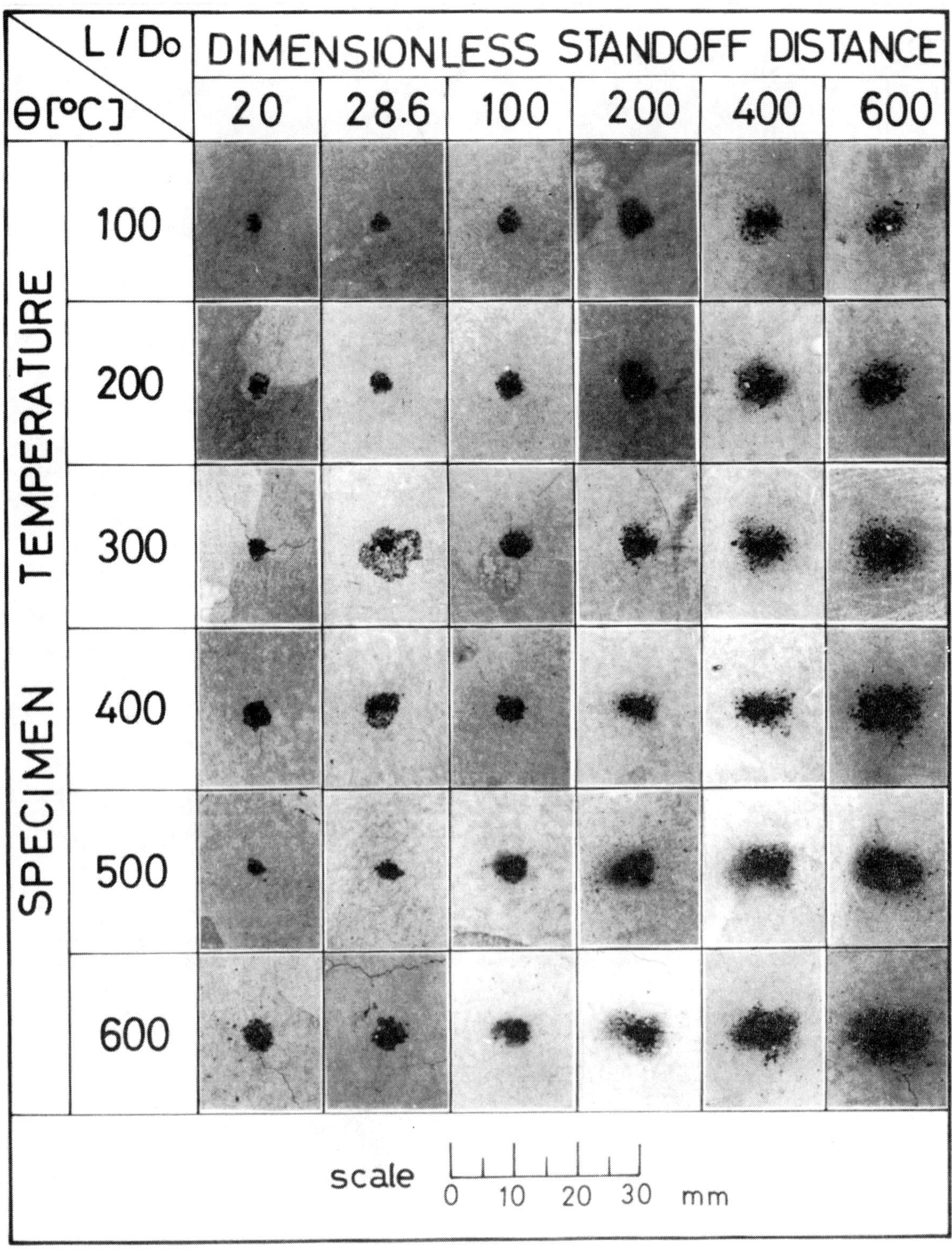

Fig. 4-1 Typical fracture patterns on the surface of the specimens broken by the impinging water jet at the different dimensionless standoff distances and the specimen temperatures tested (the results of the hot breaking test).

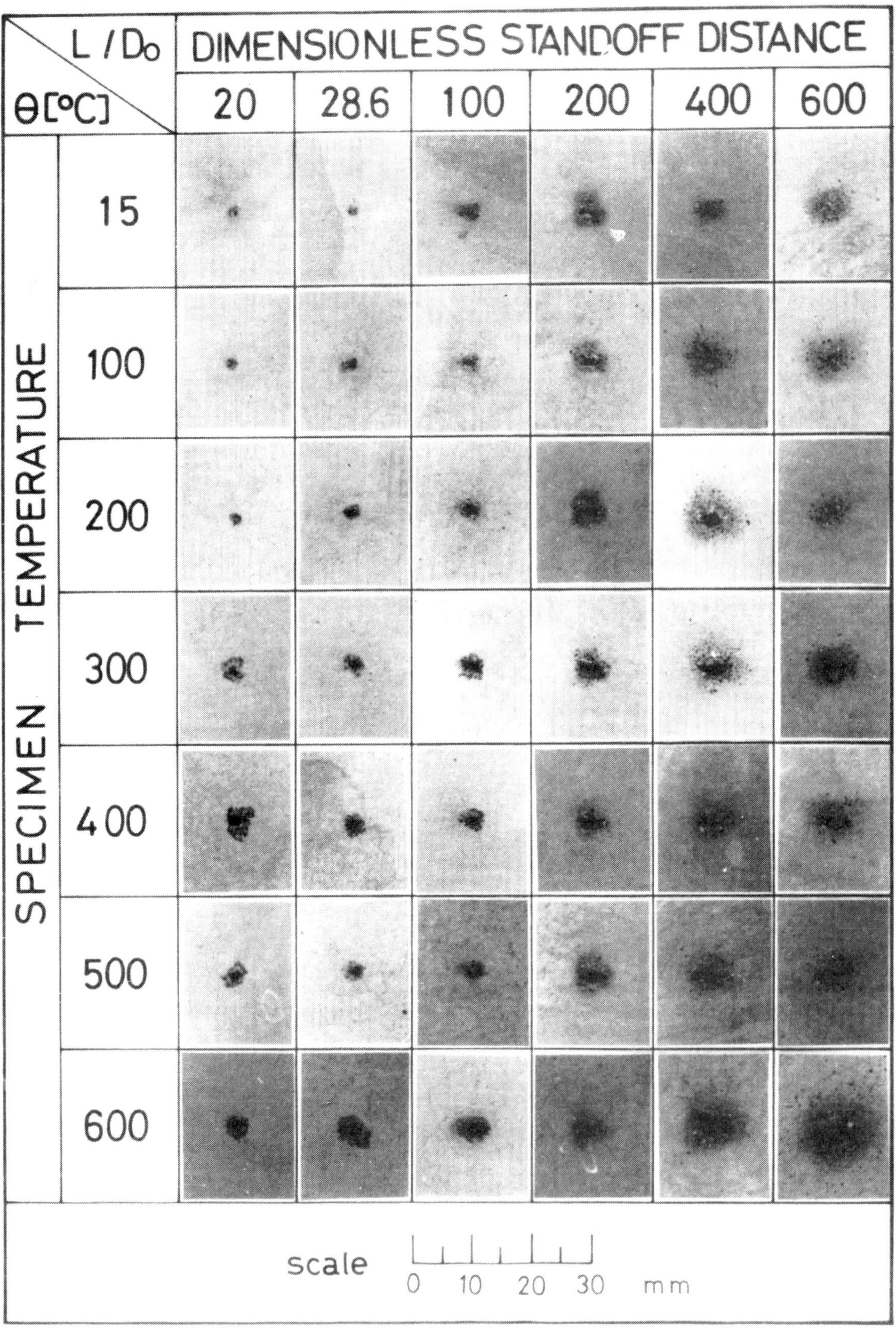

Fig. 4-2 Typical fracture patterns on the surface of the gradually cooled specimens broken by the impinging water jet at the different dimensionless standoff distances (where θ means the hot state temperatures of the specimen before being cooled).

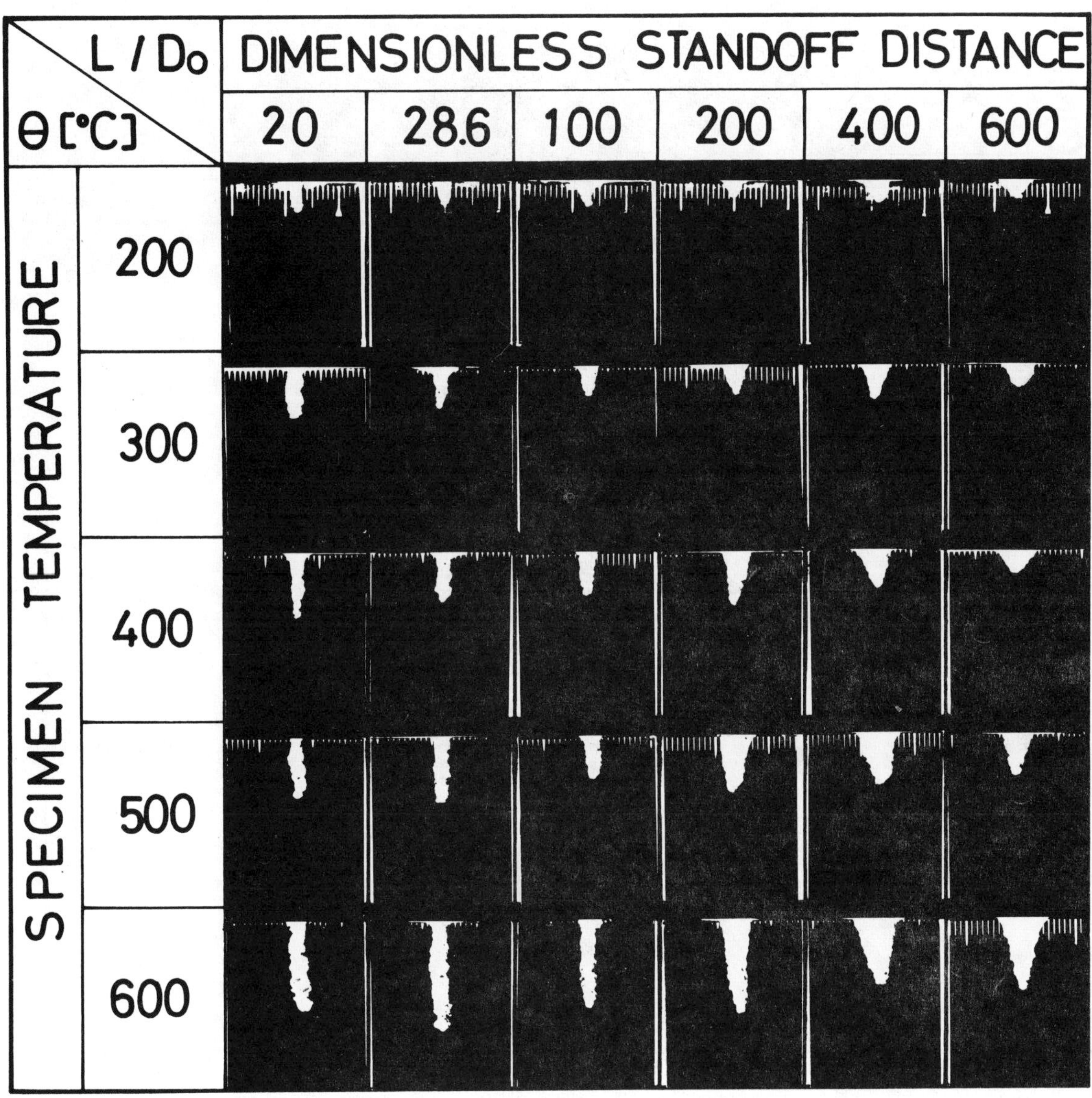

Fig. 5-1 Typical cross sections of the cavity of the specimen broken by the impinging water jet at the different standoff distances tested in cases of the specimen temperature θ=200, 300, 400, 500 and 600°C (the results of the hot breaking tests).

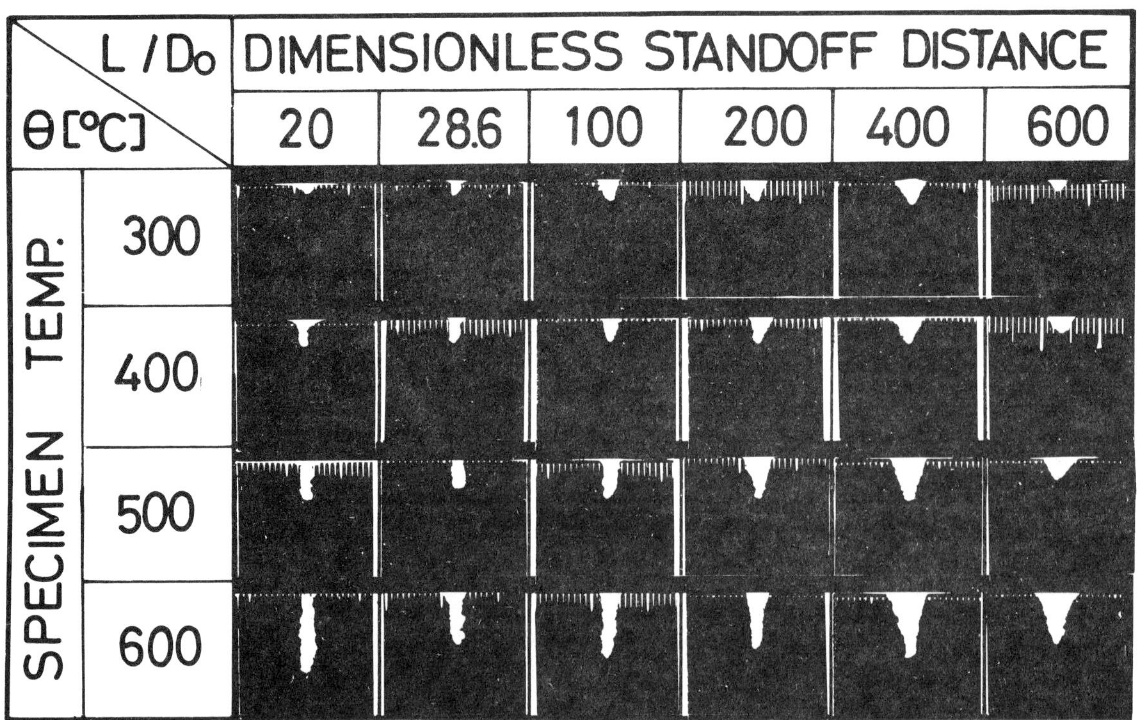

Fig. 5-2 Typical cross sections of the cavity of the gradually cooled specimen broken by the impinging water jet at the different standoff distances tested in cases of the temperature of preheated specimens, θ=300, 400, 500 and 600°C (where θ means the hot state temperatures of the specimen before being cooled).

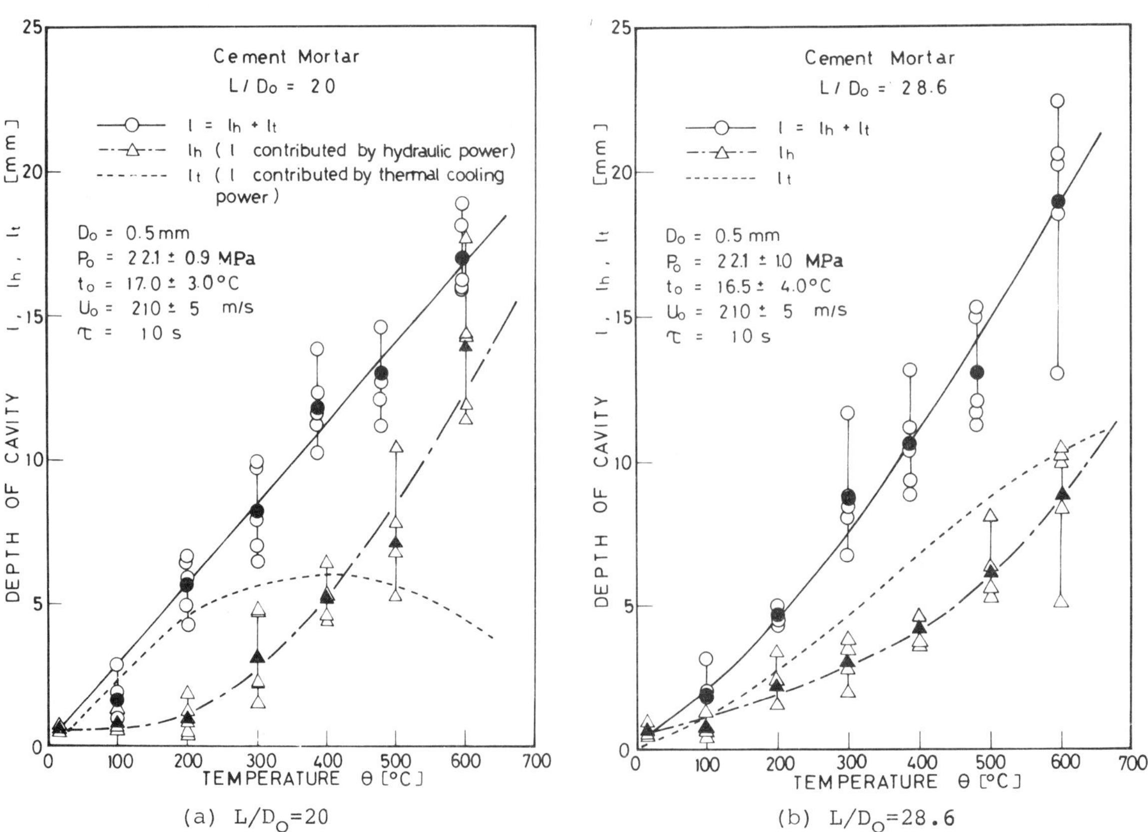

Fig. 6-1 Dependence of depth of cavity by the impinging water jet on specimen temperature in cases of L/D_O=20 and 28.6. Closed figures show mean values.

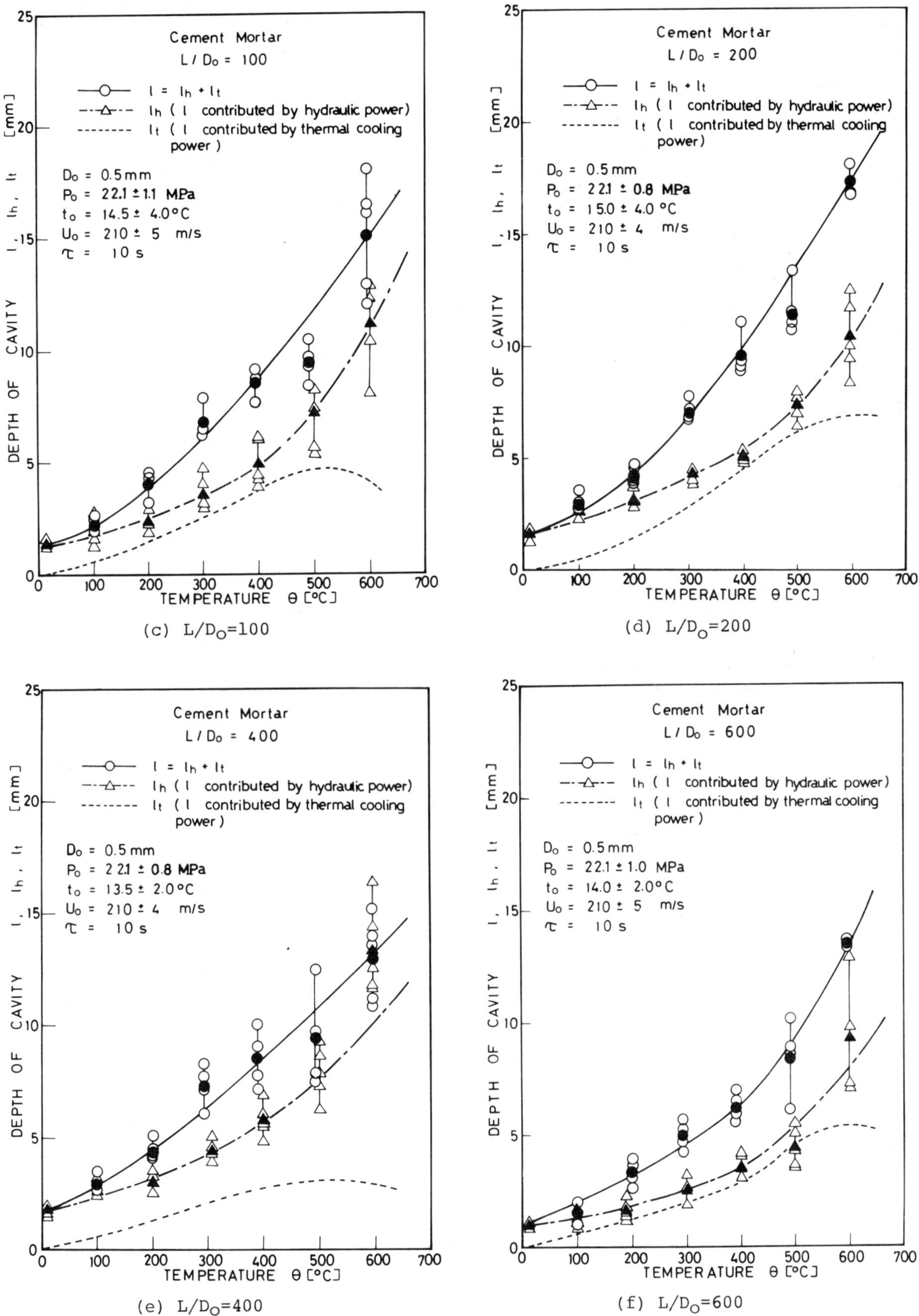

(c) L/D_O=100

(d) L/D_O=200

(e) L/D_O=400

(f) L/D_O=600

Fig. 6-2 Dependence of depth of cavity broken by the impinging water jet on specimen temperature in cases of L/D_O=100, 200, 400 and 600. Closed figures show mean values.

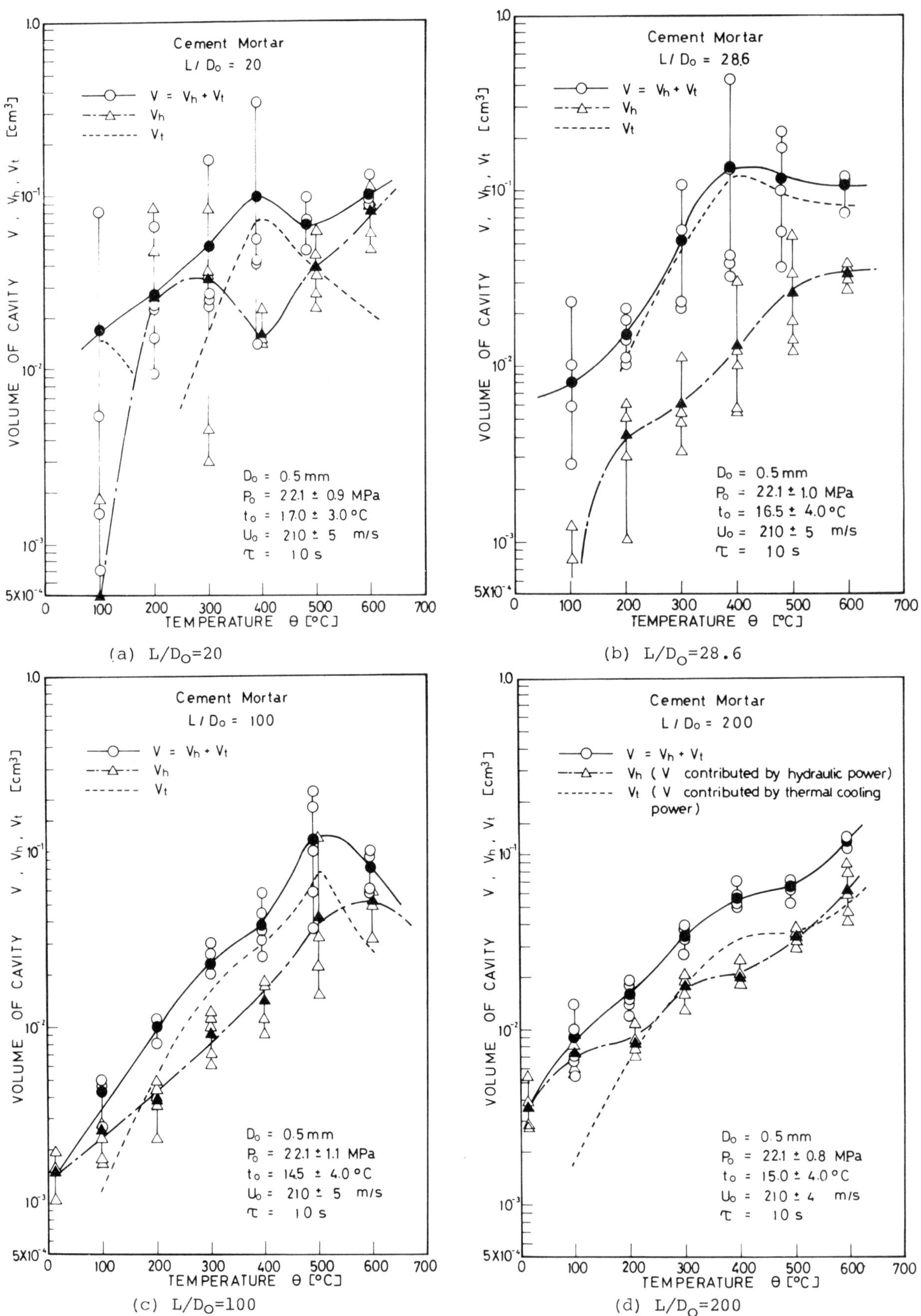

Fig. 7-1 Dependence of volume of cavity broken by the impinging water jet on specimen temperature in cases of L/D_O=20, 28.6, 100 and 200. Closed figures show mean values.

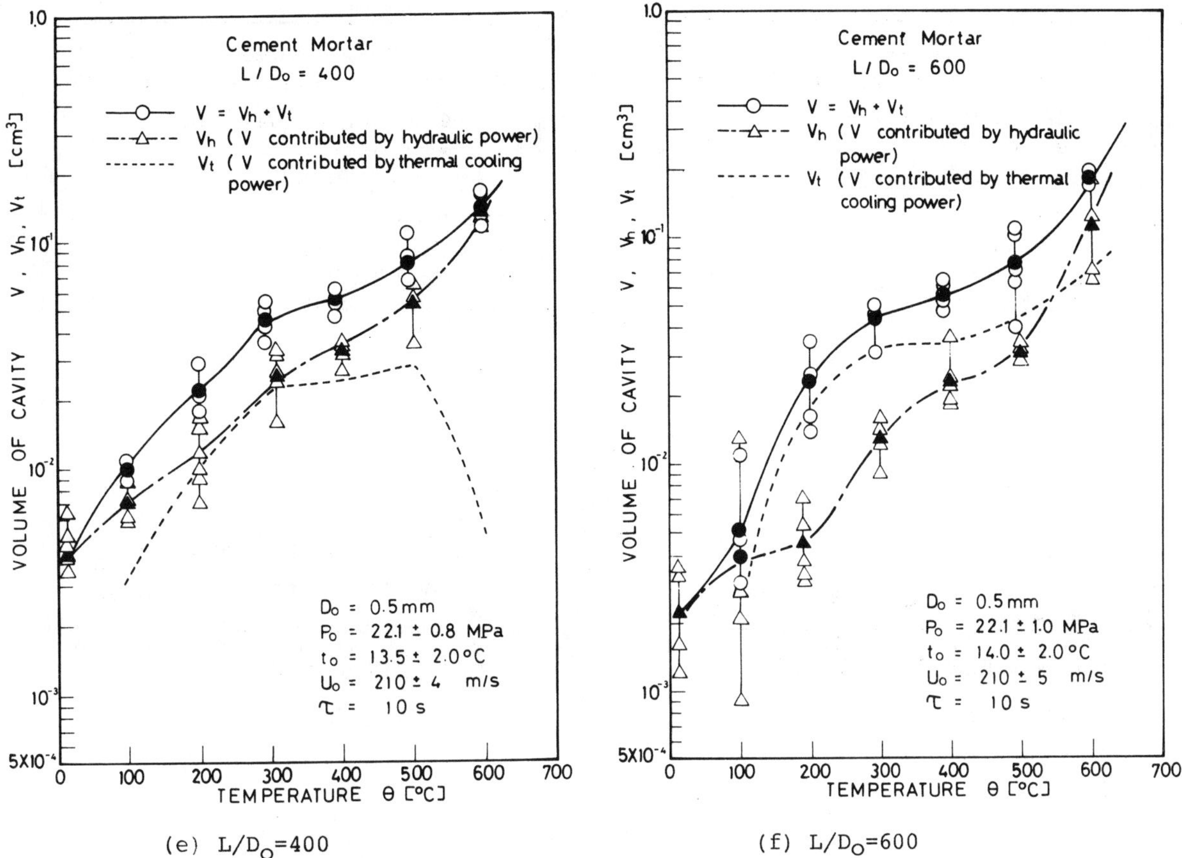

(e) L/D_O=400

(f) L/D_O=600

Fig. 7-2 Dependence of volume of cavity broken by the impinging water jet on specimen temperature in cases of L/D_O=400 and 600. Closed figures show mean values.

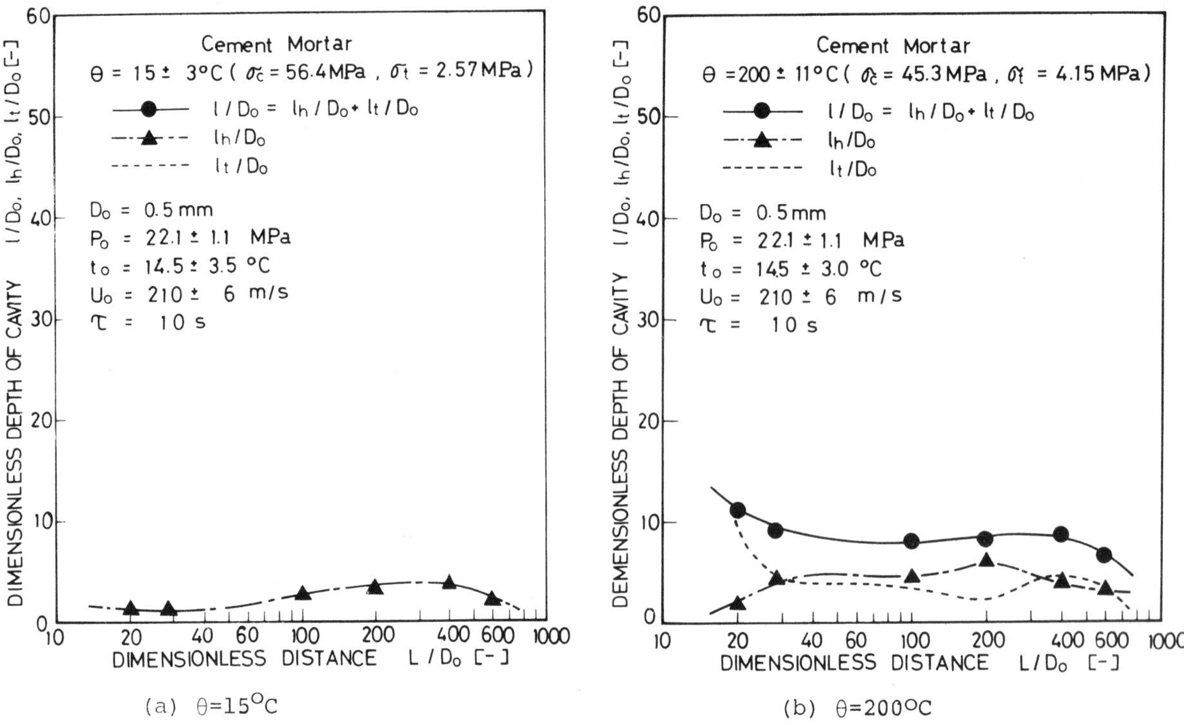

(a) θ=15°C (b) θ=200°C

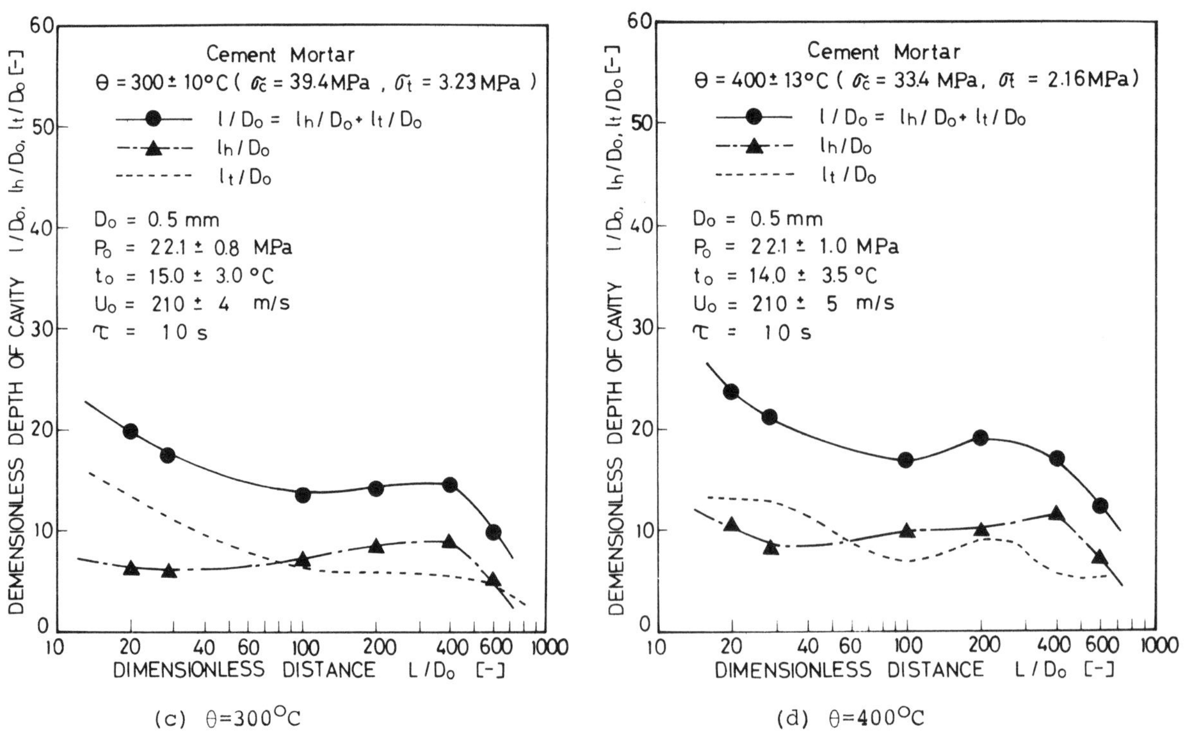

(c) θ=300°C (d) θ=400°C

Fig. 8-1 Dependence of dimensionless depth of cavity broken by the impinging water jet on dimensionless standoff distance in cases of θ=15, 200, 300 and 400°C. Each closed point shows its mean value at the same experimental conditions.

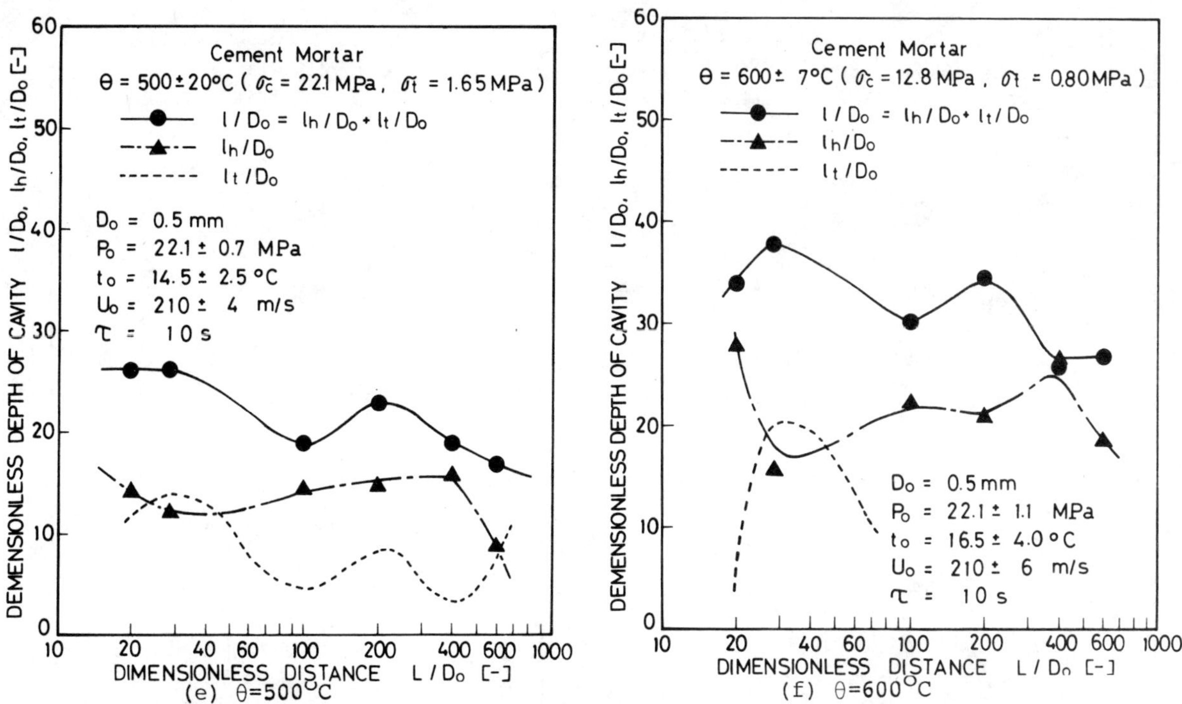

(e) θ=500°C (f) θ=600°C

Fig. 8-2 Dependence of dimensionless depth of cavity broken by the impinging water jet on dimensionless standoff distance in cases of θ=500 and 600°C. Each closed point shows its mean value at the same experimental conditions.

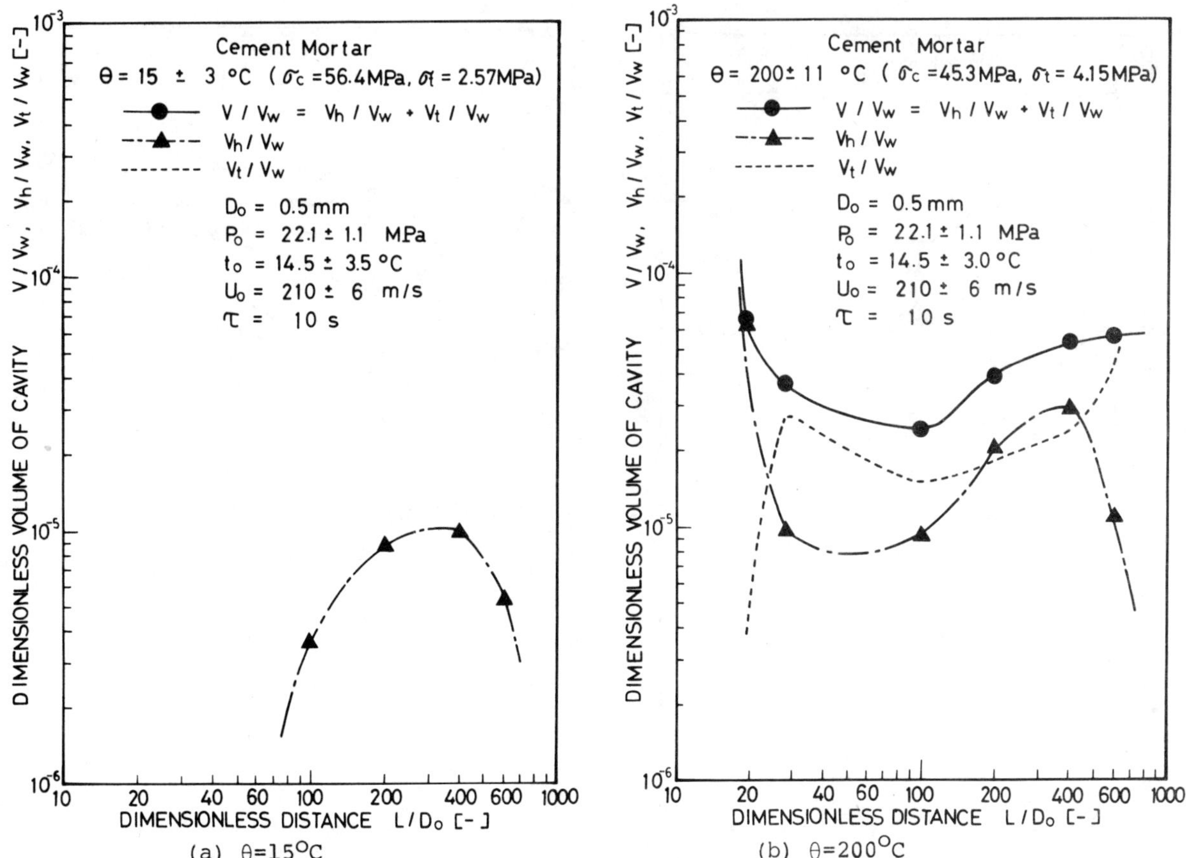

(a) θ=15°C (b) θ=200°C

Fig. 9-1 Dependence of dimensionless volume of cavity broken by the impinging water jet on dimensionless standoff distance in cases of θ=15 and 200°C. Each closed point shows its mean value at the same experimental condition.

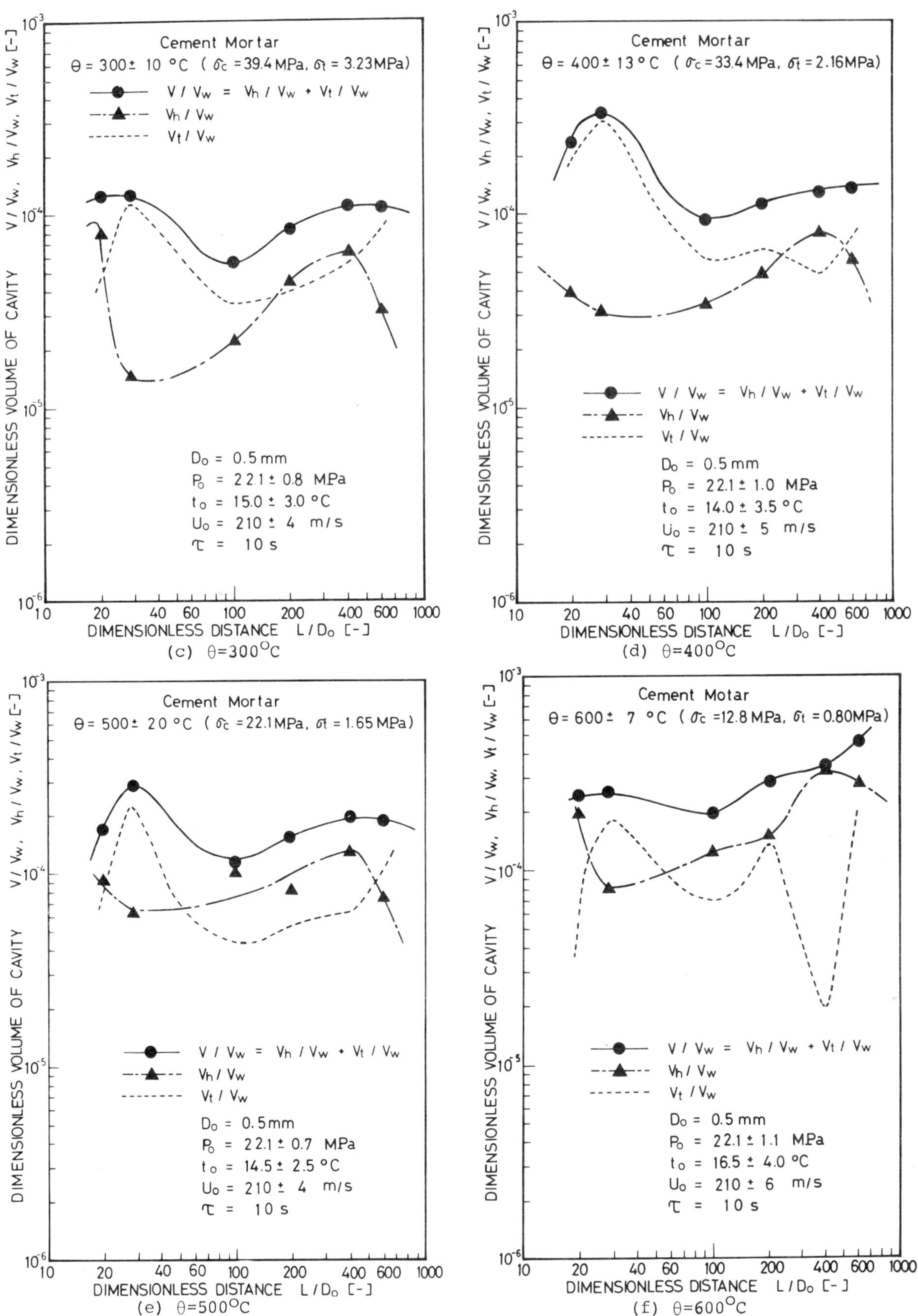

Fig. 9-2 Dependence of dimensionless volume of cavity broken by the impinging water jet on dimensionless standoff distance in cases of θ=300, 400, 500 and 600°C. Each closed point shows its mean value at the same experimental condition.

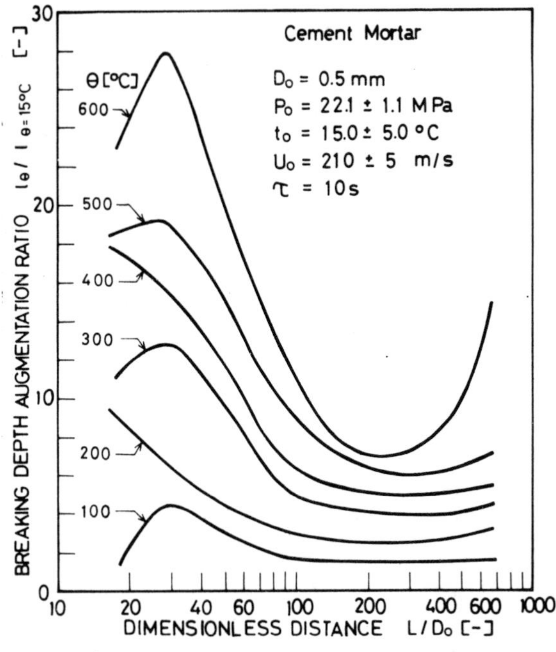

Fig. 10 Breaking depth augmentation ratio versus dimensionless standoff distance in cases of the different specimen temperatures for the preheated cement mortar.

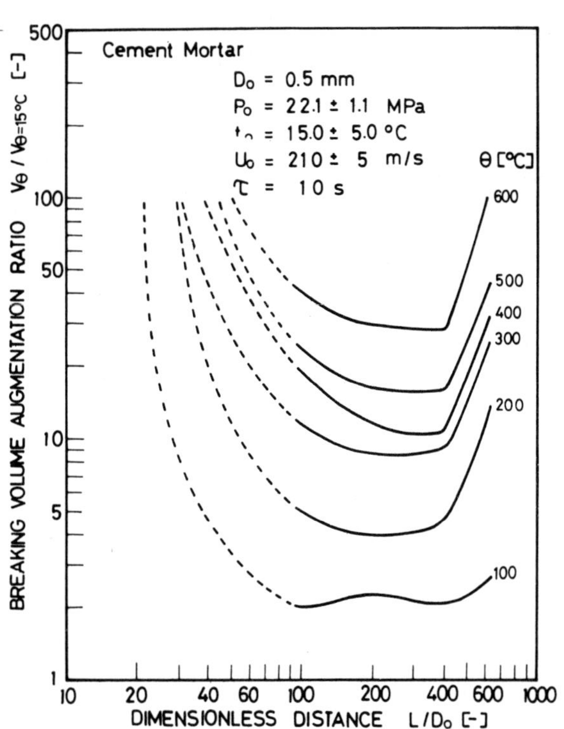

Fig.11 Breaking volume augmentation ratio versus dimensionless standoff distance in cases of the different specimen temperatures for the preheated cement mortar.

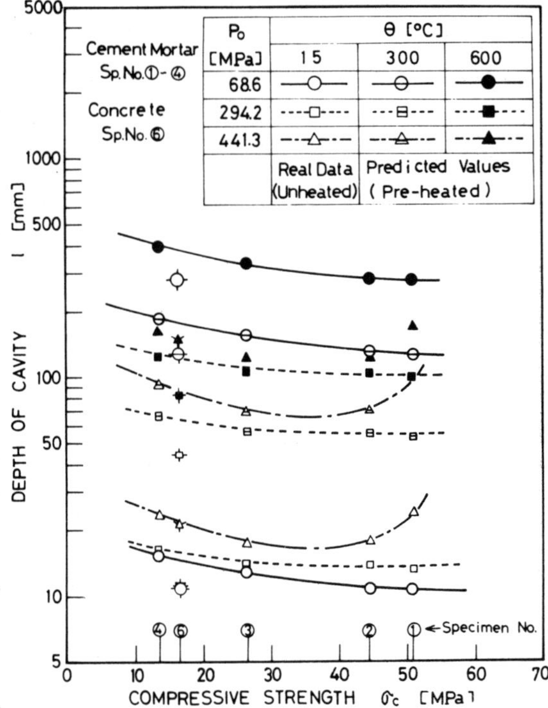

Fig. 12 Dependence of predicted depth of cavity broken by the different impinging water jets on compressive strength of the unheated specimens in cases of preheated specimen temperature θ=300 and 600°C.

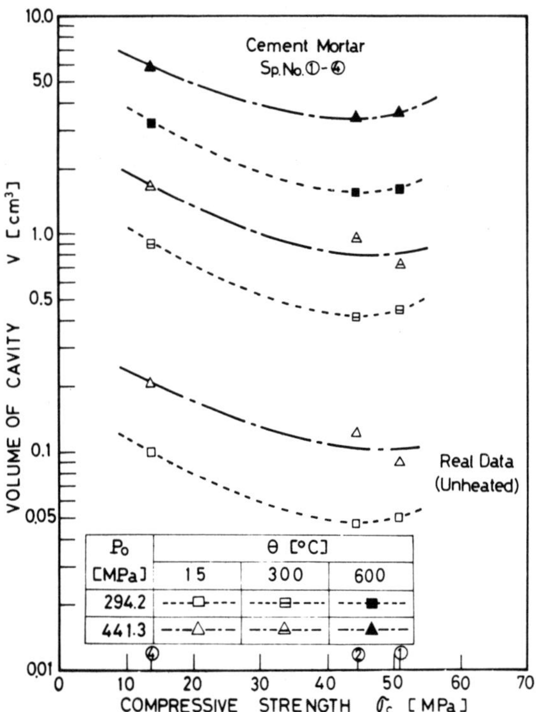

Fig. 13 Dependence of predicted volume of cavity broken by the different impinging water jets on compressive strength of the unheated specimens in cases of preheated specimen temperature θ=300 and 600°C.

6th International Symposium on

Jet Cutting Technology

6-8, April, 1982

PAPER G1

THE DEVELOPMENT OF A WATER JET SYSTEM TO IMPROVE THE PERFORMANCE OF A BOOM-TYPE ROADHEADER

N.A. Plumpton and M.G. Tomlin

National Coal Board Mining Research and Development Establishment, U.K.

Summary

At present over 600 km of underground development roadways are driven each year in UK coalmines. The preferred method of drivage is by boom-type roadheaders, but these are restricted in use by the inability of mechanical picks to cut efficiently in hard strata.

Research has shown that water jet assistance can enhance the cutting performance of mechanical picks. A prototype system for use on a roadheader has been developed in conjunction with the US Department of Energy. Successful field trials with this system have led to the current development of a preproduction version to enable full scale underground trials.

Evolution of the project to date is described, and an outline is given of the engineering problems yet to be solved before a viable system can be produced.

Held at the University of Surrey, U.K.
Symposium organised and sponsored by
BHRA Fluid Engineering

1. INTRODUCTION

Where possible, roadways in UK coalmines are driven by boom-type roadheaders which utilise mechanical cutting of the strata by rotating picks. This method is superior to the use of explosives as it is less time-consuming and laborious and does not induce weaknesses in the surrounding strata which can lead to problems in supporting the roadway and controlling roadway closure. Roadheaders are however normally restricted to use in strata of low compressive strength, owing to the inability of the picks to cut harder rocks efficiently.

Efforts to improve the performance of roadheaders have resulted in the introduction of larger, more powerful machines. These machines incorporate higher pick forces, slower cutting speeds and improved cutting head stability, all of which enable them to cut harder strata. They are, however, expensive and their cutting efficiency is impaired by pick and cutting head failure.

In recent years, research carried out in South Africa (Reference 1) and the USA (Reference 2) has shown that cutting forces on picks can be reduced and pick life enhanced by the application of high pressure water jets to assist cutting. Corresponding research and investigation, carried out by the NCB in collaboration with the United States Department of Energy, also indicated that water jets could be adopted to improve the performance of roadheaders and extend their range of use by enabling them to cut more efficiently in harder strata.

This paper describes the evolution of a project carried out jointly with the United States Department of Energy to produce a water-jet-assist system suitable for use on a standard-type roadheading machine. It covers the initial experimental research, the development and field trials of a prototype system and the current development of a preproduction version suitable for underground use.

2. EXPERIMENTAL ROCK CUTTING TRIALS

An existing test rig situated at the Safety in Mines Research Establishment was used for these trials and is shown in Figure 1. It comprises a rotating faceplate carrying the rock sample and tool holder which is traversed horizontally across the face of the rock to cut a spiral groove.

Initial experiments showed that the optimum impingement point of the jet relative to the pick tip was 2 mm and that the stand-off distance between the nozzle and rock should be as small as possible. However, for practical purposes, the nozzle was mounted within the confines of the tool holder as shown in Figure 2, result in a stand-off distance of 50 mm.

Cutting trials in Darley Dale sandstone - using water pressures between 240 and 700 bars and 1.27 m/s cutting speed - were carried out and the best results were obtained at maximum pressure. At this pressure, average reductions in normal and cutting forces of 50% and 30% were recorded. The results of a typical series of trials are shown in Figure 3.

Some indication that the use of jet assistance would reduce tool wear was also noted, but insufficient rock was available to carry out a thorough investigation.

3. DEVELOPMENT OF PROTOTYPE SYSTEM

3.1 General

The machine selected for equipping with water jet assistance was a Dosco MkIIA Roadheader, shown in Figure 4. This roadheader is by far the most predominant model operated by the NCB and over 550 of them are currently employed on underground drivages. At present their use is normally restricted to cutting in strata with a maximum compressive strength of 70 MPa, so that major benefits were envisaged in extending its cutting range to harder strata if water jet assistance proved successful.

The prototype water-jet-assist system consists of three basic units:

- A high pressure water pump
- A rotary seal around the cutting head shaft
- A cutting head fitted with high pressure nozzles.

These either replace or are fixed to existing machine components which themselves require little or no modification.

Figure 5 shows the roadheader equipped with the prototype water jet system and a description of the components follows.

3.2 High Pressure Water Pump

The type of pump selected was a hydraulically powered intensifier as used in the experimental rock cutting trials. The pump is shown in Figure 6 and has a 75 kW electric motor drive, giving a maximum pressure output of 700 bars at 45 litres/min flow. The flow can be increased at lower pressures by fitting larger output pistons and cylinders.

To avoid the potential hazard associated with the use of a flexible hose carrying high pressure water to the articulated roadheader boom, the pump was divided into two units. The hydraulic pump itself was mounted on a sledge to the rear of the roadheader, and the intensifier was mounted on the roadheader boom. Flexible hoses were used to carry low pressure oil (140 bars maximum) from the hydraulic pump; the high pressure water output from the intensifier was taken directly into the boom through a rigid steel pipe.

3.3 Rotary Seal Unit

This was designed to replace the existing seal for dust suppression water and is situated in the end section of the boom trunk as shown in Figure 7. High pressure water is fed through the pressure cylinder A into the annular area around the shaft between the two seal elements B. It then flows through the radial hole in the cutting head shaft C and axially along the shaft to the cutting head. This shaft is 140 mm diameter and rotates at 60 rev/min.

No seals were commercially available to suit this application and a development programme was instigated to produce an acceptable type. The test rig shown in Figure 8 was used to test various seal configurations with little success. Best results were obtained with a U-section Teflon seal with spring energisation, as shown in Figure 9. The life of this was however 50 hours only, compared with a minimum specified life of 500 hours. It was accepted that this short life would be satisfactory for field trials with the prototype, but that further development would be necessary for any future production version of the system.

3.4 Cutting Head

The design of the cutting head was based on the standard Dosco head for hard rock applications and is shown in Figure 10. Universal-type pickboxes were fitted to accommodate either forward-attack or radial-type picks.

To allow a reasonable flow of water from each nozzle, a maximum of fifteen nozzles are fitted. They are concentrated on the forward part of the cutting head, as the picks in this area are normally subjected to higher cutting forces than those at the rear of the head. It is in fact usual practice when cutting in hard strata to use only the forward section of the cutting head, to increase the cutting force available on each pick.

To protect the nozzles from external damage, they were mounted within the cutting head body as shown in Figure 11. The anticipated nozzle stand-off distance of 50 mm was therefore increased to 105 mm. No problems were expected from this, as synthetic sapphire nozzles were used and other research (Reference 3) had shown that with these nozzles jet efficiency would be little impaired over this greater distance.

The internal water passages in the cutting head were nickel-coated by chemical process to prevent corrosion.

4. FIELD TRIALS WITH PROTOTYPE SYSTEM

4.1 Trials Site

The site selected was a limestone mine at Middleton in Derbyshire which is used frequently by MRDE for rock cutting and drilling trials.

The limestone at this mine is a homogeneous rock having average compressive and tensile strengths of 116 MPa and 21 MPa respectively and medium abrasivity. It is tough cutting, well above the normal capability of a Dosco MkIIA Roadheader. Because of this, a mock heading was constructed from Grindleford sandstone for initial trials. This has similar physical characteristics to Darley Dale sandstone, used during the laboratory cutting trials, and has a compressive strength of 50 MPa, tensile strength of 7 MPa, and is highly abrasive.

4.2 Cutting Trials

No significant reductions in cutting rate or power consumption were noted when cutting with jet assistance in the sandstone heading. Visual observations did show, however, that dust make was considerably reduced and that frictional sparking was eliminated.

Quite dramatic effects occurred when cutting in the harder limestone. Without jet assistance it proved impossible to cut efficiently owing to excessive vibration of the machine. The tungsten carbide tips of the cutter picks suffered almost immediate destruction and the drive motor at times stalled on overload, due to the high forces imposed on the cutting head. When supporting the machine by jets however, it was possible to cut evenly at a reasonable rate and machine vibration was reduced to an acceptable level. In addition, the cutter pick tips remained intact for a much longer period.

The effect is best illustrated by reference to Figures 12 and 13 which reproduce recorder traces for a typical cutting operation. In these diagrams, the force feeding the cutting head into the rock, and the cutting rate, are directly proportional to slewing ram pressure and flow.

The roadheader prior to cutting in the limestone face is shown in Figure 14 and a comparison of pick wear is shown in Figure 15.

Different combinations of water pressure, number of nozzles and nozzle size were assessed during the trial. The optimum was found to be 9 nozzles at 700 bars pressure and 4 l/min flow per nozzle, using only the forward half of the cutting head in the rock.

At the end of the trials, an arch-shaped roadway 4.3 m wide x 3 m high x 5 m long had been driven. Typical pick consumption was 25 per metre of advance.

4.3 Equipment Performance

No problems were encountered with the water pump unit, although the intensifier mounted on the boom severely restricted the machine operator's vision when cutting to the right hand side of the roadheader.

The nozzles on the cutting head produced visually good jets which retained coherency at the pick-tip impingement point, and a well formed central core was still apparent at up to 0.75 m distance from the nozzle.

Major problems were however experienced with the rotary seals. Their life was considerably less than that achieved on the test rig, probably due to the dynamic loading on the shaft. The failure rate of the seals was not consistent and ranged from a minimum life of 1 hour to a maximum of 30 hours. Severe damage also occurred to the surface of the shaft in the seal area and attempts to rectify this situation by gun spray coating with harder materials only met with limited success.

5. PREPRODUCTION VERSION

5.1 General

Results obtained during field trials with the prototype system were considered sufficiently encouraging to warrant the further application of the technique in a coalmining environment. For this purpose the equipment needs to be reliable in operation and to meet the statutory requirements for machinery used in UK coalmines.

An improved version of the prototype system is therefore being developed. Since this is likely to be an interim development between the prototype and a future production system it has been called the preproduction version. It is expected that the development will be completed by the end of 1982, following which full scale underground trials will be conducted. This will enable full assessment to be made of the system's capabilities and equipment performance and will determine whether further exploitation is warranted.

As with the prototype, the machine chosen for adaptation is the Dosco MkIIA Roadheader. A description of the elements comprising the preproduction version follows.

5.2 High Pressure Water Pump

Two alternative types of pump are to be tried.

Because of the advantages in terms of flexibility and safety offered by a hydraulically powered intensifier, a modified version of the whole unit fitted on the prototype will be used. Instead of using one intensifier, two smaller units are to be constructed of equivalent total capacity. They will be mounted on either side of the boom as shown in Figure 16 so that they do not interrupt the operator's vision or limit the elevation of the boom. The intensifiers will be electronically controlled to run out of phase to reduce pressure fluctuations during operation. The hydraulic pump is to be mounted either on the discharge conveyor of the roadheader or on a monorail suspended from the roof of the roadway.

The alternative pump unit will be a triplex piston pump (mounted similarly to the hydraulic pump for the intensifiers). In this case a flexible hose is needed to carry high pressure water from the pump to the roadheader boom. This arrangement is more compact than the intensifier system, but the flexible hose will be extremely vulnerable to external damage and could constitute a considerable safety hazard. It is unlikely that HM Inspectorate of Mines and Quarries will acquiesce use of this hose unless it has a high factor of safety and is adequately shrouded.

5.3 Rotary Seal

The successful development of a reliable rotary seal is crucial to the overall success of the project. The seals used in the prototype system were erratic in performance and their life expectancy fell far short of the minimum of 500 hours specified for a production version. Three approaches are being investigated in parallel in an attempt to solve this major problem.

a) Research continues into the further development of seals fitted around the cutting head shaft as used on the prototype. A new design incorporating a floating seal assembly has been developed (Figure 17) which should eliminate eccentric loading on the seal caused by deflection of the cutting head shaft. In addition, several varieties of seal are to be tested and improvements are to be made to the shaft surface coating and finish. This new type of seal arrangement is considered more attractive than the other methods being investigated for reasons of economy and simplicity.

b) A layshaft configuration gearbox has been designed (Figure 18) to allow entry of high pressure water to the exposed end of the cutting head shaft via a small diameter seal. It is expected that the development of a suitable seal will be relatively straightforward and a test rig will be used to evaluate the options available. The method enables the seal to be very easily changed in situ as a unit and the specified life can therefore be considerably shorter than that of the seals fitted around the cutting head shaft.

c) The third approach is the development of a new boom with a full length water passage along its axis (Figure 19). Again a small-diameter rotary seal will be fitted at the feed end similar to that used with the layshaft gearbox. The existing boom components can be modified to accept this axial water passage, but some mechanical problems are expected regarding water passing through the epicyclic gearbox.

5.4 Cutting Head and Nozzles

The prototype cutting head gave reliable service during surface trials and there seems to be little scope for improvement in the basic design. The manufacturing technique will however have to be refined by the construction of assembly jigs to ensure accurate alignment of the nozzles vis-a-vis the cutter pick tips.

Since jet efficiency directly affects overall efficiency of the system, the configuration and performance of various types of nozzle is being investigated. The pressure profile of jets produced by various nozzles is being measured by the piezometer and the 'hole in plate' method, using the test rig shown in Figure 20. In addition, the use of polymer additives to produce a more coherent jet is being evaluated.

6. CONCLUSIONS

Indications from initial trials with the prototype system are that the use of water jets can significantly increase the cutting capability of a boom-type roadheader. The extent of these benefits cannot be fully established until full-scale underground trials have been carried out with the preproduction version.

While the technique shows extremely good potential, it does of course have its drawbacks. The overall efficiency of the system is low because the jets fire in free air when out of cut and a great deal of the installed horsepower is not realised. Since power costs represent only a very small percentage of the overall cost of driving an underground roadway this has been accepted at present. Phased operation of the jets would be necessary to maximise system efficiency. It is considered however that the sophisticated engineering techniques required to achieve this would constitute a major development which would only prove worthwhile if the system were to be exploited on a large scale.

Problems of excessive water are expected as the amount is twice that conventionally used for dust suppression, which already creates potential difficulties in operations If of course the use of water jets enables the rate of cutting to double, the existing problems would not be increased and would in fact decrease for cutting rates above this figure. One further problem noted during the field trials was the spray created around the cutting head by the high pressure jets. This severely limited the machine operator's vision when cutting. It may be reduced to an acceptable level under production conditions by the roadway ventilation system.

The development of reliable equipment, especially seals, will of course be a major criterion and most current development work is directed to this end. It is indeed likely that the future success of this work will depend entirely on successful engineering of the mechanical components to function reliably in a production environment.

7. ACKNOWLEDGMENT

Both authors thank the Director of Mining Research and Development Establishment, National Coal Board, for permission to publish this paper and the Acting Director, Division of Coalmining, US Department of Energy. The views expressed are those of the authors and not necessarily those of the National Coal Board.

8. REFERENCES

8.1 M Hood

Cutting strong rock with a drag bit assisted by high pressure water jets. - Journal of The South African Institute of Mining and Metallurgy, November 1976.

8.2 Fun Den Wang and J Wolgamott

Application of water jet assisted pick cutter for rock fragmentation - BHRA Fourth International Symposium on Jet Cutting Technology 1978.

8.3 D Ropchan, Fun Den Wang and J Wolgamott

Application of water jet assisted drag bit and pick cutter for the cutting of coal measure rocks. - Final Technical Report to the United States Department of Energy ref DOE/ET/12463-1, 1 April 1980.

FIG 1

ROCK CUTTING TEST RIG

FIG 2

PICK & NOZZLE IN TOOL HOLDER

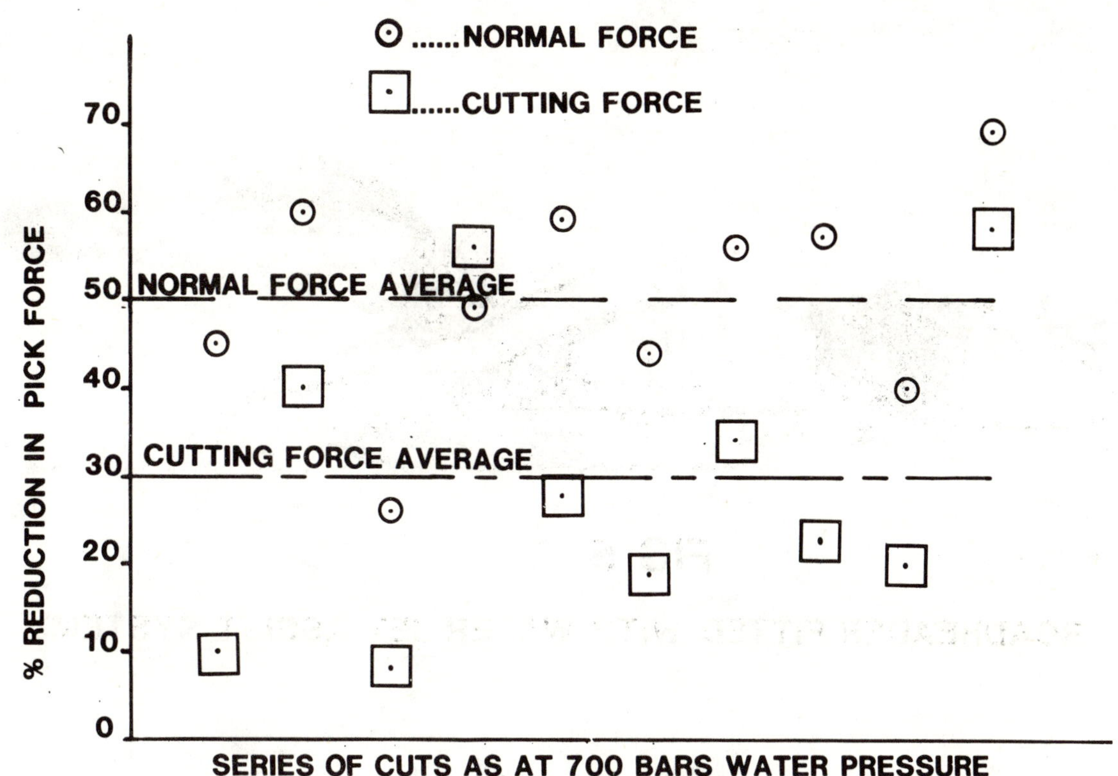

FIG 3

MkIIA ROADHEADER

FIG 4

FIG 5

ROADHEADER FITTED WITH WATER JET ASSIST SYSTEM

FIG 6

HIGH PRESSURE WATER PUMP

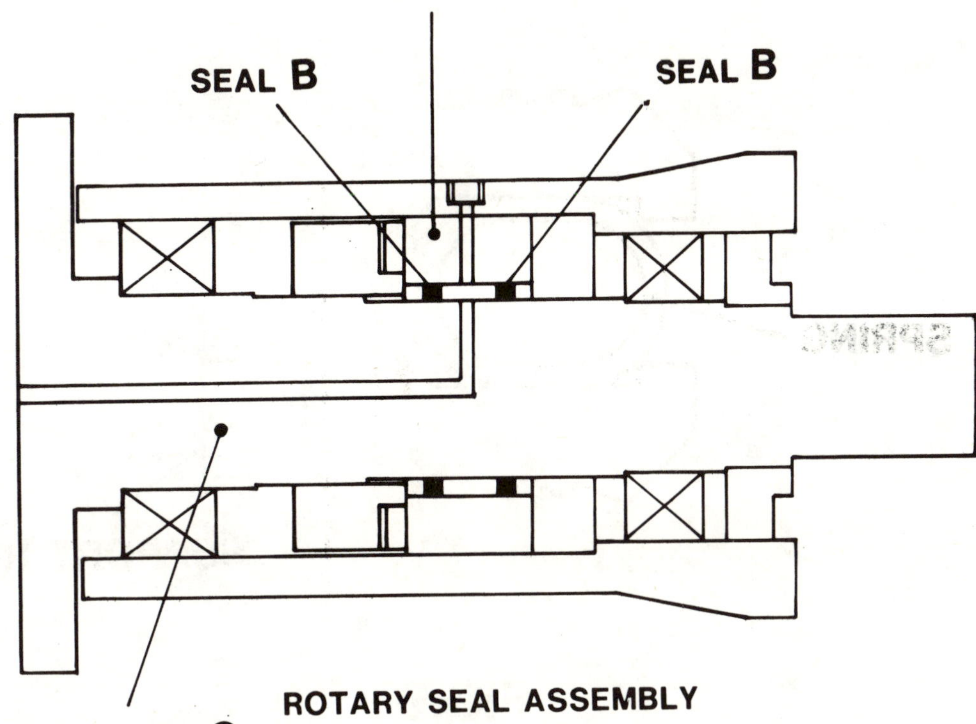

ROTARY SEAL ASSEMBLY

FIG 7

FIG 8

ROTARY SEAL TEST RIG

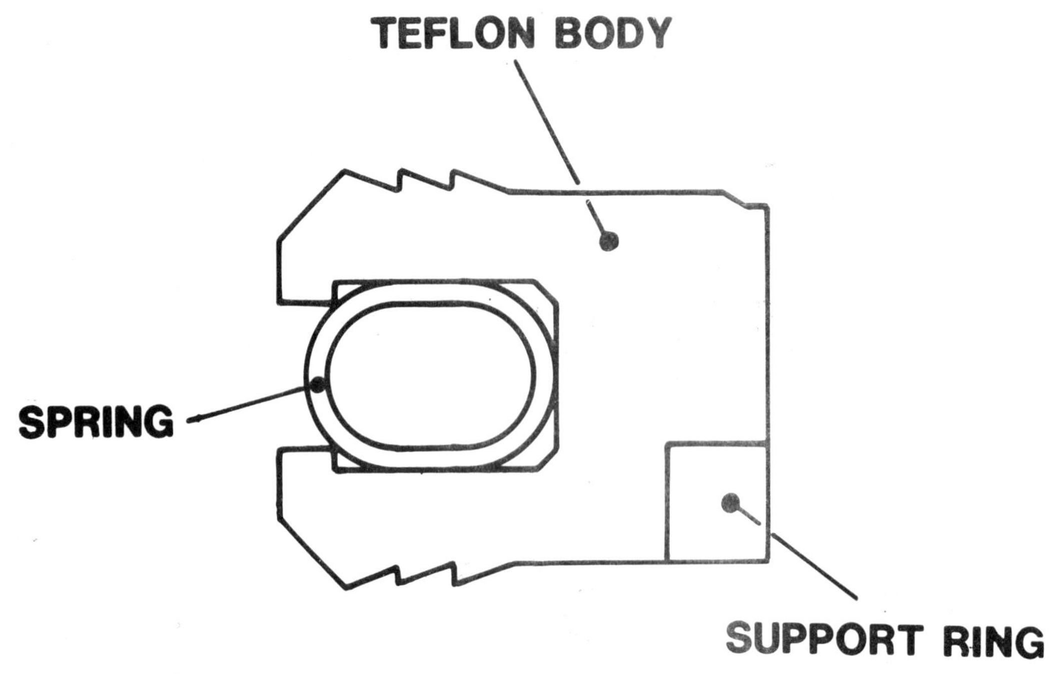

FIG.9

U-SECTION SEAL

FIG 10

CUTTING HEAD

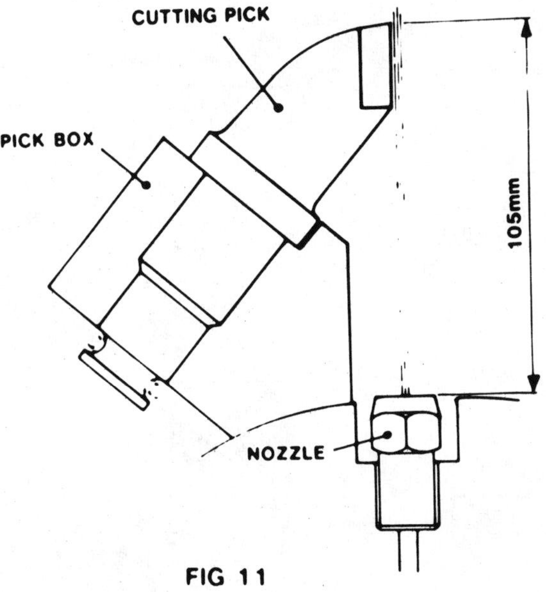

FIG 11

POSITION OF NOZZLE RELATIVE TO CUTTING PICK

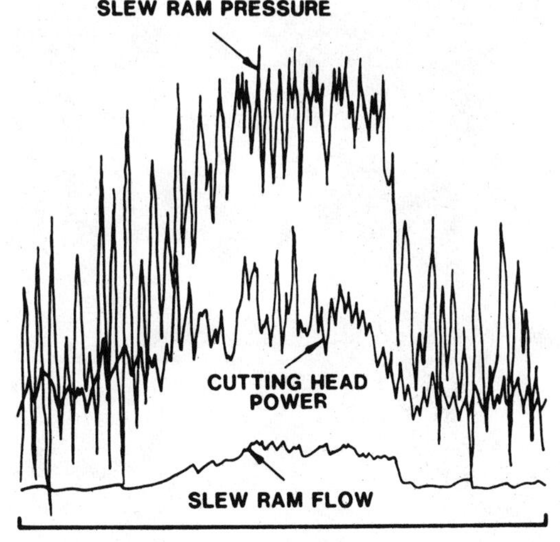

FIG. 12

CUTTING IN MIDDLETON LIMESTONE WITHOUT JETS

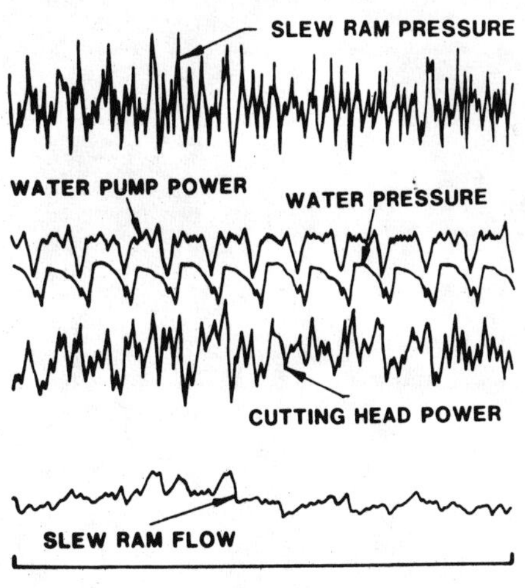

FIG. 13

CUTTING IN MIDDLETON LIMESTONE WITH JETS

FIG 14

ROADHEADER PRIOR TO CUTTING IN LIMESTONE

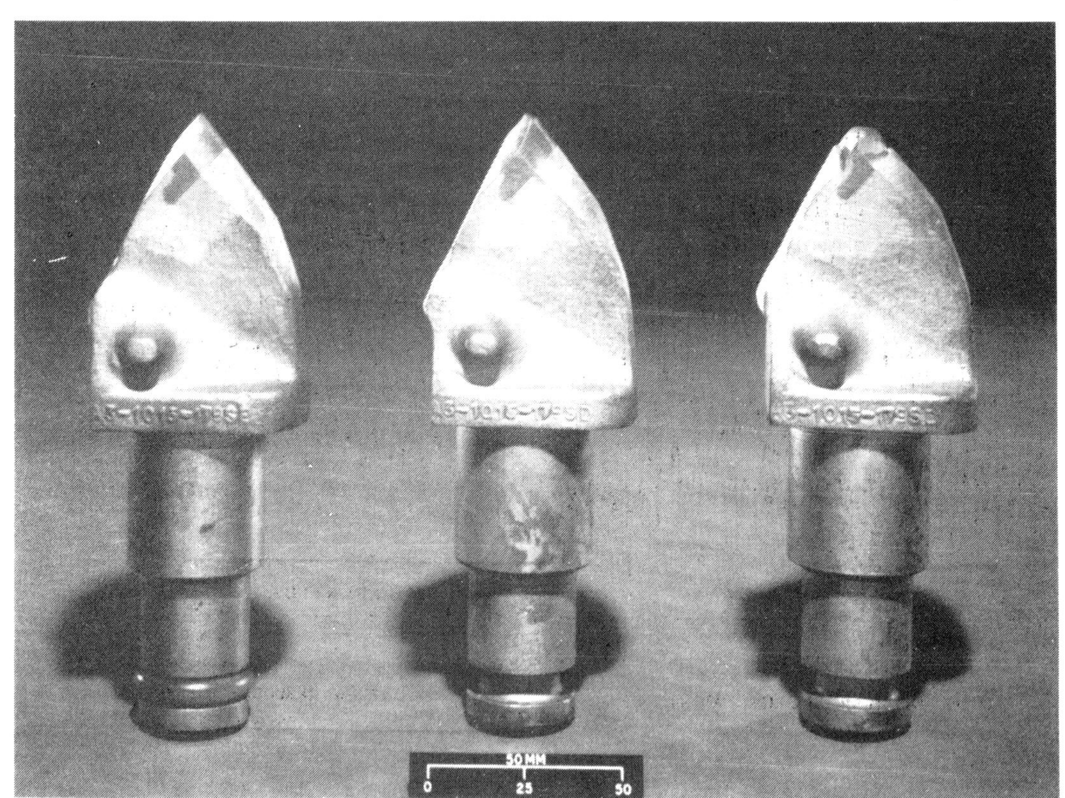

UNUSED   WITH JET   WITHOUT JET

FIG 15

COMPARISON OF PICK WEAR

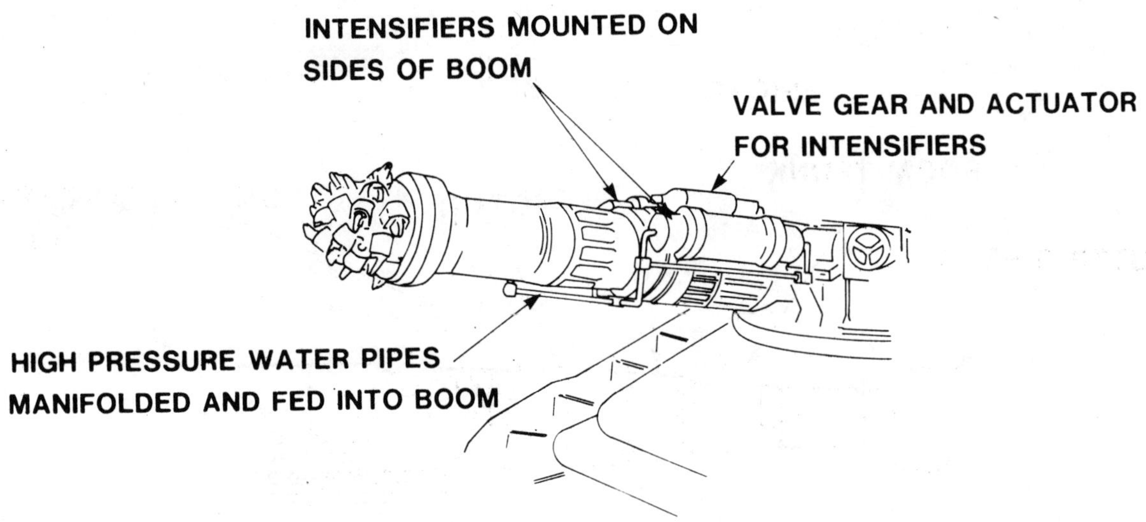

TWIN INTENSIFIERS MOUNTED ON BOOM

FIG 16

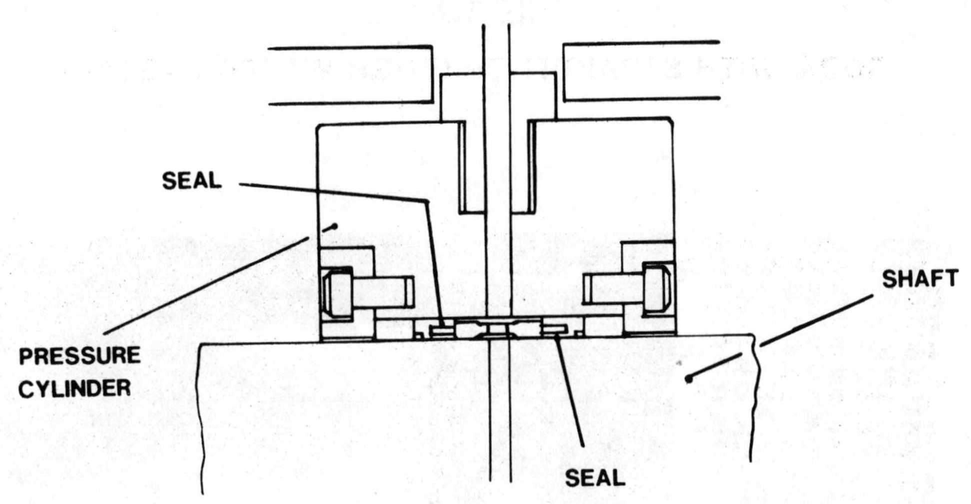

ROTARY SEAL ASSEMBLY

FIG 17

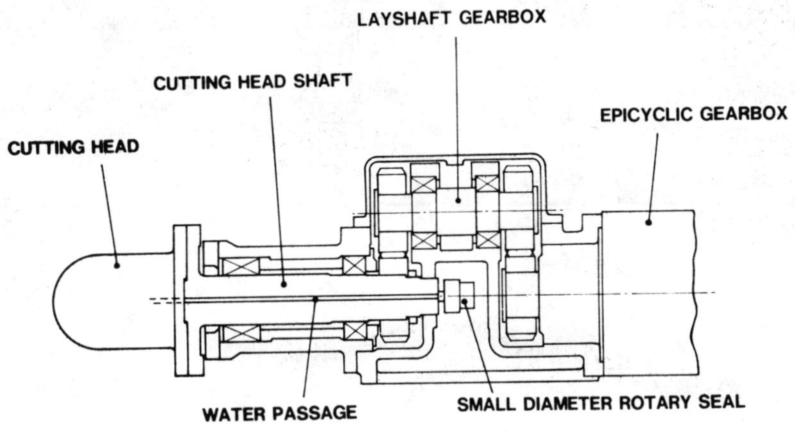

FIG 18

SECTION THROUGH LAYSHAFT GEARBOX

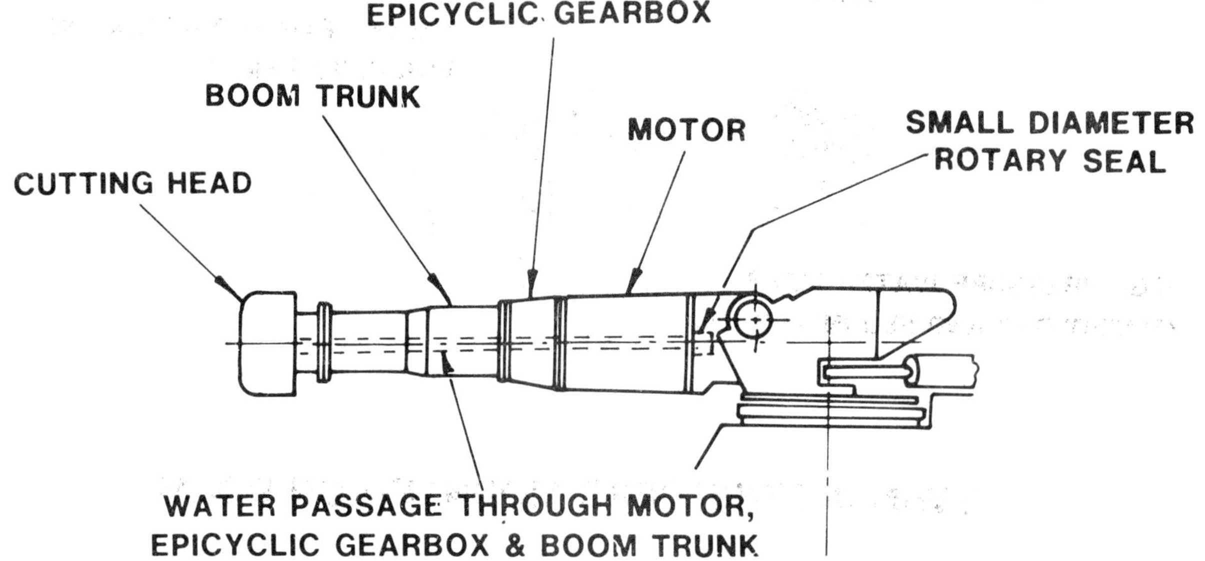

FIG.19

BOOM WITH STRAIGHT THROUGH WATER PASSAGE

FIG 20

NOZZLE PERFORMANCE TEST RIG

6th International Symposium on
Jet Cutting Technology
6-8, April, 1982

STATE OF INVESTIGATIONS ON HIGH-PRESSURE WATERJET ASSISTED ROAD PROFILE CUTTING TECHNOLOGY

L. Baumann

Bergbau-Forschung GmbH, Federal Republic of Germany

M. Koppers

Bochumer Eisenhütte Heintzmann & Co., Federal Republic of Germany

Summary

In order to widen the application range of mechanized tunnelling in the rock for application in hard-coal mining less clumsy heading systems need to be developed. For this purpose industrial-scale tests are run on the test rigs of Bergbau-Forschung as well as on one colliery. For these tests a road profile cutting machine equipped with various sensors and with a high-pressure waterjet assist in addition to the classic hard-metal tools, is used.

By systematic evaluation of test results new application possibilities for this combined heading system are found. Particular importance in this context is assigned to reduced cutting force and extension of the application range to cutting of hard rocks. Investigations also are made for finding out whether by high-pressure waterjets on their own interesting heading performances can be achieved.

A summary contains the description of a new tunnelling system incorporating the advanced profile cutting technology based on the experience gained by our tests run up to present.

Held at the University of Surrey, U.K.
Symposium organised and sponsored by
BHRA Fluid Engineering

In roadheading and tunnelling within present day mining operations predominantly full-face machines or roadheaders are used - apart from classic shotfiring - for cutting into the rock.

The mechanized systems reached a high state of technical development with respect to mechanical and cutting technology.

If, however, hard and abrasive rock is met, these mechanized systems record poor heading and tunnelling performances.

The tools presently used for cutting into the rock are approaching their technical limits. These limits are set by high tool wear and the limited possibilities of energy transfer.

For the above reasons new methods of cutting into the rock are investigated worldwide.

Since three years Bergbau-Forschung GmbH and Bochumer Eisenhütte Heintzmann & Co. run trials on application of high-pressure waterjets for roadway profile cutting as one step of tunnelling operations in hard-coal mining. This development work is sponsored by the Minister für Wirtschaft, Mittelstand und Verkehr (provincial ministry for economy and transport) of Nordrhein-Westfalen. The development work aims at improved efficiency of mechanical tools by high-pressure waterjet assist in order to reduce the high cutting forces which would allow the construction of roadheading/ tunnelling machines of less weight and higher mobility (Fig. 1).

The basic technology for roadheading with contour cutting was shown on the International Mining Fair at Düsseldorf in 1976 by Bochumer Eisenhütte Heintzmann & Co. as a component of the model "Abbau der Zukunft" (future mining) (Fig. 2).

This model exhibits a new roadheading system integrating approaches proposed for the following steps:

- Contour cutting
- Core recovery
- Clearance
- Support: transport, positioning, and setting.

The new technique was supposed to cut a slot to the required road profile large enough to place a road-support arch into it.

Then the remaining core was supposed to be recovered by known methods without affecting the structure of the surrounding strata.

Initial investigations have shown that available equipment components for contour cutting did not meet the requirements with respect to cutting accuracy, dimensions, pick wear, dust suppression, noise suppression, and spark prevention.

The use of a combined high-pressure water/hard metal tool was supposed to yield the following advantages (Fig. 3):

- high cutting accuracy
- lower cutting forces and passive forces
- less pick wear
- no dust and no sparking
- small dimensions of the machinery and equipment

The research and development work done or still to be done comprises the following phases:

1. Basic feasibility tests on lab scale and in a sandstone quarry (Fig. 4);
2. Trial runs of a full-scale experimental equipment underground;
3. Design, construction, and trial runs of a prototype of this advanced heading system with first-phase road profile cutting.

The first phase is meanwhile completed successfully. The results of the cutting trials run with very hard homogeneous rock have shown that high-pressure waterjet assisted cutting tools actually reduced the cutting forces and the passive forces considerably. These reduced values will enable smaller dimensions of the cutting tool and its carrier design and furthermore a lower rating of the drive.

In 1981, underground contour-cutting trials were run on Rossenray colliery of Ruhrkohle AG (Fig. 5).

A branch-off in "Blücher"-seam on the 703 m level was placed at our disposal for the test runs by the management. Within the framework of the colliery's long-term operations planning this branch-off, from the support side, was already allowed for, and actually the branch-off was headed approximately 1 m inby. Since the main road will not be used in the years to come, the cutting trials were run there. The test site is near the shaft and is ventilated by the downcast air stream.

EXPERIMENTAL EQUIPMENT

Profile Cutting with High-Pressure Waterjets and Hard-Metal Tools

The basic data resulting from the lab-scale work pertaining to the function of "Combined high-pressure waterjets and hard-metal tools" supplied the basis for the design of a road profile cutting unit (Fig. 6).

This experimental unit comprises essentially the baseframe for housing the forward-thrust unit moving along the longitudinal axis of the machine towards the tunnelling face. A cutting arm pivoting around the longitudinal (horizontal) axis is mounted on the thrust unit.

Both movements (pivoting of the cutting arm and forward thrust) are controlled by hydraulic rams. The pivoting angle of the cutting arm is of 90°. By bolting the piston-rod end to another bearing of the cutting arm, the pivot angle could be increased by further 90°. Thus a profile slot could be cut over 180°. Furthermore, the length of the cutting arm could be varied by means of two adapter pieces so that, for experimental purposes, three concentric profile slots could be cut. Furthermore, the possibility was given to run three different cutting speeds at identical angular speed.

For running various cutting speeds, also a hydraulic unit with a variable-delivery pump enabling infinitely variable delivery control was used.

During the trials run in Hohensyburg quarry it was found that data recording was not satisfactory.

Fig. 7 shows the basic layout of the cutting tool and the position of the strain-gage equipped transducers FS and FSR.

It is obvious that the transducer FSR measures the cutting force component FS and the transducer FR the sum of the components FP and FS. The transducers record a travel of 0.1mm at rated load. Considering the resulting deflection of the pick tip this travel could no longer be neglected and consideration had to be given to the problem of frictional losses on the joint bolts and the guidance surfaces of the tool holder. The fact that the transducer FR issued a signal representing a summed-up value including the corresponding frictional moments, made the problem even greater. When running the trials, particular attention had to be given to the frictional value on the tool holder being as much as possible constant with respect to the friction values considered when calibrating the tool on the test rig.

For these reasons the tool holder itself was equipped with a strain gage for the underground trials. Fig. 8 shows the basic array as well as the position of the strain gages on the tool holder used.

Measurement of Force P

From the drawing may be seen that the measuring point P for recording the passive force (feed force) is fitted to the tool-holder body at a place where the rectangular cross-section remains constant so that for measuring of the force P in this area of the tool holder, a homogeneous development of the tension lines can be assumed.

The measuring bridge P comprises two strain-gage sets. Each set comprises two strain-gage units arranged perpendicularly to each other. The two sets are arranged symmetrically relative to the neutral fiber run in the measuring zone. The measuring circuit is to compensate for B forces.

Measurement of Force B

The measuring bridge B comprises four strain gages arranged symmetrically relative to the tension-zero-line and they are at identical distance from the pick tip. The effect of P needs to be compensated by the bridge.

Calibration of the Measuring Equipment

The complete tool was connected to the signal amplifier and fitted to a compression and bending test rig. The test rig simulated the forces P and B acting on the cutting tool. Both forces could be adjusted stepwise up to 120 kN.

The tool was fitted to the test rig at 0°, 45°, 60°, and 90° inclination angle. The angles were measured relative to the line of force P, i.e.

0° only feed force, only P acting
90° only bending force, only B acting

The analog electric signals thus recorded are directly proportional to the forces to be measured.

Influences like

- manufacturing tolerances in the tool holder geometry
- deviations of the strain-gage position relative to the axis of symmetry
- deviations of the tension line from the axis of symmetry

just to name some of them, affect, however, the measuring results and require correction. The rated value - pre-set by the test rig - was in each case compared to the actually recorded measured value.

This resulted in a correction curve for P which changed with the inclination angle (0 - 90°). The curve for B exhibits only a very small deviation so that, for evaluation of cutting force B, a sufficiently exact proportional relation with the recorded value was obtained. The values measured are converted by means of a computer to values for cutting forces and passive forces. Fig. 9 shows the whole measuring array.

Transducers: Strain-gage bridges, 350 Ohm
Measuring amplifier: Type VE 1 (developed by Bergbau-Forschung)
Power supply for the amplifier: Type SE 1503 (developed by Bergbau-Forschung)
Signal-recorder unit: UV-beam oscillograph.

According to the regulations of the central mining inspectorate (Landesoberbergamt Dortmund) the transducers, the amplifiers, and the power supply had to be intrinsically safe. For the UV recorder unit exceptional clearance was issued for the duration of the underground testing.

Cutting Tool

The pick comprises four hard-metal tips equipped either with

- 2 leading nozzles or
- 3 trailing nozzles or
- 5 nozzles (2 leading ones and 3 trailing ones)

for high-pressure waterjet assist (Fig. 10).

According to nozzle array the forces acting on the tool as a function of the test parameters, can be controlled.

The underground trials were aimed at:

- Determination of forces acting on the tool as a function of the following parameters: pressure, cutting speed, cutting depth, nozzle array, and nozzle diameter
- Findings on the effects of the contour slot on the strata (including coal) already exposed to in terms of effects on the roadway envelope surface and on the remaining core.

In 1981 more than 900 slots were cut and the testing program for the nozzle/ hard metal tool was completed (Fig. 11).

The following statements can be made today with respect to the influence of contour cutting with high-pressure waterjet assist in underground strata (shale, clay schist):

At a water pressure of approximately 1000 bar the cutting forces and the passive forces are reduced by 50 % (Fig. 12).

The road envelope is mostly smooth (Fig. 13).

The recovery of the residual core requires relatively little effort.

Advantages of the high-pressure water/hard metal tools for roadway profile cutting:

- high cutting precision
- low reaction forces (on the tool)
- small dimensions
- low pick wear (longer useful life of the picks)
- no dust
- no sparking
- careful with respect to strata structure
- reduced time lag for support setting
- application limits of at present 800 kg/cm^2 compressive strength (crushing strength of a cube) may be extended
- increased contact interface (support arch/rock) with resulting better support behaviour
- low mechanical efforts for recovery of the remaining core (due to stress relief by the contour slot).

SUMMARY OF THE UNDERGROUND TESTING RESULTS

The gradients obtained by linear regression of the values recorded underground for cutting forces and passive forces are shown for three nozzle arrays:

- two leading nozzles
- three trailing nozzles
- two leading nozzles and three trailing ones.

Fig. 14 (2 leading nozzles) shows the cutting forces and the passive forces versus cutting depth. A parameter "water pressure" is varied from zero to 2000 bar.

Only a very slight reduction of cutting forces as well of passive forces is obtained by waterjet assist. When looking at these diagrams and also at the following ones, consideration should be given that, due to lacking homogenity of the rock strata met underground, i.e.

- changing strengths along the travel of the cutting tool
- changing directions of strata
- changing directions of cleaves and fissures

the gradients plotted for various water pressures are to be regarded as a band.

With this configuration the high-pressure waterjet assist is advantageous only in that the useful life of the cutting tool is increased since the waterjets hit the rock at the outer edges of the cutting tool where normally the wear rate is highest.

The graphs for the three-trailing-nozzle array shows clearly that cutting forces as well as passive forces are reduced with increasing water pressure. The high-pressure water supply was run at a maximum of 1500 bar. The three nozzles were of 0,63 mm diameter each (Fig. 15).

Fig. 12 shows the five-nozzle array and exhibits the previously mentioned reduction of cutting force and passive force by 50 %, recorded for a cutting depth of approximately 10 mm and a water pressure of 1000 bar. With five operational nozzles of 0.63 mm diameter each this configuration implied the high-pressure water supply being run at the limit. We may expect that with this nozzle configuration and a further pressure increase the cutting forces can be reduced even more (Fig. 12).

The differences of the values shown on Fig. 12 and those shown on Fig. 15 for the parameter

- 0 bar (no water)

may be explained by the above-mentioned lacking homogenity of the rock strata met on the underground trial site.

Furthermore, the diagrams shown represent only a few of the cutting tests run.

In addition to the described reduction of forces, the high-pressure waterjet assisted hard-metal tool exhibits the following advantages for the

- roadway profile cutting technology:

(Just to mention the most important ones)

- The roadway exhibits smooth contours; no inadverted fall of minerals from the strata.
- In less solid strata the roadway profile remains stable even though the destressed core exhibits little strength only (formation of a vault).
- Due to the exact profile cut, the support interface between arches and the rock strata is very large and consequently the support arch assumes its support function quite promptly.
- No backfilling.

PROFILE CUTTING BY HIGH PRESSURE WATERJET ALONE

In parallel to the above-described joint development work of Bochumer Eisenhütte Heintzmann & Co. and Bergbau-Forschung GmbH an alternative-design experimental unit is developed jointly with the US-Bureau of Mines. The equipment comprises a guide frame which corresponds, as to its shape, to the roadway cross-section. The tool holder is fitted to said frame (Fig. 16).

Cutting Instrument

The roadway-profile slot is cut exclusively by six high-pressure water nozzles arranged in various angles thus enabling to cut a slot of approximately 80 mm (Fig. 17).

In addition it is tried to cut the roadway-profile slot by means of two rotating nozzle heads equipped with two nozzles each (Fig. 18).

Cutting of the roadway profile by high-pressure waterjets only implies almost no reaction forces. This results in essential advantages with respect to weight reduction of the experimental equipment.

Test Results

The trials run as yet basically confirm the feasibility of this cutting principle. It was shown at the same time, however, that a cutting instrument using exclusively nozzles requires still considerable development work.

While the waterjets, due to their high performance, cut deeply into the rock, crests remain between the slots cut. These crests damage nozzles and the cutting arm.

In order to avoid such problems the nozzles previously arranged rigidly are mounted to cutting heads oscillating perpendicularly to the cutting direction. Further efficiency increase was obtained by use of rotating cutting heads with two nozzles each.

High-Pressure Water Supply

The high-pressure water supply is assured by means of so-called pressure boosters (Fig. 19) mounted to a trailer unit behind the experimental equipment and connected the cutting arm by high-pressure mains of 6.3 mm inside dia. and 14 mm outside dia. The most important components of this unit were placed at our disposal by the US-Bureau of Mines and modernized by the manufacturer of the unit, Messrs. Flow Industries, Inc. Seattle.

The unit comprises essentially the pressure generator powered by a hydraulic pump. The rating of the booster reads: 250 kW, maximum flow of 30 l/min, delivery pressure 3000 to 4000 bar. The pressure boosters work like a reversing-piston pump (Fig. 20). Their capacity and efficiency is unmatched up to present. Their operational reliability could be improved in the course of the trials so that even the particularly critical parts, like high-pressure seals, stand meanwhile several hours of operation and have a good chance of even further improvement.

FURTHER RESEARCH AND DEVELOPMENT WORK ENVISAGED

It is intended to run further complementary cutting trials on the testing site with the existing equipments until end of 1982 and furthermore to run tests with modified cutting tools and with tools for core recovery.

Besides the known methods for recovery of the core (shotfiring or impact rippers) a so-called "Wasserhammer" (water-blast unit) presently under development (Fig. 21) could be used. By a rifle-shot, like waterjet impact, successful trials for breaking rocks from carboniferous strata were run.

In parallel to these trials the advanced heading concept is further developed (Fig. 22). This system should meet the following basic performance specifications:

- The operation cycles "contour cutting", "core recovery", and "debris clearance" should be run simultaneously at a high machine utilization degree.
- Contour cutting and core recovery can be run simultaneously by one or several cutting tools moving along a frame corresponding to the roadway profile, thus cutting the contour slot, and by simultaneous disintegration of the core by a high-pressure jet disintegration unit arranged below the above-mentioned frame. The debris produced are gathered by a loading system and subsequently removed.
- Synchronously to the heading work the pre-assembled support arches are set by a manipulating equipment in a way that they assume the supporting function immediately. The time lag for support setting is reduced by inserting the arches immediately after advance of the profile-cutting system.
- The heading-system length must be limited and the system needs to be maneuverable for enabling to follow the course of the strata and seams.
- For making the system cost-effective, in particular for application in gateroads, it is to be designed in a way that it can easily be moved from heading site to the next one.
- It is of decisive importance to produce underground a smooth and dimensionally stable roadway contour for having the possibility to set the support in a way that it assumes its carrying function immediately. The support arches are locked against the roadway envelope immediately after setting, and thus the support is obtained. A stable vault is formed fairly promptly. No backfilling is required and better use of the support capacity of the arches is made. When penetrating into greater depths the cutting of exact contours will enable to allow for the increasing stresses by better support behaviour of the arches.

NB

This project is sponsored by the Minister für Wirtschaft, Mittelstand und Verkehr (provincial ministry for economics and transport) of Nordrhein-Westfalen.

The project is carried out jointly by Bergbau-Forschung GmbH, the US-Bureau of Mines/Department of Energy, and Bochumer Eisenhütte Heintzmann & Co. The high-pressure waterjet system was placed at the disposal of Bergbau-Forschung by the US-Bureau of Mines/Department of Energy within the framework of a cooperation agreement.

Fig. 1 Profile-cutting equipment underground

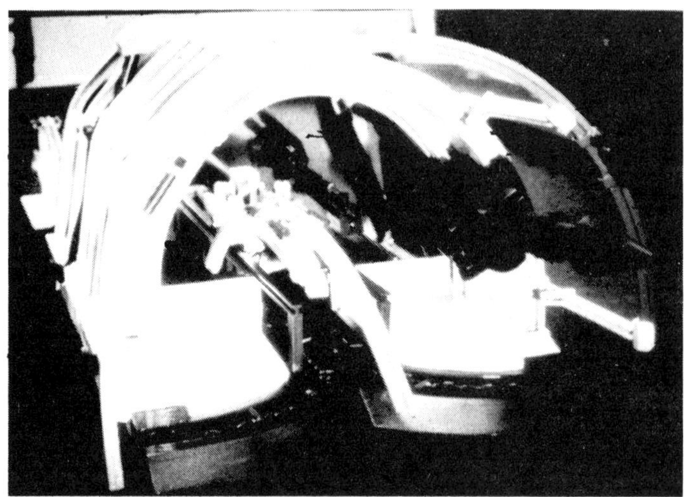

Fig. 2 Concept of a profile-cutting system, 1976

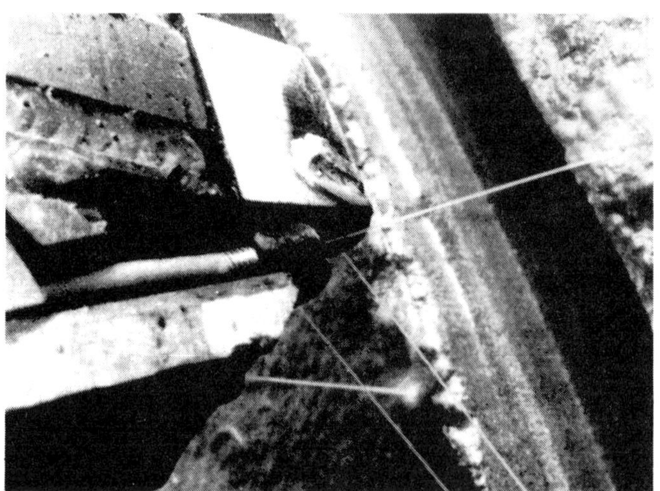

Fig. 3 Hard-metal tips with two leading and two trailing high-pressure waterjets

Fig. 4 High-pressure waterjet assisted cutting of road profiles for optimized support setting and minimized deterioration of the strata structure

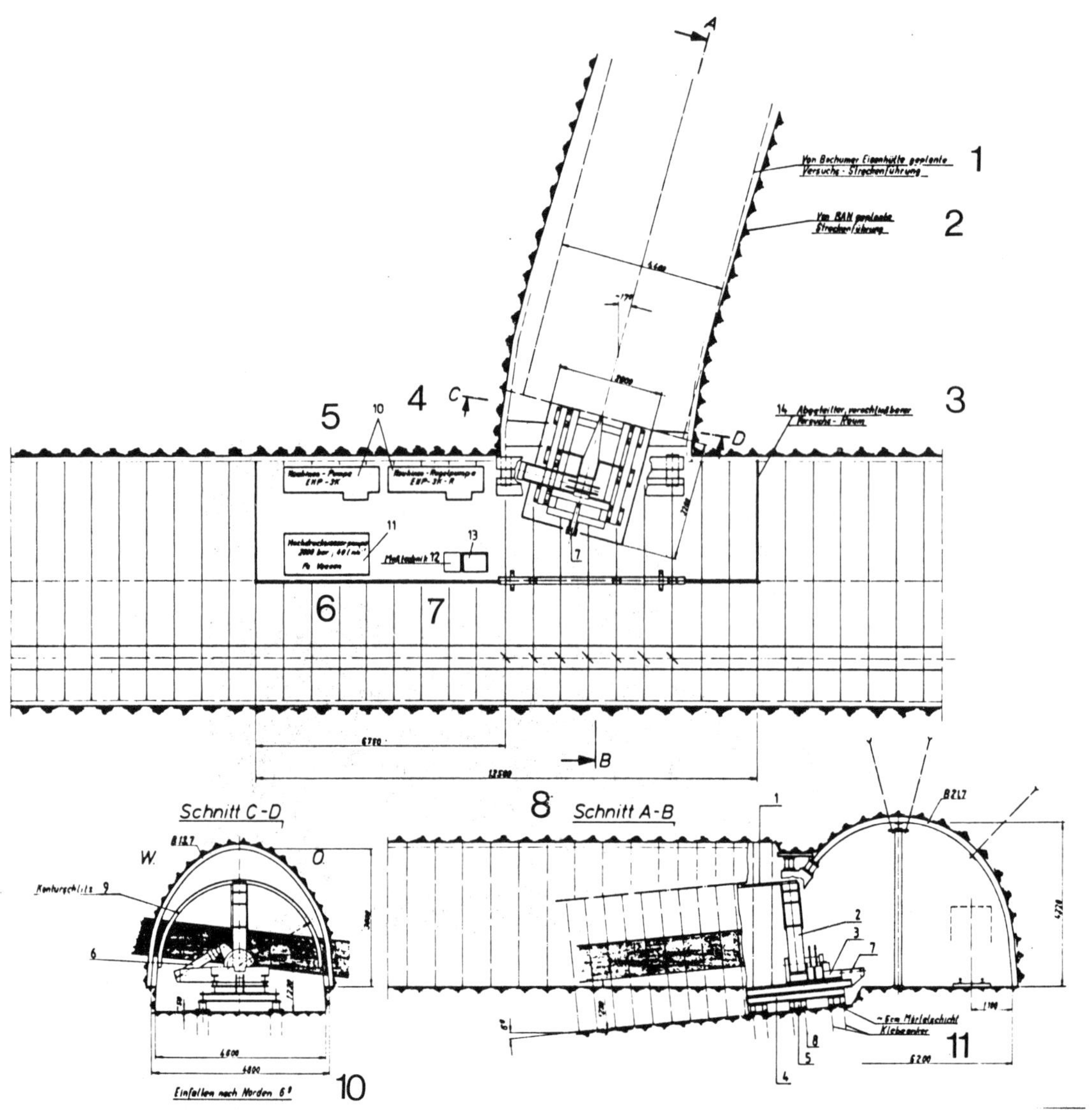

1 = Experimental heading as planned by BE

2 = Heading as planned by BAN (Western area of Ruhrkohle)

3 = Closed testing area

4 = Hauhinco Control pump, EHP- 3K - R

5 = Hauhinco pump EHP - 3K

6 = High-pressure water pump, 2000 bar, 40 l/min supplied by Vossen

7 = Measuring centre

8 = Section

9 = Contour slot

10 = Northward dip 6^g

11 = Mortar layer, approx. 5 cm thick resin-set bolts

Fig. 5 Underground testing site for profile cutting

Fig. 6 Experimental installation for testing the roadprofile cutter in a sandstone quarry

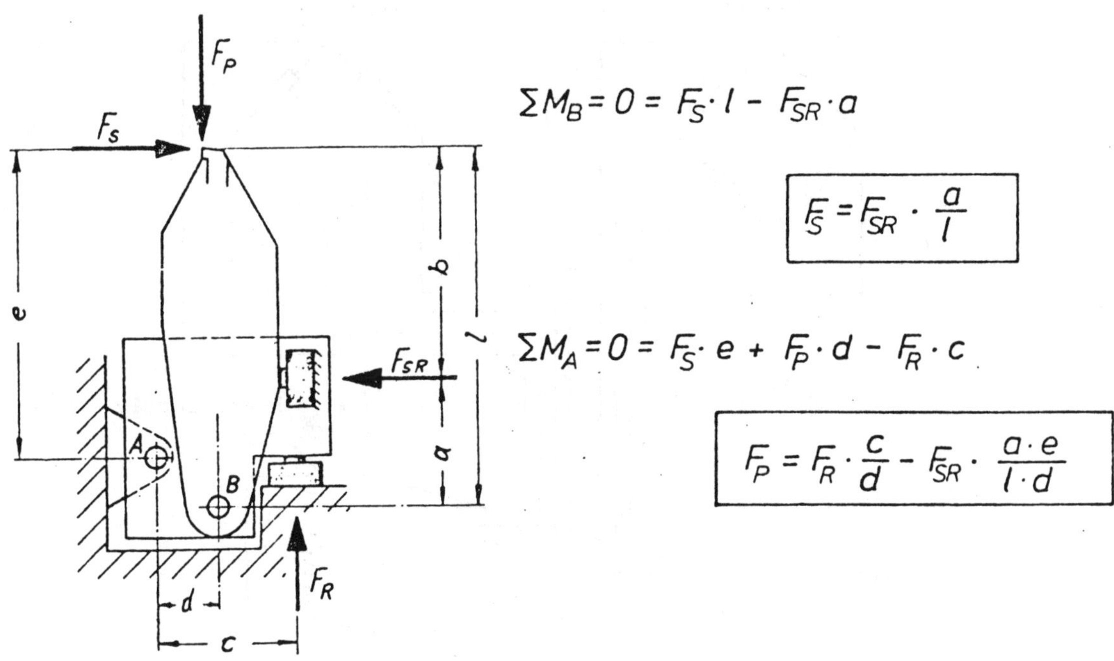

Fig. 7 Measuring scheme for cutting force determination (tests run in the quarry)

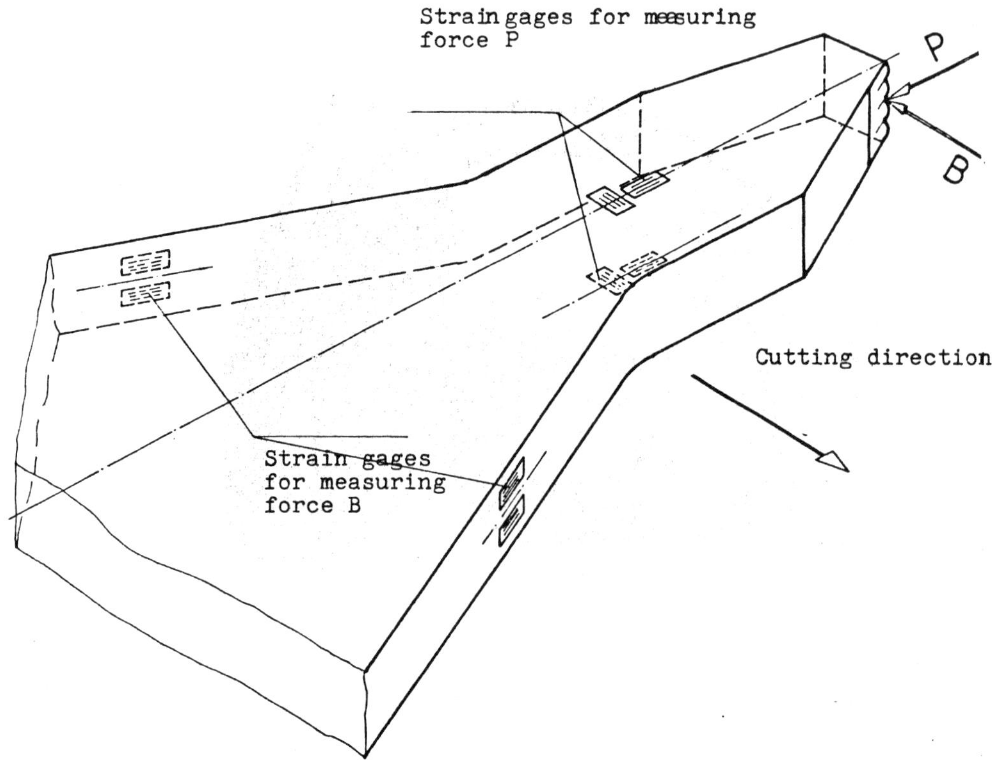

Fig. 8 Measuring scheme for cutting force determination (underground trials)

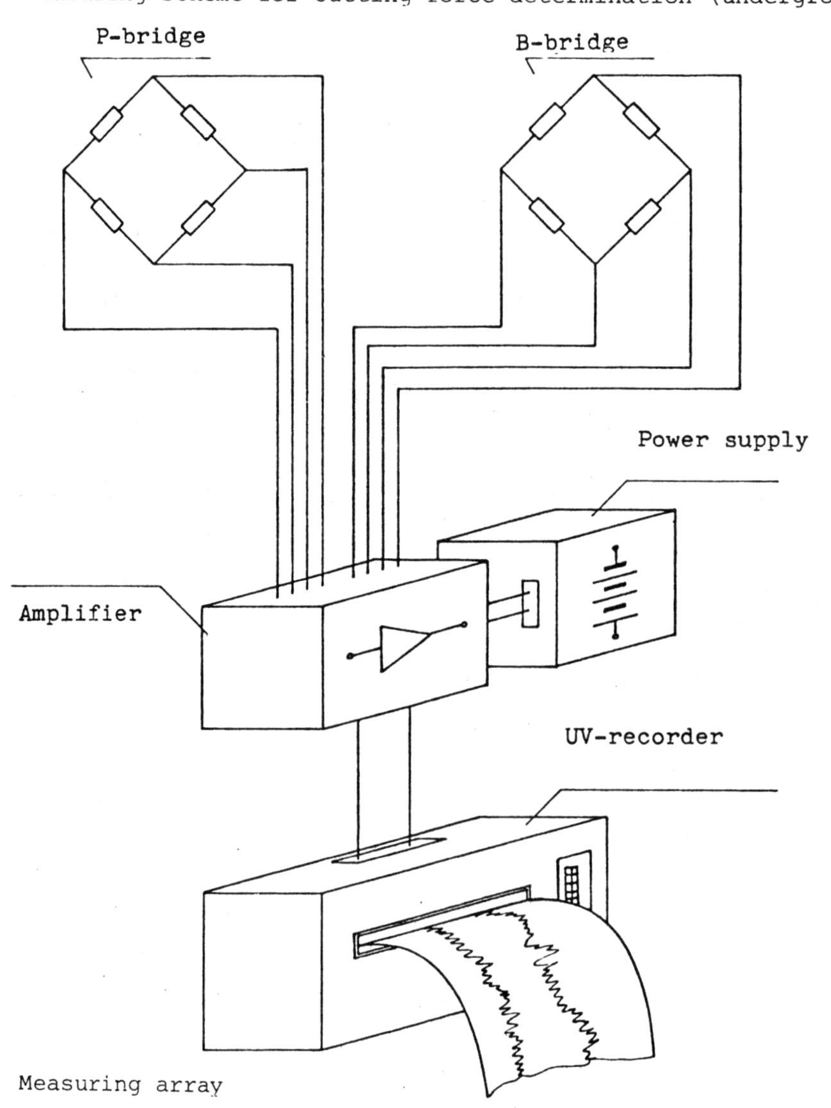

Fig. 9 Measuring array

Fig. 10 Nozzle array on the tool used underground

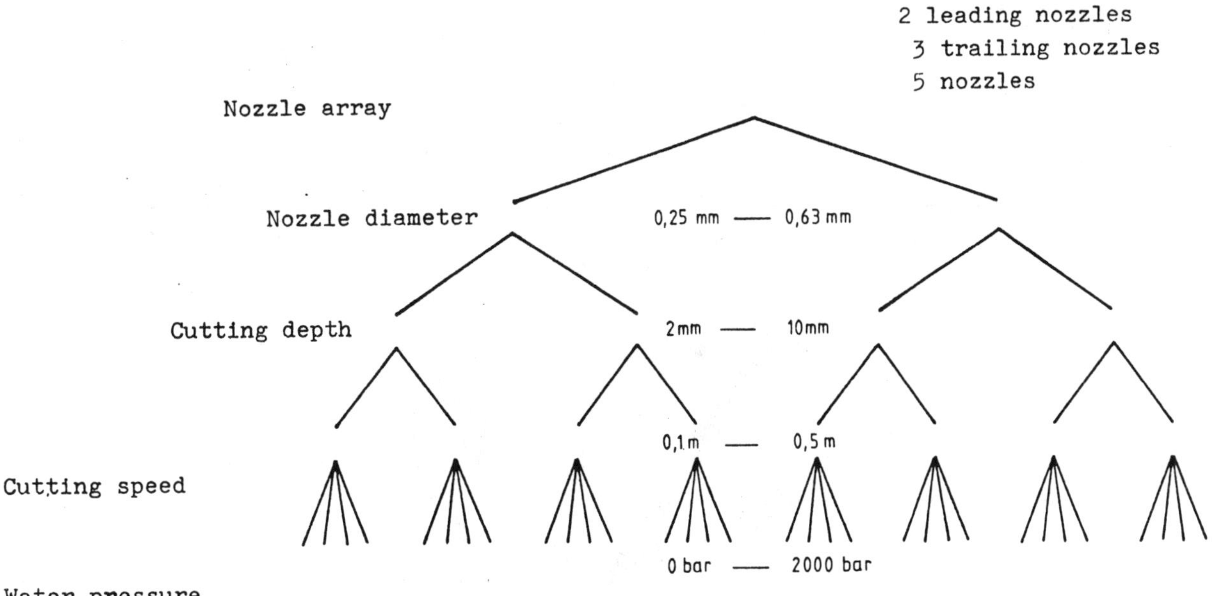

Fig. 11 Testing program scheme

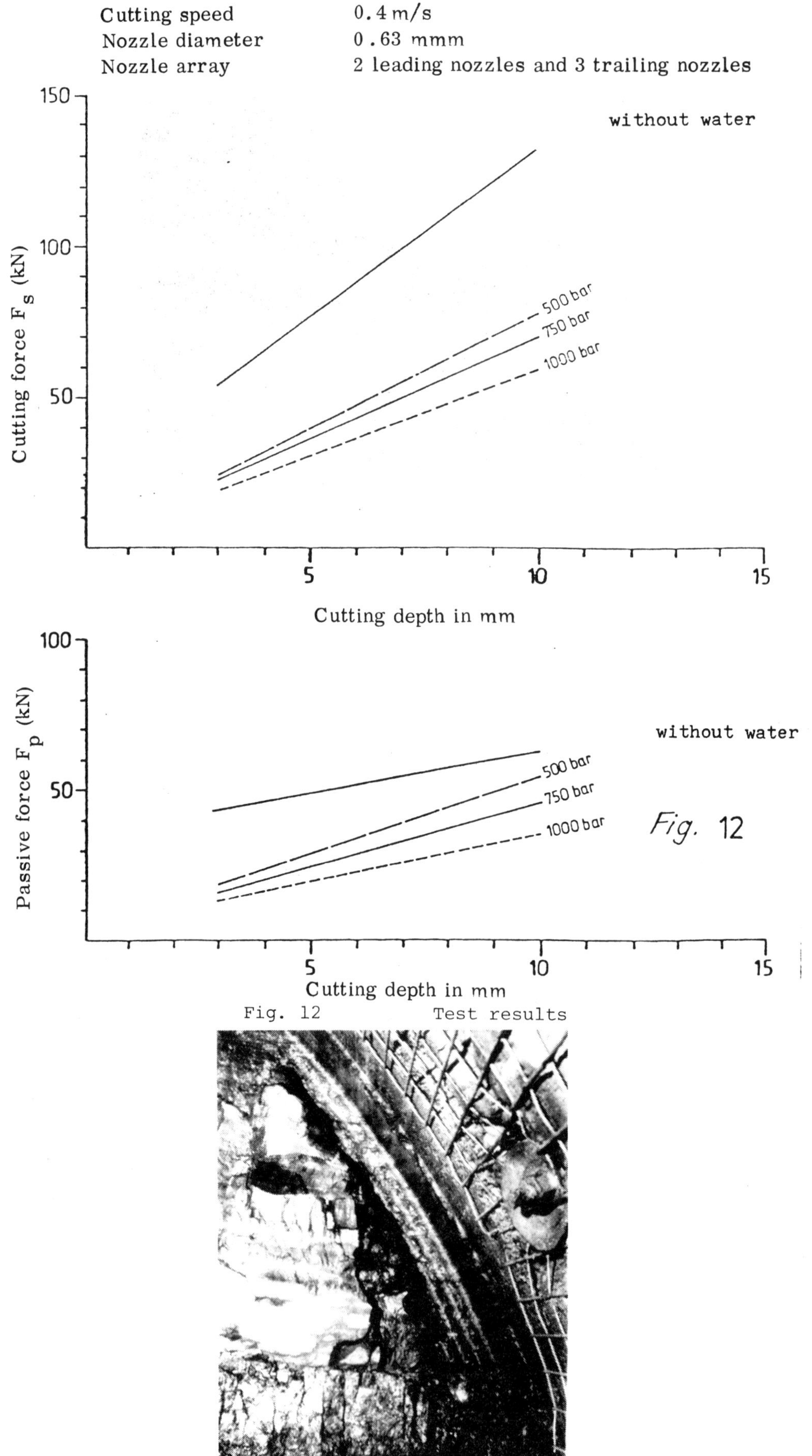

Fig. 12 Test results

Fig. 13 Roadway envelope and core as produced by profile cutting

Cutting speed	0.5 m/s
Nozzles diameter	0.63 mm
Nozzle array	2 leading nozzles

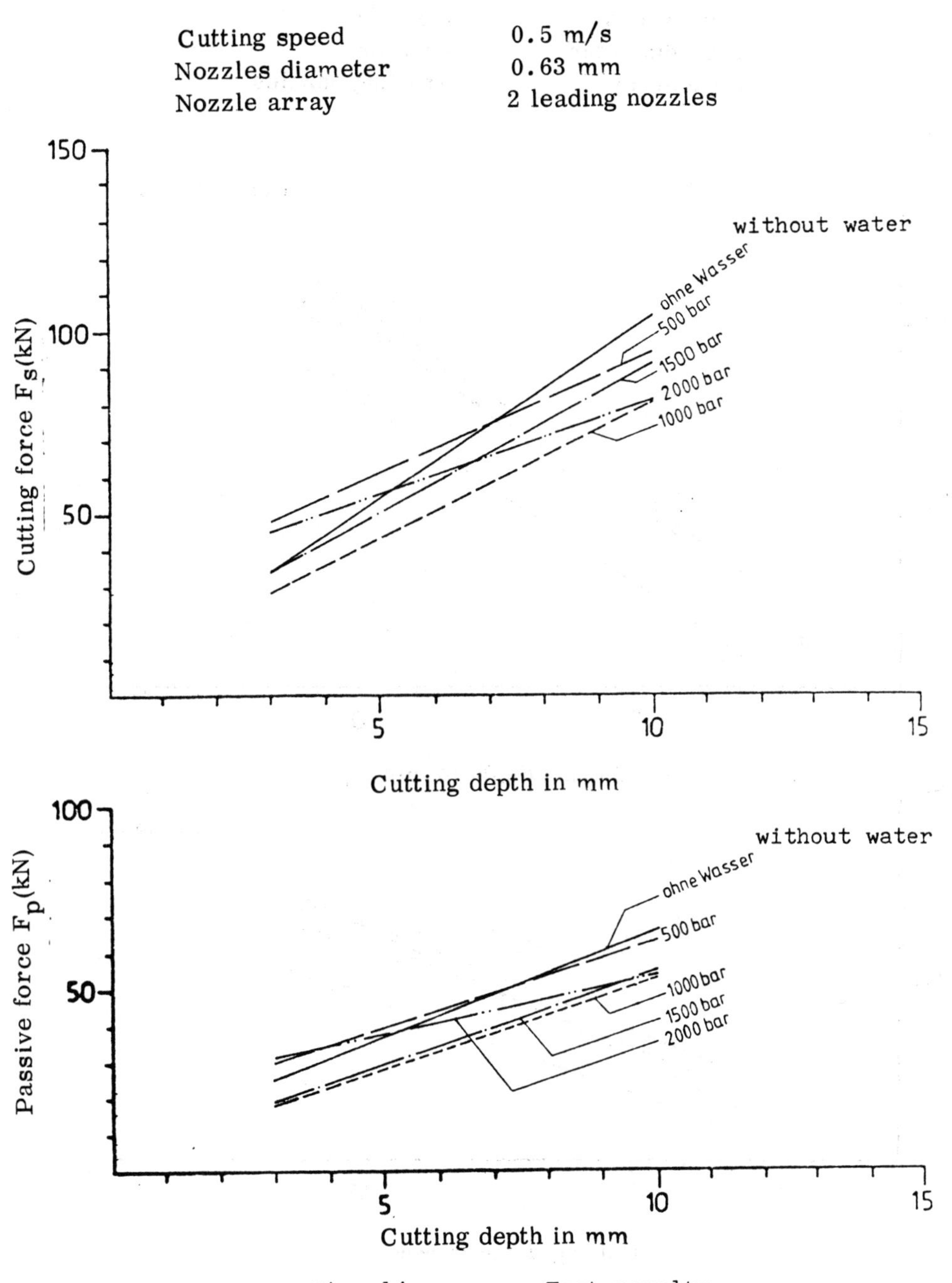

Fig. 14 Test results

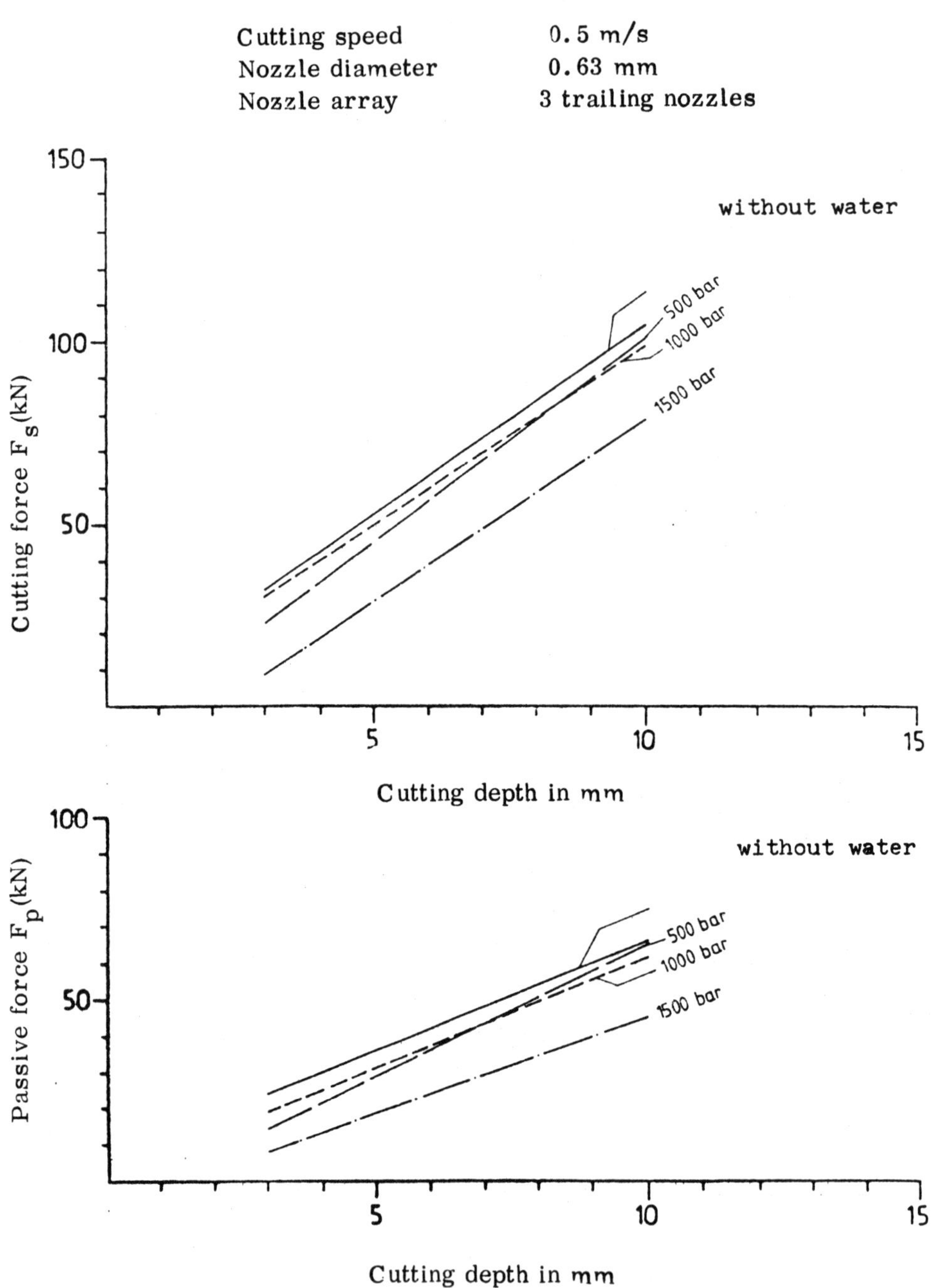

Fig. 15 Test results

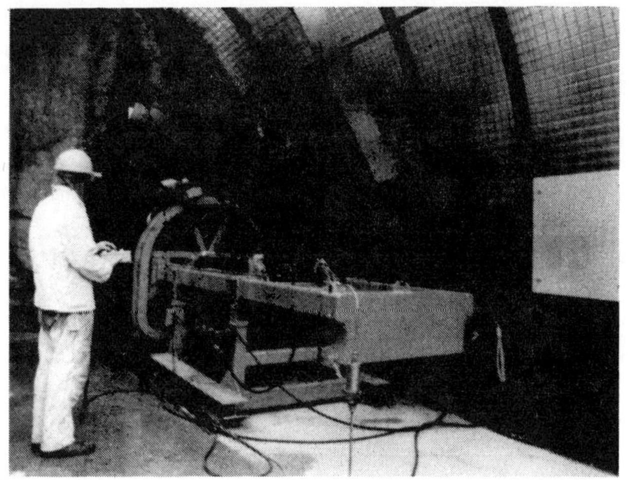

Fig. 16 Road-profile cutting bv high-pressure waterjet only

Fig. 17 Nozzle array for road-profile cutting

Fig. 18

Rotating cutting nozzles

Fig. 19

High-pressure equipment for tunnelboring tests

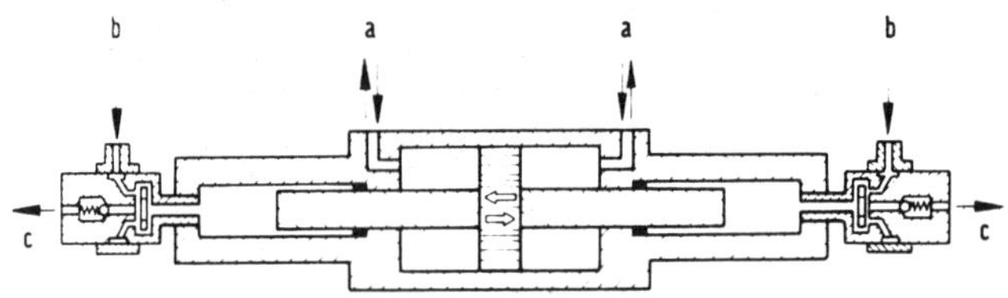

Fig. 20 Function scheme of a high-pressure booster

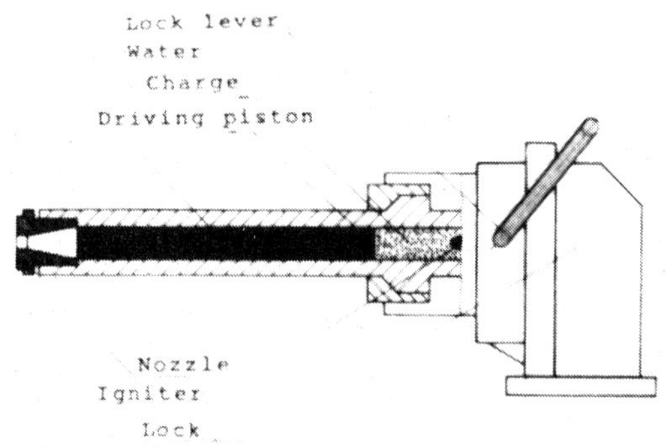

Fig. 21 Water-blast unit

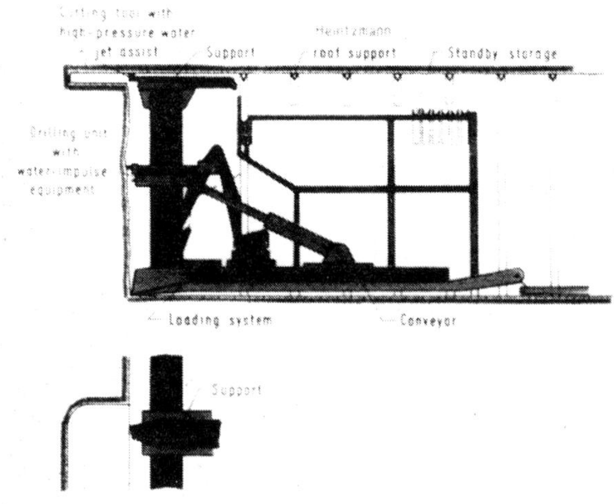

Fig. 22 Integrated profile cutting/road-heading system Bochumer Eisenhütte Heintzmann & Co.

6th International Symposium on
Jet Cutting Technology
6-8, April, 1982

A COMBINED METHOD FOR ROCK BREAKAGE

I.A. Kuzmich, Ju.A. Goldin, M.I. Ruthberg and V.S. Frolov

Skochinsky Institute of Mining, U.S.S.R.

Summary

The paper deals with the experimental results of hydromechanical breakage of rock and coal obtained both in laboratories and in-situ to develop the ways of determining the efficiency and energy consumption for hydraulic method of breakage and to form the cutting forces in hydromechanical breakage. The research helps to find optimum correlations between depth of cut, formed by thin high-pressure water jets, web width and cutting pitch for hydromechanical breakage. The investigation results of different schemes for the combined method of coal breakage by water jets and a disc cutter are considered.

The testing data of the cutting elements for hydromechanical breakages are given.

Held at the University of Surrey, U.K.
Symposium organised and sponsored by
BHRA Fluid Engineering

NOMENCLATURE

$\bar{h}_s$	= depth of slot, mm
P_o	= jet pressure, MPa
d_o	= nozzle diameter, mm
V_s	= relative traverse jet velocity, m/s
V_o	= initial jet velocity, m/s
$\bar{\sigma}_c$	= ultimate rock strength under uniaxial compression, MPa
$\bar{E}'_o$	= specific energy consumption for slot cutting, MJ/m^2
$\bar{h}$	= depth of cut (web width), mm
$\bar{t}$	= mechanical cutting pitch, mm
$\bar{t}_s$	= slot cutting pitch, mm
$(\bar{P}_z)_m$	= cutting force for mechanical breakage, kN
$(\bar{P}_z)_{h-m}$	= cutting force } acting onto the cutter for hydromechanical breakage, kN
$(\bar{P}_y)_{h-m}$	= forward force } acting onto the cutter for hydromechanical breakage, kN
$(\bar{P}_x)_{h-m}$	= side thrust } acting onto the cutter for hydromechanical breakage, kN
$\bar{P}_k$	= rock contact strength, MPa
$(\bar{E}_o)_m$	= mechanical breakage specific energy consumption, MJ/m^3
$(\bar{E}_o)_{h-m}$	= mechanical breakage specific energy consumption for hydromechanical breakage, MJ/m^3
$(\bar{E}_{min})_{h-m}$	= mechanical breakage minimum specific energy consumption for hydromechanical breakage, MJ/m^3
$(\bar{t})_{opt}$	= optimum cutting pitch for hydromechanical breakage, mm
D	= wedge contour outside diameter of disc cutter, mm
δ	= wedge rim angle of disc cutter, degrees
a''	= distance between disc cutter axis and water jet contact point with the rock massif, mm
K_α	= coefficient correcting for the angle of cutting
K_b	= coefficient correcting for the cutting rim - width of the cutter (K_b = 0.9-1.2)
K_γ	= coefficient correcting for inclination angle of the side cutting rims (K_γ = 1.0)
S	= web section, cm^2
M_g	= coefficient of dynamic friction (M_g = 0.4)
F	= blunted area of the cutter, mm^2

The combined or hydromechanical method of breakage joining the hydraulic and mechanical methods demands a justified choice of parameters to make the maximum use of the advantages of both methods. In the course of simultaneous action exerted onto the rock massif by a high-pressure water jet and a mechanical instrument the efficiency of the rock massif breakage is stipulated by the efficiency of each of these breakage methods.

I. HYDROMECHANICAL METHOD OF ROCK BREAKAGE

It has been found that in the process of hydromechanical rock breakage the depth of slot $\bar{h}_s$ mainly depends on the hydraulic parameters of water jet (the pressure and the nozzle diameter), the relative traverse jet velocity and the rock resistance to breakage.

The jet pressure in the contact plane with rock is a function of the initial jet pressure, of water physical and outflow medium constants as well as of the distance between the nozzle and the rock.

As the slot becomes deeper during rock breakage the distance between the nozzle and the face is increasing which results in reducing of the slot formation efficiency.

The cutting section of the water jet is represented by the part, within whose limits the dynamic pressure of the water jet is equal to or is higher than the minimum permissible water pressure to break the given rock. In case of negligible distances between the nozzle and the face the value of contact pressure can be taken as equal to the initial jet pressure p_o ,MPa.

The functional dependence between the depth of slot and the breakage parameters for a single jet pass can be given as follows:

$$\bar{h}_s = f\,(p_o\,;\; d_o\,;\; V_s\,;\; \bar{\sigma}_c\,), \qquad (I)$$

Using the first three primary measuring units i.e. mass, distance and time with their respective scales, expression (I) can be written as follows:

$$\frac{\bar{h}_s}{d_o} = f\left(\frac{p_o}{\bar{\sigma}_c}\,;\; \frac{V_s}{V_o} \right), \qquad (2)$$

It follows from expression (2) that the relative depth of slot $\bar{h}_s/d_o$ is dependent on the dimensionless (relative) traverse jet velocity V_s/V_o and the dimensionless parameter p_o/σ_c characterizing the breaking dynamic jet pressure, on the one hand, and the rock resistance to compression under the given breakage method, on the other.

The experimental research on estimating the breakage parameters by thin high-pressure water jets has been carried out on various rock samples. The principal parameters limits obtained during the experiments are given in Table I.

Table I

Principal parameters	Range of variations
Initial jet pressure p_o ,MPa	20 ... 140
Nozzle diameter d_o ,mm	0.7 ... 3.0
Traverse jet velocity V_s ,m/s	0.013 ... 300
Ultimate rock strength under uniaxial compression $\bar{\sigma}_c$,MPa	9 ... 195

The obtained data have been processed in accordance with expression (2) and plotted as shown in Fig. I which indicates that for various values of the dimensionless parameter p_o/σ_c a hyperbolic dependence assumes the form:

$$\frac{\bar{h}_s}{d_o} = \frac{A}{\left(\frac{V_s}{V_o}\right)^n} \tag{3}$$

The analysis of the plotted curves shows that value n for all the values of the parameter $p_o/\bar{\sigma}_c$ is equal to 0.5.

Thus, determination of the effect of the dimensionless parameter $p_o/\bar{\sigma}_c$ onto the cutting efficiency comes to determination of both its quality and quantity effects on coefficient A in expression (3).

Coefficient A is determined from the expression:

$$A = B\left(\frac{p_o}{\bar{\sigma}_c}\right)^k \tag{4}$$

where B is equal to 0.11 and k is equal to 0.75.

The values of coefficient A corresponding to different mean values $p_o/\bar{\sigma}_c$ and calculated from expression (4) are given in Table 2.

Table 2

Mean value $p_o/\bar{\sigma}_c$	Coefficient A
0.80	0.105
1.41	0.170
3.86	0.310
6.64	0.450
11.20	0.640

After substituting expression (4) for expression (3) a generalized formula for calculating the depth of rock breakage for a single jet pass can be obtained:

$$\frac{\bar{h}_s}{d_o} = 0.11\left(\frac{p_o}{\bar{\sigma}_c}\right)^{0.75}\left(\frac{V_o}{V_s}\right)^{0.5} \tag{5}$$

The experimental data plotted generally as shown in Fig. 2 have been obtained by different authors (Ref. I-7) for various parameter changes and breakage regimes (Table 3).

Table 3

Parameters	Range of variations
Initial jet velocity p_o, MPa	13 ... 1049
Nozzle diameter d_o, mm	0.203 ... 2.87
Traverse jet velocity V_s m/s	0.014 ... 3.13
Ultimate rock strength under uniaxial compression $\bar{\sigma}_c$, MPa	9 ... 200

The plotted data show different convergence degrees between the experimental data and those calculated from expression (5). The calculations have pointed out that mean variation coefficient $\bar{K}_{var}$ for every experimental value is changing within a rather wide range from 20.5 to 51.1 per cent and only the investigation data (Ref. I-2) give a systematic error with the mean value of $\bar{K}_{var}$ equal to 203 and 114 per cent.

Since most of the analysed data satisfactorily agree ($\bar{K}_{var}$ = = 30 %) with their calculated values, it is possible to recommend that expression (5) be used for preliminary calculations of the depth of rock breakage following the slot pattern.

The formula to calculate the specific energy consumption for slot cutting assumes the form:

$$\bar{E}'_o = \frac{0.007 d_o P_o^{0.5} \bar{\sigma}_c^{0.75}}{V_s^{0.5}} \tag{6}$$

The analysis of expression (6) shows that it is preferable to increase the jet pressure when choosing hydraulic jet parameters since in this case the specific energy consumption for slot cutting is not increasing so intensively as in case of a greater nozzle diameter.

Rock breakage by mechanical means is characterized by the principal parameters such as factor of rock resistance to cutting $\bar{K}$; depth of cut (web width) $\bar{h}$,mm; slot cutting pitch $\bar{t}_s$,mm; and mechanical cutting pitch $\bar{t}$,mm.

The proportion of these parameters determines the efficiency of joint action of high-pressure water jets and cutter onto the rock massif. By the changes of the cutting force one can judge about the degree of participation of the hydraulic means in the hydromechanical method of breakage.

It is reasonable therefore to take the cutting force $(\bar{P}_z)_{h-m}$ for the hydromechanical method of breakage as the principal factor of its efficiency.

So, the efficiency determination for the hydromechanical method of breakage comes to determination of the following dependence $(\bar{P}_z)_{h-m} = f\,(\bar{h}\,;\,\bar{h}_s\,;\,\bar{K}\,;\,\bar{t}_s\,;\,\bar{t})$ for two different lay-out diagrams of the high-pressure water jets and the cutter.

The first diagram (Fig. 3a) consists in arranging the water jet and the cutter along the same cutting line while the second one (Fig. 3b) consists in their arranging along different cutting lines.

Both breakage diagrams have been investigated on a stand with a planer and a high-pressure pump plant, the depth of slot varying within the limits: 0 ... 40 mm. Further breakage of the rock sample has been carried out according to the first diagram along the slot by means of I-79B and RPP-type cutters. The web width was changing within the limits $\bar{h}$ = 5 ... 25 mm. The diagram efficiency has been estimated by the reduction of cutting force $(\bar{P}_z)_{h-m}$ as compared with the cutting force necessary for rock breakage without pre-cut slots, the variations of the web width being similar.

The test results indicate the linear changes of the cutting force depending on the web width for all adopted values of slot depths. The dependence $(\bar{P}_z)_{h-m} = f\,(\bar{h})$ can generally assume the form

$$(\bar{P}_z)_{h-m} = a + b\,(\bar{h})\;, \tag{7}$$

where b and a are empiric factors.

The nature of cutting force formation as a function of depth of slots can be generally expressed by the equation:

$$(\bar{P}_z)_{h-m} = a' - b'\,(\bar{h}_s)^n \tag{8}$$

where n = 0.5 and a and b are empiric factors.

To find the optimum proportions of the parameters of hydromechanical breakage it is necessary to determine the optimum ratio between the depth of slot and the web width characterizing a decrease in cutting forces as compared with the mechanical breakage method alone for the given diagram.

The following general formula characterizing this decrease in cutting forces of the hydromechanical method as compared with the mechanical breakage method alone for the given diagram has been obtained on analysis of the experimental data:

$$\frac{(\bar{P}_z)_{h-m}}{(\bar{P}_z)_m} = 1 - 0.4 \left(\frac{\bar{h}_s}{\bar{h}}\right)^{0.5} \quad (9)$$

As follows from the analysis of expression (9), in order to lower the cutting force for hydromechanical method of breakage by 50 per cent as compared with the mechanical breakage method, it is necessary to observe the following relation between the depth of slot and the web width $\bar{h}_s/\bar{h} = 1.5 \dots 1.6$. This conclusion is supposed to be of great importance since it allows to substantially increase the sphere of application of the existing cutting elements with certain cutting and power characteristics installed in rock heading machines. It can be done by introduction of jet forming means into the cutting line of the cutters to provide the slots to be cut.

During investigation of the second diagram (Fig. 3b) both depth of slot and web width were as large as in the first case. The cutting pitch of the cutter was changing within the same range as in case of slot cutting i.e. $\bar{t} = \bar{t}_s = 0 \dots 90$ mm.

The diagrams illustrate some investigation results characterizing the changes of cutting forces as a function of depth of slots $(\bar{P}_z)_{h-m} = f(\bar{h}_s)$ (Fig.4) and of cutting pitch $(\bar{P}_z)_{h-m} = f(\bar{t})$ (Fig.5).

Analysis of the functional curves $(\bar{P})_{h-m} = f(\bar{h}_s)$ shows that the increasing depth of slot for all cutting pitches leads to a linear decrease of the average values of the cutting forces. Besides, the curve analysis makes it possible to derive an important conclusion: at substantial cutting pitches ($\bar{t} \geqslant 45$ mm) and at small depth of slot ($\bar{h}_s \leqslant 3$ mm) and the accepted web width any slot weakening of the rock massif does not influence on any change of the cutting force. In this case the obtained values of the cutting force correspond to those of the cutting force during mechanical breakage (dry cut).

The diagram (Fig. 5) gives the results of the experiments characterizing the changes of the cutting force as a function of the cutting pitch $(\bar{P}_z)_{h-m} = f(\bar{t})$, from which it follows that for different values of depth of slot, web width and rock resistance to cutting the values of cutting force have a linear increase alongside with the increasing cutting pitch.

Analysis of the experimental data for the adopted scheme of breakage has made it possible to find out that a decrease of the relation $\bar{t}/\bar{h}$ results in decreasing the cutting force, and that the intensity of this decrease depends upon the relative depth of slot $\bar{h}_s/\bar{h}$. If at $\bar{t}/\bar{h} = 1$ and $\bar{h}_s/\bar{h} = 1.1 \dots 1.2$ the cutting force is decreased by 3.0 ... 3.3 times, then, at $\bar{t}/\bar{h} = 4$ it is decreased by only 1.5 times.

To get the calculation dependences characterizing the process of hydromechanical breakage the test data were treated in dimensionless parameters (Fig. 6) which made it possible to deduce the formula to calculate the cutting force of the cutter at hydromechanical rock breakage

$$\frac{(\bar{P}_z)_{h-m}}{(\bar{P}_z)_m} = 0.18 \frac{\bar{t}}{\bar{h}} + 0.3 \left(1 - \frac{\bar{h}_s}{\bar{h}}\right) \quad (10)$$

It is worth noting that expressions (9) and (10) for the investigated schemes of rock breakage have been obtained for σ_c = 20 ... 25 MPa.

To calculate the efficiency of hydromechanical rock breakage for other values of rock strength σ_c it is necessary to introduce the quantitative connection between the cutting force and the factor of physical and mechanical rock properties into the obtained expressions.

Then by substituting expression (8) for the value of $(\bar{P}_z)_m$ in expressions (9) and (10), the following expressions are obtained to calculate the cutting force for hydromechanical method of breakage.

For the first jet-and-cutter lay-out diagram it is as follows:

$$(\bar{P}_z)_{h-m} = \left[10 \bar{P}_K K_\alpha K_\beta K_\gamma (0.25 + 0.18 S) + 2.7 \bar{P}_K M_g F \times \right. \\ \left. \times \left[1 - 0.4 \left(\frac{\bar{h}_s}{\bar{h}} \right)^{0.5} \right] \right. \tag{11}$$

and for the second diagram it is as follows:

$$(\bar{P}_z)_{h-m} = \left[10 \bar{P}_K K_\alpha K_\beta K_\gamma (0.25 + 0.18 S) + 2.7 \bar{P}_K M_g F \right] \times \\ \times \left[0.18 \frac{\bar{t}}{\bar{h}} + 0.3 \left(1 - \frac{\bar{h}_s}{\bar{h}} \right) \right] \tag{12}$$

(the depth of slot is determined by expression (5).

It is known that

$$\bar{P}_K = 0.34 \, \bar{\sigma}_c^{1.67} \tag{13}$$

In order to find the optimal parameters of hydromechanical method of breakage for the adopted diagrams the specific energy variations for the breakage process have been analysed.

It has been found for the first diagram (Fig. 3a) that specific energy consumption decreases with an increase of the depth of slot (Fig. 7) while for the adopted web width and depth of slot $\bar{h}_s = \bar{h} = 15 \ldots 20$ mm it decreases by 2 times as compared with the rock breakage without destressing the rock massif.

Thus the optimal value of the relation $\bar{h}_s/\bar{h}$ must be equal to I.

For the second diagram (Fig. 3b) it has been found that for dry cut the specific energy consumption reaches its minimum at $\bar{t} = 30$ mm (Fig. 8) depending on the cutting pitch for the adopted web width $\bar{h} = 15$ mm. The variations of specific energy consumption (Fig.9) for hydromechanical method of breakage showed that for depth of slot $\bar{h}_s = 8 \div 10$ mm and different web width $\bar{h} = 15 \ldots 30$ mm the minimum energy consumption corresponds to the cutting pitch $\bar{t} = 45 \ldots$ ... 70 mm which is by 1.5 - 2 times lower than in case of dry cut. It must be noted that an increase of web width leads to an increase of the optimal cutting pitch. The connection equation for the listed parameters assumes the form:

$$(\bar{t})_{opt} = 1.4 (\bar{h} + 18) \tag{14}$$

Analysis of the experimental data characterizing the variations of specific energy consumption depending on the relative cutting pitch $(\bar{t}/\bar{h})$ has showed that its variations have a parabolic dependence, similar to the one given in Fig. 9, with a minimum which determines the optimal ratio between the cutting pitch and the web width. For the adopted values of web width ($\bar{h} = 15 \ldots 30$ mm) the optimal ratio ($\bar{t}/\bar{h}$) varies from 2.2 to 3.6.

In order to find the dependence of optimal ratio $\bar{t}/\bar{h}$ as a function of $\bar{h}$ the data (see Table 4) characterizing the dependence $\bar{E}_{min} = f(\bar{t}/\bar{h})_{opt}$ have been analysed.

Table 4

$\bar{E}_{min}$, MJ/m^3	$\bar{h}$, mm	$(\bar{t}/\bar{h})_{opt}$
0.90	15	3.2
0.82	20	2.9

Table 4 (continued)

$\bar{E}_{min}$, MJ/m^3	$\bar{h}$, mm	$(\bar{t}/\bar{h})_{opt}$
0.66	25	2.4
0.55	30	2.2

The dependence equation $(\bar{t}/\bar{h})_{opt} = f(\bar{h})$ assumes the form:

$$\left(\frac{\bar{t}}{\bar{h}}\right)_{opt} = 4.2\,(1 - 0.016\,\bar{h}) \qquad (15)$$

Expression (15) thus makes it possible to determine the optimal ratio between the cutting pitch and web width for hydromechanical method of breakage (the second diagram).

II. HYDROMECHANICAL METHOD OF COAL BREAKAGE

Skochinsky Institute of Mining is carrying out research on coal breakage by high velocity fluid jets together with different types of cutters and disc cutters. This research has provided the basis for development of hydromechanical cutting elements.

The previous investigations (Ref. 9) have made it possible to find and study the principal factors determining the efficiency of hydromechanical method of coal breakage as well as to develop different schemes of interaction between water jets and the mechanical cutting or shearing means (Fig. 11).

During investigations of the combined schemes of coal breakage by means of water jets and cutters (Ref. 10), the depth of slot being assumed as a unifying hydraulic factor, the following results have been obtained:

- there exists an optimum jet advance to provide for a minimum cutting force and it is equal to 1.35 $\bar{h}$ for the investigated web width h = (2 ... 6)cm under the experimental conditions;
- for both developed schemes of hydromechanical breakage (see Fig. 11,e,f) - consecutive (the water jets cut the slots at the side of the cutter towards the virgin part of the massif) and chess-board (the water jets cut the slots along the lines of the adjacent cutters) - the optimal depths of slot ($\bar{h}_s/\bar{h}$) = 0.7 for the consecutive scheme and ($\bar{h}_s/\bar{h}$) = 0.35 for the chess-board scheme have been found as well as the optimal slot cutting pitches that substantially decrease the specific energy consumption for mechanical breakage;
- when breaking a coal massif according to the consecutive scheme the optimal slot cutting pitch is determined by the following expression:

for tangent cutters of IT-2S type

$$(\bar{t}_s)_{opt} = 0.8\,\bar{h} - 0.05\,\bar{h}^2 + 16\,\bar{h}_s/\bar{h} - 10\left(\frac{\bar{h}_s}{\bar{h}}\right)^2 - 3.9 \qquad (16)$$

and for radial cutters of I-90MB type

$$(\bar{t}_s)_{opt} = 8.5\,\frac{\bar{h}_s}{\bar{h}} - 7.5\left(\frac{\bar{h}_s}{\bar{h}}\right)^2 - 0.1\,\bar{h} + \bar{h}_s - 0.7 \qquad (17)$$

One of the perfection ways for hydromechanical method of coal breakage lies in the utilization of shearing wedge discs (disc cutters) as the mechanical means of breakage. In this case a high-pressure water jet cuts an advanced slot and the pillar left between the slots is sheared by a wedge disc rolling along the slot (Fig. 12).

The experimental tests of the schemes for combined breakage by means of water jets and a disc cutter were carried out on cement-coal blocks 60x90x100 cm large with the resistance to cutting A =

An analysis of the experimental data characterizing the forces of the breakage process, the energy consumption for mechanical breakage and the size composition of broken coal has showed that of the investigated combined schemes of breakage of coal massif (Fig. 13 and Fig. 11 a,b,c) the one-sided scheme of breakage by water jet and a disc cutter with one-sided sharpening angle is the most rational one.

For the optimum scheme of hydromechanical breakage the following data have been found by comparative investigation:

- the optimum distance a'' between the water jet and the disc cutter axis in terms of the cutting force and the intrusion (any decrease of a'' results in a decrease of both cutting force and intrusion);
- the optimum ratio of the cutting pitch to the web width to minimize the energy consumption for mechanical breakage ($\bar{t}/\bar{h}$ = 3.8... 0.4 h);
- the optimum ratio of the depth of slot to the web width to minimize the forces acting onto the disc cutter axis and the energy consumption for mechanical breakage ($h_s/\bar{h}$ = 0.6...0.8);
- the optimum disc cutter type and dimensions in terms of force and power indices of hydromechanical breakage process.

Fig. 16 presents the results of a comparative analysis of the scheme efficiency for hydromechanical breakage at different interaction processes between high-pressure water jets and cutters or disc cutters. For coal massif breakage by means of a disc cutter together with water jets the cutting force is by 20...25 per cent lower and the energy consumption is by 1.5-2.0 times less than those for coal breakage by water jets together with a cutter. This has provided for 70... 80 per cent of + 25 mm grade coal while dust concentration has not exceeded permissible norms.

On the basis of the stand investigations the following prototypes of hydromechanical cutting elements have been developed: hydromechanical auger-type cutter provided with cutting and jet-forming means; hydraulic cutting element provided with independent kinematics of hydraulic and mechanical units for a narrow-web shearer of KSh-3M (KSh-3G) type; hydromechanical auger-type cutter provided with disc cutters for 2K-52 shearer.

Production tests of KSh-3G shearer provided with hydromechanical and hydraulic cutting elements have been carried out in a 60 m long coal face with KM-81Э mechanized complex (Fig. 17). The experimental zone was characterized by the following parameters: seam thickness of 2.4 m, dip angle of 12° and included two claystone interlayers with overall thickness of 0.3 m. The resistance to breakage of class coal was A = 1200 N/cm, and taking into consideration the claystone interlayers it was equal to 1900 N/cm. The coal seam belongs to moderately dusty seams according to the quantity of dust produced referred to the total broken mass.

Efficiency of water jets used together with a mechanical cutter in the course of their simultaneous action onto a coal massif has been estimated by the energy consumption for breakage, by the grade of coal produced and by the dust concentration.

The design of hydromechanical auger-type cutting element envisaged different nozzle orientations versus the cutter. Water pressure (from 5 to 32 MPa) and nozzle diameter (from 1.9 to 3.7 mm) were varying for each scheme within the cited limits.

The production tests of the hydromechanical auger-type cutting element for KSh-3G shearer have proved the fact that complex utilization of water jets and mechanical cutter provides for the following:

- increased power and capacity of the shearer without any increase of its weight and dimensions (the mechanical component of energy consumption for hydromechanical breakage lowering to 0.7 ... 0.9 MJ/m^3

= 1600 N/cm. During the tests the nozzle diameter was d_o = 2.0 mm, the block displacement speed versus the nozzle and the disc cutter was constant and made V_s = 0.15 m/s, the pressure P_o to ensure the necessary depth of slot was chosen in accordance with the results obtained during a series of preliminary hydraulic cuts with the pressure changing from 10 to 35 MPa.

The efficiency of the schemes for combined breakage has been evaluated on the basis of the cutting force ($\bar{P}_z$), forward force ($\bar{P}_y$) and side thrust ($\bar{P}_x$) acting onto the axis of the disc cutter, of the specific energy consumption for mechanical breakage ($E_o)_m$, of the granulometric composition of the broken product and its specific dust production (Ref. 11,12).

The investigation results of combined breakage schemes by means of a front-action disc cutter with a depth of cut h less than the depth of slot h_s (see Fig. 11 a,b,c) showed that complete breakage of the pillar is not ensured by any of the investigated schemes of breakage, that is, shearing of the pillar is not provided for the height equal to the depth of slot.

The scheme of coal massif breakage using a disc cutter with one-sided sharpening angle for $\bar{h} > h_s$ (Fig. 13) has proved to be more efficient. In this case the pillars are mainly sheared along the base where the maximum tensile stresses are concentrated. This can be achieved by a disc cutter rolling along the slot and acting not only onto the side of the inter-slot pillar but also onto its base by pressing the top of the rim into the coal massif.

Fig. 14 shows the forces $\bar{P}_z$, $\bar{P}_y$ and $\bar{P}_x$ (acting onto the disc cutter) as a function of the slot cutting pitch (the breakage pitch) $\bar{t}$ for two schemes: with a front-action disc cutter (Fig. 11 a) and with a disc cutter with one-sided sharpening angle at $\bar{h}$ =5cm (Fig.13).

Disc cutters with outside diameter D = 200 mm and wedge rim angle δ = 30° were used during the tests. They were carried out at the ratio between the depth of slot and the depth of cut $\bar{h}_s/\bar{h}$ = 1.1 (for a front-action disc cutter) and $\bar{h}_s/\bar{h}$ = 0.6 (for a one-sided sharpening angle disc cutter). An increased cutting pitch $\bar{t}$ provides for greater forces $\bar{P}_z$ and $\bar{P}_y$, the difference between the values of P_y being substantial over the whole range of the pitch values. Application of the disc cutter with one-sided sharpening angle together with a water jet reduces the forward force $\bar{P}_y$ by 1.6 - 2.4 times as compared with that of the front-action disc cutter. This is due to the fact that in the course of intrusion in a coal massif the side surfaces of a front-action disc cutter are interacting not only with the surface of the pillar under breakage but also with the virgin part of the massif. The interaction of analogous nature takes part in forming the side thrust $\bar{P}_x$.

Specific energy consumption for mechanical breakage$(\bar{E}_o)_m$ as a function of the cutting pitch $\bar{t}$ is given in Fig. 15. As one can see, the energy consumption for mechanical breakage by a disc cutter with one-sided sharpening angle together with a water jet is much lower than that of breakage by a front-action disc cutter. It is particularly true and clear at the great values of mechanical cutting pitch $\bar{t}$ and depth of cut $\bar{h}$. When using the scheme of breakage by the disc cutter with one-sided sharpening angle together with water jets over the whole range of the cutting pitch $\bar{t}$, any increase of the depth of cut h results in a decrease of the specific energy consumption for mechanical breakage$(\bar{E}_o)_m$ while the values of the optimal slot cutting pitch increases by 1.3 - 1.5 times versus the scheme of breakage by the front-action disc cutter. Possible breakage at a greater pitch $\bar{t}$ provides for a higher coal output of bigger grades: the screen analysis has proved that over 70...80 per cent of coal belong to grade + 25 mm.

versus 1.4 MJ/m^3 for dry cut);

- an increased production (up to 70 per cent) of +25 mm grade coal;
- reduced consumption of cutters (by 10-15 times);
- provision of dust concentration sanitary norms; at hydromechanical coal breakage with water specific consumption of (47...67) 10^{-3} m^3/t and water pressure of 30 MPa specific dust production and dust concentration are equal to 2.0...2.5 g/t and 7...15 mg/m^3 respectively.

The investigation results of the hydromechanical method for coal breakage by water jets cutting the advanced slots (parallel to the floor and the roof) from the stripped coal face with further side breakage of the weakened section of the seam by means of the auger-type shearer are of special interest. It has been found that cutting of advanced slots at depths equal to the auger web width provides for a 50% decrease of shearer energy consumption necessary for mechanical breakage. Design simplicity of the hydromechanical cutting element provided with independent kinematics of mechanical jet-forming units, its durability and possibility of being installed in any shearer ensure its prospective use for the combined breakage.

The investigation results of the hydromechanical method for coal or rock breakage presented in the paper for the schemes of combined breakage make the basis for further development of hydromechanical cutting elements to be used in various heading machines and shearers.

REFERENCES

1. Moodie, K., Artingstall, G.: "Some experiments on the application of high pressure water jets for mineral excavation". In: Proc. Ist Int. Symp. Jet Cutt. Technol. (Coventry, U.K., 1972) Cranfield, U.K., BHRA Fluid Engineering, 1972, Paper E25, p.44.

2. Brook, N., Page, C.H.: "Energy requirements for rock cutting by high-speed water jets". In: Proc. Ist Int. Symp. Jet Cutt. Technol. (Coventry, U.K., 1972) Cranfield, U.K., BHRA Fluid Engineering, 1972, Paper B1, p.12.

3. Matsumoto, K., Hamada, H., Fukuda, T., Shirio, A.: "High-pressure jet cutting". In: Proc. Ist Int. Symp. Jet Cutt. Technol. (Coventry, U.K., 1972) Cranfield, U.K., BHRA Fluid Engineering, 1972, Paper B53, p.75.

4. Zelenin, A.N., Veselov, G.M., Konyashin, Yu.G.: "On regularities of rock breakage by water jet at pressure up to 2000 atm.". (Закономерности разрушения горных пород струей воды при давлении до 2000 атм.). Voprosy gornogo dela, Moscow, Ugletechizdat, 1958. (In Russian).

5. Goldin, Yu.A., Frolov, V.S., Shimanov, I.V.: "On choice of index of rock resistance to hydraulic breakage". (О выборе показателя сопротивляемости горных пород гидравлическому разрушению). Publications of Skochinsky Institute of Mining, Vol.178, Moscow, 1979. (In Russian).

6. Harris, H.D., Mellor, M.: "Penetration of rock by continuous water jets". In: Proc. 2nd Int. Symp. Jet Cutt. Technol. (Cambridge, U.K., 1974) Cranfield, U.K., BHRA Fluid Engineering, 1974, Paper H1, p.15.

7. Hoshino, K., Nagano, T., Tsuchishima, H.: "Rock cutting and breaking using high speed water jets together with "TBM" cutters". In: Proc. Ist Int. Symp. Jet Cutt. Technol. (Coventry, U.K., 1972) Cranfield, U.K., BHRA Fluid Engineering, 1972, Paper B89, p.100.

8. Baron, L.I., Glatman, L.B., Gubenkov, E.K.: "Rock breakage by heading machines". (Разрушение горных пород проходческими комбайнами). Moscow, Nauka, 1968, (In Russian).

9. Kuzmich, I.A., Ruthberg, M.I.: "Investigation of the interaction between high-speed water jet and cutter during breakage of a rock mass." In: Proc. 4th Int. Symp. Jet Cutt. Technol. (Canterbury, U.K., 1978) Cranfield, U.K., BHRA Fluid Engineering, 1978, Paper J4, p.37.

10. Ruthberg, M.I., Kuzmich, I.A.: "Investigation and choice of optimum schemes for hydromechanical coal breakage". (Исследование и выбор рациональных схем гидромеханического разрушения угля). Publications of Skochinsky Institute of Mining, Vol.160, Moscow, 1978. (In Russian).

11. Ruthberg, M.I., Khrameshkin, S.I., Merzlyakov, V.G.: "Investigation of combined method of rock breakage by water jet and shearing disc". (Исследование комбинированного разрушения горного массива струей воды и скалывающим диском). Publications of Skochinsky Institute of Mining, Vol.189, Moscow, 1980. (In Russian).

12. Kuzmich, I.A., Ruthberg, M.I., Merzlyakov, V.G.: "Some investigation results for the schemes of rock breakage by high-speed water jet and shearing disc". (Некоторые результаты исследования схем разрушения горного массива высокоскоростной струей воды и скалывающим диском). Publications of Skochinsky Institute of Mining, Vol.190, Moscow, 1980. (In Russian).

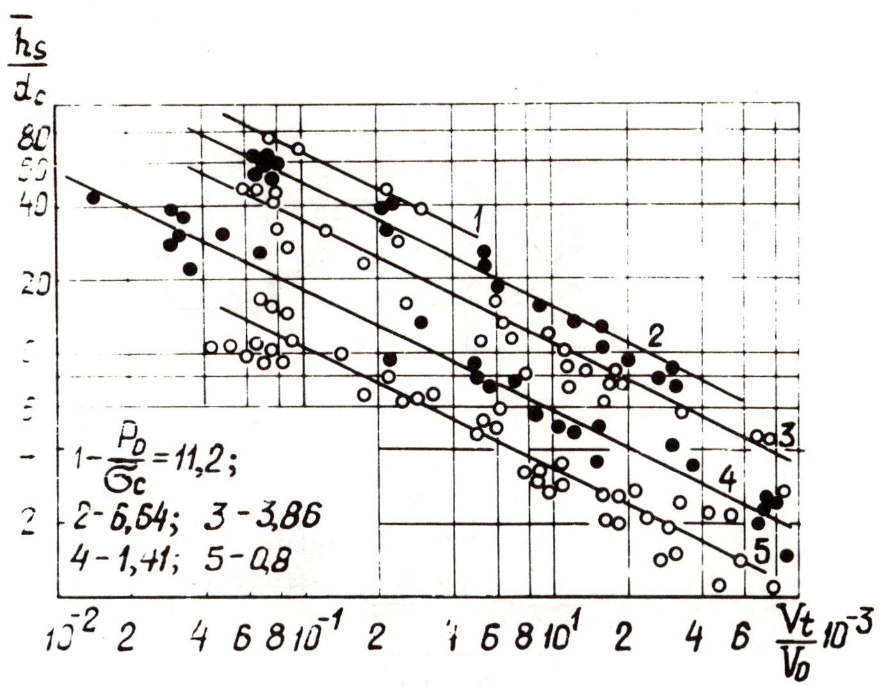

Fig. 1. Relative depth of slot as a function of relative traverse jet velocity.

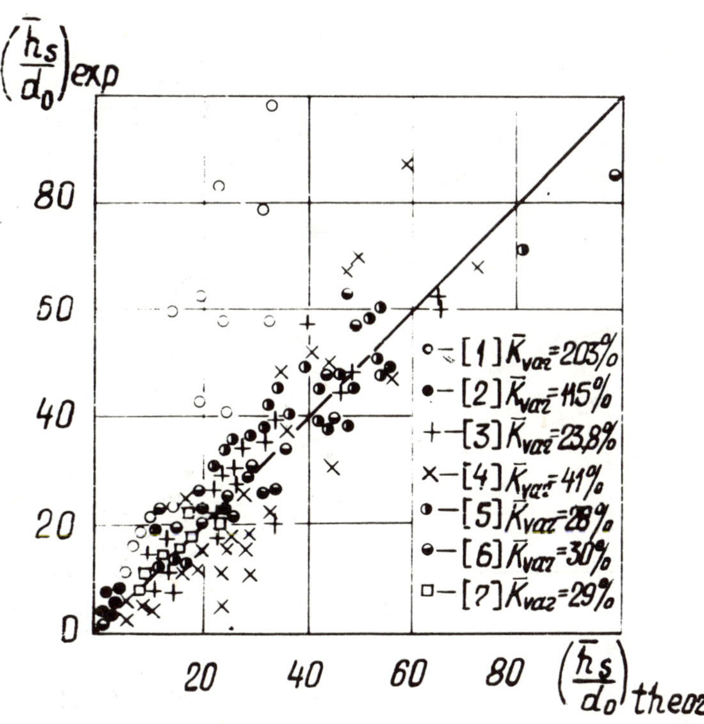

Fig. 2. Comparison between experimental and calculation data for various investigation results.

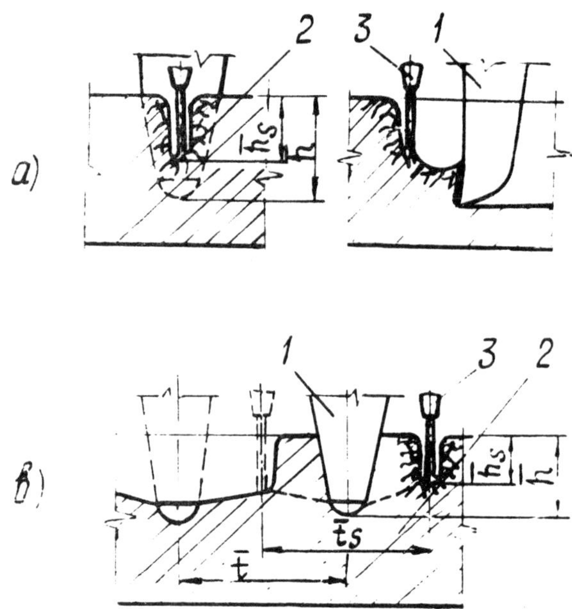

Fig. 3. Schemes of hydromechanical rock breakage:
a) jet - cutter along the same cutting line,
b) jet - cutter along different cutting lines.
1 - cutter, 2 - rock mass, 3 - slot.

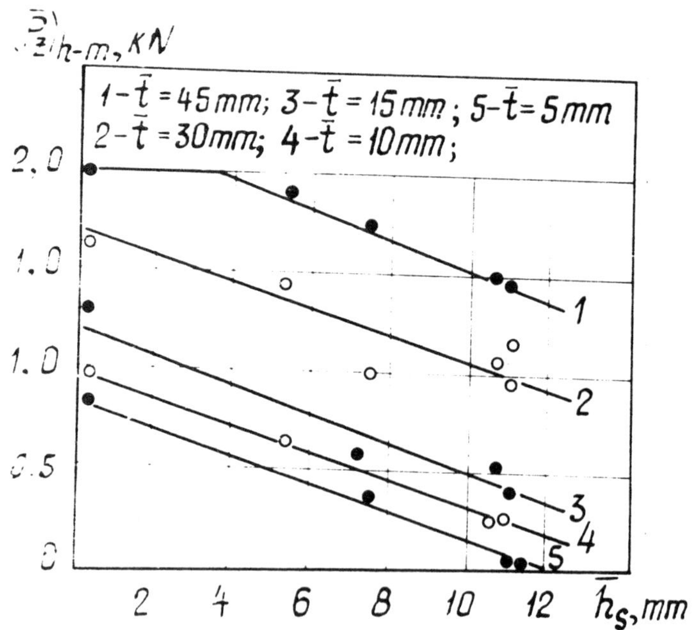

Fig. 4. Cutting force as a function of depth of slot at $\bar{h}_s$ = 10 mm and $\bar{\sigma}_c$ = 25 MPa.

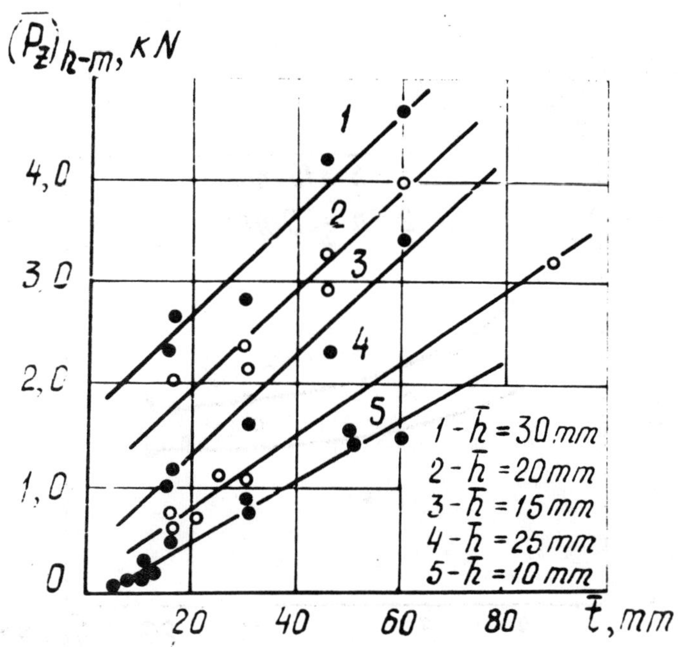

Fig. 5. Cutting force as a function of cutting pitch at $\bar{h}_s$ = 8...10 mm and $\bar{\sigma}_c$ = 35 MPa.

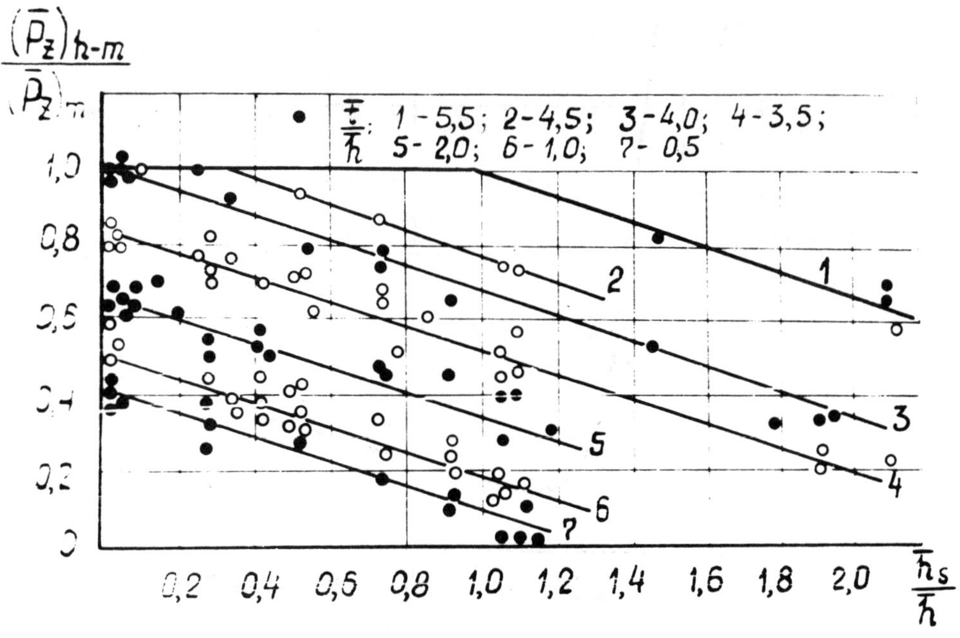

Fig. 6. Relative cutting force as a function of relative depth of slot.

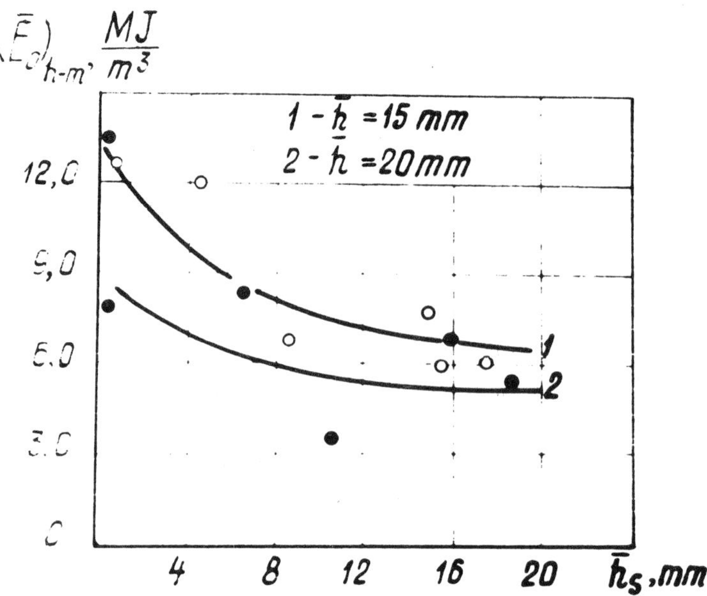

Fig. 7. Specific energy consumption for hydromechanical breakage as a function of depth of slot.

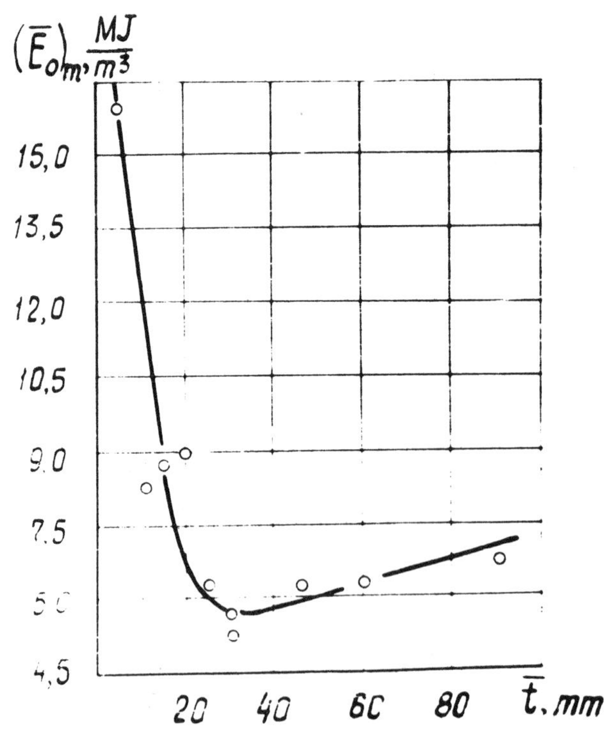

Fig. 8. Specific energy consumption for mechanical breakage as a function of cutting pitch at 15 mm web width.

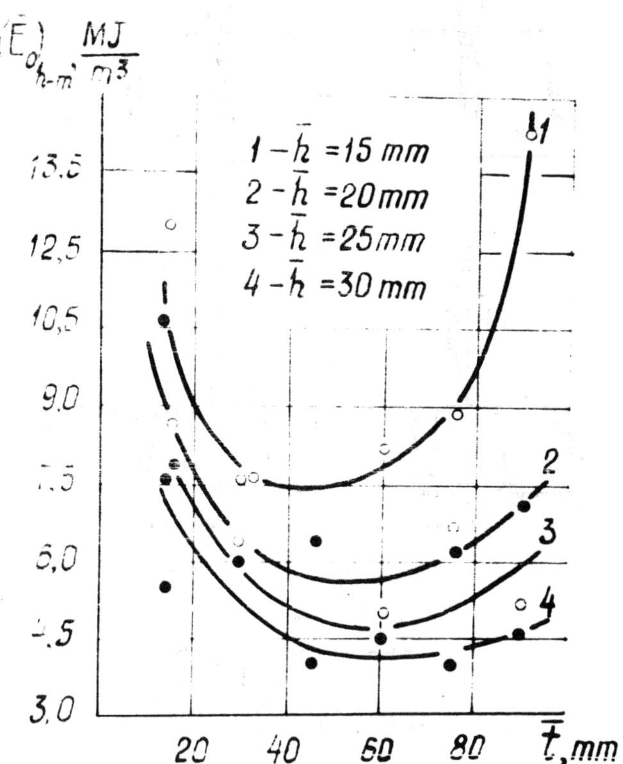

Fig. 9. Specific energy consumption for hydromechanical breakage of rock (σ_c = 35 MPa) as a function of cutting pitch.

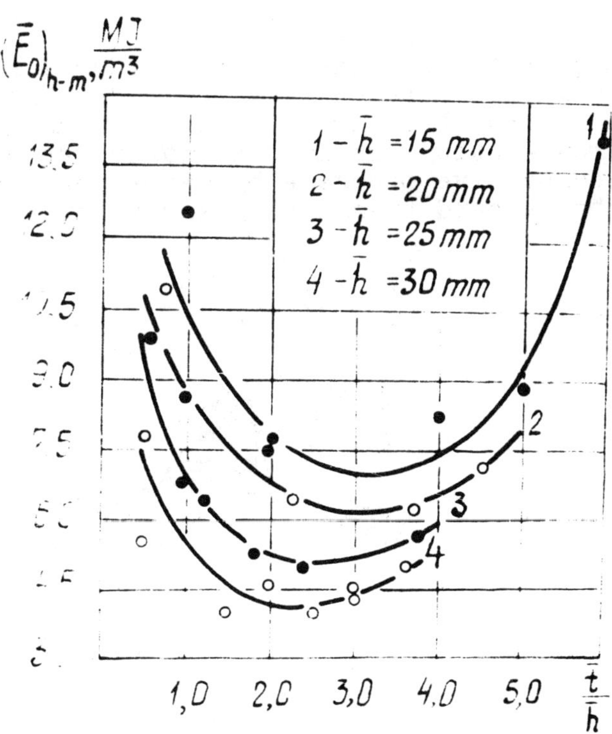

Fig.10. Specific energy consumption for hydromechanical rock breakage as a function of relative cutting pitch.

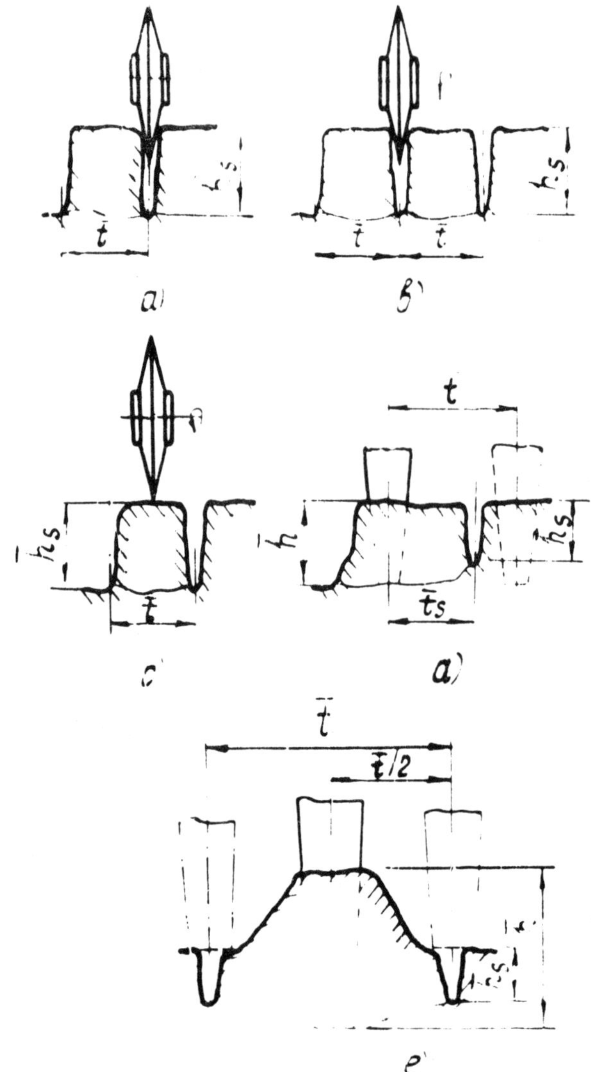

Fig.11. Schemes of the combined method of coal massif breakage:
a) unilateral breakage by water jet and disc cutter,
b) bilateral breakage by water jet and disc cutter,
c) front-action breakage by water jet and disc cutter,
d) consecutive breakage by water jet and cutter,
e) chess-board breakage by water jet and cutter.

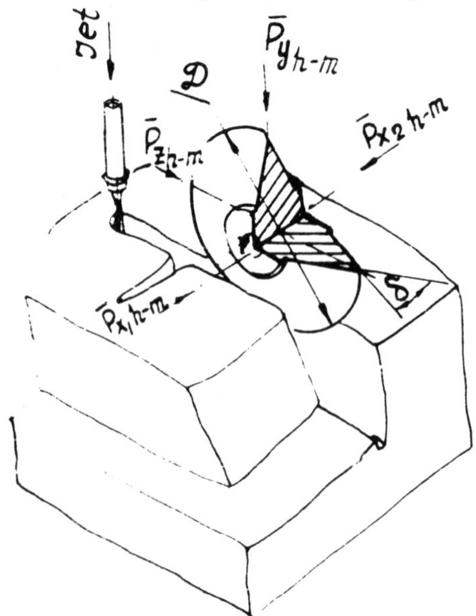

Fig.12. Forces acting onto a disc cutter at hydromechanical breakage of coal massif.

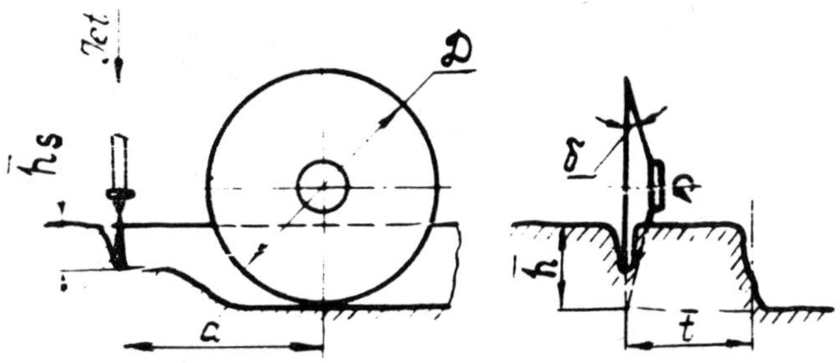

Fig.13. Combined method of coal mass·f breakage by water jet and disc cutter with one-sided sharpening angle.

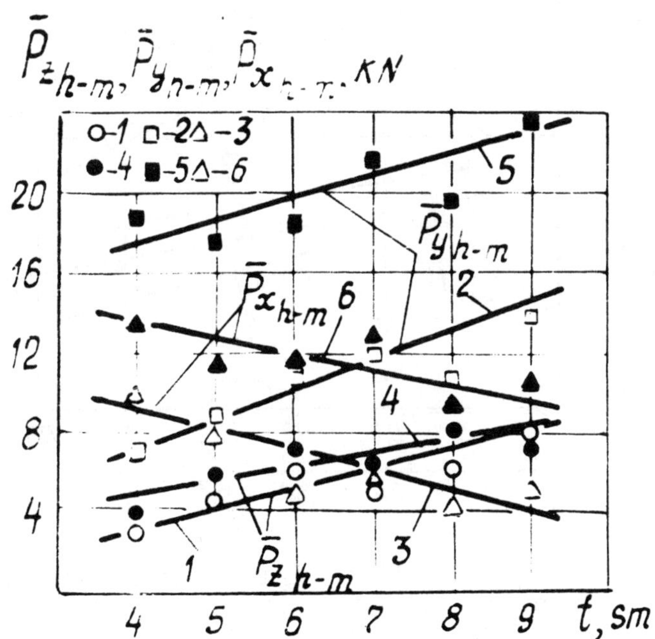

Fig.14. Cutting force $(\bar{P}_z)_{h-m}$, forward force $(\bar{P}_y)_{h-m}$ and side thrust $(\bar{P}_x)_{h-m}$ as functions of mechanical cutting pitch for coal breakage by high-speed water jet and disc cutter with unilateral (1,2,3) and bilateral (4,5,6) sharpening angles.

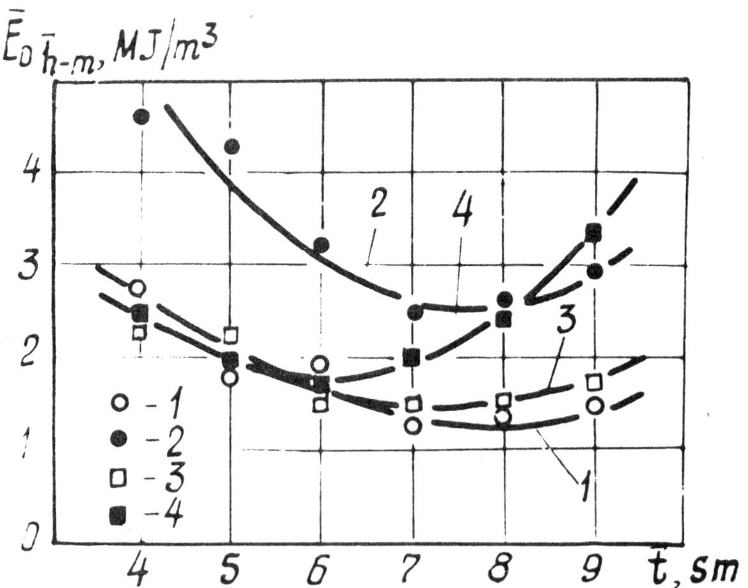

Fig.15. Effect of cutting pitch onto specific energy consumption for mechanical coal breakage by water jet together with disc cutter with unilateral (1,3) and bilateral (2,4) sharpening angle. For 1 and 2 - $\bar{h}$ = 5 cm, for 3 and 4 - $\bar{h}$ = 4 cm.

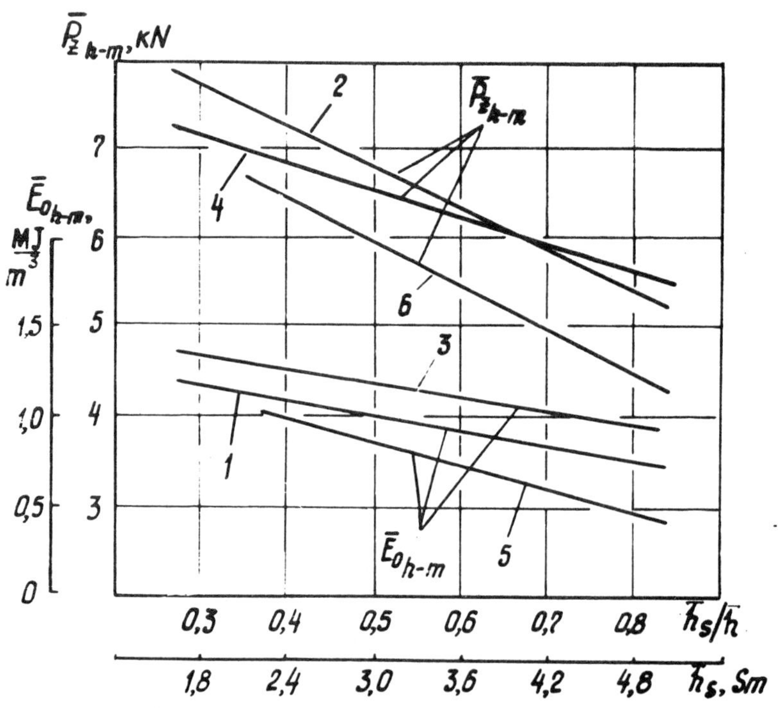

Fig.16. Cutting force $(\bar{P}_z)_{h-m}$ and specific energy consumption $(\bar{E}_o)_{h-m}$ for mechanical breakage as a function of depth of slot $\bar{h}_s$ (ratio $\bar{h}_s/\bar{h}$) for hydromechanical coal breakage ($\bar{h}$ =6cm, $\bar{A}$=1600 N/cm)

1,2) breakage by water jet and IT-2S cutter:
$\bar{t}_s = (\bar{t}_s)_{opt} = 4.5$ cm, $a/\bar{h} = (a/\bar{h})_{opt} = 1.35$

3,4) breakage by water jet and I-90MB cutter:
$\bar{t}_s = (\bar{t}_s)_{opt} = 4.5$ cm, $a/\bar{h} = (a/\bar{h})_{opt} = 1.35$

5,6) breakage by water jet and disc cutter:
$D = 200$ mm and $\delta = 25^o$:
$\bar{t}/\bar{h} = (\bar{t}/\bar{h})_{opt} = 1.4$; $a = 220$ mm.

Fig.17. KSh-3G narrow-web shearer provided with a hydromechanical auger-type cutting element.

6th International Symposium on
Jet Cutting Technology
6-8, April, 1982

JET-MINER SURFACE TRIALS RUN AT BERGBAU-FORSCHUNG, GERMANY

Th. Krämer

Bergbau-Forschung GmbH, Federal Republic of Germany

Summary

Following a detailed discussion of the underground trials run with the experimental version of the Jet-Miner, the development of the prototype is delineated.

The prototype design, the drive units for high-pressure water supply and haulage, the array of the high pressure water cutting heads, the water and energy supply system as well as integrated operation with the other coal face equipment are explained. Subsequently, initial results of surface trials are discussed and the preparations with respect to underground operation of the Jet-Miner are described.

Held at the University of Surrey, U.K.
Symposium organised and sponsored by
BHRA Fluid Engineering

Legend:

F_S : Cutting force

F_K : Pulling force (chain)

s : Web

v_M : Travelling speed

d_o : Nozzle diameter

p : Water pressure

p_H : Hydrostatic drive pressure

$\dot{V}$: Flow

P : Electric drive performance

σ_D : Compressive strength

σ_Z : Tensile strength

1. Introduction

The average worked seam thickness in German hard coal mining which today is 1.9 m will be reduced in the future to approximately 1.5 m. This requires the development of heavy-duty coal winning machinery specially designed for use in seam thickness ranging between 1.0 m and 1.5 m. The coal ploughs used today in seams of such thickness do not meet the requirements in all respects, and consequently new developments, especially on the basis of plough technology, are pushed ahead more vigorously.

The following targets are set for winning machinery to be used in hard coal and in seams of 1.0 to 1.5 m thickness:

- high winning performance in hard coal
- large-sized lumps (improved quality)
- cutting into adjacent rock
- less dust.

The development of the Jet-Miner, a winning machine using high-pressure waterjets for cutting, constitutes an important step towards the above-mentioned targets. The co-operation between M.A.N.-GHH Sterkrade, Ruhrkohle AG, and Bergbau-Forschung GmbH resulted first in an experimental unit by which the basic suitability of the techniques implied were proven (1) (2) (3).

Based on knowledge gained throughout the trials a first Jet-Miner prototype was constructed by M.A.N.-GHH-Sterkrade. This prototype is supposed to meet the requirements defined to a very fargoing extent. The following chapters will describe the development of this prototype and the surface trials run.

2. Underground Trials Run with the Experimental Version

2.1 Trial Face

Following to the good results of the surface trials run with the experimental version, an experimental coal face was established in Lohberg colliery in seam N (4). The experimental coal face was of 50 m length and of 1.6 m of seam thickness. According to its strength the coal was categorized as "difficult to plough" - Fig. 1-. In comparison to the configuration run in the surface trials, the machine was slightly modified; the key data read:

Machine under tests:	
Operation voltage	U = 1000 V
Installed power for the pumps	P = 300 kW
Flow (max)	$\dot{V}$ = 226 l/min
Water pressure	p = 70 MN/ m^2
Total mass	m = 23000 kg
Hydrostatic haulage drive	
Installed power	p = 3 x 80 kW

The measuring program covered the following parameters:

- Cutting force of a cutting head F_S
- Pulling force (chain) F_k
- Pressure of the hydro-motor P_H
- Travelling speed v_M

In addition coal samples for determination of particle-size distribution and moisture content were taken, and furthermore dust and noise measurements were carried out.

2.2 Test Results

From the coal-winning point of view the experimental version worked quite satisfactorily from the very beginning. The coal face was undercut smoothly by the oscillating high-pressure waterjets and subsequently broken off by the cutting heads. The coal face did not exhibit any cutting or grinding traces caused by the mechanical tools. This meant that the coal, as envisaged, was cut exclusively by the waterjets and then broken off by the cutting heads.

A further proof for this phenomenon is the fact that after the tests the cutting heads exhibited no wear and in particular no wear imputable to mechanical cutting work.

The hydraulic drive of the haulage system was adjusted to a value of F_K = 370 kN for the pulling force acting on the chain. The average value as recorded by measurement was of approximately F_K = 340 kN. This force may be split up into following sub-components:

- Chain	ca. 60 kN
- Loading work and machine travel	" 160 kN
- Cutting resistance of the upper head	" 5 kN
- the central head	" 20 kN
- lower head	" 120 kN

This means that approximately 40% viz. 145 kN of the available power is made use of for mechanical removal of the coal from the face. 80% of these 145 kN were picked up by the lower cutting head. The other cutting heads really undercut the coal face by the waterjet and consequently applied only insignificant mechanical break-off efforts. The following nozzle equipment turned out to be the most efficient:

- lower cutting head d_o = 1.95 mm
- central cutting head d_o = 1.75 mm
- upper cutting head d_o = 1.55 mm.

Thus the power requirement of the high-pressure pumps could be reduced from P = 300 kW to P = 220 kW and the water flow could be reduced from $\dot{V}$ = 226 l/min to $\dot{V}$ = 170 l/min - Fig. 3 - .

A main target of the underground trials was to obtain a maximum incremental area of face advance (web x travelling speed). The chain-traction effort and the web s = 0.4 m were kept constant for the experimental version. The travelling speed v_M, however, was adapted to the prevailing coal hardness. Fig. 4 shows the optimum working configuration. At a web of s = 0.4 m a travelling speed of v_M = 0.4 m/s was obtained. This corresponds to an incremental area of face advance of 9.6 m^2/min.
The corresponding values of the most efficient conventional winning machines (ploughs, shearer loaders) read 7.5 and 6.0 m^2/min respectively for the same coal hardness.

Besides these high winning performances also a considerably better particle-size distribution of the raw coal was achieved - Fig. 5 - .

The operation mode of the machine enabled a very careful handling of the raw coal which eventually resulted not only in a better particle-size distribution but also in definitely less dust production. Comparative measurements have shown that the quantity of dust released corresponds to only approximately 30% of the dust produced by a shearer loader in the same seam.

Surprisingly, the moisture content of the coal was only half as high as the one recorded for comparison in a shearer-loader face. Consequently, the moisture content of the raw coal did not cause any problem to experimental operation.

In trial operation problems were encountered above all with respect to loading. The experimental unit left only little clearance underneath the machine frame so that the broken off coal could not be removed from underneath the machine. Furthermore the electrical energy supply as well as the low-pressure water supply recorded frequent trouble. In this configuration the supply system - a modified version of the supply system used for shearer loaders - turned out to be ill-suited.

3. Development of the Jet-Miner

3.1 Concept

After successful trials run with the experimental version the development of the first Jet-Miner prototype was started in cooperation of Ruhrkohle AG, M.A.N.-GHH-Sterkrade, and Bergbau-Forschung GmbH.

Since the basic suitability of the oscillating-waterjet system was proved throughout the trials, the original concept and design features of the cutting heads were retained to a fargoing extent. As mentioned in the introduction the Jet-Miner was supposed to be used in seam thicknesses ranging between 1.0 m and 1.5 m. Accordingly, a machine body of 1000 mm height was designed. According to an array comprising two or three cutting heads, a cutting height from 1.0 to 1.5 m can be catered for - Fig. 6 - . Since during the underground trials clearing problems arose, the total machine body of the Jet-Miner was situated in the space between the conveyor and the coal face. The conveyor is just straddled by a portal structure so that a good clearance profile is assured - Fig. 7 - .

In order to eliminate problems of energy and water supply a special cable guidance system was developed which is supposed to allow trouble-free supply via cables and hoses even at travelling speeds of v_M = 0.5 m/s.

The prototype built by M.A.N.-GHH-Sterkrade subsequent to elaboration of a joint concept exhibits the following design data:

Length :	8.56 m
Height (min):	0.98 m
Pump drives:	215 kW
Water flow :	162 l/min
Water pressure :	70 MN/m^2
Total mass :	18 000 kg

- Fig. 8 -

3.2 Surface Test Rig

The Jet-Miner prototype is designed for coal winning underground. For checking the overall function, a surface testing program is run at present as was done with the experimental version as well. Except for the length, the surface test rig corresponds for all machinery components to the underground installation. In the testing facilities for coal winning and coal clearance techniques of Bergbau-Forschung GmbH, a 35 m long test rig was set up. The rig implies a mock-up coal face of 13 m length, 1.5 m height and 6 m depth. This mock-up coal face, as to its strength, corresponds to the respective values of coal prevailing in the production face envisaged (Fig. 9).

The surface as well as the underground test array comprises the following components:

- Guidance of the Jet-Miner:	Halbach & Braun Plough haulage system with chain 34 x 126 mm
- Haulage drive:	2 hydrostatic drives of 143 kW each Düsterloh)

- Conveyor: EKF 3 (Halbach & Braun)
- Conveyor drive: 60 kW (surface)
- Cable/supply system: M.A.N.-GHH-Sterkrade
- Advance system with controls: Control bracket (Halbach &Braun)
- Electrical control: Audio-frequency system (Siemens)

3.3 Surface Trials

At present only few tests actually are run. The tests were limited up to present to the determination of tractive efforts for machine haulage and to checking of the hydraulic drive system with its electric control. The tests have shown already that the required haulage efforts are much smaller with the prototype than was the case with the experimental version. This is due to the reduced mass as well as to the guidance system. In April 1982 the surface trials will be completed so that more comprehensive results can be discussed.

4. Preparation of Underground Operation

Start of underground operations on Lohberg colliery is scheduled for January 1982, following the successful conclusion of the surface trials. In these last-mentioned tests only components cleared for underground operations are used. The hydrostatic drive system e.g. is run on non-inflammable fluids only. Furthermore for control of the drives and the Jet-Miner an audio-frequency system supplied by Siemens is used by which all plant components (conveyor, Jet-Miner, machine haulage, and high-pressure pumps) are controlled centrally from a control console.
This means that the Jet-Miner needs no longer to be accompanied by an attendant which, by the way, would hardly be possible due to the relatively high travelling speed (v_M = 0.4 m/s) in the thin seams.

Furthermore measuring equipments of Bergbau-Forschung will be used which are cleared for underground operation. The following parameters will be measured and recorded by UV-recorders and magnet-tape systems:

- hydraulic pressure of the hydrostatic drives (measure for chain-traction effort)
- speed of the drives (travelling speed).

On Lohberg colliery preparatory work for the underground trials is in process. In the seam R 1 the existing plough system is scheduled to be replaced by the Jet-Miner. The coal face is 250 m long and the seam thickness is of 1.45 m - Fig. 10 - . The Jet-Miner is designed in a way that it can run on the existing plough guides. Only the drive stations need to be replaced and the special guidance system for energy and water supply needs to be fitted.

5. Summary

After the very successful underground trials run with the experimental version whereby coal-winning performances were achieved which are hardly achieved by conventional coal-winning methods, the prototype version was jointly designed by M.A.N.-GHH-Sterkrade, Ruhrkohle AG, and Bergbau-Forschung GmbH and constructed by M.A.N.-GHH-Sterkrade.

This Jet-Miner is suited for operation in seam thicknesses ranging between 1.0 and 1.6 m. At a travelling speed of v_M = 0.4 m/s webs of s = 0.4 m are scheduled to be cut. This requires a water pressure of p = 70 MN/m^2 at a flow rate of $\dot{V}$ = 162 l/min. The pumps on the machine require 215 kW of installed power. The Jet-Miner hauled by outside drives undergoes intensive testing on mock-up coal faces on the Bergbau-Forschung test rigs before being transferred to trial operation underground. After these surface trials, operation under Production conditions in seam R 1 of Lohberg colliery is scheduled.

6. Acknowledgement

The research and development work was sponsored by the "Bundesminister für Forschung und Technologie (Federal ministry for research and technology)" as well as by the "Minister für Wirtschaft, Mittelstand und Verkehr (Ministry for economics and transport)" of Northrhine-Westfalia.

The project was set up jointly by M.A.N.-GHH Sterkrade, Ruhrkohle AG, and Bergbau-Forschung GmbH.

7. References (Quellennachweis)

1. Henkel, E.H.
   Aims for a High Pressure Water Jet Assisted Coal Winning System
   Fifth International Symposium on Jet Cutting Technology,BHRA, Hanover, June 1980, p 303/312

2. Krämer, Th.
   Investigation of High Pressure Water Jet Cutting Techniques for Coal Winning
   Fifth International Symposium on Jet Cutting Technology, BHRA, Hanover, June 1980, p 313/326

3. Schwarting, K.H.
   Schneiden mit Hochdruck-Wasserstrahlen - ein neuartiges Verfahren der untertägigen Kohlengewinnung
   M.A.N.-GHH-Sterkrade, Nachrichten für den Bergbau, 1981, Heft 2

4. Schwarting, K.H., H. Goris, Th. Krämer und G. Wille
   Entwicklung und Untertageerprobung eines Gewinnungsverfahrens mit Hochdruckwasser-Schneidstrahlen
   Glückauf 117 (1981)

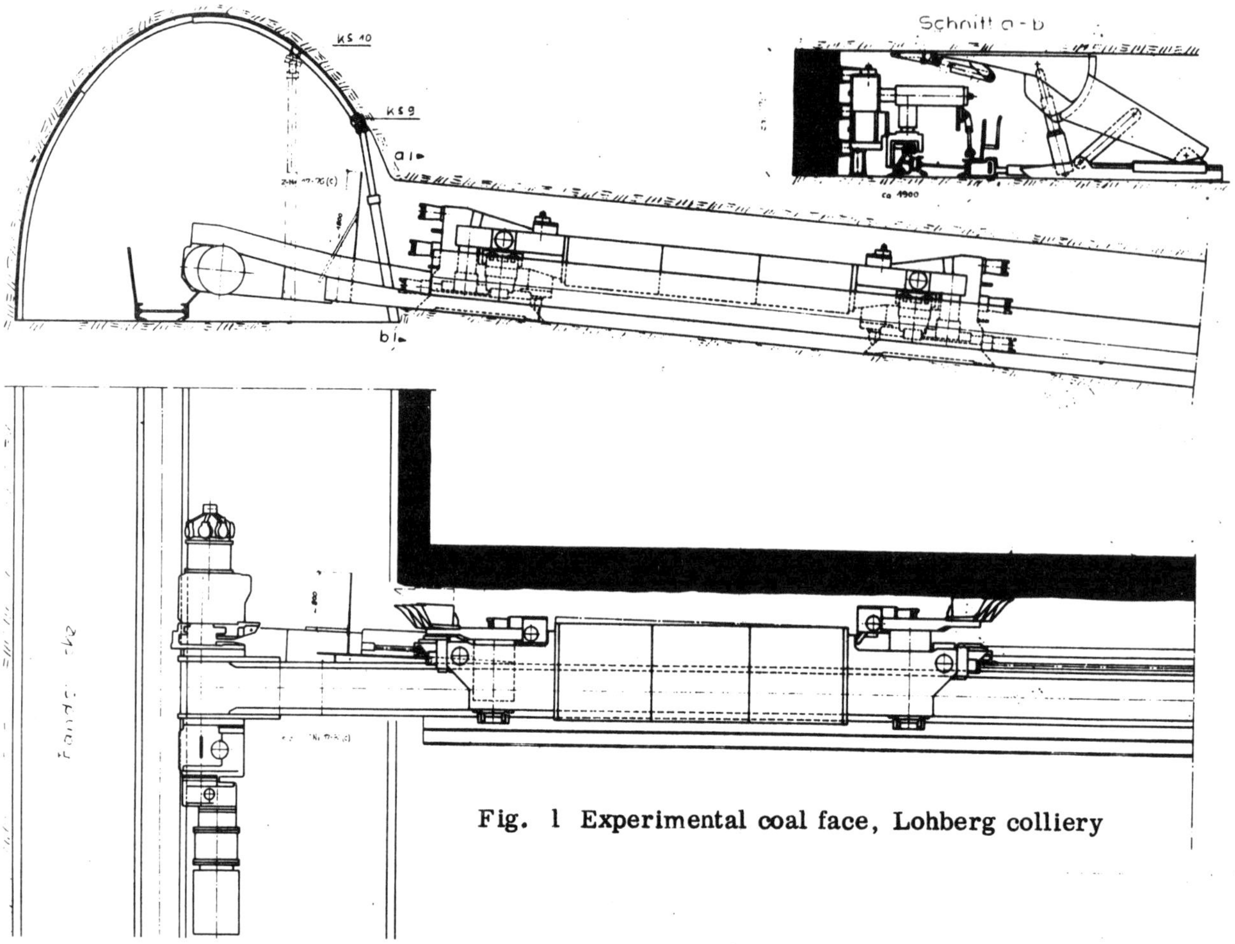

Fig. 1 Experimental coal face, Lohberg colliery

Fig. 2 Experimental Version

Fig. 3 Experimental version in the face

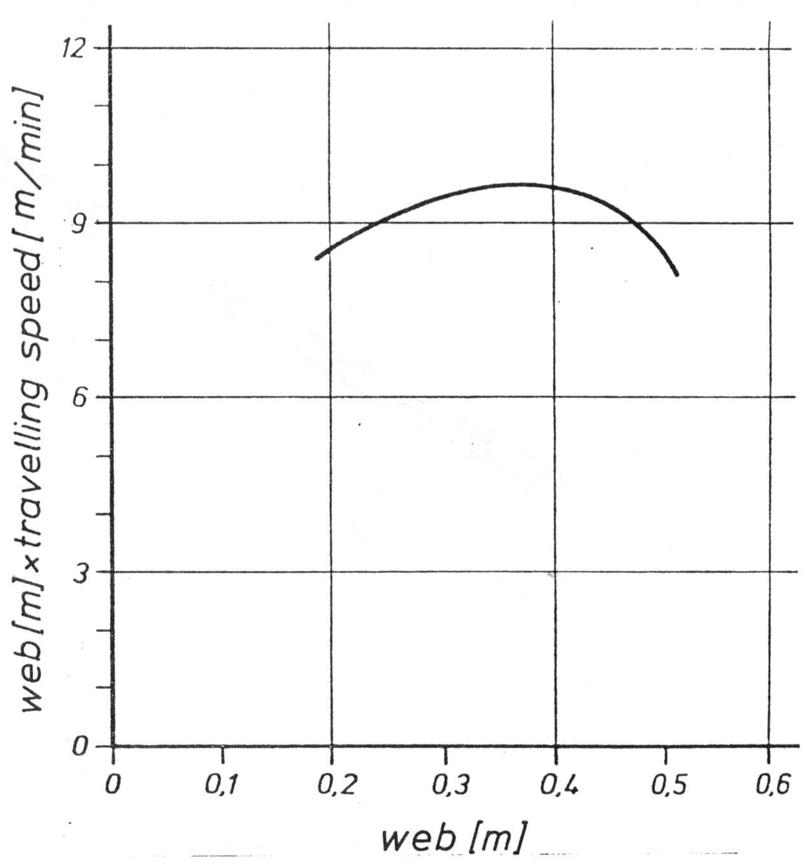

Fig. 4 Web by travelling speed versus web

Fig. 5 Coarsely sized coal

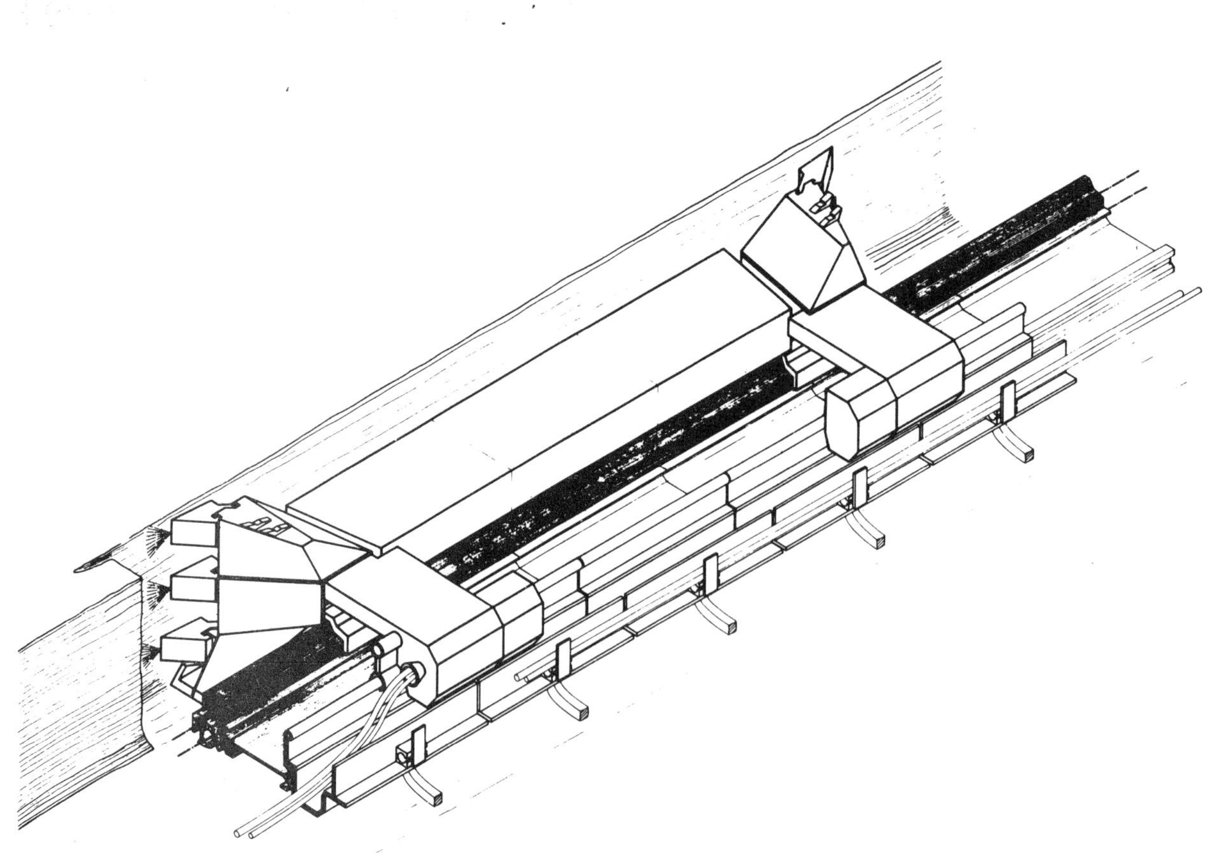

Fig. 6 Jet-Miner prototype

Fig. 7 Jet-Miner prototype

Fig. 8 Jet-Miner on the convey

Fig. 9 Test array at Bergbau-Forschung GmbH

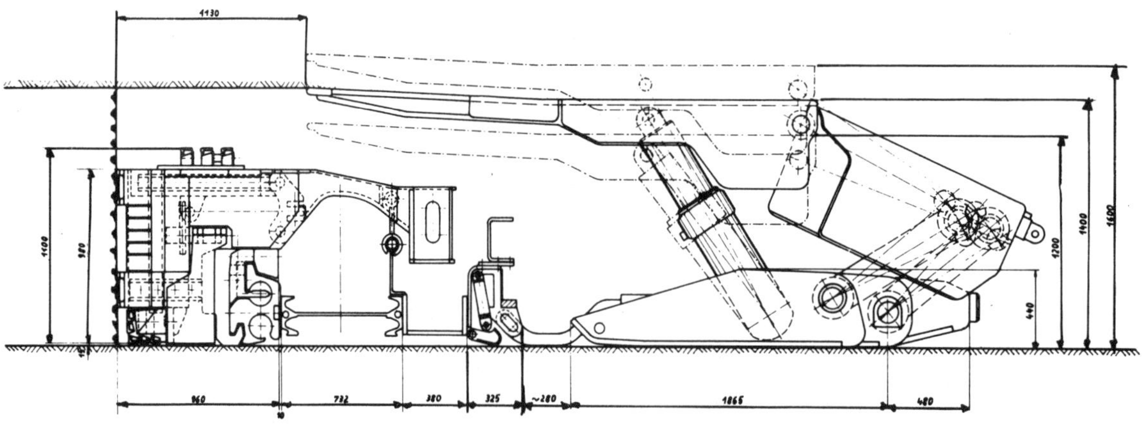

Fig. 10 Jet-Miner on Lohberg colliery

6th International Symposium on
Jet Cutting Technology
6-8, April, 1982

THE HYDRAULIC MINING OF COAL IN SOVIET COAL MINES

B. Ekber and M. Markus

Skochinsky Mining Institute, U.S.S.R.

Summary

The report outlines the present state of hydraulic coal mining in Soviet collieries. The characteristics of mining and geological conditions as well as technological schemes for hydraulic coal mining in flat and steep seams in Donetsk and Kuznetsk coalfields are given.

The equipment used for hydraulic coal mining and high pressure water systems are also described.

An analysis of the effect of the technological parameters variations onto the hydraulic coal mining efficiency is presented. The dependence of productivity of hydraulic coal face workers on variations in thickness and dip-angles of coal seams is shown.

Some recommendations concerning the method of choosing the most effective domain for hydraulic coal mining with respect to mining, geological and organization factors are given.

Held at the University of Surrey, U.K.
Symposium organised and sponsored by
BHRA Fluid Engineering

In the USSR in the Donetsk and Kuznetsk coal basins 10 mines currently utilize hydraulic techniques wholly or partially for mining. The geological conditions at these mines vary greatly and comprise:

seam thickness - from 0.7 to 10 m and higher,
dip - from 0 to 90°,
coal hardness - from soft up to extremely hard,
gas content - from gasless up to supercategorical, liable to outbursts,
seam deposition - from uniform without any disturbances up to complicated geological formations,
hypsometry - from simple to rather complicated with a wide range of angle variations,
depth of operation - from seams outcropping down to 720 m.

The main technological systems of mining operations employed in Soviet mines:

- when working flat seams: mining system by long pillars along the strike (or on dip) with the extraction of short faces without any support (Fig.1); longwall mining with mechanized complexes;

- when working steep seams: (above 35°): mining system with hydraulic mining and sub-level caving (Fig.2).

Depending on geological conditions and mining equipment being employed in the hydromines the following methods of coal mining in the working and development faces are used: hydraulic cutting (HC); mechanical and hydraulic cutting in short faces (MHC); mechanical and hydraulic cutting in longwalls (MHCL); blasting and hydraulic cutting in a short mining faces (BHC).

In 15 years at hydromines took place the process of competition of different coal mining methods which can be characterized by the following figures (in % of overall coal production at hydromines):

	1965	1970	1975	1980
Hydraulic cutting	15.7	36.6	38.4	48.8
Mechanical and hydraulic cutting in short faces	11.8	30.9	36.8	26.2
Mechanized complexes (longwalls)	-	9.5	18.8	19.9
Blasting and hydraulic cutting in short faces	72.5	23.0	6.0	5.1

These data serve as a visual demonstration of competition processes of coal cutting methods at the operating hydromines. Since 1965-1980 the coal cutting water pressure was increased from 6.0 MPa up to 12 MPa and therefore the increase of utilization of the hydraulic coal cutting becomes evident.

In the USSR, it is considered that in the nearest future the most progressive coal cutting methods in complicated geological conditions, where mostly the hydraulic mining is used, would be the hydraulic cutting and mechanical hydraulic mining in short-walls. The share of the coal mining by these methods reached in 1980 75%.

According to the above mentioned data the hydraulic coal cutting by 1980 has gained 50% and exhibits a progressive tendency to increase.

The passed years of competitive coal mining by different methods of mining permit to determine the most efficient technological systems for particular geological conditions:

- in Kuzbass at hydromine Inskaya: 100% of production is carried out by hydraulic cutting for seam thickness from 2.5 up to 6.65 m, dip angle - 18-25° and depth of operation - 320 m;

- in Kuzbass at hydromine Krasnogorskaya: 100% of production is performed by hydraulic cutting under seam thickness from 1.5 up to 13 m, dip angle - 40-50° and depth of operation - 327 m;

- in Kuzbass at hydromine Tyrganskaya: 100% of production is performed by hydraulic cutting for seam thickness from 1.6 up to 13 m, dip angle - 50-80° and depth of operation - 240 m;

- in Kuzbass at hydromine Yubileynaya: 53% of production is obtained by hydraulic cutting for seam thickness from 1.5 up to 3.75 m, dip angle - 0-35° and depth of operation - 450 m;

- in Donbass at hydromine Pioner: 100% of production is got by hydraulic cutting for seam thickness from 0.8 up to 1.1 m, dip angle - 10-13° and depth of operation - 500 m;

- in Donbass at hydromine Krasnoarmeiskaya: 42% of production is got by hydraulic working of seams from 0.9 up to 1.25 m thickness, dip angle - 10-13° and depth of operation - 400 m. Thus, 6 mines (from 10) currently utilize hydraulic coal cutting and 4 of them employ 100 per cent hydraulic mining systems.

The hydraulic coal cutting at hydromines is carried out under water pressure of 10-12 MPa and water flow rate 150-300 cubic meters per hour by remote control monitors, providing the monitor operator situation at 10-15 m from the actual production area.

Currently efforts are undertaken to pass over to the operation with 16 MPa thus providing substantial increase of hydraulic coal cutting rate. When working by shortwall mining systems the face length under hydraulic cutting depends on the jet effective action and under mechanical/hydraulic cutting - on the miner production and roof strength.

In the period of initial work of hydraulic mining system, when the hydraulic cutting was only introduced and driving of the headings was labour-consuming, the tendency was to increase the distance between the headings, that is to the maximum increase of the face length L,m. In the following years, by introduction of the continuous miners, a new tendency appeared to decrease the length of the mining face because of minimizing the mining losses.

In the future forecasting period (as it was mentioned above) because of the growth of mining machines technical parameters, the length of the mining area will increase. Studies showed that increase of the machines' autonomous operation of hydraulic and mechanical/hydraulic coal mining results in productivity rise.

Autonomous operation of the mining machines (their capability to enter the mining area independently with the remote control) enables to increase the distance between the tailoring workings and to decrease the specific volume of their driving.

Presently the mechanical/hydraulic miner K-56MG is remotely controlled which allows the operator to be situated at a distance of 15 m.

Projects considering self-propelled hydromonitors of high efficiency will be completed in the nearest future.

In the forecasting period of time the hydromonitor water pressure and water flow rate will increase and, therefore, will increase the mining face parameters.

Thus, the mentioned tendencies of parameters increase of hydraulic cutting and increase of mining machines' autonomousy will result in parameters increase of the short mining face and in corresponding decrease of tailoring works specific share at hydromines, taking into account the achievement of optimum coal losses in the mining process.

These tendencies of face increase in accordance with retrospective are being well approximated by the expression:

$$Lk = 0.03t^2 - 1.5\,t + 24,\mathrm{m} \qquad (1)$$

For longwalls this expression is accepted in conformity with (1)

$$L_{д} = 0.003\ t^2 + 5.2\ t + 32, m \quad (2)$$

Symbols here and further:

L_k - length of the short mining face, m;

$L_{д}$ - length of the mining area (longwall), m;

t - an integer denoting the ordinal number of the year, starting from 1950, to which the index is being determined.

The load onto the operating face depends on different factors and first of all on:

mining machine output;
seam thickness and coal strength;
operating face length and surrounding strata characteristic;
reliability of the machine and its maintenance system;
number of auxiliary operations in the working face incompatible with the mining operation and, after all, on the limitations, connected with the working area ventilation.

In the period of 1969-1975 the average load onto the operating face of hydromine increased 3-5 times and at the conventional mines - 2 times.

The output increase of the progressive mining methods, which will take place in the forecasting period, enables to conclude that the load level onto the hydromine operating face in forecasting period will rise.

Load variation on the operating face in the forecasting period (A_3, t/ per day) according to methods of coal cutting and coal basins can be represented by the following models:

Kuznetsk coal basin

$$A_3^{HC} = 20\ t\ \ e^{0.01\ t} \quad (3)$$

$$A_3^{MHC} = 25\ t\ \ e^{0.01\ t} \quad (4)$$

$$A_3^{MHCL} = 31\ t\ \ e^{0.01} \quad (5)$$

Donetsk coal basin

$$A_3^{HC} = 12\ t\ \ e^{0.01\ t} \quad (6)$$

$$A_3^{MHC} = 18\ t\ \ e^{0.01\ t} \quad (7)$$

$$A_3^{MHCL} = 22\ t\ \ e^{0.01} \quad (8)$$

Results of testing enable to state that the most perspective systems of mining under hydraulic technology of coal mining would be the systems of short mining faces. In particular favourable conditions mechanized complexes may be used (longwalls). Operating experience gained during the hydromines operation, scientific and theoretical studies permit to obtain methods of productivity estimation influenced by factors of hydraulic cutting, mechanical/hydraulic coal mining and mechanized complexes (2).

Rather interesting is the investigation of productivity and hydraulic cutting influenced by the main factors: seam thickness, flow rate and pressure of water, variation of operating working length. Some data on these studies are stated below.

SEAM THICKNESS

As it is clear from the above-mentioned list of hydromines, the hydraulic cutting is carried out in seams of 0.8-13 m thickness. We have made a generalization of data on hydraulic cutting productivity at all hydromines for the last 5 years from 1976 to 1980 which enables to show a diagram of hydraulic cutting productivity variation per man during mining operations, see Fig.3. In the conditions of the USSR hydromines the team on extraction operations consists of a hydromonitor operator, his assistant and, besides, workers on driving the tailoring workings providing the extraction area.

As it can be seen from the previously mentioned figure, the man productivity in the mining area is greatly influenced by the seam thickness. This can be described for flat and steep seam depositions, respectively, by the following expressions:

$$\Pi_n^z = 11\ m \tag{9}$$

$$\Pi_K^z = 17 + 3\ m \tag{10}$$

where

Π_n^z and Π_K^z - workman productivity of the winning face when cutting hydraulically under flat and steep seam depositions, respectively, tons per shift

m - seam thickness, m.

Presently the hydraulic cutting technology is considered to be the most effective when working seams of steep deposition (above 35°), but working flat seams of 1 to 3.5 meter thickness by mechanized complexes in favourable geological conditions (uniform thickness, dip, absence of faults and dirt inclusions etc.) results exceeding production rates of hydraulic cutting can be obtained.

The analysis of longwall's mechanized complexes operation for the last five years has allowed to make the following generalization:

- in moderate geological conditions, constantly deteriorating, the workman productivity of the winning face (Π_n^K) (mechanized complexes) can be expressed by the following expression:

$$\Pi_n^K = 4 + 4\ m \tag{11}$$

- in the most favourable geological conditions, providing the operation of the mechanized complexes, the man productivity of the winning face ($\Pi_n^{K\Pi}$) should be:

$$\Pi_n^{K\Pi} = 16\ m \tag{12}$$

Fig.4 illustrates graphically that on the one hand, the operation of complexes in moderate, still deteriorating, geological conditions is less effective than employment in similar conditions of the hydraulic cutting and, on the other hand,

- the employment of the complexes in the conditions favourable for their operation is rather effective than hydraulic cutting.

Likewise, the general conclusions regarding to the employment of hydraulic cutting of coal presently can be itemized:

- hydraulic coal cutting in the winning face presently is mostly effective in the following conditions:
- steep seam deposition (above 35°);
- flat seam deposition (to 35°), seam thickness more than 3.5 m;
- flat seams deposition (to 35°) of 0.8 to 3.5 meter thickness, in in the conditions, where it's impossible to use the mechanized

complexes in the most effective way.

WATER FLOW RATE VARIATION

As studies have shown, the increasing of hydromonitor water flow rate results in its output rise and in extension of the jet effective range thus, in turn, providing the decreasing of coal mining time from the extraction face under its constant parameters or enables to increase the extraction face parameters and decrease the volume of cutting operations. As a result, all this influences the decreasing of labour-consuming operations and the increasing of labour productivity, respectively. Besides, mentioned factors permit to optimize the winning face parameters, hydraulic monitor performance and to reduce the coal losses in the mining area.

In example being considered we adopt the average pressure presently used at the hydraulic monitor nozzle H = 10.0 MPa and H = 15.0 MPa. Dimensions of the mining face are considered to be constant B = 6 m, L = = 10 m.

Fig.5 shows a diagram plotted according to the calculation data of hydraulic cutting productivity variations (t/h) due to flow rate variations.

Fig.5 shows that increasing the water consumed by hydromonitor 4 times results in increasing hydraulic cutting productivity 4-7 times. Consequently, the necessity of such a measure is obvious and the possibility of its realization depends on putting into operation the hydromonitors 12 GD providing flow rate up to 450 m^3/h and by this time being manufactured in series.

Analysing the diagrams the following conclusions can be stated:

- water flow rate increasing under pressure of 10 MPa from 100 m^3/h up to 400 m^3/h results in hydromonitor output increasing 4 times, the hydromixture consistency (by weight) will be 1 : 5 + 6;

- water flow rate increasing under pressure 15.0 MPa from 100 m^3/h up to 400 m^3/h results in hydromonitor output increasing 7 times, the hydromixture consistency (by weight) will be 1:2.5.

Above-mentioned items permit to make a conclusion: the effectiveness of flow rate intensifying grows simultaneously with pressure increasing from 10.0 to 15.0 MPa. Besides, its clear that under pressure 10 MPa the hydromixture consistency is low and, consequently, the water energy is used not completely. Only under pressures about 15.0 MPa the consistency obtains the optimum dimension and uses completely the water energy for cutting and hydraulic transportation of coal.

WATER PRESSURE VARIATION

Water pressure increasing at the nozzle of the hydromonitor (the same authors' studies) results in hydraulic cutting productivity increasing and in extending the range of effective impact onto the coal solid. Further we shall analyse the extent of this impact.

For the analysis, water flow rate through hydromonitor nozzle is considered to be 150 and 300 m^3/h. The rest of parameters are analogous to the previous ones.

Fig.6 presents a diagram of hydraulic cutting production variation against the water pressure at hydromonitor nozzle; the diagram is plotted in accordance with data of calculation.

As it can be seen from Fig.6, the increasing of water pressure four times results in increasing hydraulic cutting productivity increases effectively with the pressure increase, at a definite increase level it obtains such values which prevent the further pressure growth. Such an obstacle is the flushing of the dislodged coal out of the face which is carried out with the solid-to-water ratio of S : W = 1 : 3 (2.5).

When keeping such a ratio the cutting and flushing away processes are carried out continuously. When this ratio is not maintained due to water volumes decreasing, the process is disrupted, that is the cut coal is gradually stored in the mining area and needs high pressure water to flush the dislodged coal (cutting at this time is not performed) or additional supply of low pressure water to flush away the coal.

From this picture it can be also seen that under given conditions it is advisable to increase the pressure only up to 15.0 - 16.0 MPa which provides continuity of the technological process without additional water supply.

So, by analysing Fig.5 and 6 it can be stated that for the given seam geological conditions the water pressure at hydromonitor nozzle is most advisable to be 15.0 - 16 MPa and water flow rate up to 300 - 400 m^3/h.

WINNING FACE PARAMETERS VARIATION

When working with the recommended for the forecasting periods systems of mining by short faces, the main specific share in the total headings driving falls on the tailoring operation, its volume, other conditions being the same, depends on the length of the extraction face (increment) "L". Therefore, by variation of this parameter it's necessary to remember and take into account that its decreasing results in rising of the tailoring operations volume and, correspondingly its increasing - in decreasing of this volumes. Besides, as it was above-mentioned, variation of this parameter influences the level of the operational losses.

Consequently, for this parameter optimization it is necessary to study its joint influence on the variation of the extraction operation labour-consuming character and labour-consuming character of the tailoring workings driving.

Initially let's determine the influence of "L" variation on the hydraulic cutting productivity.

After making all the necessary calculations let's plot the diagram of variations of extraction operations labour-consumption at different combinations of "Q", "H" and the data of tailoring operations labour-consumption variations against the working face length (Fig.7). It can be seen from the figure that for adopted parameters the optimum face lengths will be:

under H = 10.0 MPa and Q = 300 m^3/h - L = 10.0 m
under H = 15.0 MPa and Q = 150 m^3/h - L = 8.0 m
under H = 15.0 MPa and Q = 300 m^3/h - L = 15.0 m
under H = 10.0 MPa and Q = 150 m^3/h - L = 5.0 m

Operation with parameters H = 10.0 MPa and Q = 150 m^3/h, which presently is generally used at hydromines with hydraulic cutting, is far beyond the optimum range.

Cited, according to the method, calculations show that the recommended methodology of fore-casting of mining operations labour-consumption enables not only to prognose it but also to optimize all basic parameters of the mining system and methods of coal cutting.

Cited calculations of parameters optimization permit to make a conclusion about realization possibility according to the periods of forecasting.

It should be also noted that obtained parameters of the winning face are optimum not only because of operations labour-consumption but also because of the condition of minimum coal losses obtained in the mining area outline. Thus, it can be stated that there are two peaks bringing of to the highest effectiveness of the hydraulic coal mining engineering perfection: intensifying mining machines and means of mechanization of auxiliary operations implemented at hydromines.

CONCLUSIONS:

The performed analysis of hydraulic cutting basic parameters variation and optimizing of mining face parameters taking into account the variations of mining and preparatory operations labour-consumption permits to make the following conclusions:

- water consumption increase through one hydromonitor provides hydraulic cutting productivity rise, increases its autonomousy, lowers the mining and tailoring operations labour-consumption;

- water pressure increase at hydromonitor provides the rise of hydraulic cutting production, increases its autonomousy and lowers the labour-consumption of mining and tailoring operations;

- pressure water increase extra 150 - 220 MPa results in hydraulic cutting productivity rise, exceeding the transporting capability of the water flux which evokes the necessity of additional labour-consumption the expediency of which requires definite bases;

- to the definite hydraulic cutting parameters (Q and H) are corresponding optimum parameters of the mining face according to the criterion of the overall costs for carrying out mining and driving operations.

From the above-mentioned the following basic trends of NIOKR in the sphere of improving and developing the technology of mining coal in a short face:

- to realize hydromonitor water pressure increase up to 150 - 200 MPa;

- to realize water flow rate through one hydromonitor (up to 450 m^3/h);

- to intensify sequentially hydraulic machines output and increase their autonomousy;

- to design hydromonitor units operating at distances 75 - 100 m;

- sequentially, with the rise of cutting parameters and autonomousy, increase the mining face length;

- to improve available and create new means of the auxiliary transport, providing lowering of the volumes of driving headings, by 1000 t of extraction.

For carrying out the further development of technology and techniques of the mining operations it is foreseen by the main trends: to develop, to create and manufacture (serial) the following units of equipment:

- centrifugal pumps with delivery up to 450 m^3/h and pressure 16.0 MPa;

- self-propelled hydromonitors with autonomousy radius 15-40 m under pressure 16.0 MPa and consumption up to 450 m^3/h;

- hydromonitor unit with the feeding pipeline up to 50 + 150 m, at pressure 16.0 MPa and consumption up to 450 m^3/h;

- self-propelled hydromonitor with incorporated pressure intensifier;

- hydromonitor unit with incorporated pressure intensifier;

- light protecting power-supports for short winning faces for operation in especially complicated conditions to decrease coal losses;

- programme-remotely and automatic controlled hydromonitors and mechanical/hydraulic machines.

All above-mentioned machines should be universal, that is suitable to Donets and Kuznetsk basins particular conditions or for standard line of machines for working seams of different thickness and dip angle.

The use of new developed machines and mechanisms, when mining operations are performed, will allow to obtain data characterizing a coal mining enterprise of a new technical/economic level with labour productivity exceeding by 4-6 times the productivity in analogous conditions in 1975.

REFERENCES

1. Dokukin A.V., Rudinkin Ju.A. and others. Principal methods for long-term forecasting of the basic parameters of underground coal face. Moscow, Skochinsky Institute of Mining, 1971.
2. Tsyapko N.V. and Chapka A.M. Hydraulic coal cutting in underground operations. Moscow, Gosgortekhizdat, 1960.

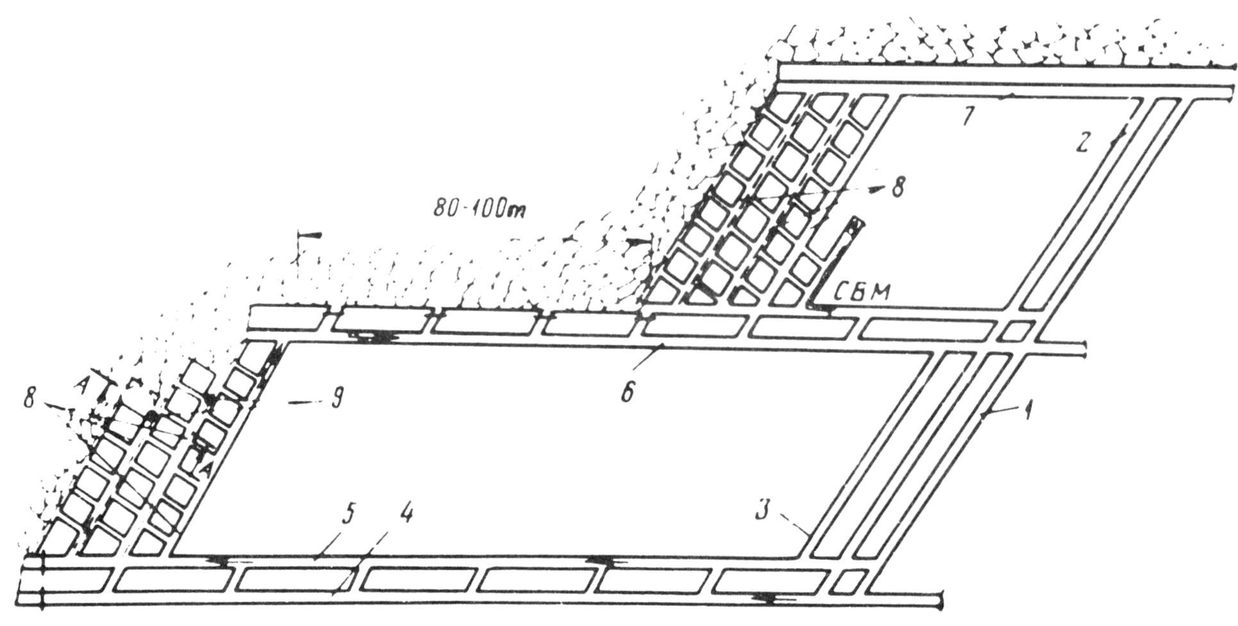

Fig.1 Short-wall system of mining with full caving

1 - slope; 2 - slope for slurry flushing; 3 - heading on rise; 4 - accumulating entry; 5 - parallel road; 6-7 - ventilating entries; 8-9 - mining headings

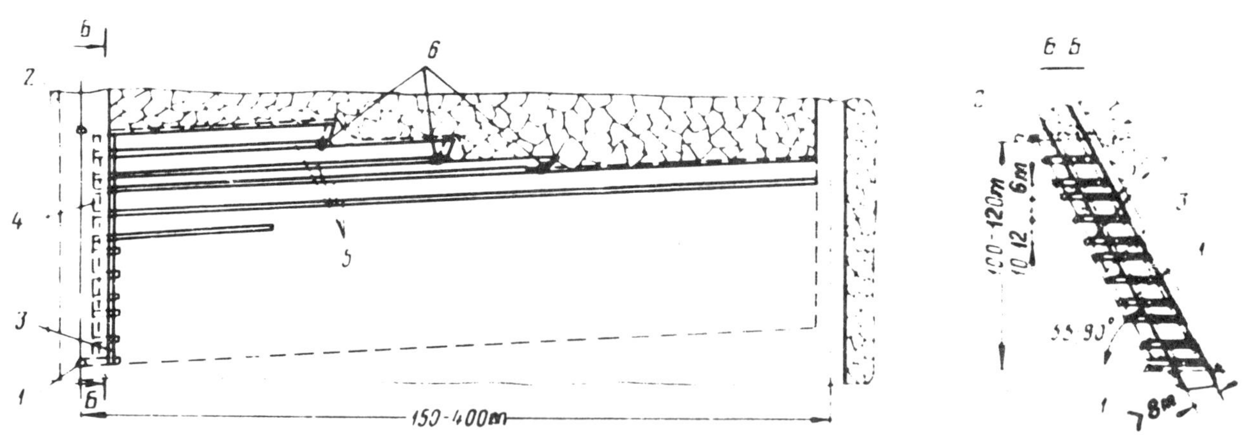

Fig.2 Sub-level mining system

1 - accumulating crosscut; 2 - ventilating crosscut; 3 - ventilating heading; 4 - haulage slope; 5 - sublevels; 6 - hydromonitors; 7 - sublevel crosscuts

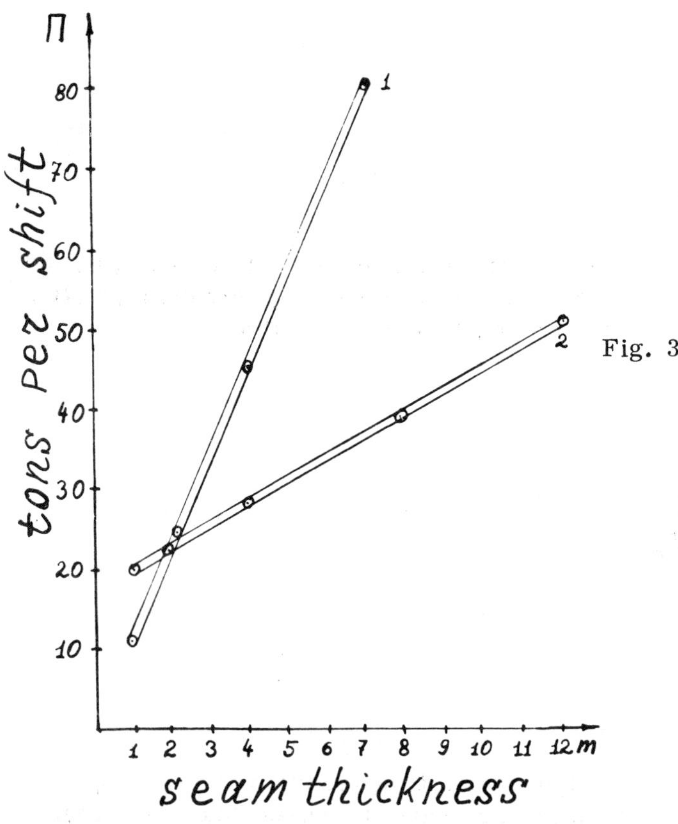

Fig. 3 Production of the mining area man under hydraulic cutting
1 - labour productivity when working flat seams; 2 - labour productivity when working steep seams;

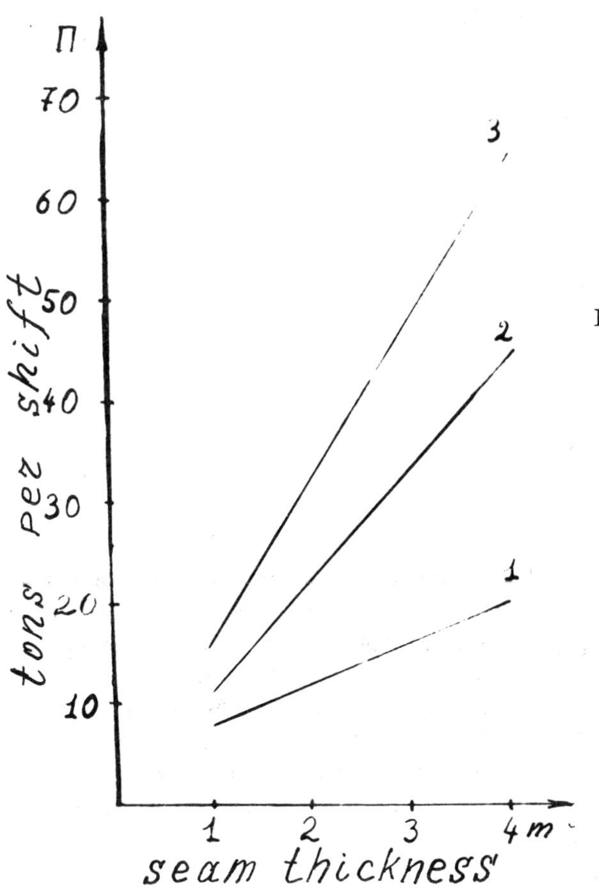

Fig. 4 Production of the working face man
1 - labour productivity in complex/mechanized faces at medium geological conditions; 2 - labour productivity at hydraulic cutting; 3 - labour productivity in favourable conditions.

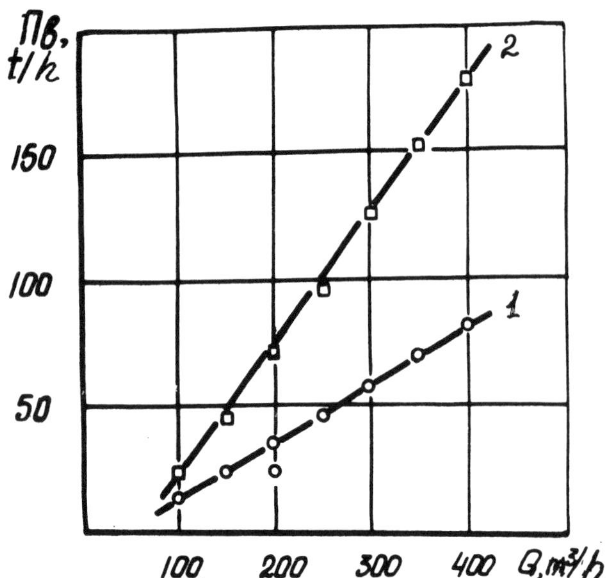

Fig. 5

Hydraulic cutting productivity varies against water flow rate; 1 - water pressure 10.0 MPa; 2 - water pressure 15.0 MPa.

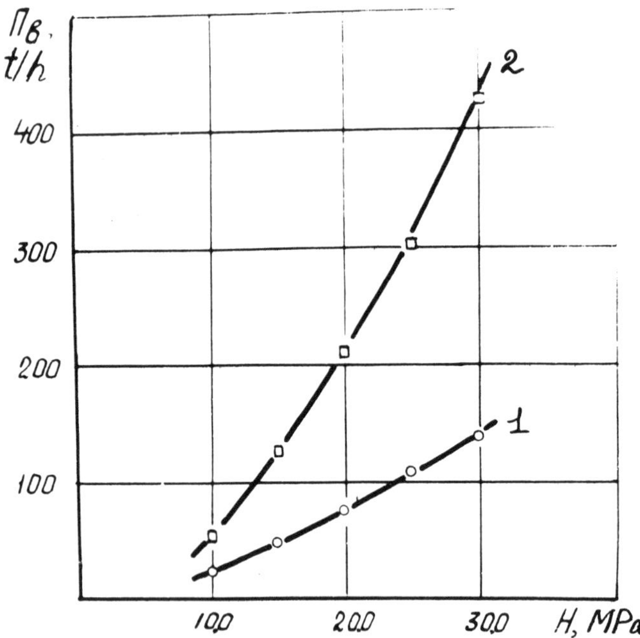

Fig. 6

Hydraulic cutting productivity varies against water pressure; 1 - water flow rate through hydromonitor nozzle - 150 m^3/h; 2 - water flow rate through hydromonitor nozzle - 300 m^3/h.

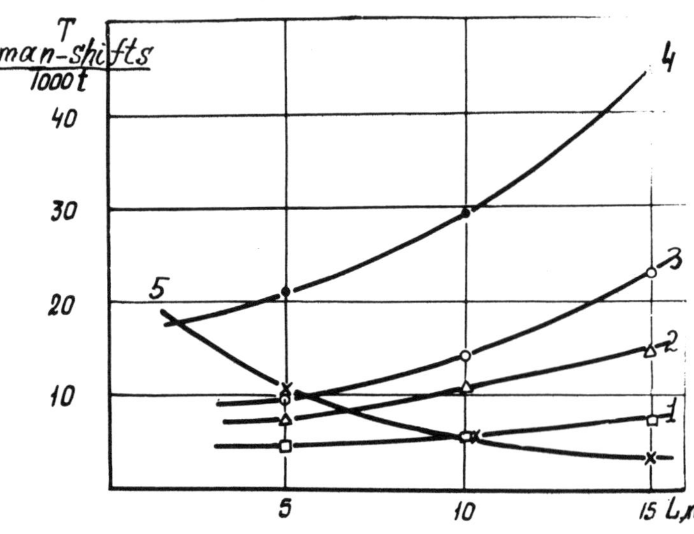

Fig. 7

Variation of mining and tailoring operations labour-consumption; 1-mining operations 1 - mining operations labour consumption at: H = 10.0 MPa, Q = 150 m^3/h;
2 - H 10.0 MPa, Q 300 m^3/h;
3 - H 15.0 MPa, Q 150 m^3/h;
4 - H 15.0 MPa, Q 300 m^3/h;
5 - labour consumption of tailoring operations

6th International Symposium on
Jet Cutting Technology
6-8, April, 1982

CLEANING COKE OVEN DOORS

D.J.H. Odds

F.A. Hughes & Co. Ltd., U.K.

Summary

This paper will report on the work carried out over the last six years on cleaning coke oven doors, using high pressure water jets.

On a coking plant the doors are regularly removed to enable the re-charging process to take place. The doors have to be cleaned within one and a half minutes during the "pushing cycle". Build up of hard carbonaceous, bitumastic layers on the doors and seals prevent easy removal and replacement.

Additional benefits of clean door seals are the prevention of polluting the surrounding area and a health hazard to the coke oven operators.

Investigation into optimising the cleaning operation will be reported, also a report on various jet shapes traverse speeds, etc. will be discussed.

Held at the University of Surrey, U.K.
Symposium organised and sponsored by
BHRA Fluid Engineering

CLEANING COKE OVEN DOORS

INTRODUCTION

Coke is used in the iron making processes and to make the coke, coal is baked in ovens for about 18 hours.

The gases released from the coal are used for firing the steel furnaces. There are many by-products, such as tar benzine and the residue coke is discharged, and cooled, prior to being fed into the hungry blast furnaces, as required, for the steel making processes.

The ovens used for coke making are tall and thin; they are up to 10 metres high, 0.5 metres wide and 30 metres deep. In a large coking plant there are several batteries and each battery comprises 50 ovens, see fig. 1.

The charge of coal is fed into the top of the ovens and after cooking for about 18 hours, vertical doors placed at each end of the oven are removed, the red hot coke is pushed through into purpose-built railway trucks. The coke is then taken to a 'quencher' fig. 2 for cooling and then fed by conveyors directly to the blast furnaces.

After the coke has been pushed through and the oven is empty, the two doors are replaced. There has been an increasing problem with existing coking plants where the doors have suffered from fouling with tar deposits from the coal. During the 'cooking process', bitumen separates out mainly on the bottom of the oven and if there are any gaps in the door seal, coal tar oozes out of the door. One can imagine that with large narrow doors making a metal to metal seal is quite difficult if not impossible. The original face of the oven door jamb is machined flat and the original new door with the metal seal is produced and is made as reasonably flat as possible, but with the heat of the cooking process, some distortion of the doors is inevitable.

The doors are held onto the oven with two latching catches; as these are turned from the vertical to the horizontal, the latch tightens onto the door jab, pushing it towards the oven. At times, it is impossible to get the door back onto the oven through having a build up of bitumen on the faces, thus a spare door has to be collected from the end of the battery and placed onto the open oven. This disruption wastes time and the target of pushing at least 6 ovens per hour falls sadly behind. On the side of the oven where the coke is deposited into the railway trucks, the door extracting machine is known as the 'Coke Guide', see fig. 4 & 5. The coke guide removes the door, it moves along, positions the guide at the open oven and when the coke is pushed through, it prevents the red hot coke from falling onto the bench or walkway around the oven.

On the other side of the oven, the door extracting machine is very different and is known as a 'pusher'. It is large enough in size to carry a large ram up to 30 metres long, the ram face being the size of the oven. The pusher removes the door and then lines up a large ram which is approximately the size of the oven and this ram pushes all red hot coal through the oven and out of the other side. This car is known as the 'Pusher' see fig. 6. Usually on the pusher side of the oven the other side of the oven, the track, trolley car and most spares spaces are hotter and dirtier, as every ten minutes a charge of red hot coke is pushed out of the side of the oven.

THE ENVIRONMENT

Leaky doors allow coal gas and sulphurous fumes to escape to the surrounding countryside. When coal gas escapes from a leaky door, usually the gases ignite, causing a flame on the outside of the oven. These flames allow local overheating of the outside skin of the door and door jamb and this uneven heating can cause distortion. So, in addition to the buildup of carbonaceous deposits on the door and the door jamb, a distorted door presents an even greater problem in making a good seal.

So the benefits of clean doors which are well-sealed are many and are listed below, but not necessarily in order of their importance:-

1. The operators have an improved working environment in that there is a reduction of obnoxious fumes.

2. The surrounding countryside is protected from pollution and the Alkali Inspector keeps a watchful eye on this.

3. The doors can be removed and replaced as required and do not hold up the production of coke.

4. The doors are sealed against the ingress of air into the oven.

5. The oven temperatures can be stabilised at the required temperature.

6. A reduction of distortion due to flaming gases, which would overheat the doors and frames.

7. A reduction in the burning or wearing of the doors, seals and frames.

TASKS TO BE CARRIED OUT

The only time that is available for cleaning the coke oven doors is during the pushing cycle. The time allowed here is about 1 minute. Hitherto mechanical means have been tried by using hand-held scrapers, but with the size of the door and the time available, only one small part could easily be cleaned by an operator. This is sufficient in places such as India, where labour is cheap and their ancient coke ovens are only 3 M high, but, of course, modern oven doors are up to 7 M high and are impossible to hand-scrape.

This led to the development of a tracing motion which contained a large number of spring-loaded scrapers. This tracing motion was offered up to the door which had been taken off. Pressure was then applied to the scrapers and the scrapers were transported around the door seal with an hydraulic chain drive conveyor. This system allowed buildup of tar on the scrapers, lumps of tar fell onto the chain drive, jamming up the system, causing chain breakages and the scraper jumped over hard deposits. This led the maintenance and operating staff to look for an alternative method of cleaning doors.

The first approach of the idea of using a water jet was rejected on the grounds that the water would cause thermal shock to the door which has a ceramic insultation cladding attached to it. It was also thought that the cold water jetting would cause further distortion to the metal seal. Early trials with hand-held lance showed that if the water was directed at the seal and not the door plug (the ceramic insulation) then thermal shock cracking did not take place. It was concluded that the amount of water used in high pressure water jetting was dispersed in evaporation before sufficient heat loss of the door plug occurred.

Trials were carried out to establish the cleaning rates and other information such as stand-off distance, water jet pressure, volume of water, effective spray pattern, velocity of traverse of water jet, etc. etc. Initial ideas also included the use of a double water jet so that the door seal was cleaned twice in the one cycle. The object was to replace the scrapers with a lance carrier.

There was also the problem of hose pipe snagging. With high pressure water jetting, the hose pipe remains in a fairly stiff condition and therefore, if there was to be a rotation of the water jet lance around the door seal avoiding the sprocket wheels and chains, then hose pipe trials had to be carried out.

The possibility of a water jet lance with three nozzles attached to it was considered against a single jet of high pressure water a greater stand-off distance see figures 7 & 8. The next problem to tackle was the fact that it was found that a new or recently completely cleaned door could be kept clear if a water jet cycle was carried out at every door removal. Whereas doors which were already encrusted with hard bitumous deposits had to have a more intense cleaning operation carried out on them until bare metal was achieved. Always remembering that the time allowed for the door cleaning cycle had to not interfere with the 'pushing time'.

One idea was to water jet the seals at 700 bar for one traverse of the periphery, knowing that it would not completely clean the door. After three cleaning operations, the door seal should then be down to bare metal. After this, the cleanliness could be maintained with a single pass carried out at a lower pressure - say 300 bar.

The methods that were considered for changing the pressure of water jets are listed as follows:-

1. The high pressure water is supplied by a triplex plunger pump and the size of the plungers could be increased or decreased to give higher or lower volume of water by changing the size of the plungers.

2. The nozzle could be changed in the cleaning head to increase or decrease the orifice size.

3. One could vary the number of nozzles in the cleaning head.

4. The pressure could be changed by re-setting the unloading valve which has the effect of matching the jetting nozzles with the pump plungers used. The unloader is an adjustable spring-loaded annular orifice which can be increased or decreased to achieve any pressure desired. This could be wasteful of power in dumping water not required, and although it is convenient in other water jetting situations, in a fixed installation this is not always found to be desirable.

5. The speed of the pump could be increased or decreased with a variable speed motor, thus increasing or decreasing the volume of water. Eventually, the last solution was used, but as only two pressures were required, then a double wound stator was used and it was found that without changing the nozzles, simply by changing a connection on the motor winding, that the motor would rotate at either 1450 rev/min or 960 rev/min. The pressure being approximately inversely proportional to the flow, therefore the higher speed rotation of the pump gave 690 bar and lower rotational speed of the pump gave a pressure of 450 bar. This simplified the operation of the speed changes or pressure changes for the operator. He could merely move a lever on the starter which was indicated 'high pressure' or 'low pressure'.

   So, as a door was removed, it was examined and the pump started at the required speed. Eventually, after the ovens had been used for about 3 times and all doors were kept clean by the lower pressure of 450 bar, this saved pump power, use of water, pump wear, gland wear, nozzle wear and was generally thought to be the most efficient way of using the jet cleaning system.

DESIGN SOLUTIONS

The first cleaning machines were to be used on the coke side which generally has a more difficult problem with door sealing, as the hot coke is passed through this side of the oven. The coke guide machine is fairly small as it merely has the door extracting facility; a separate trolley car is towed behind the door extractor, which is quite a simple openwork structure. There was difficulty to find space to accommodate quite a large pump, 125 hp unit and a 400 gallon tank. The size of the tank was determined by the number of cleaning operations or pushing cycles required per shift and in order not to interrupt the flow of coke pushing, it was planned that the tank would be filled at the beginning of each shift. The tank was placed as high as possible in the coke guide machine to give a positive head to the triplex plunger pump, low level alarm and shutdown facility ensured that low level of the tank would either stop or inhibit the starting of the pump, this protected the pump from dry running.

High pressure water is fed through a pulsation damper and a high pressure hose to the nozzle cleaning lance. Special attention had to be paid to the gland of the pump to ensure that the coal dust and ash dust which penetrates every corner of the door extracting machines did not affect the plunger seals and crankcase oil seals. see figure 9.

The tracing motion which originally carred the scrapers had to be redesigned to take the water jetting lance. The first design of the tracing motion carried two cleaning lances each on cleaning one half of the track around the periphery of the door seal. Eventually, this was replaced by one single lance, which rotated completely around the periphery of the door seal.

Early use of the door cleaner in production had some problems with the lance becoming bent due to the buildup of tar on the door plug. This led to refinements on the hydraulic drive of the tracing motion. A relief valve was fitted so that if an obstruction was encountered, then the rotation would stop, giving a stall situation, after running the first machine for 15 months, it was then decided to instal a water jet cleaner on the pushing side. The problems of installation were much simpler as the pushing machine is very large having to accommodate the hydraulic ram and there is much more space available for the water tank and the high pressure pump. Although this side of the coke ovens is a good deal cleaner than the coke guide side, provision was made for protecting the pump unit from coal and ash debris. figure 9.

FINANCIAL ASSESSMENT

The installation of the water jet door cleaner will reduce the number of ovens lost due to the inability to latch doors after pushing because of their dirty condition. The number of ovens lost before the water jet cleaner became operational was 30 per month. This has now been reduced to 10. A saving of 20 ovens a month was made and this will increase the coke output by 4,080 tons per annum, assuming a wet coke output of 17 tons per oven. There are other expected savings, but these cannot be evaluated:

A. Reduction in maintenance costs because of the reduced oven door jamb and refractory damage due to firing around ovens caused by overheating and distortion.

B. Slight increase in gas (by-product) production because they are recovered and not burned around the door seals.

20 ovens per month x 12	=	240 ovens per year
17 x 240	=	4,080 tons
4,080 tons x £46	=	£187,680 per year

C. Capital Expenditure:

Material i.e. Pump unit	
Tracing **Motion**,tank etc.	£70 000
Installation	£20 000
Total	£90 000

The payback time is about 6 months as this is the first machine on the battery.
The second machine on the pusher or cooler side has a longer payback period, say 15 months as the problems of fouling are not so great and thus the saving returns are less.

CONCLUSIONS

High Pressure Water Jetting has proved to be a practical facility which assists the production of coke. It improves the environment of the coke plant operators and reduces the atmospheric pollution.

The cleaning of doors is completed within the pushing cycle by a machine which is both uncomplicated and reliable. Clean doors vastly improve door extraction and maintenance downtime on door machines has been reduced. Gas emissions have been cut down to a minimum, thus preventing oven door fires, figure 10 which lead to oven door and jamb damage. There is more gas available for by-products recovery. A reduction of air ingress, which gives damage to the oven structure (silica cracking and door jamb burning) also air ingress leads to high temperature within the oven chamber and give free carbon in tar which, in other words it causes poor separation. This, in turn, leads to roof carbon and stickier ovens. Maintenance downtime has improved machine availability owing to lack of door extraction replacement problems, easier to maintain a water jet cleaner than conventional mechanical cleaners. Water jet cleaner has no contact with the door trough or seal, therefore, no seal damage is incurred.

NOZZLE TRIALS

TEST NO	NOZZLE TYPE	TIME SECS	FLOW AT 350 BAR L/MIN	FLOW AT 700 BAR L/MIN	NOZZLE DISTANCE MM	TEST CONDITIONS & COMMENTS	RESULTS
1	Fan 30°	-	20		Various	Hand Held	Poor
2	Fan 30°			29.5	230	Hand Held	Poor
3	Fan 30°			29.5	50	Hand Held	Good
4	Fan 30°			29.5	50	Hand Held Moving with Carriage	Effective
5	Solid				Various	Static	Not Effective
6	Stream				200	Hand Held	Small Area
7	Hollow Cone		77		50	Static	Not Effective
8	Solid Cone		11 at 103 Bar		100		Some Removal
9	Solid Cone		19 at 103 Bar		50		Some Removal
10	Solid Stream		12		200	Lance Moving with Carriage	Poor
11	Fan 35°		77 at 550 Bar		100		Very Good
12	Twin Fans			59	75		Good
13	Three Jets			104	200	60° Angle	Good
14	Three Jets		70		200		Not Very Effective
15	Star			30	100		Area Not Covered
16	Star			30	100	Improved Aim	Very Good

NOZZLE TRIALS

TEST NO	NOZZLE TYPE	TIME SECS	FLOW AT 350 BAR L/MIN	FLOW AT 700 BAR L/MIN	NOZZLE DISTANCE MM	TEST CONDITIONS & COMMENTS	RESULTS
17	Star				250		OK for Washing Hard Deposit Remains
18	Cone 60°			30	160	2 Passes Bake Deposit	Some Removal
19	Star 40°			45	160	Hot Door ONE Pass	Some Areas Left
20	Star 40°			40	100	Hot Door ONE Pass	Well Cleaned
21	Cone 60°			30	100	Hot Door ONE Pass	Well Cleaned
22	Fan 40°			45	100	Hot Door Very Dirty	Some Dirt Left on Bottom
23	Fan 35°			68	155	Cold Door Very Dirty	Required 5 Passes Very Good
24	Star 40°	45		68	155	Half Clean	Moderate
25	Star 40°	30		68	155		Some Deposit Left
26	Star 40°	35		68	155	Part Cleaned	Stubborn Deposit Remains
27	Fan 35°	45		68	155	Part Cleaned	" "
28	Fan 35°	45		68	210	Very Dirty Door	" "
29	Fan 35°	45		68	210	Standard Production Door	" "
30	Fan 35°	45		68	100	Very Dirty Door	Very Clean
31	Fan 35°	45		68	100	Dirty Door	Very Clean
32	Fan 35°	45		68	100	Dirty Door	Very Clean

NOZZLE TRIALS

TEST NO	NOZZLE TYPE	TIME SECS	FLOW AT 350 BAR L/MIN	FLOW AT 700 BAR L/MIN	NOZZLE DISTANCE MM	TEST CONDITIONS & COMMENTS	RESULTS
33	Fan 35°	42		68	120	Heavily Encrusted	Poor because of bad Alignment
34	Fan 35°	42		68	120	Heavily Encrusted	Better (33)
35	Fan 35°	42		68	120	Heavily Encrusted	Good
36	Cone 40°	42		45	120	Heavily Encrusted	Poor
37	Cone 40°	42		45	120	Heavily Encrusted	Poor
38	Cone 40°	100		45	120	Soft Tar	Very Good
39	Cone 40°	100		45	120	Heavily Encrusted	Poor
40	Star 25°	100		55	80	Heavily Encrusted	Small line only cleaned
41	Star 40°	100		55	120	Very Dirty	Good Improvement on 40
42	Fan 35°	100		55	120	Very Dirty	Very Good
43	Fan 35°	100		55	120	Very Dirty	Very Good
44	Fan 35°	100		55	200	Already Cleaned before on 30	Very Good
45	Fan 35°	100		55	200	Soft Tar Very Dirty	Unsatisfactory
46	Fan 35°	100		55	200	Baked Hard	Some Erosion
47	Fan 35°	100		55	240	Baked Hard	Some Erosion
48	Fan 35°	100		55	240	Same Door (47) 2nd Pass	More Erosion

NOZZLE TRIALS

TEST NO	NOZZLE TYPE	TIME SECS	FLOW AT 350 BAR L/MIN	FLOW AT 700 BAR L/MIN	NOZZLE DISTANCE MM	TEST CONDITIONS & COMMENTS	RESULTS
49	Fan 35°	100		55	240	3rd Pass On (47)	Clean
50	Fan 35°	100		68	240	Very Dirty	Satisfactory
51	Fan 65°	100		68	140	Very Dirty	Excellent
52	Fan 65°	105		68	105	Heavy Soft Deposit	Good
53	Fan 65°	100		68	90	Dirty	Good
54	Fan 65°	90		68		Oblique Angle. Dirty Door	Poor
55	Fan 65°	60		68	90	Oblique Angle. Dirty Door	Poor
56	Fan 65°	60		68	90	Square on Dirty Door	Good
57	Fan 65°	60		68	90	Dirty Door 90° Jet	Good
58	Fan 65°	60		68	90	Very Dirty	Good
59	Fan 65°	60		68	90	A Previously Cleaned Door	Very Good
60	Fan 65°	45		68	90	Dirty Door	Some Dirt Left
61	Fan 65°	45		68	90	Very Dirty Door	Good
62	Fan 65°	45		68	90	Very Dirty Door	Good
63	Fan 60°	45		28	60	Hard Baked Door	Poor
64	Solid	45		50	90	Dirty Door Mixture of Hard and Soft	Very Good but Scribble

Fig. 1 Coke oven battery seen from 'pusher' side.

Fig. 2 Quencher

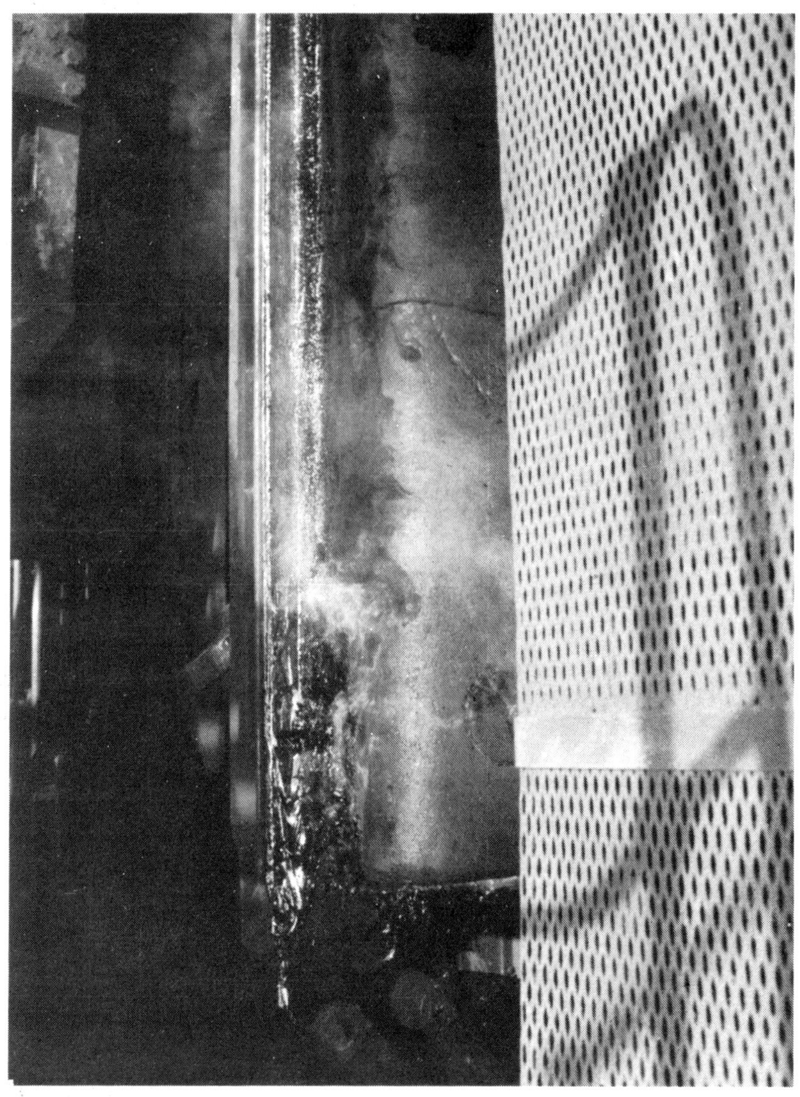

Fig. 3 View of bottom of door with tar deposit hanging from seal.

Fig. 4 Tracing motion lance carrier

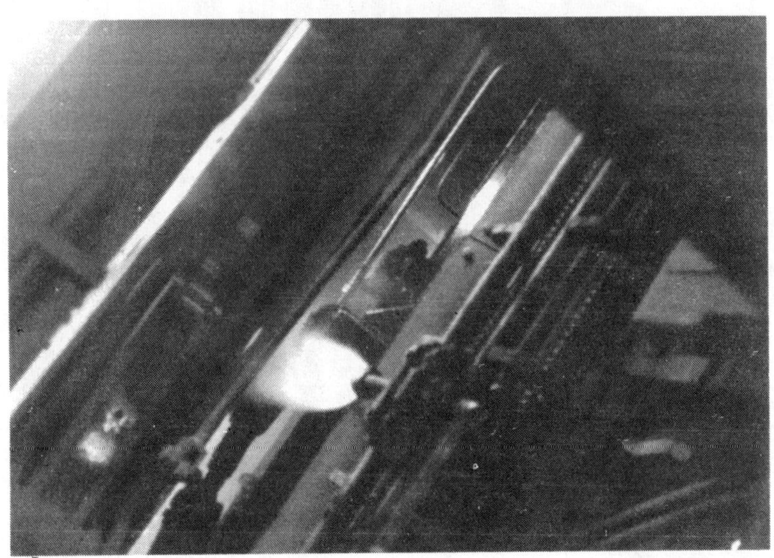

Fig. 5 Jetting lance with single nozzle in operation

Fig. 6

Water jetting unit in its normal environment

Fig. 7

View of dirty coke oven door, showing leakage

Fig. 8 Cleaned up battery 12 months after installation of water jetting equipment

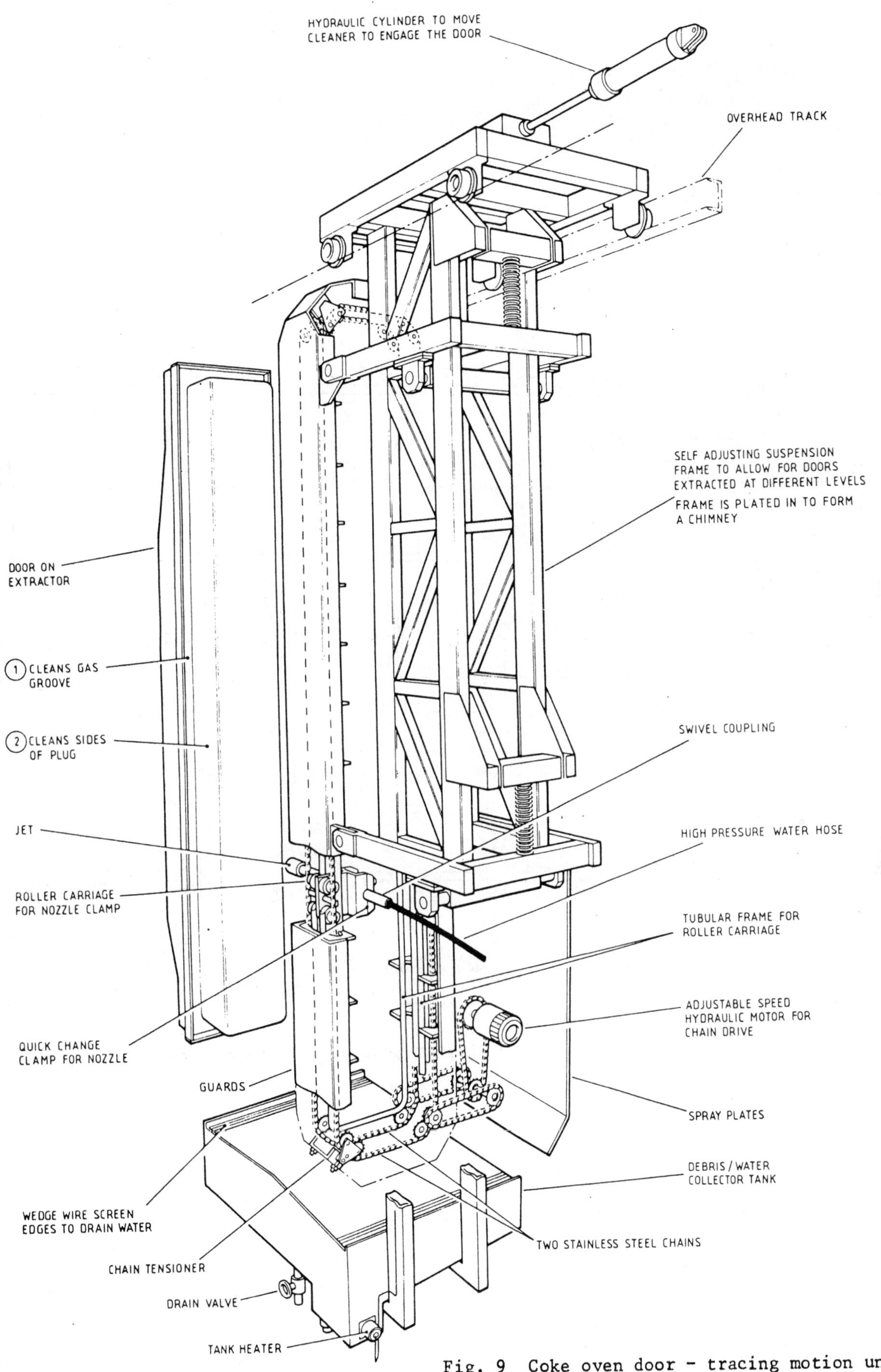

Fig. 9 Coke oven door - tracing motion unit

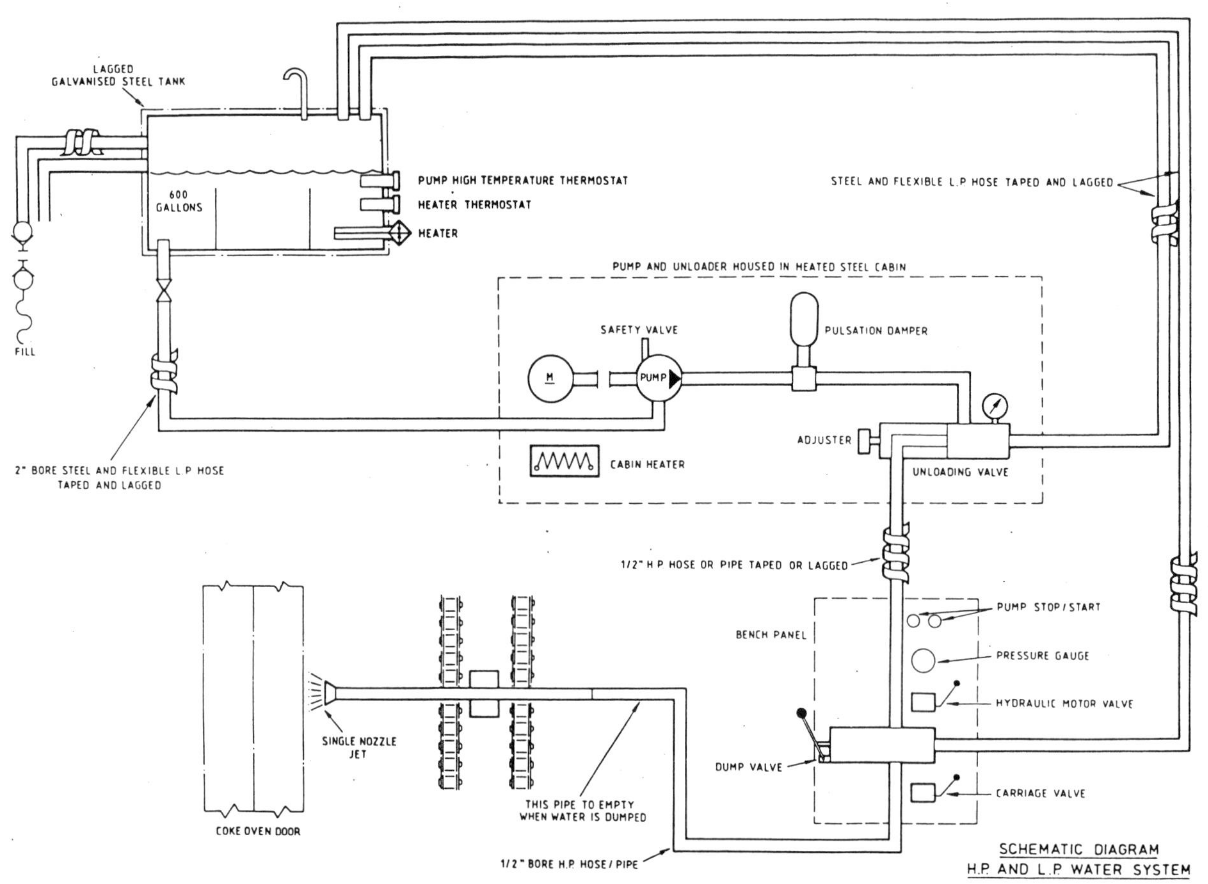

Fig. 10 Schematic diagram HP and LP water system

6th International Symposium on
Jet Cutting Technology
6-8, April, 1982

APPLICATION OF WATER JET CUTTING IN SOLID PROPELLANT TECHNOLOGY

S. Ganapathi, M.K. Venkitakrishnan and T.S. Ram

Indian Space Research Organisation, India

Summary

This paper explains the use of jet cutting in solid propellant technology. The facility available at VSSC and method of cutting propellant and insulation are described. Results of the tests conducted to prove the soundness of the hardware subjected to jet are shown. Experimental trials done during the optimisation of propellant cutting are explained and the results are discussed.

Held at the University of Surrey, U.K.
Symposium organised and sponsored by
BHRA Fluid Engineering

1. INTRODUCTION

Solid propellants are used for booster rockets of launch vehicles. Diameter and weights of todays rocket motors have increased remarkably and acceptance standards are rigorous. It is understandable therefore the problem of disposal of defective motors become all the more important. A way has to be found out to dispose of the propellant in a safe manner which also reclaims the expensive motor case for further use. A high pressure water jet is used for this application. The water jet slices the rubbery solid propellant without friction and heat and the motor case is totally unaffected. The extensive use of water - the best medium to fight propellant fires - assures complete safety.

It was found out that various systems using pressure from 40 to 700 MN/m^2 were used for disintegrating different materials. Low pressure jets were found suitable for our use since they have insufficient energy to affect the hardware which is salvaged. Though abrasive jets at still lower pressures are able to cut the propellants, they are not preferred considering the friction at propellant surfaces, erosional problems in nozzle, piping etc.

2. JET CUTTING FACILITY AT VSSC

This facility was established for salvaging big rocket motors produced in our country for the space programmes of ISRO at Vikram Sarabhai Space Centre (VSSC), Trivandrum.

Our solid propellants are highly filled polymers having the following physical properties:-

Density	:	1.6×10^3 to 1.8×10^3 kg/m^3
Tensile strength	:	0.2 to 1.2 MN/m^2
Elongation	:	20 to 300%
Hardness (shore A)	:	50 to 80
Auto-ignition	:	520 to 550°K

2.1 System Description

The system consists of power pack, lance unit, rocket motor supporting system, waste disposal arrangements etc.

2.2 Powerpack

Considering the properties of propellants, hardware, safety aspects and equipment availability we decided on an equipment which could develop -

Maximum pressure	:	70 MN/m^2 approx.
Maximum volumetric discharge	:	0.08 m^3/min. approx.

The power pack is horizontal, quintuplux, single acting, reciprocating plunger pump equipped with air cushion chamber for smooth flow pattern. Pump is driven by a 150 H.P electric motor. This is supplied by M/s Tritan Corporation, U.S.A.

2.3 Lance Unit

This consists of high pressure pipings, nozzles and its supports. The lance comprises of 0.1 m diameter x 6 m long pipe with end fittings through which the high pressure piping was taken into the front end with which nozzle is attached. The lance is mounted over a swivelling mechanical jack so that the nozzle may be positioned at different heights and orientations as shown in Figure-1. The lance also acts as a safety device in case of high pressure line failure. Provision is made for fixing one or more than one nozzle at the tip. Diameter of nozzle is 1.16×10^{-3} m and material is stainless steel. The nozzle pressure was controlled by means of a bleed-off valve which returns the excess water back to reservoir.

There are three possible arrangements in cutting:

1. A stationary material to be cut with a moving jet;
2. A moving material with stationary jet;
3. Combination of 1 and 2.

We have followed the 2nd method in our system.

2.4 Rocket motor supporting system

As shown in Fig.1 the defective rocket motor is kept over the rollers which in turn is fixed over the trolley. The trolley is movable on rails grouted on the floor. Independent drives are provided for rollers and trolley. Thus the rocket motor has rotational as well as longitudinal movement in either directions. The peripheral speed and longitudinal speed of the rocket motor may be infinitely varied from 2.2 to 15.2 m/min. and from 12.5×10^{-3} to 65×10^{-3} m/min. respectively. All the movements are controlled from a remote control room.

2.5 Method of cutting

Inside configuration of solid rocket motor is generally as shown in Fig.2. This motor with handling harness weighs 4500 kgs. For an initial cut, a 'T' nozzle is used with twin jets (diametrically opposite) on the propellant port. Simultaneous rotation and axial movement of the motor make a helical cut on the propellant. The pitch of the helix is adjusted to be equal to or less than the depth of cut. The helically made radial cut is followed by an axially oriented circumferential cut which uproots a layer of propellant throughout. This gives a spiral of propellant which breaks by itself into pieces during the operation. This procedure is repeated until all the propellant is removed. In all the cases the maximum stand off distance is kept around 5 mm. Fig.3 and 4 show the increase in propellant surface roughness with depth of cut. This can be due to the tearing off of propellant. Material removal rate is approximately $2 \times 10^{-3} m^3$/min.

The rocket motor is lined with some special rubber insulation as shown in Fig.2 whose thickness varies from 3 to 20 mm. This insulation is also removed by selecting the proper cutting speed. Fig.5 shows this operation. Fig.6 shows the cut pieces of rubber insulation.

2.6 Waste disposal system

The oxidiser, ammonium perchlorate, is soluble in water. After cutting the propellant, lumps are collected in a tray fitted in the front end of the rocket motor. The collection becomes easy because of spent jet water and rotation of the motor. These lumps are removed at regular intervals either for disposal or for recovery of ammonium perchlorate.

2.7 Safety aspects

Since the propellant is susceptible to fire, the salvaging operation of rocket motor is classified as hazardous operation. Hence all safety precautions are taken into consideration.

The high pressure system is provided with relief valve to avoid the pressure shootup. The controls for all operations are remote and located in a control room.

In the facility, instructions as shown in Fig.7 and 8 are displayed for the observations of operators and supervising personnel (Ref. 1).

3. SOUNDNESS OF HARDWARE

A number of trials on different types of metals are carried out to find out the effects of high pressure jet on the metal during cutting. Also the effect of number of nozzles as well as orientation of jet are studied. Table-1 compares the test results of the specimens of 15 CDV 6 steel subjected to jet with those which are not subjected to jet. From the test results it is confirmed that in no way jet affects the soundness of the rocket motor hardware.

4. EXPERIMENTAL STUDIES

During the optimisation of method of cutting of propellants in casebonded rocket motors a good amount of tests are conducted. Those results are presented and discussed here.

4.1 Experimental set up

A general arrangement of the experimental set up is as shown in Fig.9. The liquid used is plain water at room temperature. The nozzle having size as discribed earlier is used. The depth of cut is measured with a standard depth gauge. A motor having solid propellants as shown in Fig.9 is used for cutting experiments. The physical properties of the propellant is as given earlier. Time of jet application is measured by a standard stop watch. The pressures are measured by an on line pressure gauge fitted near the pump.

Different experiments are conducted to study the effects of parameters such as stand-off distance, traverse speed, water jet pressure on the penetration depth in propellant. In all the tests jet axis is kept perpendicular to the propellant surface. The penetration depth is an average of ten measurements of depth of cut at equal distance along the cut.

5. RESULTS AND DISCUSSIONS

5.1 Jet pressure

The relation between the pressure and penetration depth obtained from the tests is shown in Fig.10. Constant stand-off distance of 3.5 mm is maintained for all the tests. Fig. 11 shows the appearance of cuts in propellant after the penetration tests. The linear variation of penetration depth with pressure is fortuitous and varies very much with traverse speed.

5.2 Jetting time

Fig.12 shows the relationship of penetration depth with jetting time for stationary jet. From the figure it is clear that major cutting takes place in the initial phase.

5.3 Stand-off distance

Fig. 13 shows the relation between the stand-off distance and penetration depth for different pressures. The tests are conducted at constant surface speed. From the test results it is found that penetration depth decreases linearly with stand-off distance. This decrease is found to be almost independent of pressure. Also it is noticed that as the stand-off distance increases, the opening of the groove cut (kerf width) increases and becomes irregular.

5.4 Traverse speed

Penetration depth for different pressures and traversing speeds are measured at constant stand-off distance and the results are shown in Fig.14. Study shows that the penetration depth increases inversely with traverse speed and almost flattens at lower speeds. Trials at still lower traverse speeds is not carried out due to certain limitations in our facility.

5.5 Specific energy

The specific energy of propellant removal from rocket motors for different pressures is given as in Fig.15. The tests are carried out for a traverse speed of 30 mm per sec. at a constant stand off distance of 3.5 mm. Propellant removal rate is calculated as (Ref. 2)

$$\dot{V} = \frac{S\,h^2}{2}$$

where V is volume of propellant removed, h is penetration depth and S is traverse speed. From the results it is found that good efficiency is attained beyond 42 MN/m^2.

6. CONCLUSIONS

From the operating experience gained over the past years, water jet cutting is found to be a very convenient and safe method for salvaging the rocket motor hardwares.

7. REFERENCES

1. C.M.Ward: "Safety considerations arising from operational experience with high pressure jet cleaning". In:Proc. 1st International Symposium on Jet Cutting Technology, BHRA Fluid Engineering, Cranfield. April, 1972, Paper F1. 15 pp.
2. K.Moodie, G.Artingstall: "Some experiments on the application of high pressure water jets for mineral excavation" : In:Proc. 1st International Symposium on Jet Cutting Technology, BHRA Fluid Engineering, Cranfield. April, 1972, Paper E3, 29 pp.

TABLE - 1

Specimen Identification	Details		U T S Kg mm^{-2}	% of Elongation	Hardness BHN
(B)	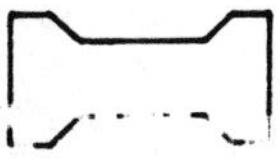	NO JET	112	12	345
(D)	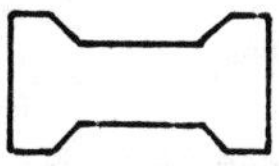	NO JET	112	13·1	345
(F)	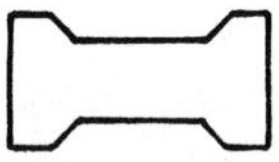	NO JET	110	13·1	329
(A)	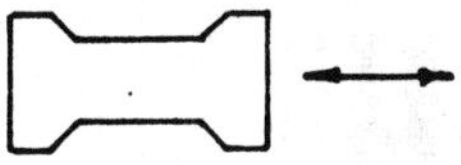	⟷	112	—	354
(C)	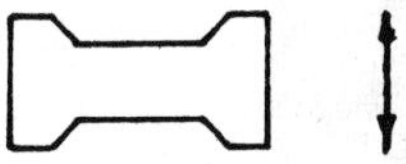	↕	112	12·6	345
(E)	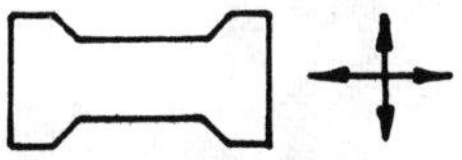	✛	111	13·2	345

Remarks:—

1. Arrows indicate the jet movements.
2. Jet action is perpendicular to the face of the specimen.
3. Stand-off distance is 1mm.
4. Operating pressure is 70 $MN\,m^{-2}$.
5. Nozzle diam. is 1·16mm.
6. Jet action time is 5 min in each specimen.
7. Material: 3 mm. thick 15 CDV 6 sheet.

Fig.1 - Set up showing the defective rocket motor in position.

Fig. 2 - Solid rocket motor ready for jet cutting.

Fig. 3

Fig. 4

Fig.3 & 4- Jet cut propellant pieces.

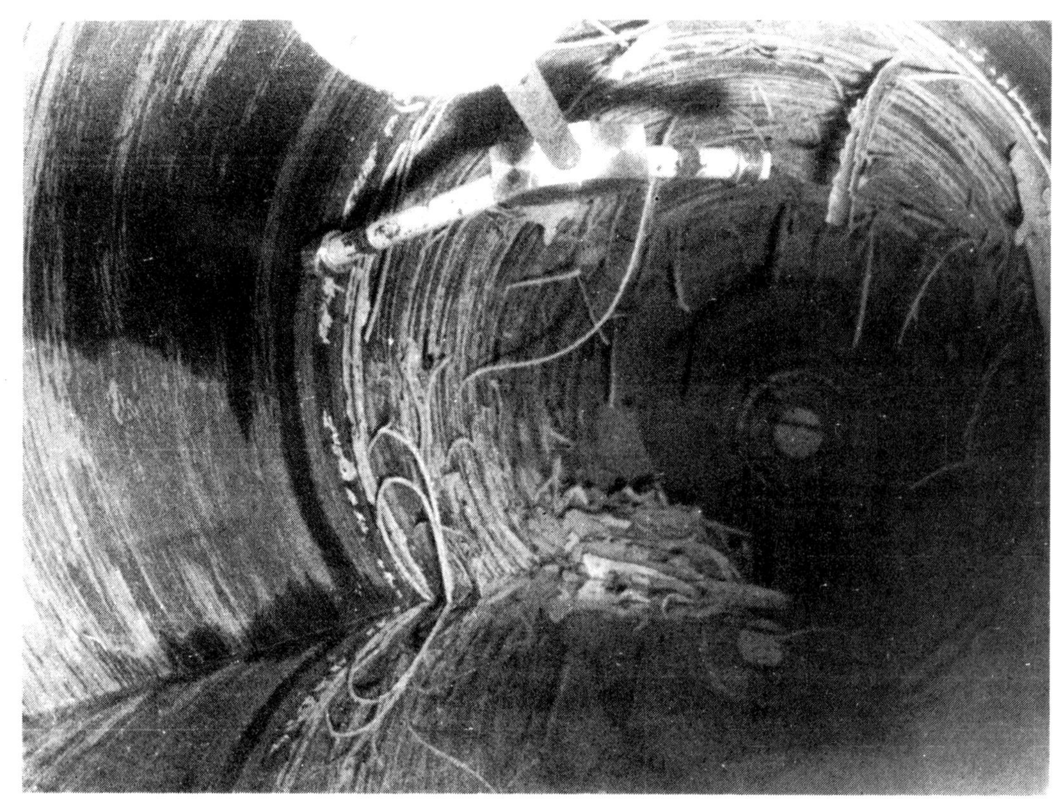

Fig.5 - Removal of Insulation

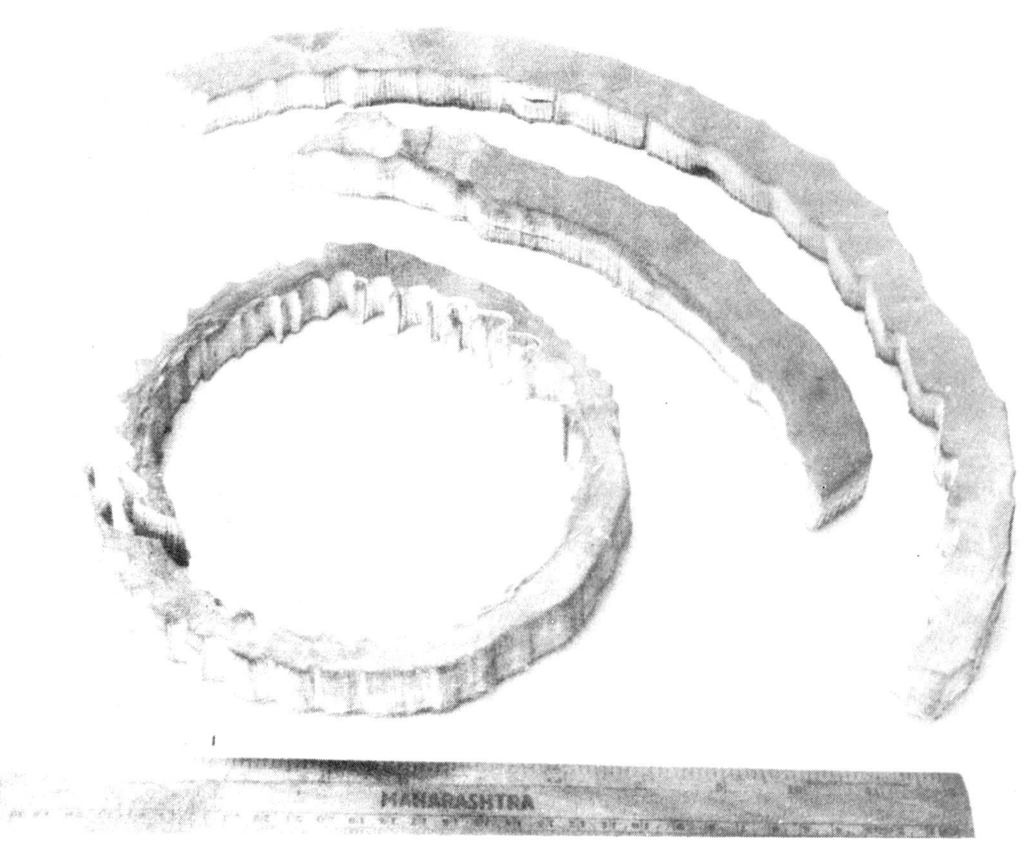

Fig.6 - Jet cut insulation pieces.

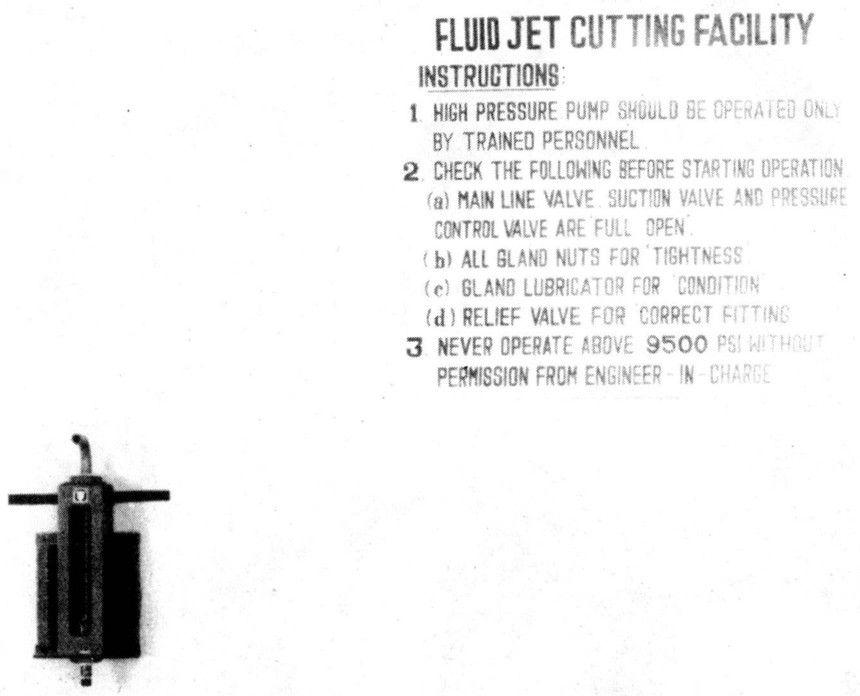

Fig.7 - Display in the control room showing the safety precautions.

FLUID JET CUTTING FACILITY
INSTRUCTIONS
1. MAKE SURE ALL NOZZLES, LANCES, HOSES AND FITTINGS ARE IN SOUND CONDITION. CHECK WITH YOUR ENGINEER IN DOUBT.
MAKE SURE YOUR PROTECTIVE CLOTHING IS IN FIRST CLASS
TION AND THAT YOU ARE WEARING ALL OF IT.
3. MAKE SURE ALL GUARDS ARE IN POSITION.
4. DO NOT START WORK WITHOUT HAVING A STANDARD PROCEDURE COVERING THE ACTUAL JOB. CHECK WITH YOUR ENGINEER IN DOUBT.
5. ENSURE YOUR WORKING AREA IS FREE FROM ALL UNWANTED ITEMS.
6. STOP THE JET BEFORE ALLOWING ANY OTHER PERSON TO ENTER OR LEAVE THE AREA
7. DO NOT OPERATE ABOVE 9500 PSI WITHOUT PERMISSION ENGINEER-IN-CHARGE.

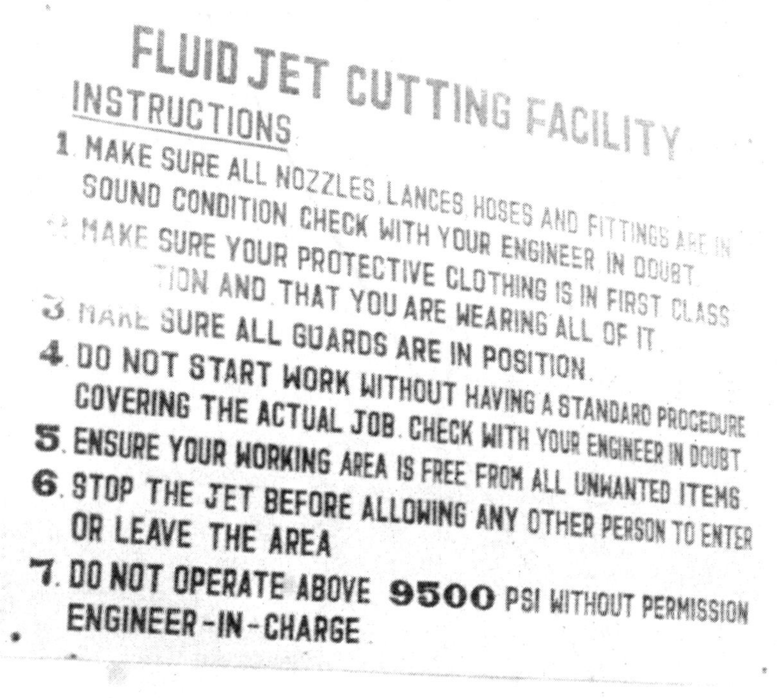

Fig.8 - Display in the working hall showing the safety precautions.

Fig.9 - Rocket motor used in the jet cutting experimental studies.

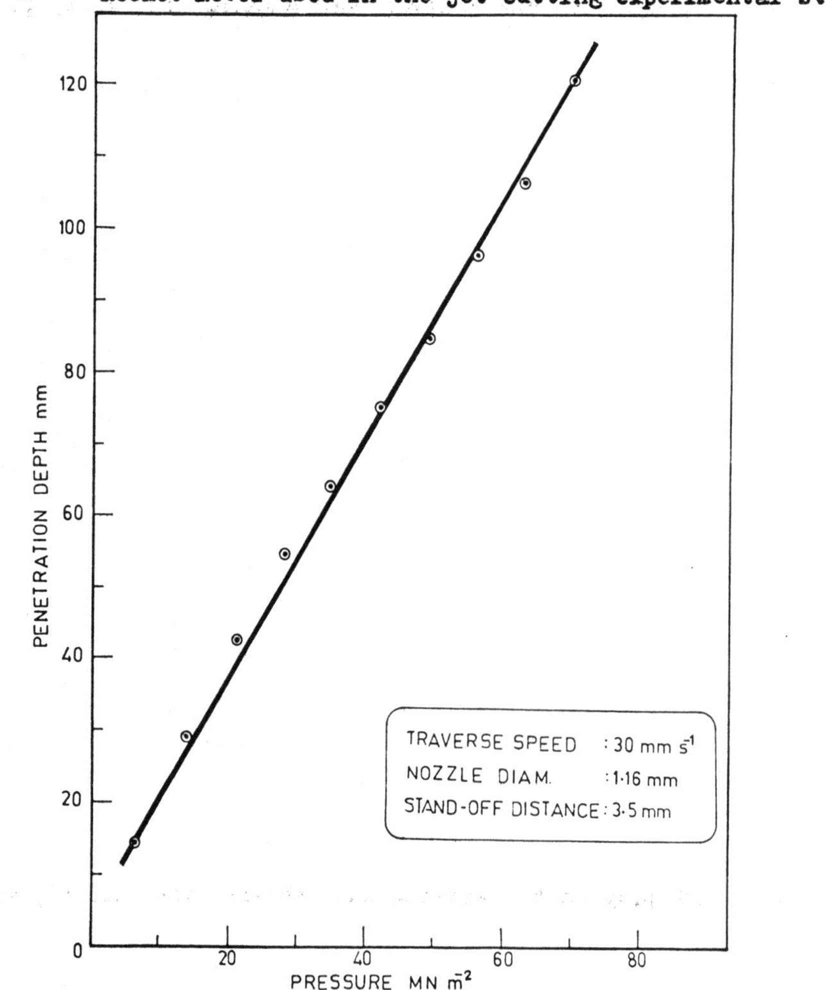

Fig.10 - Relation showing the penetration depth and jet pressure.

Fig.11 - Appearance of cut at different pressures.

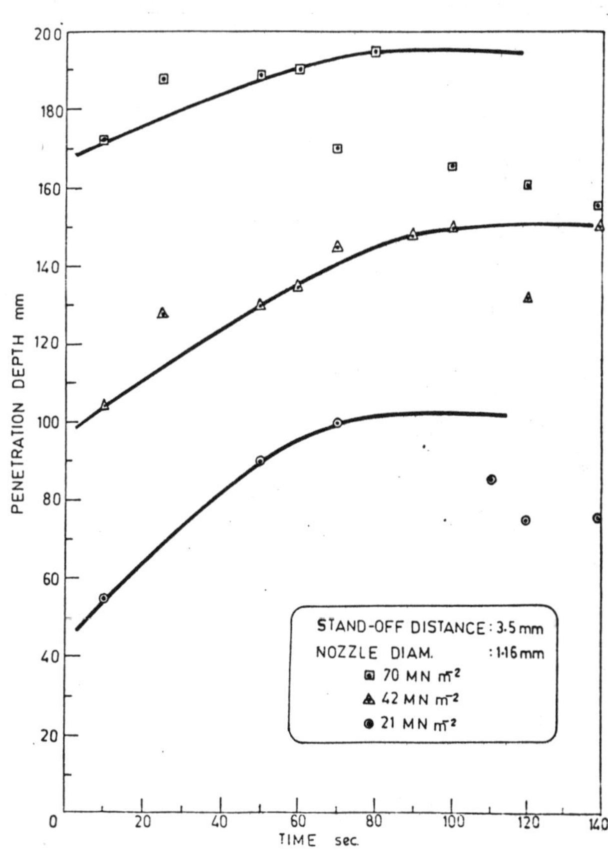

Fig.12 - Relation showing the penetration depth and jetting time for stationary jets.

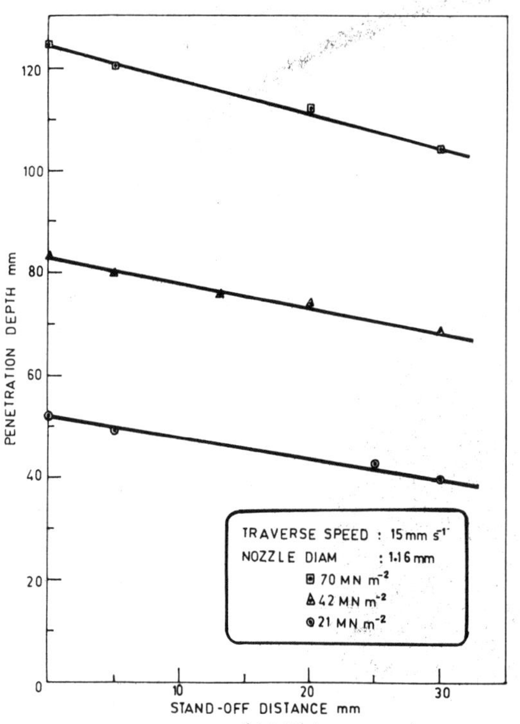

Fig.13 - Relation showing the penetration depth and stand-off distance.

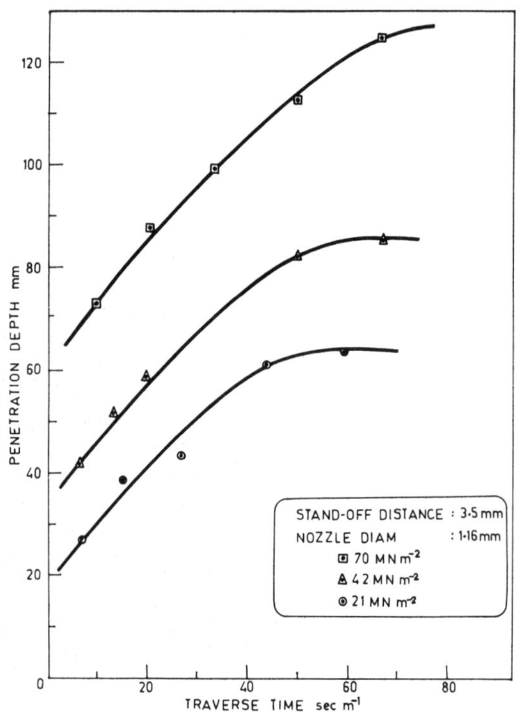

Fig.14 - Relation showing the penetration depth and traverse speed.

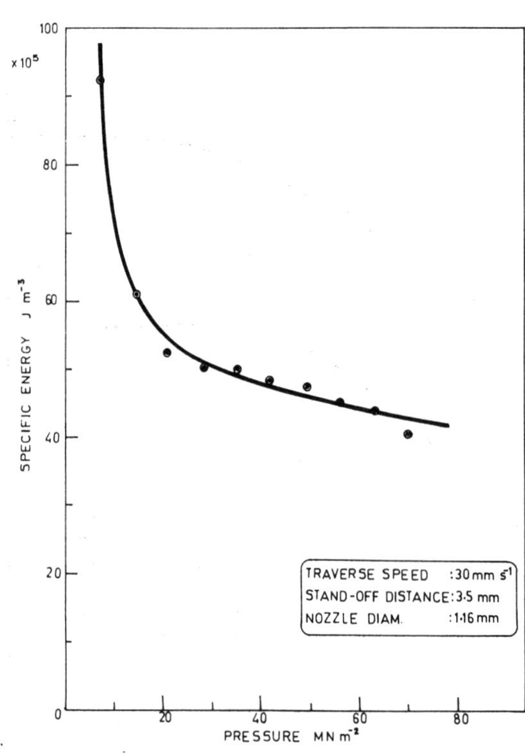

Fig.15 - Relation showing the specific energy and the pressure.

6th International Symposium on
Jet Cutting Technology
6-8, April, 1982

DEVELOPMENT OF WATERJET TOOLS FOR THE INSTALLATION AND REPLACEMENT OF BURIED UTILITIES

J.M. Reichman and D.P. Kelley

Flow Industries Inc., U.S.A.

Summary

A study sponsored by the Electric Power Research Institute was undertaken to develop waterjet soil-boring and cable-removal tools for the installation and replacement of directly buried electrical distribution cable.

The soil-boring tool is provided with a waterjet-assisted impactor that increases its rate of penetration two to four times over that of a tool with an unassisted impactor. The tool also incorporates a novel steerable, self-advancing system that greatly increases its directional control capabilities.

The cable removal tool advances along a failed cable and clears away the soil surrounding it with an array of waterjets, leaving a stable hole as it passes. This tool reduces cable replacement costs by eliminating the need for surface excavation and subsequent restoration.

Both tools have been fabricated and are now undergoing extended field testing by various electric utility companies.

Held at the University of Surrey, U.K.
Symposium organised and sponsored by
BHRA Fluid Engineering

1. INTRODUCTION

The use of underground distribution cable is becoming increasingly more common as utility companies, municipal governments, and developers seek to reduce the clutter of overhead wires and minimize outages caused by storm damage. Installing underground cables is readily accomplished at relatively low cost in rural areas and new subdivisions by means of such conventional methods as trenching and cable plowing. However, installation costs increase significantly when conventional methods are used in urban areas since they require cutting through existing pavement and/or landscaped properties. Therefore, underground soil penetrating tools for use in urban areas have been developed as alternatives to conventional surface cutting tools. Unfortunately, the present generation of underground soil penetrating tools are restricted by their lack of positive directional control and limited range to bore lengths of no more than 23 m, which makes them suitable only for special applications. Clearly, the range and directional control capabilities of the underground soil penetrating tool must be improved if it is to realize its potential as an economically viable underground distribution cable installation device.

The increasingly frequent use of underground distribution cable in urban areas presents another problem in addition to that of initial installation, i.e., the economical replacement of failed cable. The replacement of failed cable by conventional methods is prohibitively expensive due to the cost of excavating the cable and then restoring the surface to its original condition after the cable replacement operation is complete. Underground soil penetrating tools could help minimize this cost by creating a bore parallel to the failed cable. A replacement cable would be guided through this new bore and assume the function of the failed cable, which could then be abandoned in place. However, a tool capable of following a failed cable and removing the soil surrounding it while leaving a stable hole for new cable installation would provide even greater savings. The replacement cable could then be attached to one end of the failed cable and pulled into place as the failed cable is removed from the hole. This would simplify the operation and reduce cost by eliminating the tool's guidance system requirement and allowing the failed cable to be salvaged.

The problems associated with increasing requirements for underground distribution cable installation and replacement led the Electric Power Research Institute (EPRI) to fund a program at Flow Industries entitled "Development of Improved Boring Equipment for Installing Underground Distribution Cable and Conduits". The original program objective was limited to the development of a steerable boring tool capable of overcoming the technical and economic difficulties encountered in the installation of underground distribution cable. The cable-following tool portion of the program was added later as cable replacement problems and the advantages of using a failed cable for tool guidance became better understood.

This paper describes both waterjet-assisted steerable boring tool and waterjet cable-following tool development. It includes separate design reviews for both tools. Plans for further development and long-range testing are also discussed.

2. WATERJET-ASSISTED GUIDED BORING TOOL

Several soil boring tools are now commercially available for use in underground distribution cable installation. These tools are essentially impacting devices that create stable holes by compacting the soil as they penetrate it. As such, they are restricted to operating in uniform, compactible soils that are free of obstructions. Applications for these tools are also limited by their 23-m maximum boring range and lack of directional control capability. Therefore, an improved tool with greater general utility must be able to operate in non-uniform soils, hard and sandy soils, and soils in which obstructions may be encountered. It must also have a longer boring range and a directional control capability including an electronic tool detection system. The addition of the directional control and tool detection capability will allow the tool to move along a predetermined path and thus avoid obstacles. The specific design requirements for the tool are to:

1. Produce bores suitable for either directly buried cable or conduit.

2. Bore a hole 305 m long.

3. Create a stable hole in all types of soils, including rocky soil.

4. Penetrate average soils (loamy clay) at a rate of 10 to 5 mm/sec.

5. Deviate no more than 300 mm in any direction from specified path.

6. Change direction to avoid obstacles.

7. Meet all pertinent safety standards.

8. Operate from an easily transportable, integral power source.

The tool developed for this program is shown schematically in Fig. 1. The tool performs by:

1. Using a waterjet to pilot the hole and reduce friction by lubricating the soil.

2. Using a special impactor to form the hole and minimize water requirements.

3. Using a steering section to control hole direction.

4. Using a waterjet to enhance steering control by jetting in the desired direction.

5. Using a self-advancing mechanism to move the tool through the soil.

The effect of adding waterjet assist to the tool is of particular interest in this instance. This subject will be presented in some detail herein, together with brief descriptions of system components other than those used to provide waterjet assist.

2.1 Jet-Assisted Impactors.

A basic requirement of the boring tool is that it produce a stable hole. Existing soil-boring tools satisfy this requirement by compacting the soil surrounding the tool. Thus, a stable hole is formed and no slurry recovery is necessary. In view of its effectivenss, it was decided to retain this basic hole-forming method and incorporate it into the development of a new, reasonably-priced tool capable of overcoming the limitations of conventional soil-boring tools. However, it was postulated that assisting the impactor could potentially improve tool performance. Additionally, since there was reason to believe that the characteristics of soil-borer impactors might not be optimum, a study was undertaken to determine the effect of using waterjets to assist impactors in penetrating a variety of soils.

The types of tools tested are listed in Table 1. The tools were fitted with a head 100-mm in diameter which could either be run dry or with waterjet assist. The parameters that were varied included pressure and flow rate. The measured parameter was penetration rate. This rate was measured for a fixed "hold-down" force. The impactor head and experimental set-up are shown schematically in Fig. 2. The tests

Table 1. Pneumatic Impactor Characteristics

Characteristic	Jack Hammer	Paving Breaker	Post Driver	Clay Spade	Vibrator
Piston Weight	2 Kg	2.16 Kg	6.9 Kg	0.59 Kg	6.6 Kg
Piston Bore	67 mm	63.5 mm	68 mm	44 mm	70 mm
Piston Length	115 mm	76 mm	178 mm	49 mm	203 mm
Piston Velocity	6.5 m/s	8.1 m/s	4.8 m/s	12.6 m/s	3.7 m/s
Blow Energy	44 J	72 J	81 J	47 J	46 J
Blows Per Minute	2096	1120	1400	2328	2350
Power Level	1.55 Kw	1.35 Kw	1.91 Kw	1.91 Kw	1.9 Kw

were conducted in various soil types including clay, sand, topsoil, and pit run (soil and rock mixture). The soils were contained and compacted in large cylindrical vessels 1.2 m in diameter. This was done to prevent groundwater from softening the soils. The tests were performed in the center of the vessels to avoid any wall effects.

The results of the tests demonstrated that waterjet assist of impacting tools greatly enhanced their performance in all soil types. The penetration rate of the waterjet-assisted tool ranged from two to more than five times that of the dry tool. The test results for each soil type and tool are presented in Table 2. The jet pressure for these tests was 54.4 MPa and the flow rate was 5 ℓ/min through a 0.635-mm diameter nozzle. This represents a delivered hydraulic power of 4.4 Kw.

The ability to increase penetration rate without increasing the power delivered to the tool offers distinct advantages for boring. These advantages are that:

1. Smaller tools can be used to achieve an acceptable penetration rate, thus adding maneuverability to the system.
2. By limiting the power to the impactor, the chances of penetrating buried pipe are reduced.
3. A wider variety of soil types can be penetrated by a given tool and increased even further by increasing the power to the jet.

An additional advantage of adding waterjets to an impacting tool is its subsequently increased ability to penetrate and split rock. This was proven by testing waterjet-assisted and dry tool penetration through concrete blocks. In each case, the waterjet-assisted tool's penetration rate was three to four times that of the dry tool. The reason for this is believed to be the ability of the assisting jet to penetrate into cracks formed by the impactor and "hydrofracture" the rock. The ability to go through rock will increase the applicability of impactors to more soil types and minimize the chances of tool deviation from its prescribed path.

2.2 Soil Penetration Selection.

Conventional boring tools operate pneumatically with the tool driven through the soil by means of a large reciprocating piston. The tool advances as the piston impacts an anvil that is an integral part of the tool's body (Fig. 3). The soil surrounding the tool holds it in place during the piston's return stroke. It may be seen that a great deal of impact energy is lost as the force imparted to the anvil by the piston is partially dissipated into the tool's body. A tool with a smaller piston striking a separate anvil would transmit more impact energy to the soil. Such an arrangement would have the effect of increasing the tool's rate of penetration and the number of soil types in which it could be used while decreasing its total size and weight. This concept has been incorporated into the impactor design of the steerable boring tool developed by Flow Industries, Inc. (Fig. 4). Conventional and improved impactor specifications are compared in Table 3.

Table 2. Penetration Rates for Unassisted and Assisted Tools in Various Soils

Tool	Penetration Rate, m/min							
	Sand		Topsoil		Pit Run		Clay	
	Dry	Jet	Dry	Jet	Dry	Jet	Dry	Jet
Jack Hammer	1	2.6	3.1	10.8	2.9	5	2	5
Paving Breaker	2.9	4.9	2.7	8.8	1.7	3.1	7.3	7.3
Post Driver	0.24	0.73	0.82	3.8	0.24	0.48	0.42	2.1
Clay Spade	0.18	0.6	0.58	2.2	0.15	0.15	0.42	1.1
Vibrator	0.21	0.42	0.12	1.4	0.1	0.1	0.20	1.1
Chipping Hammer	0.18	0.45	0.03	0.1	0.1	0.13	0	0.06

Table 3. Tool Specifications

Specifications	Hole Hog	Flow Impactor
Piston Weight	6.80 kg	1.02 Kg
Piston Bore	51 mm	40 mm
Piston Area	2026 mm^2	1258 mm^2
Piston Length	762 mm	203 mm
Piston Velocity	5.46 m/s	12.95 m/s
Blow Energy	101.69 J	85.42 J
Blows Per Minute	390	1570
Power Level	6.7 Kw	2.2 Kw

2.3 Steering Section.

Conventional soil boring tools must be long and straight to provide for straight-line directional stability as the tool advances. However, such a design makes it difficult to apply the forces necessary to correct the tool's course if it is deflected during the boring operation or started at an incorrect angle. Therefore, a new approach to tool steering was conceived and developed for this program.

As shown on Fig. 5, the head and back section of the steerable boring tool, unlike a conventional tool, are larger than the impactor body. This allows the tool to be directed by applying force to the back of the impactor body to pivot the head in the $\pm y$ and $\pm z$ directions as desired. A separate hydraulic control line is provided to lock the steering mechanism for straight-line advance, in which case the directional stability of the steerable boring tool is approximately equivalent to that of a conventional tool.

The tool steering section and head are both 101.6 mm in diameter, the tool body is 60.3 mm in diameter, and the tool length from the head to steering section is 914.4 mm. This configuration permits the tool to make a 2-deg course correction, which corresponds to a 15.2-m radius of curvature.

2.4 Advancing Section.

A conventional soil penetrating tool advances as its piston impacts the integral head; it is held in place during the piston's return stroke by the force of the surrounding soil on its body. Such a design has the advantage of simplicity, but its inherent lack of steering capability and limited pulling ability decrease the tool's effective operating range. The steerable boring tool, which advances by means of an in-hole or external pusher, has an increased range capability. However, it lacks the conventional tool's large hole-contact area since its impactor is separate from the main body. Therefore, it requires some positive means of counteracting the force of the piston's return stroke. The necessity of satisfying this requirement led to the development of the steerable boring tool's advancing section.

The in-hole advancing section, shown on Fig. 6, comprises two grippers interconnected by a double-acting piston. Tool advance is effected by the cyclic action of this advancing section as follows: (1) the rear gripper is activated, (2) the cylinder pushes the tool forward, (3) the rear gripper is deactivated, (4) the front gripper is activated, and (5) the cylinder retracts to pull the rear gripper forward. The tool's penetration rate is adjusted to the prevailing soil conditions by varying the length of this cycle, with the maximum penetration rate being 25 mm/sec.

3. WATERJET CABLE-FOLLOWER TOOL

The present high cost of failed cable replacement could be greatly reduced by a tool that uses the failed cable as a guide while it creates a hole for replacement cable installation. The replacement cable could then be installed either in conjunction with tool operation or by attaching it to one end of the failed cable and pulling it into place as the failed cable is removed. In order to perform its function, such a tool must be capable of advancing along a failed cable while negotiating corroded concentric neutrals as required, removing soil surrounding the cable, and creating a hole with sufficient stability to permit conduit or cable installation. In addition, the tool must be portable and operate from an integral power source. A tool that satisfies these requirements is shown on Fig. 7. The tool achieves the desired performance by:

1. Using an array of waterjets to remove the soil around the cable.
2. Incorporating an internal flushing system to minimize wear.
3. Streamlining the body so as not to damage adjacent cable.

3.1 Waterjet Removal of Soil.

When cable is directly buried, the material placed around it is chosen to minimize the chance of cable damage. For this reason, sand or screened native soil is placed underneath and on top of the cable. The trench is then filled and compacted. This general burial procedure assures that the soil surrounding the cable is weak and thus simplifies removal of the soil by waterjet action.

Waterjet cutting tests were conducted in three typical soil types: sand, clay, and topsoil. As would be expected, the threshold pressure for this type of cutting is quite low. A curve of the depth of cut versus pressure is shown in Fig. 8. A curve of depth of cut versus power is shown in Fig. 9. From these curves, it is evident that soil cutting by waterjet is very effective. The crucial portion of the design, therefore, is the placement of nozzles to cut and efficiently slurry the cut material out of the hole. this was accomplished by operating the tool at 34 MPa and 15 ℓ/min. The choice of this pressure and flow rate results in efficient washing but minimum damage to surrounding cables. For opertion in more difficult soils, excess pumping capacity should be available.

3.2 Advancing.

The tool advances along the cable by means of a double-acting internal gripping system. The tool's operating cycle is as follows: (1) The internal grippers are activated, (2) the advancing side of the cylinder is pressurized causing the nozzles to press forward into the soil, (3) the internal grippers are deactivated while the rear gripper activates to hold the tool in place, and (4) The retraction side of the cylinder is pressurized to pull the tool body forward. This cycle is illustrated on Fig. 10. This tool is capable of a maximum speed of 20 mm/sec and has actually attained speeds of 7.6 to 10 mm/sec during initial testing.

4. CONCLUSIONS

All steerable boring tool components have been tested either individually or in combination and found to perform as designed. Also, the results of limited testing with an integrated tool indicate that the present configuration is capable of satisfying performance requirements with respect to soil penetration rate and directional control. The tool's maximum penetration range will be determined by more testing as will other performance characteristics. Incorporation of the electronic guidance system and the subsequent extended field testing program will probably begin in the spring of 1982.

The cable-following tool has been more extensively tested than the steerable boring tool. Controlled testing over several thousand meters of single primary distribution cable has been conducted at penetration rates of 7.6 to 10 mm/sec. Tests with primary and random-lay secondary cable and in various soils are now in progress.

It is expected that the experience gained during these tests will lead to tool modifications prior to the extended field tests. These latter tests will be performed in cooperation with selected electric utility companies.

Data obtained from the testing of both tools will be used in the preparation of a detailed cost analysis. The cost-per-meter of using these tools to install and replace cable will be determined and compared to the cost of performing the same operations using conventional tools. This comparison will be used to determine the tool's commercial viability. However, at this point, initial test data and a preliminary cost analysis appear to indicate that both tools offer significant potential cost savings.

Acknowledgement

The authors wish to gratefully acknowledge the helpful ideas and insights given them by Mr. T. Kendrew, EPRI Project Manager, during the course of this work.

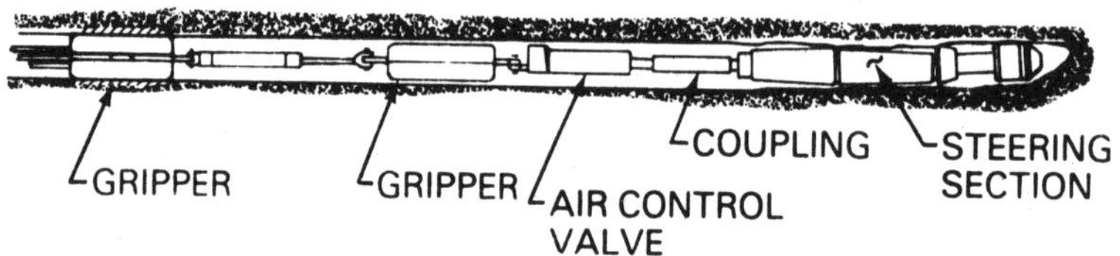

Fig. 1. Steerable Boring Tool Design

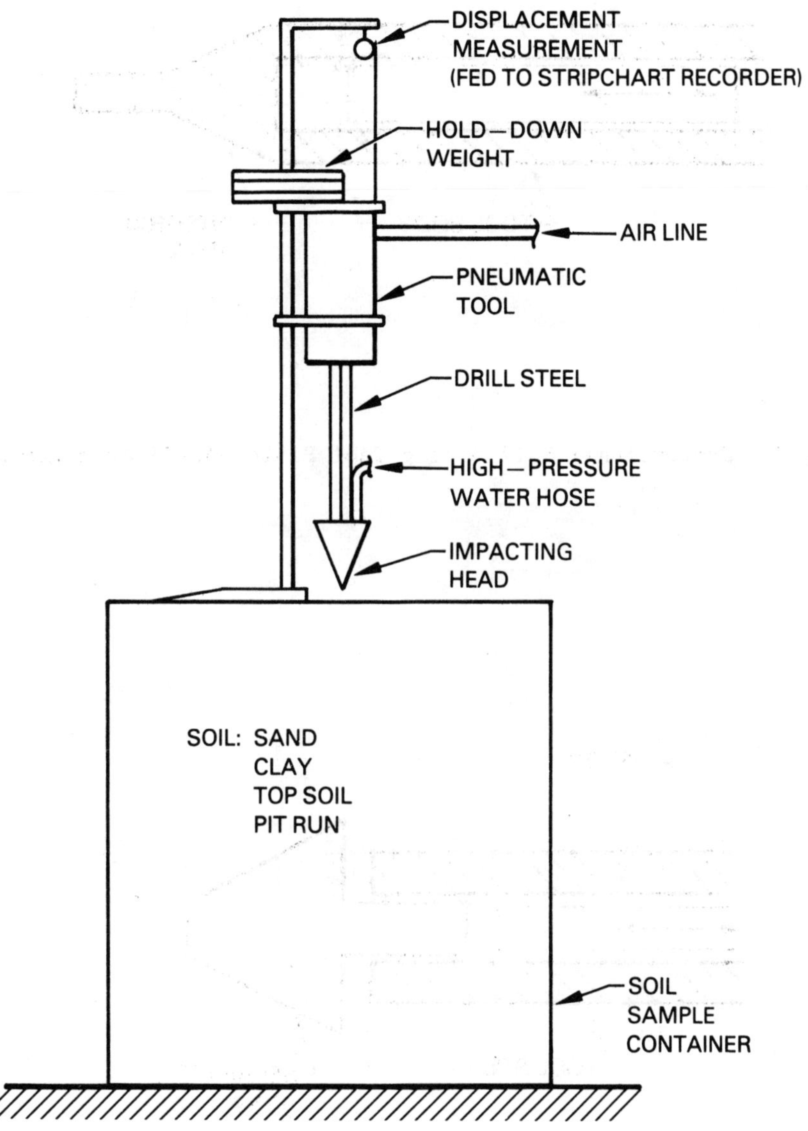

Fig. 2. Experimental Schematic

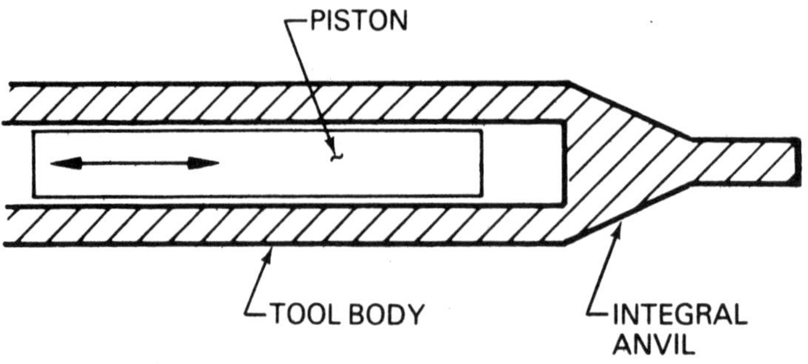

Fig. 3. Conventional Soil Boring Tool Piston-Anvil Arrangement

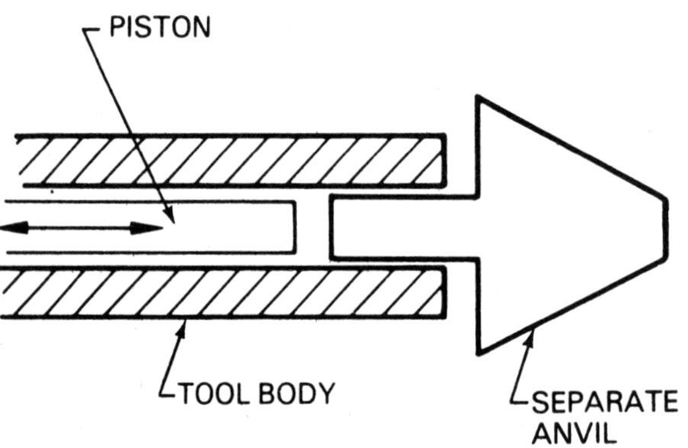

Fig. 4. Steerable Soil Boring Tool Piston-Anvil Arrangement

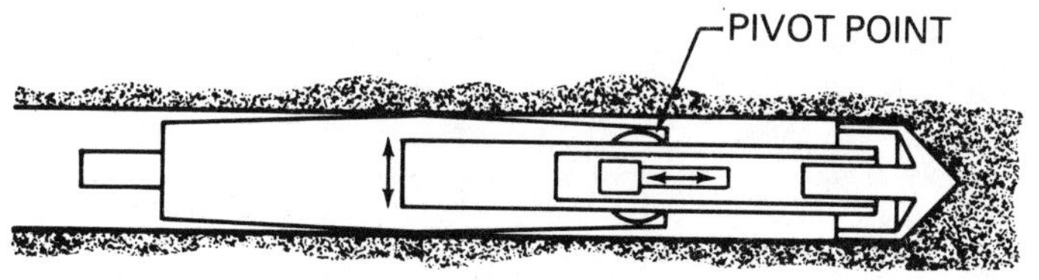

(a) Straight-line advance.

(b) Maximum correction.

Fig. 5. Steerable Soil Boring Tool Steering Section

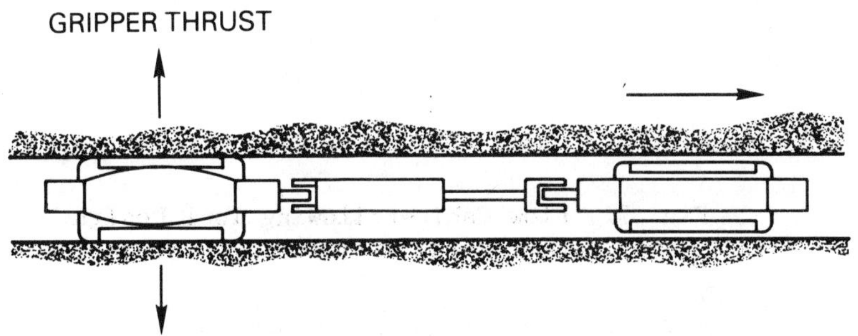

(a) Rear gripper activated (cylinder advancing stroke).

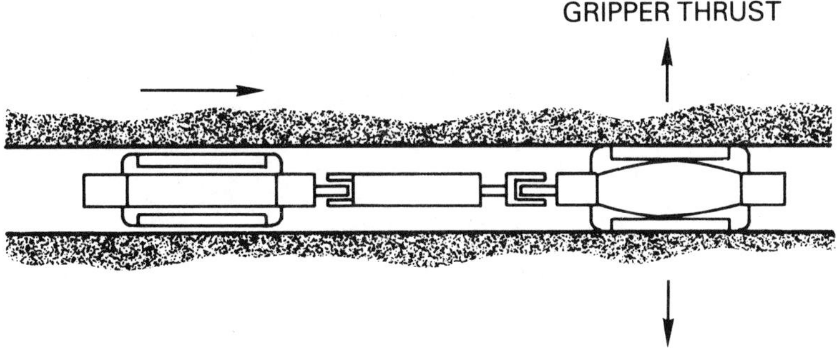

(b) Front gripper activated (cylinder retraction stroke).

Fig. 6. Steerable Boring Tool In-Hole Advancing Section

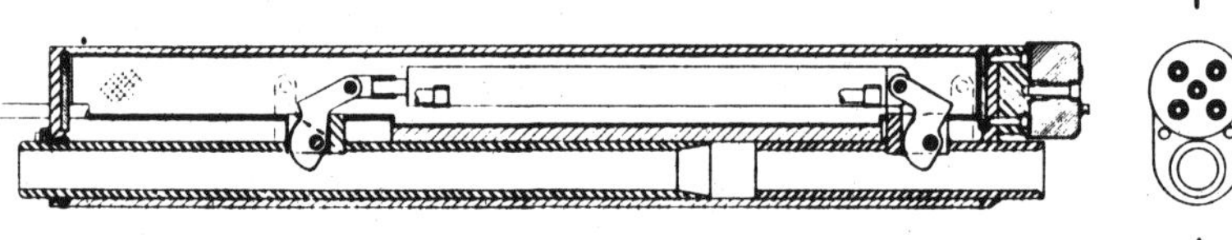

Fig. 7. Flow Cable-Following Tool Design

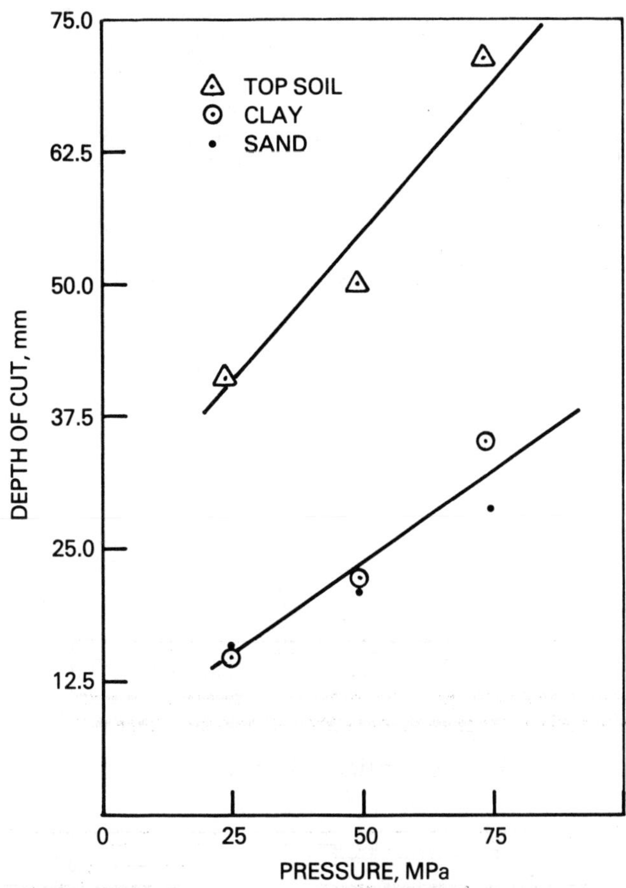

Fig. 8. Depth of Cut vs Pressure for Various Soils

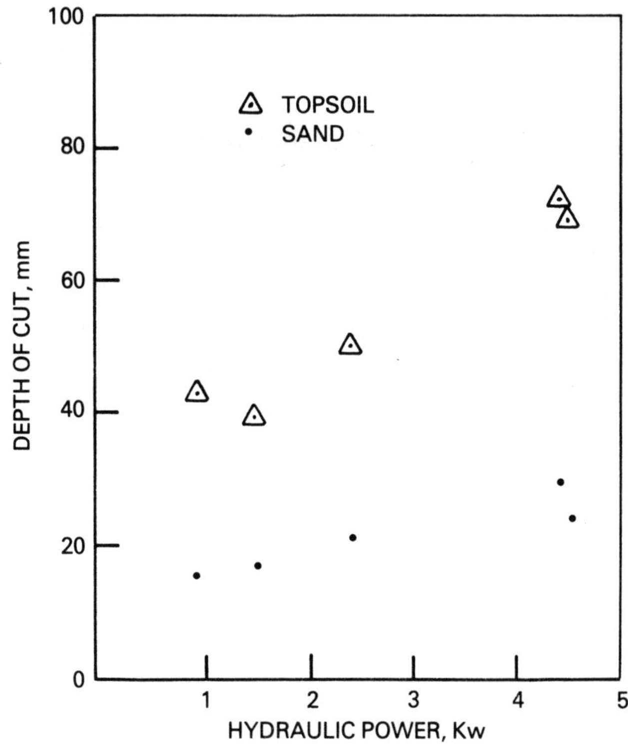

Fig. 9. Depth of Cut vs Power for Various Soils

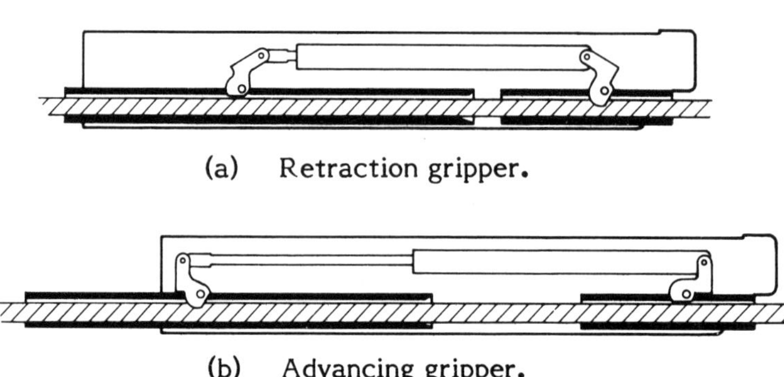

(a) Retraction gripper.

(b) Advancing gripper.

Fig. 10. Cable-Following Tool Advancing Cycle

6th International Symposium on
Jet Cutting Technology
6-8, April, 1982

COMPARATIVE USE OF INTERMEDIATE PRESSURE WATER JETS FOR SLOTTING AND REMOVING CONCRETE

D.A. Summers

University of Missouri-Rolla, U.S.A.

R.J. Raether

Dakota Grante Company, U.S.A.

Summary

The U.S. Army Corps of Engineers have engaged two competing organizations to demonstrate the effectiveness of intermediate pressure (< 130 MPa) water jets for cutting concrete.

This paper describes the trial that was carried out, and the equipment used therein.

Potential improvements in performance as a function of the parameters of test are incorporated in the discussion of the results.

Held at the University of Surrey, U.K.
Symposium organised and sponsored by
BHRA Fluid Engineering

1. INTRODUCTION

The application of water jets for cutting many rock types has, over the last 10 years, moved from laboratory testing through prototype development to commercial sale. However, during all this time the application of high pressure water jets for cutting concrete, although the subject of many research investigations, has not met with a similar degree of success for immediate field application. While it is not the purpose of this paper to review in depth the results achieved elsewhere, many of which have in fact been reported through these BHRA Symposia, nevertheless it is perhaps pertinent to discuss the work which has gone before in some detail, since, in reviewing this material, a rationale for the work, carried out by the University of Missouri, can better be understood.

2. BACKGROUND

It is now recognized that high pressure water jets have particular advantage over conventional mechanical devices for material removal in the lack of a permanent wearing tool surface and in the very high energy levels which can be transmitted into a very small area, thereby allowing very high cutting rates to be achievable. This horsepower is transmitted directly from the pump, through the cutting nozzle, to the work surface, eliminating many of the mechanical problems that currently exist in high horsepower transmission. This power transmission is also achieved without many of the noise and other environmental problems which arise with mechanical systems.

Many of the studies that have been undertaken on the excavation of concrete have been carried out at relatively high jet pressures. This, in part, may have been due to the earlier (and since shown erroneous) belief, that high pressure water jets must of necessity, operate at a pressure of 10 to 30 times the material compressive strength in order to be effective (Ref. 1). However, this pressure regime may also have been adopted because of the conceived need to cut through the aggregate contained within the concrete.

Concrete, as it is generally understood, consists of relatively hard aggregate particles, ranging typically in size from .5 cm to 8 cm in length, and chosen from rock ranging from river washed gravel to crushed sandstone or limestone.

In seeking to create a general "cutting machine" it has been presumed necessary to design a machine capable of cutting all aggregates which can be found within the concrete. To this end, therefore, systems have been developed operating in excess of 400 MPa and even at these pressures, it has been shown by Olsen (Ref. 2) that it is not always possible to cut through aggregate. Other problems also arise, where such high pressure equipment is used. At generally available horsepowers, use of such high pressure requires that a very small diameter nozzle be used. The material which is cut through at this high pressure is directed within a very thin slot, which characteristic of the system is of particular advantage when cutting softer materials such as cloth and cardboard. In the large volume removal of geotechnical material, however, this is a disadvantage, since adjacent cuts can be made in such material as close as 2 mm apart, without removing the intervening rib. Further, since the slot width is of the order of twice the nozzle diameter then, for complete material removal, a large number of adjacent passes must be made over the concrete surface in order to effect complete material removal. These disadvantages, in themselves, do not completely argue against the use of the very high pressure technology and, in fact, Flow Industries has successfully demonstrated in a quasi-commercial trial (Ref. 3) that water jets could compete advantageously over existing jack hammers in removing weather damaged concrete from road bridge decks. The high cost of the equipment and the problems in maintaining and operating such a device, together with our feeling that this pressure is not necessary, led us to disregard this approach.

In similar manner in seeking to evaluate the best system to cut the concrete, we examined the systems developed around the pulsed water cannon device. The major developer in the United States of this technology has been Dr. Cooley of Terraspace who built a device initially based on the Voitsekhovskii design (Ref. 4). This unit has been observed firing pulses of water at up to 4,000 MPa. This work, more recently, has been directed at concrete cutting at lower pressures, by Dr. Burns at the University of Waterloo (Ref. 5). While this technology has considerable potential for blasting off solid lumps to a free surface, certain limitations appear to exist, currently, on the viability of the device. At the present time it would appear that sequential water cannon blasts must be placed within approximately 10 cm or less of the free face of a concrete block in order for the fracture to break out to that free face. Secondly, it does not, as yet, appear possible to break concrete samples greater than some 22.5 cm in thickness with this technique (ibid).

In the particular program which we were asked to look at with this contract, the concrete would be at least 30 cm thick and would be solid (or partially weakened). One could not anticipate that free surfaces would be available to which the concrete could break or from which to create a slot with the device. Further it was recognized that the water cannon technology still exists as more of a laboratory device than a commercially viable technology and these factors combined, led us to conclude that water cannons would not at the present time, be a practical method of concrete removal for the particular purpose designated.

3. PRESSURE SYSTEM SELECTED

The two systems discussed to date have considered the use of very high pressure water jets, in order to cut through any aggregate which exists in the concrete. This is despite the fact that, even at 400 MPa, it has been shown that it is not possible to completely guarantee that this purpose can be achieved. However these approaches are predicated on a false assumption, namely that it is necessary to cut through the aggregate, in order to remove it. The aggregate is located in a relatively soft matrix. Conventional concretes comprise aggregate set in a matrix of compressive strength in the range of 30 to 40 MPa (this use of compressive strength is for identification purposes only, it has long been the point of view of the principal author that compressive strength is a meaningless parameter in relation to water jet cuttability).

It has been shown, in work reported by Dr. Carolyn Preece (Ref. 6) that the addition of carbide particles, while improving the resistance of a metal to conventional wear, has very little effect on the water jet erosion susceptabilities of the material. The major reason for this is that the water jet, instead of cutting through the carbide, eats out the soft phase of the material leaving the carbide particles intact. Unsupported, however, they fall out of the material without particular damage. This model seemed a relatively useful one to apply to the water jet cutting of concrete.

Most specifically, if we could wash out the aggregate by removal of the surrounding paste, one should be able to achieve a much more rapid concrete removal, at much lower pressures, than is conventionally the case. The experiment that is described in this report realistically only verifies that the concept is a viable one because, with the limited amount of time and material available, we were unable to optimize the process. However, we can, based on this initial test, make several recommendations.

4. INITIAL TEST PROGRAM

The University of Missouri-Rolla was invited by the U.S. Army Corps of Engineers to demonstrate the use of high pressure water jets for slotting concrete in a demonstration, together with the Colorado School of Mines, at the Corps facility in Vicksburg in July of 1981. The test was carried out on a concrete test pad, which had previously been poured, and which contained river gravel of a chert-like consistency in a cement whose strength was reported as being on the order of 55 MPa.

Prior to taking test equipment to Vicksburg, the University requested and received a small block of such concrete for evaluation. The concrete was placed in front of a rotating nozzle assembly which was traversed across the concrete block, at a jet pressure of 70 MPa. On the initial traverse, approximately 1 cm of material was removed from the surface of the material, all of it being cement paste from around the aggregate (Fig. 1). The water jet, however, was inclined perpendicular to the surface being cut so that the water jet was unable to penetrate underneath the pebbles thus exposed. Successive passes of the water jet nozzle over the surface, therefore had relatively little effect on undercutting the particles and penetration progress rapidly diminished after the initial pass, although a final slot depth of the order of some 10 cm was achieved. These tests were carried out at approximately 70 MPa pressure and some 80 l/min flow rate.

The basic viability of the process was demonstrated in this test, insofar as aggregate was removed from the surface of the concrete without being cut.

At this stage a pre-commercial scarrifying device for cutting concrete pavement was rented from North American Product Development Corporation of Milbank, South Dakota (Fig. 2). This device consisted of a traversing water jet nozzle system fitted on the back of a small cart, and so arranged that as the nozzle traversed across the back of the cart, perpendicular to the direction of the motion, it was interdependently rotated and could be raised and lowered, again interdependently by use of hydraulic jacks (Fig. 3). The device was driven by a single hydraulic pump which fed separate hydraulic motors for advance of the unit, rotation of the nozzle, traverse of the nozzle, and also which raised and lowered the nozzle traverse bed by means of two independent hydraulic jacks, one in each end of the equipment.

This unit was connected to the University high pressure pump, which, in order to improve performance, was changed so that the water jets would be produced at a pressure of some 100 MPa by reducing the flow rate and changing the plungers in the pump. Unfortunately we were not aware at that time that a weakened inlet spring in the pump body would not allow the pump to operate at pressures in excess of 80 MPa. The equipment was accordingly set up with an angle of diversion between the jets of some 45 deg (22.5 deg angle from the perpendicular). It was anticipated that this would be sufficient to direct the jets underneath the existing aggregate, when a traverse was made. The equipment was taken down to Vicksburg and assembled at the test site the day before a public demonstration was to be made. The requirement of the contract was that the system be capable of cutting a slot some 5 m long through the 30 cm concrete pad and the previous day it was quickly demonstrated that the equipment could very easily cut the 5 m length (this was demonstrated simply by supplying high pressure fluid to the nozzle, which was rotated at a relatively high speed as the vehicle advanced at a rate of approximately 0.7 m/min. The requirement that the system cut through the entire concrete pad was similarly demonstrated over an approximately 45 cm length prior to the test (Fig. 4).

In this preliminary test, one problem was discerned with the nozzle assembly being used. This was that the slot was only approximately 5 cm wide while the aggregate being removed was on the order of 6.5 to 7 cm maximum dimension. This meant, among other things, that large particles were sticking out of the side being cut and getting in the way of the nozzle as it was lowered. To overcome this, sequential passes were made with two adjacent slots which overlapped. In this way it was possible to gradually deepen the cut, to the point that the entire 30 cm cut through the concrete could be made. From that point it was then possible to continuously remove material down to the 30 cm level as the cart moved forward.

In the actual demonstration itself the nozzle assembly was set up across a corner of the block of concrete and a cut made down through approximately 15 cm of the concrete surface to demonstrate that it was possible to operate the cutting jet to cut through the concrete in the mode proposed (Fig. 5). Further by advancing the cart, between passes, it was possible to remove the surface layer of concrete quite effectively.

5. CONCLUSIONS AND RECOMMENDATIONS

The work to date has shown that it is possible to use relatively low pressure water to cut through concrete potentially at relatively high cutting levels using water jet pressures of the order of 70 MPa. It is not however been possible to achieve high cutting rates as yet, but only to demonstrate that the basic concept of removing the aggregate by cutting around it, is practical. In order to better achieve the cutting rates which are required for commercial use, it is recommended that the cutting angle between the two balanced and inclined jets which rotate be increased to 60 deg or greater and that the feed rate of the device as it moves down through the material be held as a relatively consistent rate of approximately two times to three times of the nozzle diameter per pass over the rock surface. It is further recommended that the system be set up so that the traverse rate and the rotational rate of the nozzle be controlled together so that the jet makes a fixed number of passes over the concrete surface/cm of advance.

This particular point is stressed since in a number of different instances it has been the primary authors observation that where the rotational speed and feed rate of the equipment are separately controlled by different motors, both of which are fed by the same hydraulic pump, that as one changes one parameter (for example increasing the advance rate) then one indirectly effects the other (slowing down the rotational speed). While this may not have any particular damaging effect with a conventional mechanical tool, what happens where this occurs with a water jet system is that as the advance rate is increased, if the gearing of the nozzle is not correct, then the jet will leave ribs of material between adjacent passes over the surface. The result is then that the nozzle cannot advance because the material still impedes its passage. It is therefore critically important in the design of this equipment that the nozzle be rotated such that it sweeps out in adjacent rotations the entire face area of the slot being cut.

It is our concluding consideration based on this work that a piece of equipment can very easily be constructed, based on existing high pressure water jet technology, nozzles, couplings, pumps, motors, already in existence, and of proven capability. With very little further effort, these can be adapted to improving the quality of the life of us all by removing the drumming beat of the jackhammer from our existence.

In conclusion, it is perhaps pertinent to quote from the very first paper on water jet cutting of concrete read before this body at the First Water Jet Symposium some ten years ago in Coventry. At that time it was stated that water jets would never be adopted by the construction industry and that progress in this area should be halted because it would cost approximately $200,000 to develop such a piece of equipment, which was way beyond the budget of all construction companies (Ref. 7). It is understood that the Flow Industries equipment offered at higher pressure (400 MPa) now costs approximately $200,000 (Ref. 3). It is the opinion of the authors that the equipment, herein described, can also be assembled for a price on the same order of magnitude, or lower. The fact that in the water jet industry ten years of progress has led to no inflation in price should perhaps be a matter of some congratulations.

6. ACKNOWLEDGEMENTS

This work was carried out with the assistance of Mr. J. Blaine, Mr. R. Robison, and Mr. J. Tyler of the Rock Mechanics and Explosives Research Center staff and made use of equipment originally developed by Dr. Clark R. Barker in the early evaluation of viability of the concept. We are indeed grateful for this assistance, without which this work could not have been carried out.

We also gratefully acknowledge the grant from the U.S. Army Corps of Engineers under the personal monitoring of Dr. Carl Pace and the considerable interest which has been shown by the U.S. Army Corps of Engineers in this program which has given us sufficient encouragement to bring us to you today.

Figure 6 is provided with compliments of the Rolla Daily News (Suzi Alexander, photographer) in which it appeared on October 4, 1981.

7. POSTSCRIPT

For the last thirteen years, while research into water jet cutting has been carried out at the University of Missouri-Rolla, as far as the University has been concerned it has been an academic experience only.

In September of 1981 the high pressure jet laboratory was however to be moved, necessitating the removal of several 15 cm thick concrete walls, and the cutting of a door through a 35 cm thick re-inforced concrete wall. The contract price was based on jackhammering all concrete.

By later agreement we removed the walls (Fig. 6) and cut the doorway using high pressure jetting - making single cuts at the rate of 0.2 m/min, and cutting to a depth of 10 cm/pass with a single, rotating, 0.063 mm diameter nozzle, operated at a pressure of around 85 MPa. While this rate was not optimized, it has certainly drawn the attention of the University administration. I might also add that it brought us a "free" concrete loading dock and working surface as an equitable exchange for our effort.

8. REFERENCES

1. Singh, M.M. and Huck, P.J., "Correlation of Rock Properties to Damage Effected by Water Jet," Proc. 12th Symp. on Rock Mech., Univ. of MO-Rolla, Rolla, Missouri, pp 681-695, (Nov. 16-18, 1970).

2. Olsen, J.H., "Jet Slotting of Concrete," Proc. 2nd Int. Symp. Jet Cutting Tech., BHRA, Cambridge, UK, April 1974, paper G1.

3. Reichman, J.M., Kirby, M.J., and Rodenbaugh, T.J., "The Development of a Water Jet Cutting System for Trenching in Concrete," Proc. 5th Int. Symp. Jet Cutting Tech., BHRA, Hanover, Federal Republic of Germany, June 1980, Paper D1.

4. Cooley, W.C., "Rock Breakage by Pulsed High Pressure Water Jets," Proc. 1st Int. Symp. Jet Cutting Tech., BHRA, Coventry, UK, April 1972, Paper B7.

5. Burns, D.J., et al, "Relationship Between Jet Penetration in Concrete and Design Parameters of a Pulse Water Jet Machine," Proc. 5th Int. Symp. Jet Cutting Tech., BHRA, Hanover, Federal Republic of Germany, June 1980, Paper D2.

6. Preece, C., Seminar on Cavitation Erosion, New York, 1979.

7. McCurrich, H. and Browne, R.D., "Application of Water Jet Cutting Technology to Cement Grouts and Concrete," Proc. 1st Int. Symp. Jet Cutting Tech., BHRA, Coventry, UK, April 1972, Paper G7.

Figure 1. Concrete cut by the water jet, showing the amount of aggregate exposure.

Figure 2. Rig developed to demonstrate the concept of water jet cutting and scarifying.

Figure 3. Detail of the nozzle drive and movement system.

Figure 4. Slot cut through the pavement.

Figure 5. Scarified pavement surface and 15 cm slot cut during the demonstration.

Figure 6. Ron Robison, a senior research specialist at UMR, operates a slotter, a tool that shoots pressurized water at high speed to cut through rock, while cutting away part of a concrete wall at the Rock Mechanics Building. Robison and Jim Blaine, a senior lab mechanic at UMR, developed the slotter and modified it for use in the current project (caption from the Rolla Daily News page 1, Vol. 1, No. 245, Sunday, October 4, 1981).

6th International Symposium on
Jet Cutting Technology
6-8, April, 1982

SOIL IMPROVEMENT METHOD UTILIZING A HIGH SPEED WATER AND AIR JET ON THE DEVELOPMENT AND APPLICATION OF COLUMNAR SOLIDIFIED CONSTRUCTION METHOD (COLUMN JET METHOD)

T. Yahiro, H. Yoshida and K. Nishi

Kajima Institute of Construction Technology, Japan

Summary

The significance of foundation works has been increasing in recent years as construction projects have become larger and more complicated. Especially ground water control and soft ground improvement govern the success of a project, playing a vital role in the safety and durability of the structure.

Therefore, we have been endeavouring in the development of an underground construction method using a high speed water jet. The endeavour has been rewarded by the development of the column jet pile method (construction method of columnar solidified body) aimed at soil improvement.

This method is intended to form a consolidated column by injecting grout to fill the voids in the ground formed by high velocity air surround water jet sent simultaneously from above the ground. Therefore the method may be regarded as a kind of grouting method. However, since the voids are formed before grout is injected, the resulting consolidated mass can be located at an accurate position in the ground. This method also excels in environmental protection to prevent the ground from land subsidence and under ground water from being dried up or contaminated.

This report introduces a programme of research work starting from the development and ending in actual application of this method.

Held at the University of Surrey, U.K.
Symposium organised and sponsored by
BHRA Fluid Engineering

1. PREFACE

As part of developing an underground construction method using a high-speed water jet, an induced injection method primarily aimed at soil improvement has been developed.

In this method, ground is broken with a high speed water-air jet to form a cavity and a material such as cement milk is injected therein.

This method enables construction of columnar solidified body with accuracy at a desired position.

2. CONSTRUCTION METHOD OF COLUMNAR SOLIDIFIED BODY (Column Jet Pile Method)

2.1 Purpose of Development

Roughly classified, the mechanism of ground improvement methods generally used at present may be grouped as follows: 1 Reduction of buoyancy by dewatering 2 Vibration and 3 Consolidation.

Sand drain method, chemico-pile method etc., using these principles are used to their respective advantages.

However, where the depth at which underground construction is carried out is large or water contents are large as in humus or where the ground supports an existing structure which is to be improved, these methods entail various problems, often resulting in little improvement.

Therefore, as one of the methods of improving ground by which the above-mentioned problems may be solved, a method of constructing a columnar solidified body by application of a wall-shaped solidified construction for building a cutoff wall has been developed.

In this method, high speed water and air jets are simultaneously shot from a monitor rotated and lifter in the ground to form a cavity. At the same time injection material is sent into the cavity to construct a columnar solidified body.

Study on this concept has resulted in: 1 this method can be effectively used for improvement of soft ground 2 the method can be used to improve clay, humus and sandy ground, without much affecting the property of the original ground by that of improved ground 3 Solidified body can be constructed as deep as 40m below the surface.

2.2 Application of the Method and Equipment

(1) The typical procedure is as follows. (Fig. 1)

1) Drilling of guide holes
By use of a boring machine or earth auger, guide holes of not less than 15 cm in diameter are drilled.

2) Insertion of a monitor
A monitor is inserted into the guide hole and lowered to the buttom of the hole.

3) Breaking of ground with high-speed water and air jets and injection of grouting material.
A high pressure pump and an air compressor are started to make highspeed concentric water-air jet by which the ground is broken. At the same time, a grout pump is operated to inject cement milk into the cavity formed by broken ground, rotating and lifting the monitor from the bottom of the hole.

(2) Table 1 shows the list of equipment requires.

Depending on the nature of the ground at the site, size of the project and environmental conditions, the capacity of the equipment required varies. The principal items of the equipment, however, are as follows.

1) High-pressure pump
   Outlet pressure 300 - 500 (kg/cm^2)
   Discharge 50 - 70 (/mm)
2) Air compressor
   Outlet pressure 7 (kg/cm^2)
   Discharge 4 (m^3/min)
3) Monitor, triple pipe (See Fig. 2)
   Water jet nozzle 1.6 mm dia. 1 ea
   Ring air nozzle, slit width 1 mm.
4) Ground pump
   Outlet pressure 35 - 60 (kg/cm^2)
   Discharge 125 - 200 (/min)
5) Rotating and lifting apparatus for the monitor
   Boring machine, 16 ton truck crane.

2.3 Specific Characteristics and Various Application

The following is the specific characteristics of this method.

(1) By use of small diameter guide holes as much as 15 cm across, large diameter columnar solidified bodies 2 to 3 m across can be constructed.

(2) By use of appropriate injection materials and method columnar solidified bodies of required shape and strength can be constructed.

(3) This underground construction method is free from noise and vibration.

(4) Since the area of injection is confined by the area of breakage of soil caused by high speed air-water jet, there is no fear of overflow of grouting material. Moreover, because cement milk is used as grouting materials, no pollution such as contamination of ground water by injection fluid is expected.

(5) Since an air lift mechanism is used for removal of slime, the larger the depth at which a solidified body is constructed, the more advantageous is the method.

Judging from the result of past performance, underground construction as deep as 40 m from the surface is possible by this method.

Further, construction of solidified body at a part of the guide hole (for example at the bottom or at the midpoint of the guide hole) is possible.

(6) This method is applicable to soft ground consisting of clay or humus of high water contents and even to soft rock with compressive strength as high as 40 kg/cm^2.

Thanks to the above-mentioned specific characteristics, this method can be used to construct a row of colunm composed of columnar solidified bodies for cutting off ground water or for preventing earch from being washed away in addition to increasing the strength of ground by conventional ground improvement method.

Examples of application are shown in Fig.3.

(a) Prevention of heaving and boiling of ground caused by excavation work.
(b) Increase of anchoring capacity.
(c) Construction of a continuous under ground wall by building solidified bodies where continuity of steel sheet piles and continuous underground was is disrupted.
(d) By constructing an independent columnar solidified body at appropriate spacing, bearing capacity of the ground as a compound foundation can be increased.
(e) By constructing the solidified bodies below the existing underground structure, strength of supporting ground can be increased, preventing uneven settlement of ground.
(f) For prevention of slope failure, shear strength of sliding surface can be increased.

3. IN-SITU EXPERIMENT FOR DEVELOPMENT OF THE METHOD

From the above-mentioned fundamental experiments, we believe in the realization of this method. But, when it is carried out in-situ applying the execution equipment, there would be some cases where expected results can not be obtained due to unforeseeable factors.

So, from the execution experiment applied to in-situ ground, we have tried to collect data on the design and execution for practical use.

3.1 Purpose of the Experiment, Test Ground and Method

We decided to make the following matters clear:1 understanding the factors which influence the diameter and strength of solidified mass: 2 characteristics of solidified mass of soil difference:3 understanding the factors which influence the cost. Experimental ground was made in mainly soft soil of alluvium, sandy and clayey layer Fig. 4 shows the Geological log of test site. In Koto-ku, Tokyo, sandy layer contains mainly silt and fine sand which are under 10 in N-value and partly clay layer whose thickness is from 10 cm to 20 cm. The clayey layer in Shiki-city, in Saitama Prefecture is the clay and humus, whose N-value is zero. In these ground, constructed depth of solidified mass is from G. L. - 2.0 m to G. L. - 5.0 m in sandy ground, and from G. L. - 1.0 m to G. L. - 4.0 m in clayey ground. Experimental equipment and procedure of experiment were the same, which are shown in Fig.1. and Table.1. But in the experiment, as the Table - 2 shows, we changed some factors such as the outlet pressure of high speed water jet and grouting pressure of grout and made 9 pieces of columnar solidified mass in sandy ground and 13 pieces in clayey ground. Measuring items were as follows;

During experiment 1 Variation of experimental factor 2 Specific gravity of slime. After experiment 1 physical and mechanical test of core 2 Elastic wave velocity and density log performed in situ. After these measurements were completed, the shape of excavated solidified mass was examined.

3.2 Outline of the Results of the Experiment

Summary of the result of the experiment is as follows:

(1) Shape of solidified mass is almost columnar and photo-1 and 2 show them constructed in sandy ground and clayey ground. From them, maximum diameter is 3.5 m in sandy soil and 2.5 m in clayey soil.

(2) Judging from the relationship between outlet pressure, flow of high speed water jet and shape of accomplished solidified mass, the larger the outlet pressure and flow changes, the larger the diameter of solidified mass becomes. Its increasing rate is 1.6 in sandy ground and 0.5 in clayey ground in the case of outlet pressure of 200 kg/cm^{2}and 400 kg/cm^2.

(3) Judging from the relationship between outlet pressure of grout and the diameter of accomplished solidified mass, there is little difference in sandy ground. Contrary to that, in clayey ground, the higher the outlet pressure rises, the larger the diameter becomes, and when outlet pressure changes from 0 to 30 kg/cm^2, the diameter increases by 40 to 50 %.

(4) Judging from the relationship between flow of grout and the diameter and strength of accomplished solidified mass, in the case of sandy ground there is no effect on the diameter and strength increases as flow of grout increases. In clayey soil, if the grout is less than the volume of solidified mass, it solidifies imperfectly. When the flow of grout changes from 200 /min, to 300 /min, the diameter increases by 20 % and strength increases by 10 %.

(5) Judging from the relation between rotation speed of hydraulic monitor and shape of solidified mass, the less, the rotational frequency is, the

diameter becomes larger, and in the sandy soil, when rotation speed changes from 5 to 10 r.p.m., the diameter of column increases by 15 %.

(6) The lifting speed of hydraulic monitor influences very much the shape of solidified mass. In sandy soil, when the lifting speed changes from 5 to 10 cm/min., the diameter decreases by 10 - 30 %. In the case of clayey ground, the trend is found as shown in Fig. 5.

(7) In the case of construction by the stepped system in which the hydraulic minitor rotates at 5 r.p.m. for one minute each time it is raised by 5 cm at the same depth, the diameter increases to 3.2 m even in the clayey soil.

(8) The compression strength of solidified mass is from 30 kg/cm^2 to 50 kg/cm^2 in sandy soil, and from 13 kg/cm^2 to 50 kg/cm^2 in clayey soil. The relationship between elastic wave velocity (S-wave) and compression strength is as shown in Fig. 6.

4. PRACTICAL APPLICATION EXAMPLE

4.1 Application of Improvement of Foundation Ground, in the Case where a New Building was Constructed after the Break Up of the Structure above the Ground

(1) Purpose of Application

The building which was built up on 1935, with 2 stories, below ground, and 8 above, was not bearable on its faculties and social demands, so it was forced to be rebuilt. In the new building (with 2 stories under the ground, 13 above, and 3 tower house), we had the following problems on the underground construction.

1) As the site was almost filled with the building it was difficult to transfer the buried pipe around the site by dismantling the breast wall.
2) The 1600 pine piles which had supported the building became the obstacles to the underground construction.
3) Concerning cost and execution time, it was difficult to adapt the caisson method and benoto method.

As a means of solving these problems, retaining wall, foundation slab and pine piles were retained and this method was applied to the ground improvement from foundation slab to bearing layer, according to the designer.

(2) Construction

The ground is made up by the following.

The top - G.L - 7.5 m ..... Sandy layer N value is 10 - 20
- G.L - 12 m ..... Silt layer N value is 2 - 10
- G.L - 27 m ..... Clay layer N value is 2 - 5
- G.L - 28 m ..... Sand layer
- below ..... Sand and gravel layer N value is 50

Column solidified pile which aims to improve the ground, has improvement diameter 2 m, total improvement area 1,089 m^2, total construction piles 285, total improvement volume 18,397 m^3, as the Fig. 7 of excution plan and Fig. 8 of execution section show. After the dismember of the building above the ground, the construction was carried out with the Bl floor as a execution floor to the following execution procedure.

a. Bore concrete floor of B 1 and B 2 to cobble stone layer with ϕ200 mm core tube.
b. Bore the ground in the execution zone to 1 m under Temma sand and gravel layer of bearing layer with ϕ150 mm wing bit.
c. Set the base machine (R8H320) which is an equipped 24 m flying leader with rotation and lifting equipment. And then, hydraulic monitor is inserted to the bottom of guide hole.

d. Jet the high speed water and air and then cement suspension too. After this work, rotate and lift up the hydraulic monitor.

e. A day later, grout cement suspension to the space between breezed column solidified pile and underground base floor.

The specifications of this execution is as follows.

- High speed water jet;
  - pressure - 400 kg/cm^2,
  - discharge - 50 /min
- Compressed air;
  - pressure - 7 kg/cm^2,
  - discharge - 1 m^3/min.
- Cement suspension;
  - water-cement ratio - 100 %,
  - pressure - 15 - 30 kg/cm^2,
  - discharge - 150 - 200 /min.
- Hydraulic monitor;
  - rotary speed - 5 r.p.m.
  - lifting speed - 5 - 10 cm/min.
- Secondary grout;
  - cement 983 kg, aluminum powder 83 kg and water 688 kg for mixing cement suspension 1,000 .

Table 1 shows the used machine and Fig. 9 shows the execution system.

(3) Execution Result and its Examination

1) The actual execution result

We can see the machine transference and results of each work in Table 3 and Table 4. For the execution an average was as follows. Boring for guide hole excavation was 1.5 holes/day. Average execution time for jetting and grouting was about 6 hours. Their standard execution cycle is shown in Fig. 10.

2) Secondary grouting

As the grouting material is a cement suspension, it is sure to cause a space between a top of column solidified pile and underside of the underground base floor. Breezing ratio shows value of the space divided by execution length.

The frequency curve of this breezing ratio is shown in Fig. 11 and relation between breezing ratio and grouting volume is shown in Table 5.

3) Exploration of the result of execution

For the exploration of the result of execution, we examined as the following.

- Core boring - 15 holes. After 3 weeks from the execution, all core boring was put in the position of 90 cm from center of executed pile and its center.
- Standard penetration test - 5 holes. It was done every 1 m after 4 weeks from the execution.
- Physical logging - 2 holes
- Core test

In the results of core observation of core boring, we can see no difference between outer and inner solidified mass, and their were the mixture of soil and cement.

Standard penetration test as examined for 5 piles of all the executed 285 piles. From the result, N-value of column solidified piles was over 50 with sandy layer and over 30 with clayer layer (see Fig. 12). Sample for properties of core were picked in the position of 90 cm from the center and the center, too, as to the optional 15 column solidified piles among 285 piles. These samples were examined for physical and chemical properties. The results were as follows. On the column solidified piles executed in sandy layer, apparent specific gravity was 1.4 - 2.0 g/cm^3, compressive strength was 13 - 75 kg/cm^2, cohesion was 4.5 kg/cm^2, internal friction angle was 21 - 30°, P-wave velocity was 2.3 - 3.0 km/sec. and S-wave velocity was 1.3 - 1.5 km/sec. On the column solidified piles executed in clayey layer, apparent specific gravity was 1.4 - 1.8 g/cm^3 compressive strength was 12 - 50 kg/cm^2, cohesion was 7 - 10 kg/cm^2, internal friction angle was 21 - 30°, P-wave velocity was 1.5 - 1.9 km/sec. and S-wave velocity was 0.8 - 1.0 kg/ sec. As to geophysical logging, examined about electric, density and sound velocity, of which the result is shown in Fig. 14. From this, density was 1.6 - 1.9 g/cm^2 and elastic wave velocity was 2.2 - 3.6 km/sec. Judging from characteristics of solidified body by the analysis of sonic log, there were parts with lower compressive strength, but, as a whole, we should judge that the column solidified piles were homogeneous and good.

4.2 Utilization for the Purpose of Earthquake-Proof Reinforcement of Constructed Building

Shimizu-city in Shizuoka Prefecture which has Fuji Seito Co. head office and factory, is under the very probable conditions of occurrence of a large earthquake. It is pointed out as a Tokai-earthquake countermeasure area on the large scale earthquake countermeasure special measure law.

The building was researched and examined on the foundation ground, material and structure point of view, to find out its poor earthquakeproof. So soil improvement was executed on column jet method about ground as the most effective earthquake-proof measure without stopping (operation) of the productive facilities of the factory for a long time.

This chapter describes an earthquake proof reinforcement method by the improvement of foundation ground.

(1) Purpose of Application

The ground of this site tends to be subject to liquefaction with a uniform fine sand and high ground water level. So we examined liquefaction of building site in the simple method of H.B. Seed with the maximum acceleration of the ground surface α max = 200 gal. Equivalence mean shear stress is as the following.

$$\tau d = 0.65 \frac{rh}{g} \alpha \max rd$$

In this formula, r : bulk density of soil, h : depth, g : acceleration of gravity, rd : stress damping coefficient. And shear stress amplitude τen is as follows:

$$\tau en = \left(\frac{\sigma d}{2\sigma_o'}\right)_{50} \frac{Cr\ Dr}{50\%} \sigma v'$$

In this formula, $(\sigma d/2\sigma_o')_{50}$: ratio between shear stress amplitude $\sigma d/2$ (σd : axial differential stress) which causes liquefaction by the times repetition and initial effective stress σ_o' on the repeated triaxial test to the sand of relative density 50 %, $\sigma v'$: intitial vertical effective stress, Cr: correct modulus of relative density. Results of calculation is shown in Table 6. It shows liquefaction at every point. Liquefaction sand layer has a large connection with particle size, density, under ground water level, so they should be changed to the different direction from the direction which causes liquefaction, as a liquefaction measure.

o Devisable measures are as follows:

a) Change for the soil whose particle size doesn't tend to be subject to liquefaction.

b) Increase of density by compaction.

c) Lowering of ground water level.

d) Direct support of bearing layer.

e) Grouting, which solidified ground.

But, it is difficult to transfer the manufacture machine in the building during operation, so a machine for this execution should be of a compact type. And in addition, considering schedule, cost, and execute condition, it was impossible to improve the ground of foundation.

And then we decided to form columnmar solidified in the ground directly under the pillar on the column jet method which has high strength and reliability. Because, in case of a big earthquake when liquefaction phenomena occurs, it is necessary to prevent a sudden collapse and the damage to the men and facilities should be as little as possible, so the building should be supported for as long a time as possible.

(2) Outline of the Ground

Describing from the coast direction, reclaimed ground, muddy ground and sandy ground made up the out skirts of this site in parallel with the coast line. And they cross with the soft ground which is an ancient river and it is in parallel with the Tomoe river. The sedimentary conditions of soil layers are shown in geological log (Fig. 13). Soil properties of each layer are as follows.

1) Silt Mixed with Sand and Gravel
Upper part is supposed to be buried layer or dredge soil, water content is very high, very soft ground and contains sand and gravel in-homogeneous. Lower part is supposed to be ancient marine soil, and places sand and gravel between band like or a striped pattern.

2) Sand and Gravel with Silt Layer
Totally, it is supposed to be buried layer, sand and gravel is main and it places silt between pocket like here and there, and some places are mixed with sand and gravel. During work the water content increases and it makes the layer quite loose, as N=2 penetration depth 52 cm per a blow.

3) Alternation of Silt and Fine Sand
Each layer is composed of silt mainly and it places sand layer of 5 - 15 cm thick between silt at the position of No. 1 and 2 is clayey and has much clayey, but that at No. 3 has little clayey. As for sand, they are fine sand whose particle size is regular and partly medium sand.

4) Sand
Layer at the position of No. 2 is composed of medium sand mainly, which contains pebbles whose diameter is under 10 m/m. Layer at the position of No. 3 is made up from medium sand whose particle size is regular, and has the same property with medium sand which composed of alternation with silt.

5) Silt
This silt layer contains a little shell flake and the upper part of it is composed of clayey silt which has much cohesive and supposed to be non-liquefaction layer.

(3) Execution Order

Execution order is based on standard order but there are some parts because of the execution to inner and surround of existing building.

1) Begin from the slab of existing building, core boring this concrete (slab, footing) first, and then bored soil of execution zone.

2) Insert of hydraulic monitor
Insert hydraulic monitor to guide hole and setting on the bottom of the hole, and then pull out the casing-pipe. But in the case where the construction was done inside the building, with a low ceiling, used hydraulic monitor rod of short length (=1.0m) and added one for each and to set.

(4) Design and Control

1) Design

Allowable bearing capacity for a improvement pile is found, as a sort of cast-in-situ pile by the following formula with the skin friction P = 0.

$$PA = \frac{15\ N\ Ap}{Fs} - W$$

In this formula,

N : Mean N-value of end of pile (Fig. 13)
Ap : Section area of pile
W : Weight of pile

Calculating with safety coefficient Fs=1.2 is as follows.

$$PA = \frac{15 \times 3.5 \times 3.14}{1.2} = 126 \text{ ton}$$

Divide surcharge load by allowable bearing capacity for a pile and placed as Fig. 14 shows.

2) Control

(a) Grout Control
During execution, keep a lookout for the pressure and flow of high speed water, air, and grout, to satisfy regulation.

(b) Combination Control
Perform calculated cement and water to confirm the combination W/C = 100 % of cement milk, grout and controlled on the specific gravity of cement milk. As specific gravity 1.50 in the time of W/C = 100 %, carry out the extract inspection once for every 4 - 5 batch (approximately 1 m^3) and controlled specific gravity between 1.48 and 1.52.

(c) Control of Slime
Density of discharged slime to ground surface is owing to the soil properties of object ground and cutting speed of ground. So, we observed slime under execution and measured its specific gravity at the same time. When the specific gravity was small or sand particle was not discharged, we repeated the execution in the same depth and found the certainty.
Specific gravity of slime was between 1.52 and 1.66 and it was supposed to be a good value.

(d) Observation of Breezing, Secondary Grout
As long as cement milk is used for grout, we are concerned about breezing. As breezing of 2 - 3 % was found in the project example of past execution, observed the phenomena of breezing and, the next day, hard mixed cement milk of grouted (W/C = 60 %) to the same hole and made the effort to elevate execution accuracy.

(5) Test of Execution Effect and Consideration

As the quality control of after the execution, physical and mechanical test of sampling core, standard penetration test, geophysical well logging were carried out.
The results are as follows.

1) Physical & Mechanical Test of Sampling Core
Table 7 shows totally. Compression strength c is between 37.6 kg/cm^2 and 68.2 kg/cm^2 and its average is 53.6 kg/cm^2. Static poisson's ratio ( ) is from 0.20 to 0.38 and static young's modulus (Es) is from (1.64 to 2.77) x 10^4 kg/cm^2. Tensile strength (t) is from 3.93 to 6.20 kg/cm^2 and it is about 1/10 of compression strength. As to elastic wave velocity,

P-wave (Vp) is from 1.92 to 2.44 km/sec., S-wave (Vs) is from 0.87 to 1.27 km/sec. As to share coefficient, cohesion (C) is 20.6 kg/cm^2, angle of internal friction ( ) is 28.5°, besides, measuring cement volume in columnar solidified mass. It is shown in Table 8. According to it, the cement volume of the lower part is rather little, but todally, well balanced.

2) Standard Penetration Test

Standard penetration test was carried at the point of 80 cm far from the center of columnar solidified bodies. The result is shown in Fig. 15. From the top to around G. L-5.0 m is N-30 - 50, good result, but on the other hand, N-value becomes lower as it comes closer to the lower part. The reason of this, is supposed to take much time to generate of strength and there conditions much cement, so increase of compression strength will be expected.

3) Geophysical Logging

As geophysical well logging, sonic logging, density logging and neutron logging were examined. As the result, elastic wave velocity, intensity, density, porosity were found, which are shown in Fig. 16 and Fig. 17. Elastic wave velocity (Vp) of improved ground is from 2.0 to 2.8 km/sec., density ( ) is from 1.8 g/cm^3 to 2.2 g/cm^3, porosity (n) is from 20 to 40 %.

4) Consideration

Columnar solidified body formed by column jet method satisfied each condition of compression strength, diameter, coefficient of adherence with foundation slab etc. on design spec. and accuracy of execution. And it fully attained the purpose of earthquake-proof reinforcement a countermeasure.

5. CONSIDERATION OF RESULTS OF EXECUTION

As we can see in this method, it is no longer at that of expansion and propagation. However, there are some problems in the mater of in-situ execution which should be improved.

Looking at previous execution, the import nt methods of execution are improvement of foundation ground, liquefaction prevention of sand layer, and increasing of horizontal resistance of pile etc. The objects of execution are made up by soft rock, sand, silt and clay. Consideration about some problems on design, and execution which we found are as follows:

1) In order to deal with the design which is the expansion of the applied field, collect and analyze the data on compression strength of solidified mass, diameter and properties of soil, pressure and flow of jet, etc.

2) On the execution, the way of present condition is shown in Table 9. In this case, outlet pressure of water jet is 200 - 500 kg/cm^2, in flow to /min. For future accuracy of execution and reduction of cost, it is necessary to get the wider execution data on outlet pressure and flow of water jet and to strive for the re-utilization of slime.

3) We gave a judgement for execution effect of 1 excavation of solidified mass 2 physical and mechanical test of sampling core and 3 In-situ test.

In-situ test of these methods was done by physical and mechanical test, N-value, and pressiometer test. In the case of sampling core, in addition to the physical and mechanical test, we made a thin section and to took the microscopic photograph.

Photo 3 - photo 6 show them in sand layer and humus layer.

From them, we can see anorthosite, quartz, pyroxene, and olivine in sand layer. Those mineral particle are fresh and uncontaminated. Each particle is surrounded with cement and made up to be well solidified mass. Humus was made up with cryptocrystalline mainly with clay, and contains a few quartz particle which mixed into it as coming from the outside mineral. There are some parts which is lack of solidified mass in the plate. And so the humus is made up inhomogeneity rather than sandy soil is.

4) For equipment, we now use boring machine, earth auger to excavate the guide hole, and tower of piling machine, rotary table for rotation and lifting equipment of hydraulic monitor. But for the future, we expect for wide site like seaside industrial zone and exclusive machines which can be used in any narrow working area like an old buildings. For example, Photo. 7 shows hydraulic monitor of down-the-hole driving monitor, which was developed for those expectation. This equipment used reaction force of water jetting when it drives, and as the point of driving part and rotation part are almost the same situation, it doesn't lose the energy when it transmits.

6. POSTSCRIPT

For popularization and further development of this method, several problems have to be solved by putting this method into more frequent practice.
For example,
Items related with apparatus;

1. Development of the high speed drilling machine for drilling guide holes.
2. Control equipment capable of grasping the process of constructing solidified bodies should be developed.
3. Control equipment by use of geophysical exploration should be developed.

Items related with executed method;

(1) Economical means for disposal and re-use of slime.

(2) Selection and application of the appropriate injection materials costwise and strengthwise.

(3) Most appropriate execution method for the purpose intended, from the standpoint of configuration and strength.

REFERENCE

1. T. Yahiro, H. Yoshida & K. Kishi:
On the Characteristics of High Speed Water Jet in the Liquid and its Utilization on Induction Grouting Method, KICT Report No. 16, May '74. Kajima Institute of Construction Technology.

2. T. Yahiro, H. Yoshida:
Induction Grouting Method Utilizing High Speed Water Jet, VIII International Conference on Soil Mech. and Foundation Engineer, Moscow 1972.

3. T. Yahiro, H, Yoshida & K. Nishi:
On the Characteristic of High Speed Water Jet in the Liquid and its Utilization on Induction Grouting Method, II International Symposium on Jet Cutting Technology, Cambridge 1974.

4. T. Yahiro, H. Yoshida & K. Nishi:
The Development and Application of a Japanese Grouting System, Water Power Dam Construction, February 1975.

5. J. Wakabayashi:
Grouting System Places 100-ft-deep Cut off wall, ENR, May 2, 1974.

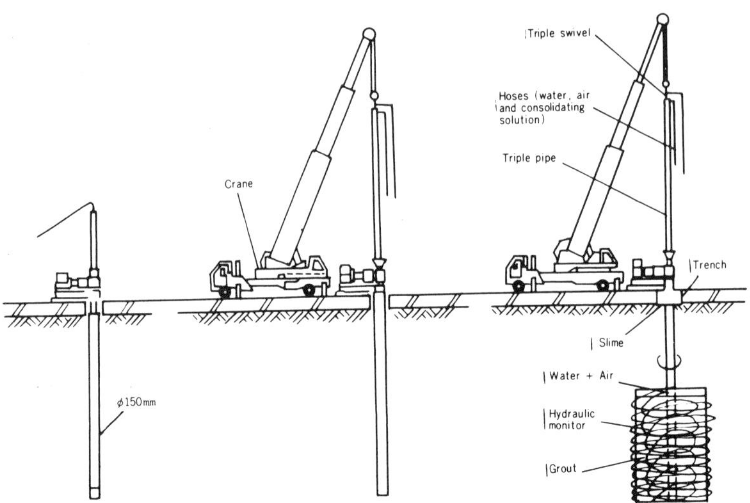

(1) Guide Hole Excavation

A guide hole of 15 - 20 cm in diameter is sunk to the depth where the structure is to be built.

(2) Monitor Setting

A monitor is placed down in the guide hole.

(3) Rotation Jetting and Lifting of Monitor

The monitor is rotated and lifted simultaneously at the specified speed as the jetted stream breaks the soil.

Fig. 1 Execution Procedure of Column Jet Pile Method

Table 1 Machine and Equipment

Application	Nomenclature of machine	Kind of machine	Maker	Capacity	Scale Width	Scale Length	Scale Height	Power	Weight
Guide hole for installation	Boring machine	Megaro - 150	Japan Longyear	50 - 500 M	1,170	2,280	1,450	11kW	1,500 kg
	Boring pump	MG - 10	Mineral Research Drilling Co., Ltd.	30 kg/cm² 105 ℓ/min	540	920	1,000	7.5	250
Jet pile	Ultra high-pressure pump	PG - 75 BB	Mineral Research Drilling Co., Ltd.	100 kg/cm² 67 ℓ/min	1,200	2,300	1,100	55.0	2,500
	Grout pump	MG - 25	Mineral Research Drilling Co., Ltd.	63 kg/cm² 200 ℓ/min 60 kg/cm² 125 ℓ/min	840	2,720	1,485	18.5	1,000
	Grout mixer	PM - 18	Mineral Research Drilling Co., Ltd.	500ℓ x 2 tanks	1,295	2,520	2,212	15.0 (7.5 x 2)	1,400
	Compressor	PDR - 370	Hokuetsu Industries, Co., Ltd.	7 kg/cm² 105 m³/min	1,690	4,200	2,000	PS 110	2,800
	Rotary equipment	Boring machine	Japan Longyear	50 - 500 m					
	Lifting equipment	Megaro- 150							
	Truck crane			16 t					
Water supply and discharge equipment	Turbine pump	TK - 100	Takasago Machine Co., Ltd.	ϕ100 Two stage 1000 ℓ/m				11 kW	
	Water pump	OS - L	Takasago Machine Co., Ltd.	ϕ 50 200 ℓ/mℓ					
	Bond pump	T - 610		ϕ150 200 ℓ/mℓ ϕ100 1000 ℓ/mℓ					
	Water bath			20 m³/Right					
Others	Cement silo		Koyo Machine	30 t 500 kg/min	3,000	4,500	12,000	2.7	6,000
	Truck Crane			16 t					

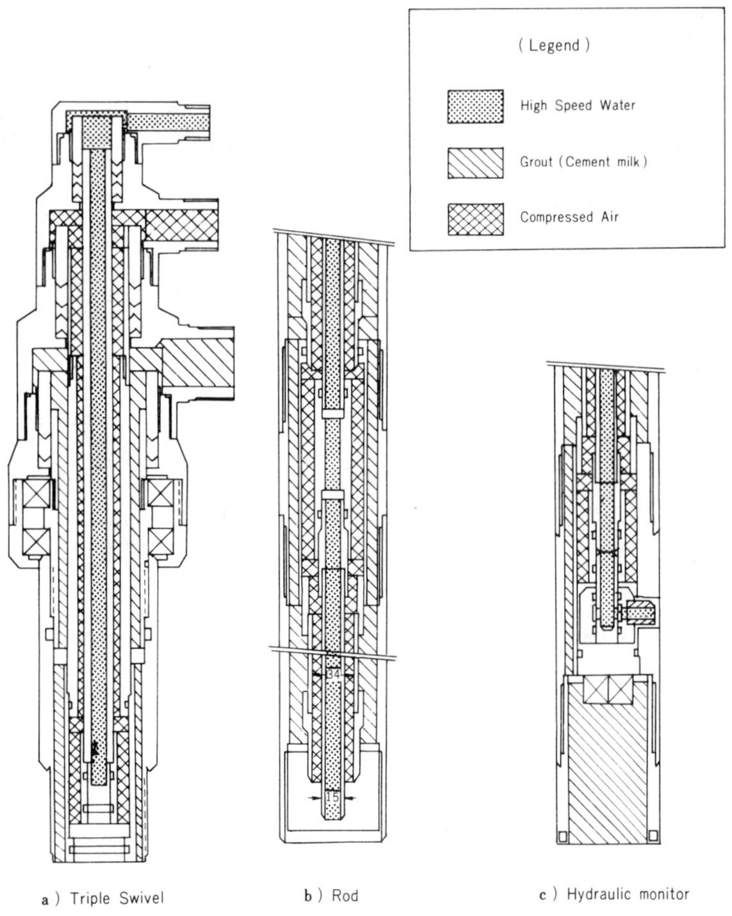

Fig. 2 Execution Apparatus for Column Jet Method

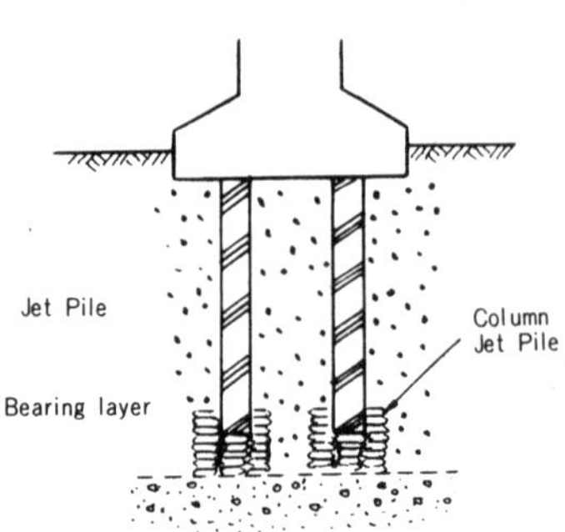

Improvement of bearing layer for piers or existing piles

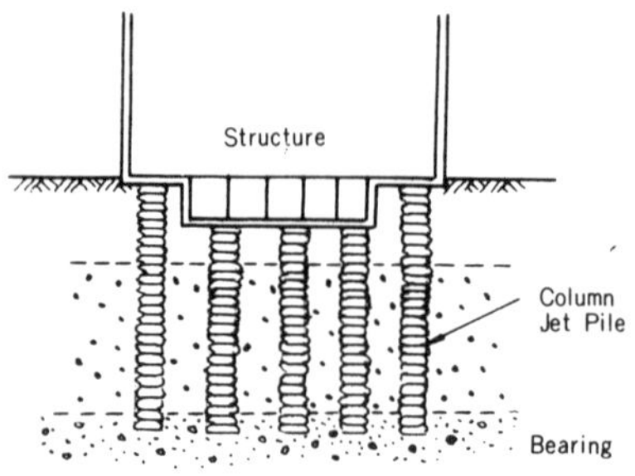

Improvement of ground under the existing structure

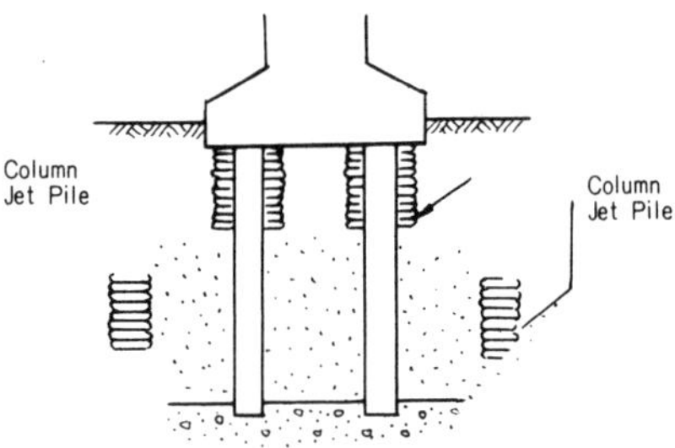

Improvement of resistance capacity of piers and existing piles against horizontal force

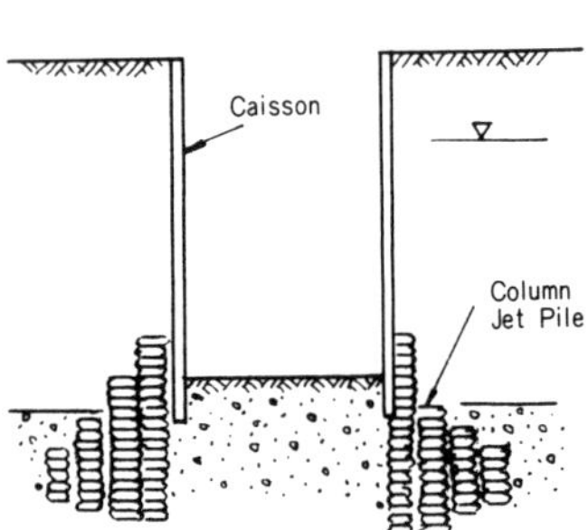

Reduction of up-lift

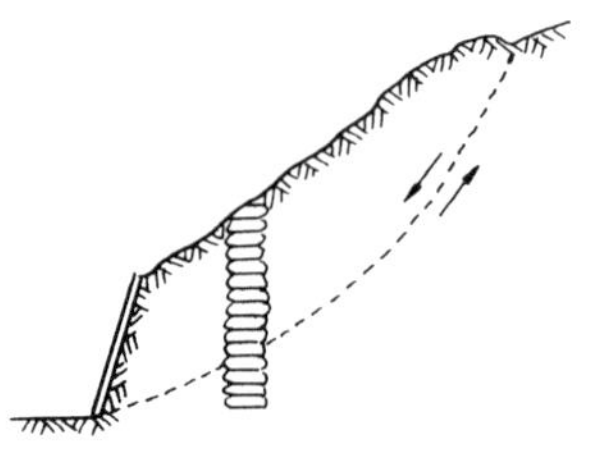

Prevention of slope failure

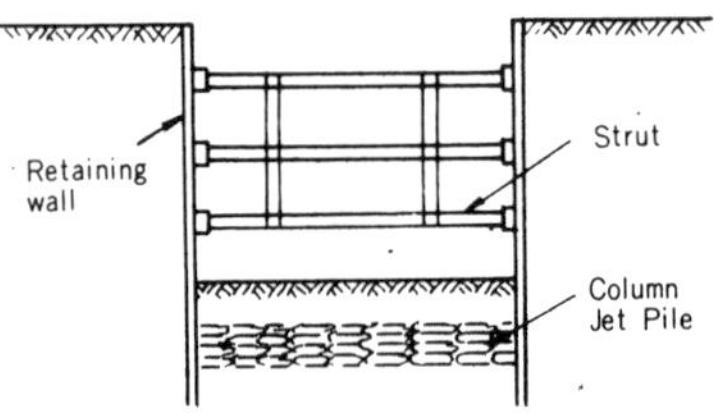

Soil improvement below excavated bottom

Fig. 3 Application of Column Jet Pile Method

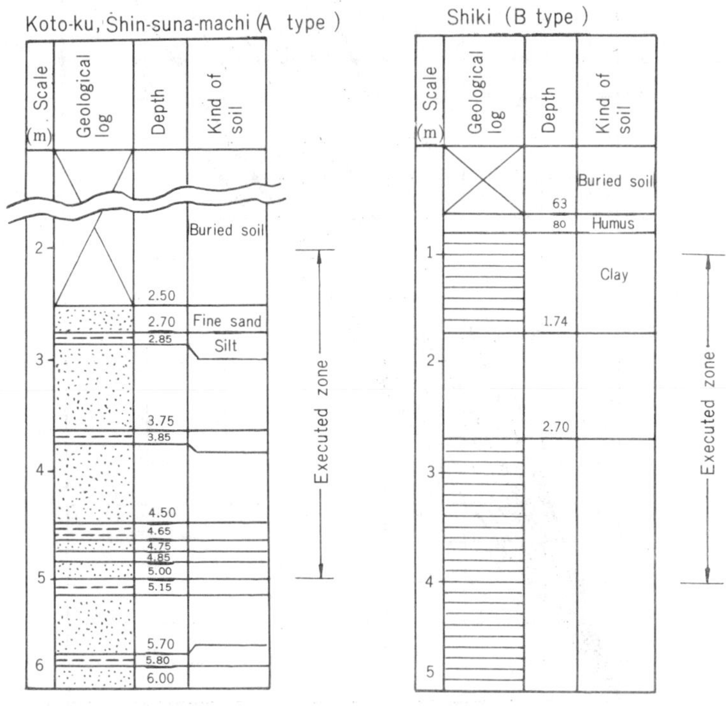

Fig. 4 Geological Log of Test Site

Table 2 Test Parameter and Variable Value

Experimental factor / Variable value soil	High speed water		Grout			Execution		Compressed air	
	Outlet pressure (kg/cm^2)	Flow (ℓ/min)	Outlet pressure (kg/cm^2)	Flow (ℓ/min)	Grout	Revolutions (r.p.m.)	Lifting speed (cm/min)	Outlet pressure (kg/cm^2)	Flow (m^3/min)
Variable value	0, 200	0, 35	0, 30	100, 150	Cement milk	5,	5, 7.5		
						5, 10		0, 5	0, 10
	400	70	50	200, 300	Mortor		10		
Sandy soil	400	70	0	100	Cement milk	5	5	5	1
Clayey soil	400	70	30	200	Cement milk	5	5	5	1

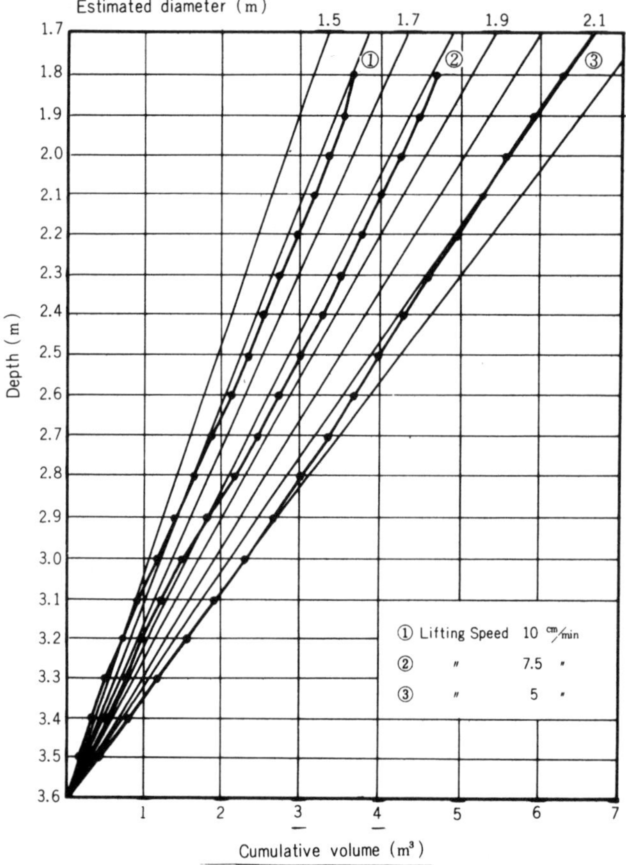

Fig. 5 Relation between Execution Depth and Cumulative Volume

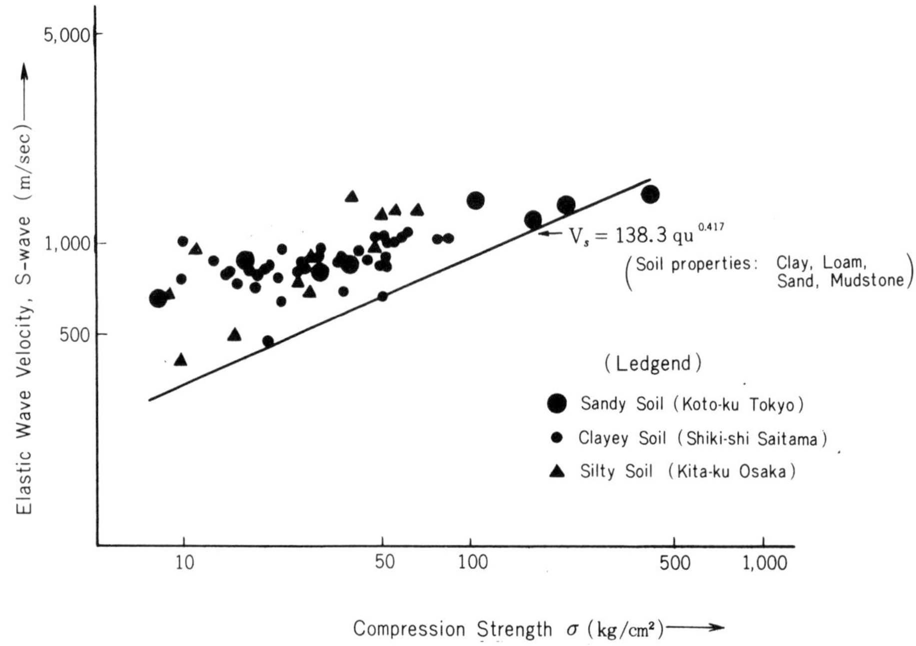

Fig. 6 Relation between S-wave and Compression Strength

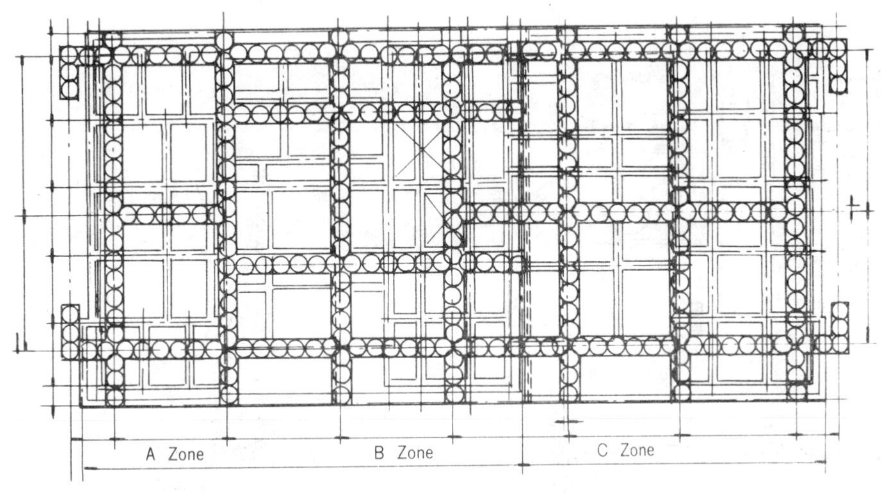

Fig. 7 Plan of Execution

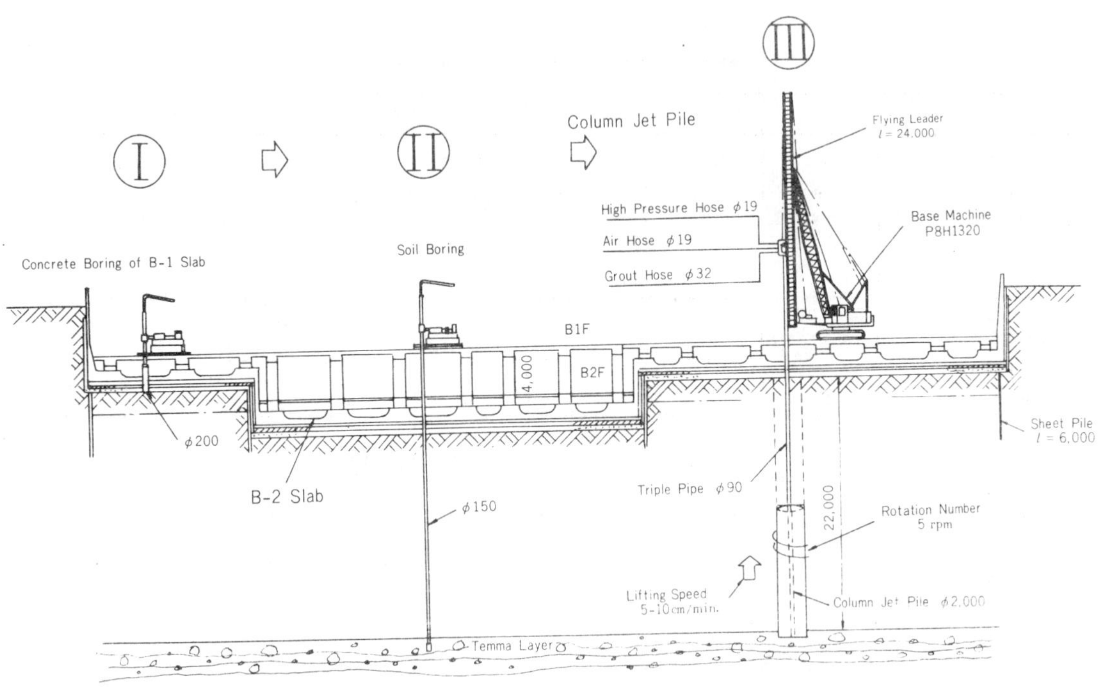

Fig. 8 Section of Execution

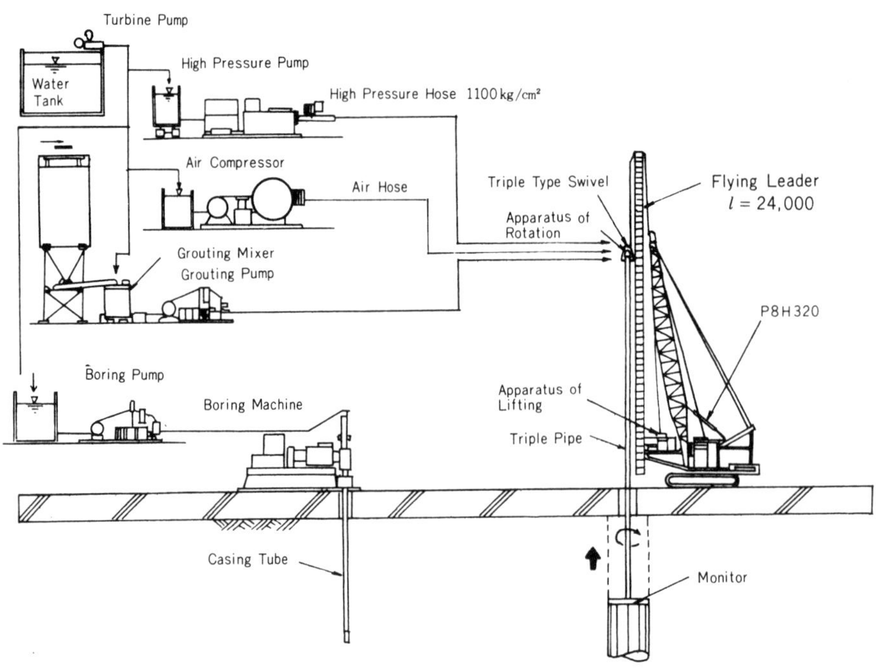

Fig. 9 Execution System of Used Machine

Table 3 Operation Result of Machine

	No. 1	No. 2	No. 3	No. 4	Boring Machine	Total	Average
Type of Machine	P-H-320	P-H-330	P-H-335	P-H-335	M-150		
A (days) Period of Execution	12/20-3/2 74	12/20-3/2 74	12/20-2/28 71	12/20-2/1 44	11/20-3/1 103	366	65.8
B (days) Working Days	41	31	34	24	21	151	32.5
(days) Waiting Days	21	31	25	10	68	155	21.7
(days) Holiday	12	12	12	10	14	60	12
C (hours) Working Time	427.5	451.0	437.0	144.0	307.0	1,756.5	362.4
D (number) Number of Execution	A 28 B 50	B 58 C 23	B 16 C 57	C 22	A 20 B 2 C 9	285	63.5
E (m) Length of Execution	1,516	1,550	1,542	484	767	585.9	1,273
Working Ratio B/A (%)	55	42	48	55	20		50
Working Ratio D/B (No./day)	1.9	2.6	2.1	0.92	1.5		1.9
Working Ratio E/B (m/day)	37.0	50.0	45.4	20.2	36.5		38.2
Working Ratio C/D (hour/No.)	5.30	5.30	6.00	6.30	9.55		5.50
Working Ratio C/F (Min./m)	17	17	17	18	24		17'15"

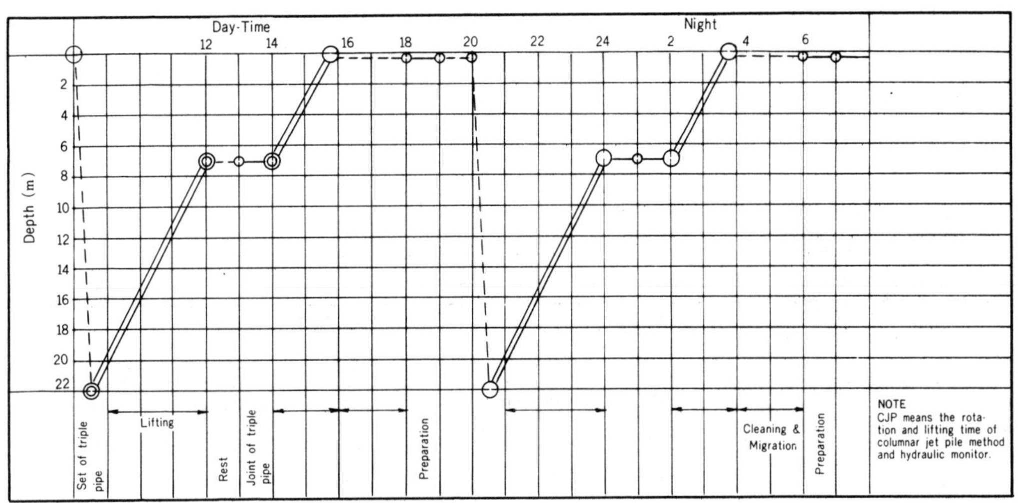

Fig. 1.0 Standard Execution Cycle

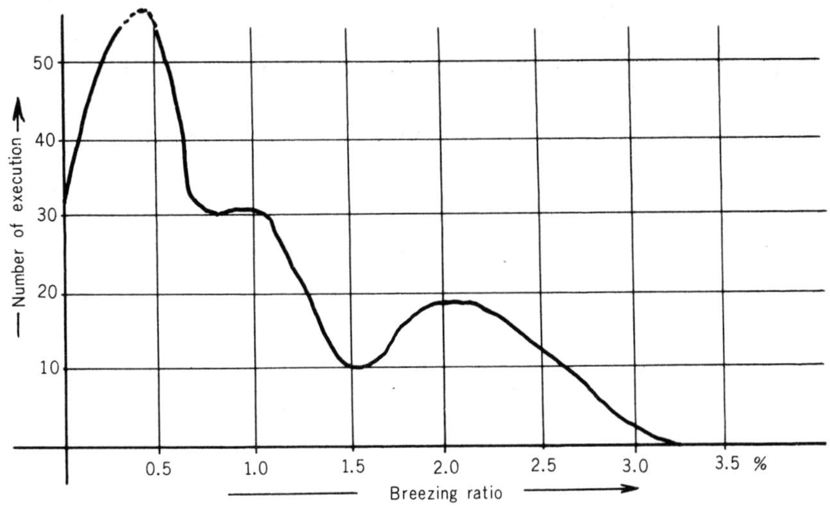

Fig. 11 Frequency Curve of Breezing

Table 4 Result of Execution

	Table of Execution	Total Magnitude		Unit
Magnitude of Execution	Column Jet Pile	Numbers	285	
		Lengths (m)	5,859	
	Boring	Concrete Lengths (m)	641.19	2.25 m/a Column Jet Plle
		Lengths of Chestnut Stone (m)	327.2	1.15 m/a Column Jet Pile
		Lengths of Soil	6,263	21.98 m/a Column Jet Pile
Material of Execution	Column Jet Pile	Magnitude of Cement (t)	11,340.53	1.94 t/Length of Column Jet Pile (m)
	Boring	Magnitude of Bentonite (t)	342	0.005 t/Length of Boring (m)
Discharge	Column Jet Pile	Soil (m^3)	18,122.5	3.08 m^3/Length of Column Jet Plle (m)
		Solidified Soil (m^3)	3,310	0.60 m^3/Length of Column Jet Pile (m)
Labor Supply	Column Jet Pile	Jetting & Grouting (persons)	1,281	0.22 persons/Length of Column Jet Pile
		Mixing & Supply (persons)	1,045	0.18 persons/Length of Column Jet Pile
		Odd Job (persons)	3,396	0.58 persons/Length of Column Jet Pile
		Electoric Job (persons)	184	0.03 persons/Length of Column Jet Pile
		Operator of Machine (persons)	913	0.16 persons/Length of Column Jet Pile
	Boring	Operator of Machine (persons)	1,823	0.31 persons/Length of Boring (m)
		Odd Job (persons)	153	0.03 persons/Length of Boring (m)
Power and Water Supply	Column Jet Pile and Boring	Electric Power (kWH)	245,810	42 kWH/Length of Column Jet Pile (m)
		Water Supply (m^3)	22,978	3.9 m^3/Length of Column Jet Pile (m)
		Light Oil (ℓ)	28,500	4.9 ℓ/Length of Column Jet Pile (m)

Table 5 Results of Breezing

Zone	Numbers of Execution	Length of Breezing (cm)	Breezing Ratio (%)	Grouting Volume (ℓ)
A	47	7.3	0.33	224
B	126	10.3	0.61	272
C	112	7.3	0.33	293
Total	285	Ave. 8.9	Ave. 0.43	Ave. 280

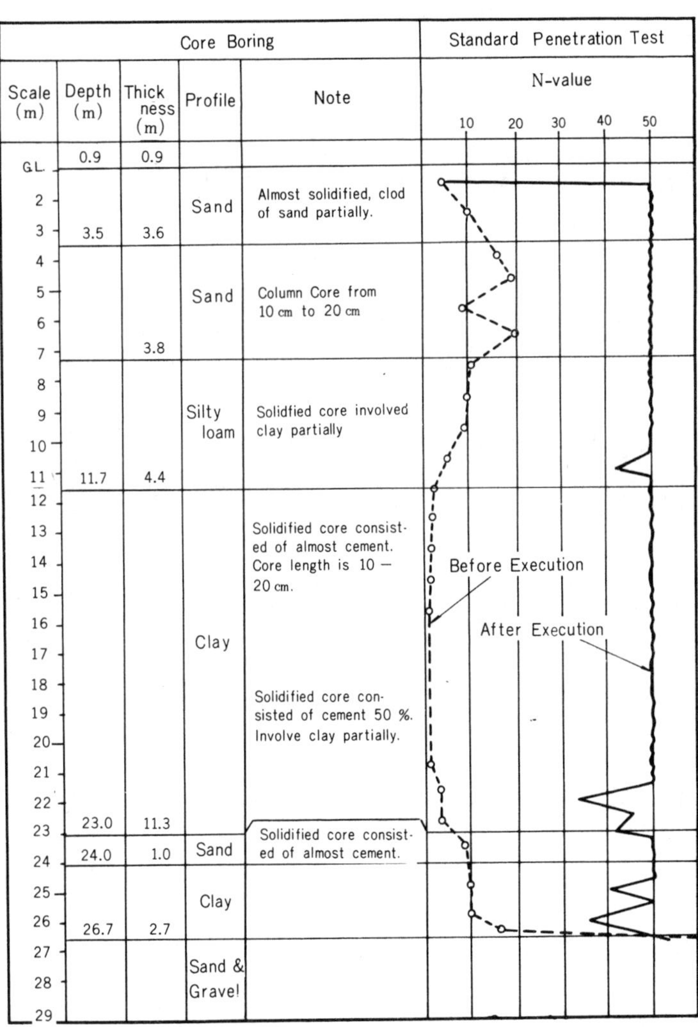

Fig. 1.2 Result of Standard Penetration Test before and after Execution

Table 6 Consideration Result of liquefaction

Items Point No.	Equivalent mean shear stress d (t/m^2)	Shear stress amplitude en (t/m^2)	Judgement
No. 1	0.745	0.541	liquefaction
No. 2	1.080	0.927	"
No. 3	1.524	1.284	"

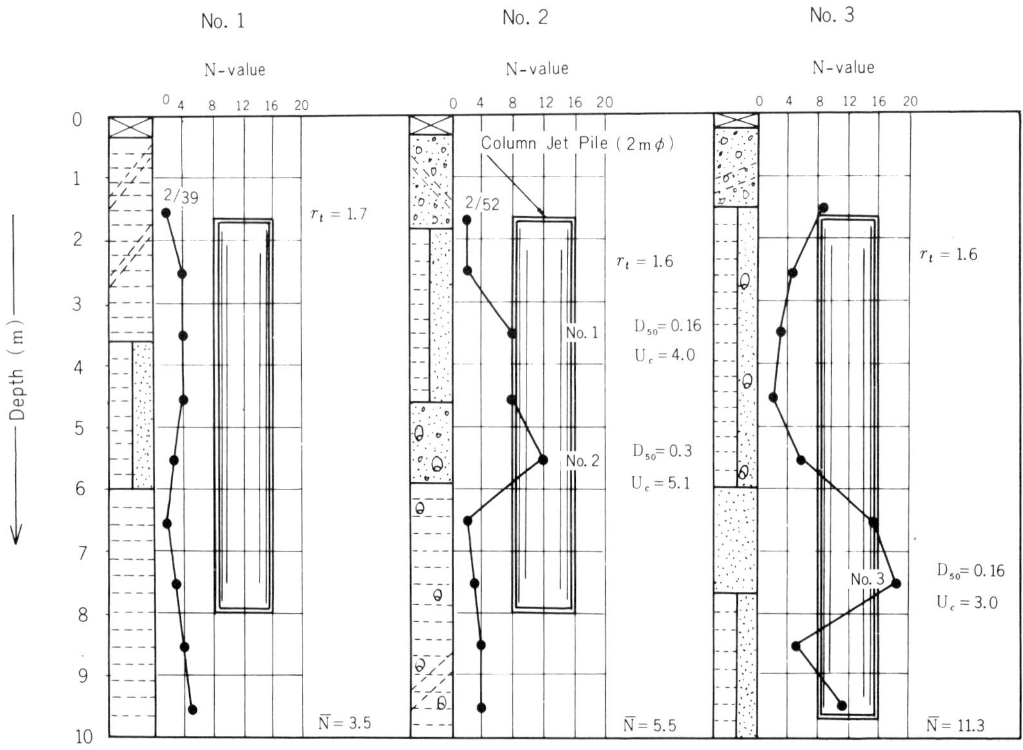

Fig. 13 Geological Log and N-value

Table 7 Result of Physical and Mechanical Test on Sampling Core

Items \ No.	1	2	3	4	5	6	7	8	9	10	11
Density (g/cm^3)	2,200	2,218	2,230	2,153	2,362	2,377	2,125	2,249	2,301	2,318	2,354
Water content (%)	15.0	17.2	20.5	23.1	21.0	25.9	29.4	26.7	24.6	25.7	30.6
Water absorption (%)	33.2	28.0	30.6	36.2	33.8	35.4	38.1	34.7	31.9	38.2	43.0
Void Ratio (%)	42.2	38.3	40.6	43.8	44.4	45.7	44.8	43.8	42.3	47.0	50.8
Tensile strength (kg/cm^2)								4.80	6.20	3.93	
Compression strength (kg/cm^2)				40.7	68.2	67.7	37.6				53.8
Static poisson's ratio				0.20	0.33	0.38	0.26				
Static young's modulus (kg/cm^2)				2.53×10^4	2.77×10^4	2.13×10^4	1.64×10^4				
P-wave (km/sec)	2.04	2.10	2.05	1.93	2.07	2.25	1.92	1.95	2.44	2.20	2.07
S-wave (km/sec)	1.27	1.17	1.12	1.03	1.14	1.02	0.90	0.94	1.02	0.91	0.87
Dynamic poisson's ratio	0.18	0.27	0.29	0.30	0.28	0.37	0.36	0.35	0.39	0.40	0.39
Dynamic young's modulus (kg/cm^2)	8.55×10^4	7.87×10^4	7.36×10^4	6.06×10^4	8.02×10^4	6.91×10^4	4.78×10^4	5.47×10^4	6.79×10^4	5.48×10^4	5.05×10^4
Cohesion (kg/cm^2)		20.6									
Angle of internal friction (°)		28.5									

Table 8 Result of Cement Content Test

Content material / Depth (GL-m)	Water (%)	Cement (%)	Soil (%)
2.0 - 3.0	11.2	37.8	51.0
3.0 - 4.0	10.0	34.3	55.7
4.0 - 5.0	11.3	38.9	49.8
5.0 - 6.0	28.9	42.3	28.8
6.0 - 7.0	28.2	27.1	44.7
7.0 - 8.0	30.1	27.4	42.5

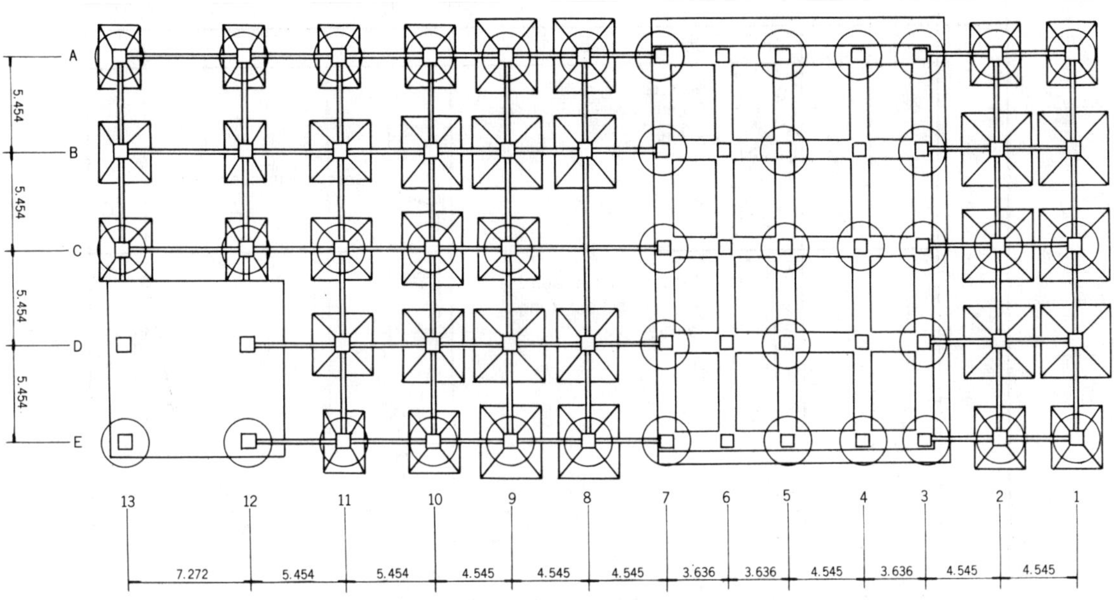

Fig. 14 Arrangement of Column Jet Pile

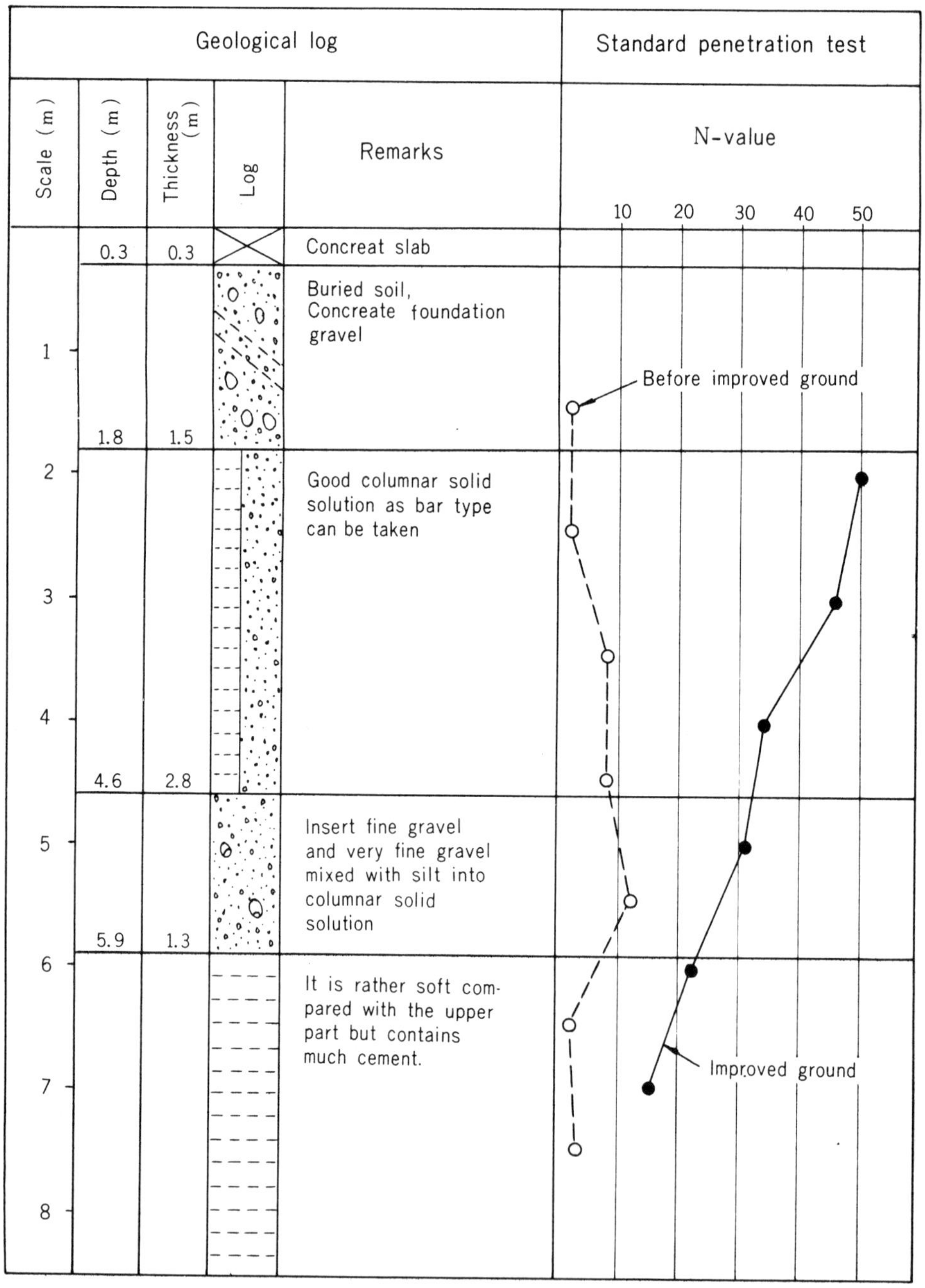

Geological log					Standard penetration test
Scale (m)	Depth (m)	Thickness (m)	Log	Remarks	N-value (10, 20, 30, 40, 50)
	0.3	0.3		Concreat slab	
1	1.8	1.5		Buried soil, Concreate foundation gravel	Before improved ground
2, 3, 4	4.6	2.8		Good columnar solid solution as bar type can be taken	
5	5.9	1.3		Insert fine gravel and very fine gravel mixed with silt into columnar solid solution	
6, 7, 8				It is rather soft compared with the upper part but contains much cement.	Improved ground

Fig. 15 Result of Standard Penetration Test

Table 9. Result of Execution

Execution Procedure	• High Speed	• Outlet Pressure } High → Diameter (Large) • Flow } Diameter (Large)
	• Grouting Pressure	• Sandy Soil • Clayey Soil → High → { Homogeneity, Large Diameter
	• Grouting Flow	
	• Fit Rotation Velocity of Hydraulic Monitor	• 5 - 10 r.p.m. Homogeneity, Strength → High
Shape and Properties of Consolidated Mass	Shape	• Clayey Soil : 2.5 mφ • Sandy Soil : 3.5 mφ
	Properties	• γt : 1.8 - 2.0 (g/cm^3) • σ : 30 - 50 (kg/cm^2)
Cost		• Lifting Speed • Grouting Flow • Displacement of Slime • etc.

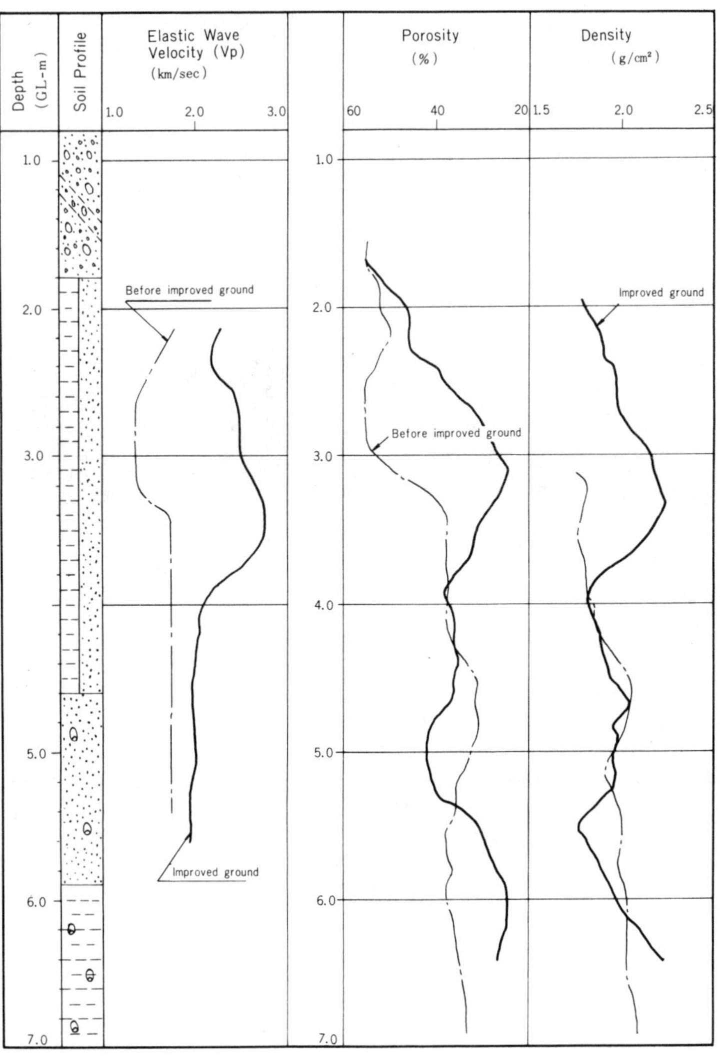

Fig. 16 Result of Geophysical Logging

Photo - 1 Taken from consolidated mass formed in sandy layer

Photo - 2 Taken from consolidated mass formed in clayey layer

Photo 3 - Microphotograph of consolidated mass section formed in sand layer (Parallel Nicols, X80)
Photo by T. Yahiro et al 1980

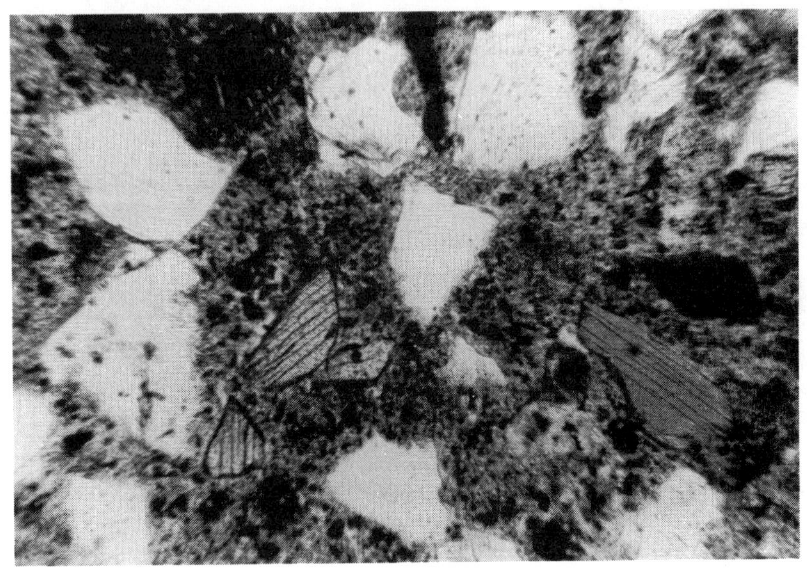

Photo 4 - (Crossed Nicols, X80)

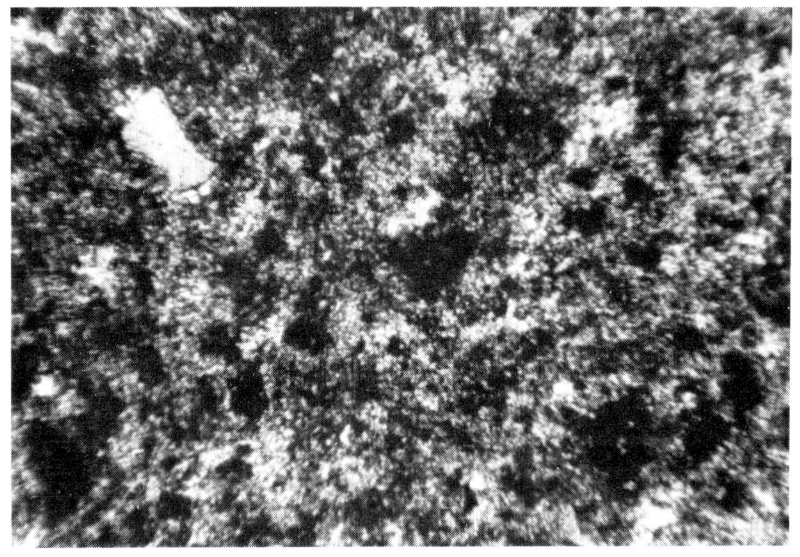

Photo - 5 Microphotograph of consolidated mass section formed in humus layer, (Parallel Nicols, X80)
Photo by T. Yahiro et al 1980

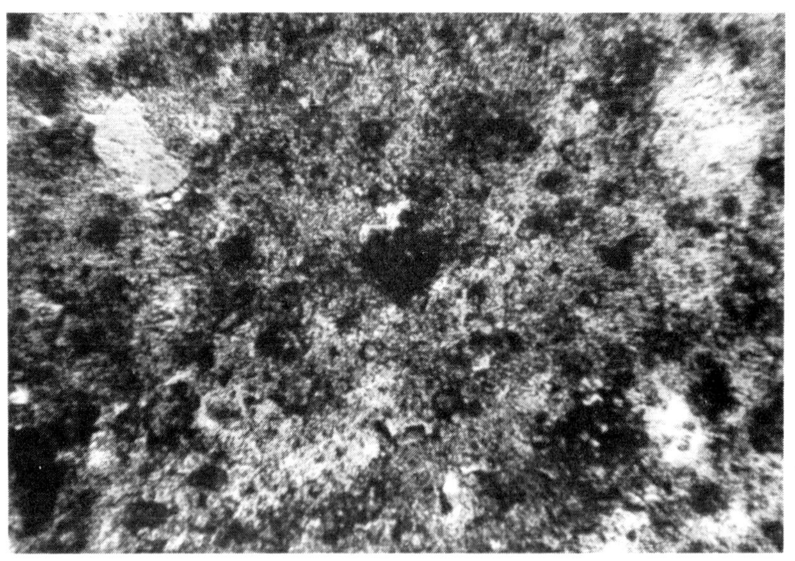

Photo - 6 (Crossed Nicols, X80)

Photo - 7 Hydraulic monitor of down-the-hole-driving type

6th International Symposium on
Jet Cutting Technology
6-8, April, 1982

EVALUATION OF AN ABRASIVE CLEANING SYSTEM

C.R. Barker and M. Mazurkiewicz

University of Missouri-Rolla, U.S.A.

M. Anderson

Partek Corporation, U.S.A.

Summary

Cutting tests using glass samples were run using a production abrasive cleaning system to relate the sample weight loss to standoff distance and sand consumption for a constant power input. Four different water nozzles were tested with and without a cone mixing chamber to assist in accelerating the sand to cutting speed.

Equations were fitted to the cutting data to project predicted cutting performance outside the region where data were taken, and to compare the performance of the four nozzle arrangements tested. This comparison reveals that the best arrangement of those tested was the six jet water nozzle. The most likely explanation for this result is that the abrasive can more easily penetrate into the smaller jets, and the area where the water is exposed to the abrasive is increased by using more water jets.

Held at the University of Surrey, U.K.
Symposium organised and sponsored by
BHRA Fluid Engineering

NOMENCLATURE

$CW_b^{(n)}$ = weight loss of glass sample in grams using the cone mixing chamber

$SW_b^{(n)}$ = weight loss of glass sample in grams without the cone mixing chamber

$a_o \rightarrow a_5$ = constants

b = a subscript of CW and SW to denote a specific standoff distance in meters

d = standoff distance in meters

n = an integer superscript on CW and SW to denote the number of water jets used in the nozzle

s = sand consumption rate in kg/hr

1. INTRODUCTION

In some cases the use of water alone does not produce satisfactory cleaning rates. When metal surfaces must be cleaned of tight rust, mill scale, paint, and other coatings it is common practice to use an abrasive material such as sand in conjunction with the high pressure water. Fig. 1 illustrates a typical cleaning application using equipment manufactured by Partek Corporation.

Many abrasive cleaning systems are currently available, and their principle of operation seems to be quite similar. In each case the high pressure water is used to accelerate the particles of the abrasive material to a speed sufficient to remove the deposit and clean the parent surface for further treatment. Frequently, the location at which the abrasive particles are accelerated is just before they are expelled from the nozzle to strike the surface being cleaned. The abrasive is generally transported from a holding tank to the point where it is injected into the high pressure water stream by a flow of air which may be provided by an auxiliary air compressor or be the result of the suction induced by the water jet.

This process on the surface appears to be quite simple, but in reality requires some careful engineering to produce an efficient cleaning apparatus. One of the difficulties which must be overcome is to mix the high velocity-high density stream of water with the low density-low velocity stream of air and abrasive without a large mixing energy loss. The mixed stream must provide a pattern which is of adequate size and distribution to provide a good cleaning pattern. Therefore, the distribution of the abrasive within the jet stream must be carefully controlled to use the abrasive most effectively. The materials from which the mixing chamber are manufactured must be chosen carefully in order to provide an acceptable service life.

In order to evaluate the performance of an existing abrasive cleaning system and compare it to other designs, a test program was performed using a pressurized sand pot supplied by Partek Corporation. Compressed air was used to transport sand from the sand pot through a rubber hose to the mixing chamber where the sand was mixed with a stream of high pressure water. Samples of glass were exposed to the abrasive mixture for a fixed length of time and the weight loss of the glass was recorded. This paper will describe the various nozzle configurations tested and present the results of the glass cutting tests.

2. DESCRIPTION OF TEST ARRANGEMENT

2.1 Test equipment

Air was supplied from a Henke Model 452TV air compressor capable of producing 18.9 ℓs^{-1} (40 CFM) at 0.689 MNm^{-2} (100 psi) to an auxiliary tank shown in Fig. 2. The air flow passed through a water trap and then an air flow meter before dividing into two flows at the sand tank. A portion of the air travels directly to the line feeding the ABRAS-I-JECTOR. The remainder of the air passes into the sand tank and flows with sand through the regulating valve to combine with the air flowing directly to the ABRAS-I-JECTOR.

A 7.62 m length (25 ft) of 1.27 cm (0.5 in.) I.D. hose was used to connect between the sand pot and the top stainless steel tube shown in Fig. 3. The device on the right side of Fig. 3 is the ABRAS-I-JECTOR and the lower tube is the high pressure water supply line. The pressure transducer shown is a BLH Type DV Pressure Cell and it was used to record the water pressure at the entrance to the water nozzle.

2.2 Details of the ABRAS-I-JECTOR

Referring to Fig. 4, the high pressure water flows into an annulus in the ABRAS-I-JECTOR where it is then discharged through individual passages to form jets which are directed along the surface of the cone in the tungsten carbide insert. The air and abrasive flow is directed along the axis of the cone to mix with the high pressure jets before being expelled into the outside air.

2.3 Test stand

The test stand shown in Fig. 5 was used to clamp the ABRAS-I-JECTOR and direct it toward a glass sample held against a steel plate as shown in the figure. The distance from the ABRAS-I-JECTOR to the sample could be varied by moving the steel plate along the test stand frame. The glass samples used were 26 cm x 26 cm (10.25 in. x 10.25 in.) with a thickness of either 0.64 cm or 1.27 cm (.25 or .5 in.).

2.4 Sand pot instrumentation

One of the important parameters in evaluating different abrasive cleaning equipment is the rate of abrasive consumption. In order to accurately measure this parameter, the sand pot was suspended from a strain gauged force transducer which employed semi-conductor gauges to achieve the required sensitivity. The output of the force transducer was recorded on a strip chart recorder so that a continuous record of the weight of the sand pot was produced. This arrangement is shown in Fig. 6.

3. TEST PROCEDURE FOR GLASS CUTTING TESTS

3.1 Procedure followed

With the glass sample clamped to the test stand and protected by a steel shield, the high pressure water pump was started and the pressure set at the pump. The pressure signal from the transducer at the ABRAS-I-JECTOR was recorded on a strip chart recorder as well as the weight of the suspended sand pot. While the pump adjustments were being made, the sand control valve was set to the desired rate and the shut off valve opened in the air line leading to the ABRAS-I-JECTOR. The sand pot weight versus time trace was observed on the strip chart to ensure that a constant slope was being generated. When steady state conditions were achieved, the glass sample was uncovered and after 5 seconds of exposure the valve was closed stopping the flow of sand and air to the ABRAS-I-JECTOR. The air flow and air pressure readings were taken during the tests and recorded along with the other data.

3.2 Sample trace from strip chart

A typical trace for tests #131 and #132 is shown in Fig. 7. The top trace in the figure is the pressure trace and the weight of the sand tank is the lower of the two. Notice in the upper left hand corner of the figure on the #131 pressure trace the point at which the pump is turned on and the bypass valve closed to raise the water pressure to the operating point. At the left side of Fig. 7, the weight trace shows a small increase which occurred when the chart was stopped and the air compressor was recharging the tank after the previous run. When the chart was started again, the weight trace shows the effects of adjusting the sand flow control and the stable condition achieved before the test was started. The actual cutting of the sample takes place in the 5 second interval before the closing of the air and sand valve which appears on the chart as a spike followed by a "ringing" trace. Shortly after this, the pump was turned off and the chart stopped to prepare for test #132.

4. GLASS CUTTING TEST RESULTS

4.1 Six jet water nozzle with cone chamber

The first water nozzle arrangement tested had six water passages of diameter 0.572 mm (0.0225 in.) spaced at 60 deg intervals around the annulus depicted in Fig. 4. The initial tests were run using the cone mixing chamber and a 30-60 mesh sand. With this particular nozzle configuration the operating pressure of the water was maintained at 68.9 MNm^{-2} (10,000 psi) and the water flow measured as 0.464 ℓs^{-1} (7.36 gpm) which results in an input power level of 32 Kw (42.9 hp) from the high pressure water. The power supplied from the compressed air was calculated from the air flow and tank pressure data using the equation

$$\text{Air Power (Kw)} = \text{Air Flow } (\ell s^{-1}) \times \text{Air Pressure } (MNm^{-2}) \qquad (1)$$

The power input from the water was essentially constant from test to test. However, the power input from the compressed air varied depending upon the air compressor cycle and the actual rate of sand flow. Therefore, the cutting data presented in Fig. 8 was taken with a power input ranging from an extreme low of 34 Kw to an extreme high of 39.5 Kw excluding, of course, the data presented at the operating pressure of 34.5 MNm^{-2} (5,000 psi).

Figure 8 shows that for an operating pressure of 68.9 MNm^{-2} (10,000 psi) the weight of glass removed in 5 sec of exposure increases significantly as the sand flow rate is increased and decreases rapidly as the standoff distance is increased. The data appear to fit quite well a second degree polynomial function which passes through the origin. This agrees well with the fact that the water alone would not cut the glass at this pressure. Hence, if a least squares fit is used for the data of Fig. 8 the resulting equations relating weight loss in gms to sand consumption in kg/hr are:

$$CW_{0.5}^{(6)} = 0.31154\ s - 7.3677 \times 10^{-5}\ s^2 \quad (2)$$

$$CW_{0.75}^{(6)} = 0.20748\ s - 6.5546 \times 10^{-5}\ s^2 \quad (3)$$

$$CW_{1.0}^{(6)} = 0.13333\ s - 3.1612 \times 10^{-5}\ s^2 \quad (4)$$

where

s = sand consumption in kg/hr

CW = weight loss in gms

the subscript on CW is standoff distance in meters, and the superscript is the number of water jets used. Using these equations it is then possible to predict a weight loss for a specified sand consumption rate at the three distances of 0.5, 0.75, and 1.0 m. For example using sand rates of 100, 200, 300, 400 and 500 kg/hr should result in the predicted weight loss shown in Table 1.

Table 1

Predicted Weight Loss for Constant Sand Rates

	Weight Loss - g		
s (kg/hr)	0.5 m	0.75 m	1.0 m
100	30.4	20.1	13.0
200	59.4	38.9	25.4
300	86.8	56.3	37.2
400	112.8	72.5	48.3
500	137.4	87.4	58.8

These predicted values were then plotted as shown in Fig. 9 and a second degree polynomial fitted to them in a least squares sense with the assumption that the weight loss at a distance of 2.0 m or more would be minimal.

The resulting equations for constant sand consumption using the cone chamber are:

$$CW_{100}^{(6)} = 55.1 - 57.2\ d + 14.8\ d^2 \quad (5)$$

$$CW_{200}^{(6)} = 107.8 - 112.2\ d + 29.2\ d^2 \quad (6)$$

$$CW_{300}^{(6)} = 157.4 - 164.1\ d + 42.7\ d^2 \quad (7)$$

$$CW_{400}^{(6)} = 204.6 - 214.1\ d + 55.9\ d^2 \quad (8)$$

$$CW_{500}^{(6)} = 249.4 - 261.8\ d + 68.6\ d^2 \quad (9)$$

where the subscript on CW indicates a constant sand rate in kg/hr, d is distance in meters, and the superscript denotes the number of jets used.

Having the equations relating weight loss to sand consumption at a fixed distance, and another set relating weight loss to distance at a fixed sand rate, permitted constructing Fig. 10 which shows the predicted performance of the six jet system with a constant power input of 37 $\pm$ 2 Kw. The surface depicted in Fig. 10 can be approximated by the function whose form is:

$$CW(d,s) = a_0 s + a_1 s^2 + a_2 ds + a_3 ds^2 + a_4 d^2 s + a_5 d^2 s^2 \quad (10)$$

where

CW = weight loss in gms

d = standoff distance in meters

s = sand consumption in kg/hr

a's are constants

The form in Eq. (10) was chosen so that CW = 0 when s = 0 and the values of the constants "a" were chosen by fitting the function in a least squares sense to a set of 54 data points generated from Eqs. (2) - (9). The result is:

$$CW(d,s) = (5642.5s - 1.3117\ s^2 - 5843.2\ ds + 1.2152\ ds^2 + 1.5114\ d^2 s - 0.27893\ d^2 s^2) \times 10^{-4} \quad (11)$$

Eq. (11) fits the data points very closely with a maximum error of 5 percent in the region $0 \leq d \leq 2$ and $0 \leq s \leq 500$.

Table 2 shows a comparison between the actual cutting data and the values predicted by Eq. (11). Again, the agreement is very acceptable.

Table 2

Comparison Between Actual Cutting Data and Equation (11) Values

Test	d	s	Actual CW	Predicted CW
4	0.5 m	133.9 kg/hr	41.5 g	40.1
5	0.5	141.2	42.0	42.2
41	0.5	185.7	56.0	54.9
6	0.5	284.2	80.5	81.8
49	0.5	435.4	123.8	120.2
7	0.5	504.8	137.5	136.7
11	0.75	148.0	31.0	30.0
10	0.75	171.2	30.0	34.5
13	0.75	193.9	37.5	38.8
9	0.75	277.8	56.0	54.3
14	0.75	397.7	69.5	75.1
8	0.75	480.3	85.5	88.5
15	1.0	177.1	22.0	22.0
16	1.0	294.6	37.0	35.4
17	1.0	507.1	60.0	56.8
18	1.0	532.5	61.5	59.2

Eq. (11) can therefore be used to predict the cutting performance in the region represented by Fig. 10 for the six jet system using the cone chamber if the power level is the same as that used in the original cutting tests.

4.2 Six jet water nozzle without cone chamber

The cutting tests just described were repeated without the cone mixing chamber. The results without the mixing chamber are shown in Fig. 11 and generally resulted in a removal of a smaller weight of glass for the same conditions as the previous test. The largest difference in cutting occurred at the standoff distance of 0.5 m where generally the weight removed without the cone was only 75%-85% of that removed with the cone. A smaller difference was noted with and without the cone at 0.75 m and 1.0 m where the weight loss was 84%-86% at 0.75 m and 88%-95% at 1.0 m of the corresponding value using the cone chamber. No significant difference in cutting performance occurred at 34.5 MNm^{-2} (5000 psi) for low sand rates. At rates of sand consumption above 400 kg/hr, the system with the mixing chamber performed slightly better than the one without the chamber.

4.3 Discussion of damage zones for six jet nozzle

In order to compare the cutting performance of the various nozzle arrangements, it is important to examine the area of the sample in which glass has been removed. If a very small depth of penetration is spread over a large enough area, the resultant weight loss may be relatively high. On the other hand, coverage of a small area to a greater depth would also produce a large weight loss.

The procedure followed to qualitatively compare the damage zones between samples was to place the sample on a level surface beneath a copying camera. Then, knowing the sample weight loss and the density of the glass, the total volume of the damage zone was computed. A quantity of black latex paint equal to one-third of this volume was then injected into the damage zone from a hypodermic syringe and the paint was allowed to reach its own level. An exposure was taken followed by a second injection of paint equal again to one-third of the volume of the damage zone. After the second exposure, the last third of the volume was added and a third exposure taken.

The result of following this procedure can be viewed in Fig. 12-14 which are photographs of the samples 6, 9, and 16 cut with the 6 jet nozzle and the cone mixing chamber. The scale is such that the full frame of the photographs corresponds to 14.4 x 18.7 cm (5.66 x 7.375 in.). Notice that the area of the three damage zones is nearly the same although sample 6 was cut at a distance of 0.5 m, sample 9 was cut at 0.75 m, and sample 16 was cut at 1.0 m. Each of the three samples was cut using a sand flow rate of approximately 285 kg/hr (630 lb/hr) and the same power input level was used in each case. The volumes represented by the three figures are however, quite different. Sample 6 lost 80.5 g or a volume of 36 μm^3 while 9 and 16 have volumes of 24.8 and 16.4 μm^3, respectively. Hence, to evaluate the three photographs, it must be understood that in Fig. 12 the darkest zone has a depth of 0.635 cm (0.25 in.) and represents a volume of 12 μm^3 while the darkest zone in Figure 13 is only 0.4 cm (0.158 in.) deep and represents a volume of 8.3 μm^3. In Fig. 14, the depth of the darkest zone is 0.2 cm (.080 in.) with a volume of 5.5 μm^3. These photographs suggest that as the target moves back from the nozzle to a larger distance the area of the damage zone does not change significantly but the depth of penetration is sharply reduced. Hence, since it is known that the jet and abrasive spread significantly with distance, much of the abrasive that reaches the target is not moving with sufficient speed to cut the glass.

4.4 Other nozzle arrangements

Cutting tests were also performed using water nozzles with 2, 3, and 4 jets in each case spaced evenly around the circumference. Tests were run with and without the cone mixing chamber using the same procedure described earlier. Standoff distances of 0.5 and 0.75 m were used with the same type of sand used in the case of the six jet nozzle. The power input for the 2 and 3 jet nozzles was nearly identical to that used with the 6 jet nozzle. However, due to a slightly larger total nozzle area with the 4 jet nozzle, the operating pressure of the water was lower and the resulting power input was 83% of that used with the 2, 3, and 6 jet nozzles. Perhaps an adjustment factor should be applied to the cutting data for the 4 jet nozzle to compensate for the reduced power input. However, an insufficient number of trials were run to accurately relate cutting performance to power input. Therefore the results presented for the four jet nozzle have not been modified.

If equations similar to Eq. (2) and (3) are fitted to the cutting data from the 2, 3, and 4 jet nozzles, the resulting equations are for the case with the cone chamber:

$$CW^{(2)}_{0.5} = 0.24593\ s - 1.4875 \times 10^{-4}\ s^2 \qquad (12)$$

$$CW^{(3)}_{0.5} = 0.29126\ s - 1.7332 \times 10^{-4}\ s^2 \qquad (13)$$

$$CW^{(4)}_{0.5} = 0.19612\ s - 2.5485 \times 10^{-5}\ s^2 \qquad (14)$$

$$CW^{(2)}_{0.75} = 0.15800\ s - 8.0000 \times 10^{-5}\ s^2 \qquad (15)$$

$$CW^{(3)}_{0.75} = 0.18326\ s - 9.5700 \times 10^{-5}\ s^2 \qquad (16)$$

$$CW^{(4)}_{0.75} = 0.17420\ s - 9.7980 \times 10^{-5}\ s^2 \qquad (17)$$

The corresponding equations for the case of the straight end without the cone mixing chamber are:

$$SW_{0.5}^{(2)} = 0.14203\ s - 4.0563 \times 10^{-5}\ s^2 \quad (18)$$

$$SW_{0.5}^{(3)} = 0.18901\ s - 6.5230 \times 10^{-5}\ s^2 \quad (19)$$

$$SW_{0.5}^{(4)} = 0.17486\ s - 4.2691 \times 10^{-5}\ s^2 \quad (20)$$

$$SW_{0.5}^{(6)} = 0.22764\ s + 2.6218 \times 10^{-6}\ s^2 \quad (21)$$

$$SW_{0.75}^{(2)} = 0.10950\ s - 5.8453 \times 10^{-5}\ s^2 \quad (22)$$

$$SW_{0.75}^{(3)} = 0.11474\ s - 1.3933 \times 10^{-5}\ s^2 \quad (23)$$

$$SW_{0.75}^{(4)} = 0.13695\ s - 4.5363 \times 10^{-5}\ s^2 \quad (24)$$

$$SW_{0.75}^{(6)} = 0.17425\ s - 4.7423 \times 10^{-5}\ s^2 \quad (25)$$

A comparison between the various arrangements can then be made by examining the ratio of the weight removal equations for the two systems being compared. For example, if the six jet system with the cone chamber were to be compared to the four jet system with the cone chamber when the standoff distance in both cases was 0.5 m, the appropriate ratio would be:

$$\frac{CW_{0.5}^{(4)}}{CW_{0.5}^{(6)}} = \frac{0.19612s - 2.5485 \times 10^{-5}\ s^2}{0.31154s - 7.3677 \times 10^{-5}\ s^2} \quad (26)$$

which is shown plotted in Fig. 15. The results for the two and three jet configurations compared to the six jet are also shown in Fig. 15 for a standoff distance of 0.5 m in all cases. Fig. 16 shows the performance comparison at 0.75 m standoff distance, again with the cone chamber.

The results for the two standoff distances without the cone chamber are shown in Figs. 17 and 18.

4.5 Comparison of damage zones for the various nozzle arrangements

The characteristic shapes of the damage zones in the glass samples can be compared by using the same photographic technique described earlier. When the samples were cut with a standoff distance of 0.75 m without the cone chamber the damage zones produced by the 2, 3, 4, and 6 jet systems are shown in Figs. 19-22 respectively. Notice in Fig. 19 the two distinct lobes cut by the two jet system. Fig. 20 shows a damage zone with three lobes as produced by the three jet nozzle while in Fig. 21 the pattern is nearly square when a four jet nozzle was used. Fig. 22 is similar to Fig. 20 although a six jet nozzle was used in the case of Fig. 22.

The sand consumption rate for each of the Figs. 19-22 was approximately 375 kg/hr (826 lb/hr) and the input power was equal to 37 kw (49.6 hp) for the 2, 3, and 6 jet cases. The power input for the 4 jet case was 30.7 kw (41.2 hp).

5. DISCUSSION OF RESULTS

The influence of using the cone mixing chamber can be clearly seen from the cutting data. In every case regardless of the specific parameters more glass was removed by using the cone than without it. The cone chamber must assist in the acceleration of the sand particles and help to concentrate the sand pattern to increase the effectiveness of the cutting.

The importance of the sand rate is illustrated by Fig. 10. No glass was cut from the samples using the water alone. The weight loss of the glass varied almost linearly with sand consumption for a constant standoff distance. Since the input power was essentially constant, the specific energy required to cut the glass would be lowest at the highest rate of sand consumption and would approach infinity as the sand consumption rate approached zero. The standoff distance also has a significant effect on the specific energy requirements when the sand rate is held constant. A significant reduction in specific energy is achieved by decreasing the standoff distance.

The six jet water nozzle provided superior cutting performance compared to the two, three, and four jet nozzles. In Figs. 15-18 the cutting ratios generally were in the range of 0.6 to 0.9 for the lowest sand rate and in the range of 0.5 to 0.8 at the highest sand rate. This suggests that a great deal of the sand is not being accelerated to a speed of sufficient magnitude to damage the glass. As the number of jets is reduced, the diameter of each individual jet must be increased to maintain the same power input. Therefore the total region where the sand can mix with the high pressure jets is reduced by decreasing the number of jets in the nozzle. In addition, it is probably more difficult for the sand to penetrate into the interior of a large jet than a small jet. The consequence of both of these effects is that less sand is accelerated to cutting speed with fewer jets and the amount of glass cut from the sample decreases.

6. CONCLUSION

This work represents the initial phase of a study to evaluate abrasive cleaning and attempt to define the most significant aspects of the operation. No conclusions should be drawn from the results presented here in regard to projected cleaning rates for various surfaces. Instead, the importance of a clear understanding of the mixing process in which the abrasive is accelerated to cutting speed has been emphasized.

It is clear that there is a strong correlation between cutting rate, sand consumption, and standoff distance for a given level of power input. It is also obvious that more cutting can be accomplished with the same amount of sand if a better means can be found to accelerate the sand, or more of it, to cutting speed. The tests of the water nozzles indicate that smaller jets can accelerate the abrasive more effectively than larger jets. The difference may be due to the inherent problem of the inability of the abrasive to penetrate into the very fast moving water stream. The answer to this and other questions about the mixing process will require further study.

7. ACKNOWLEDGEMENTS

The authors wish to acknowledge the assistance of Mr. Mike Ginn, President of Partek Corporation and Mr. Amos Pacht, Manager of Engineering, for suggesting this project and for the financial support which they provided. In addition, Mr. L. John Tyler and Mr. James Blaine assisted in conducting the cutting tests and fabricating the experimental apparatus. Thanks should be given to Mr. Way King for his assistance in preparing the computer data plots, and Mrs. Mujde Erten for drawing the figures.

Figure 1. Typical abrasive cleaning operation.

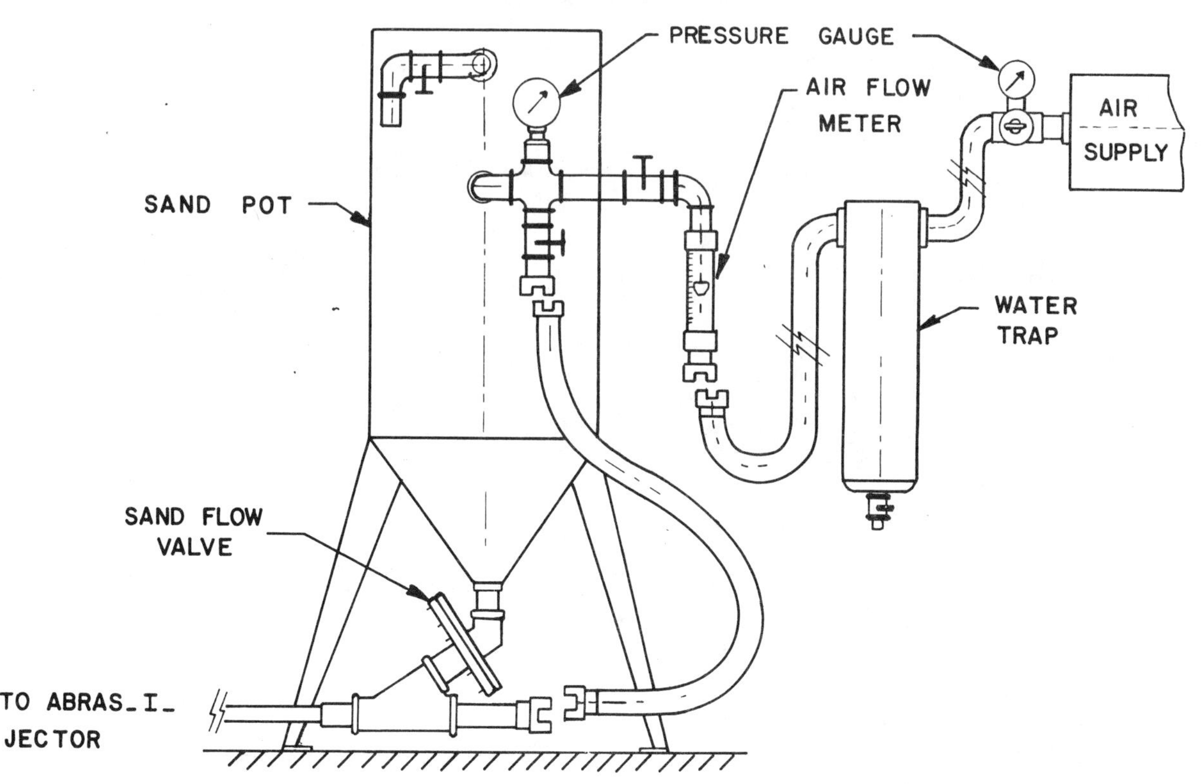

Figure 2. Arrangement of the pressurized sand pot.

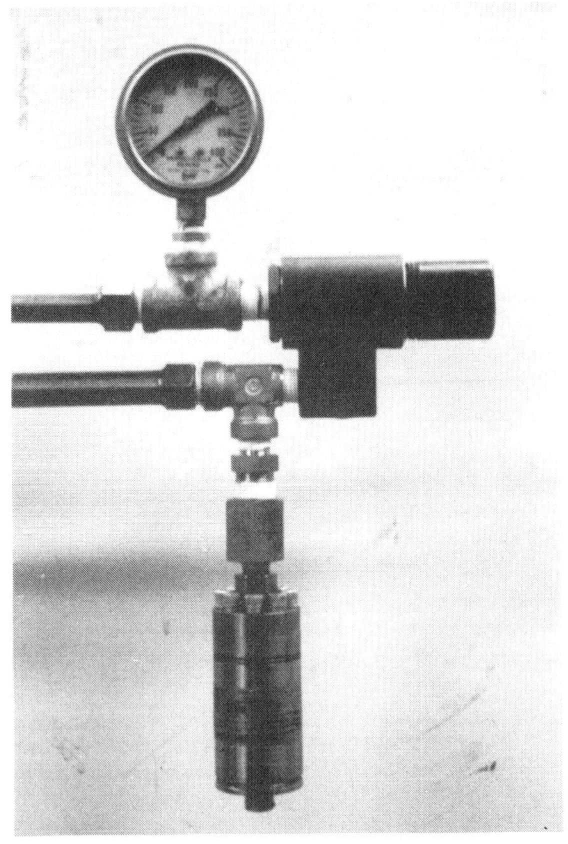

Figure 3. Photograph of the Abras-I-Jector.

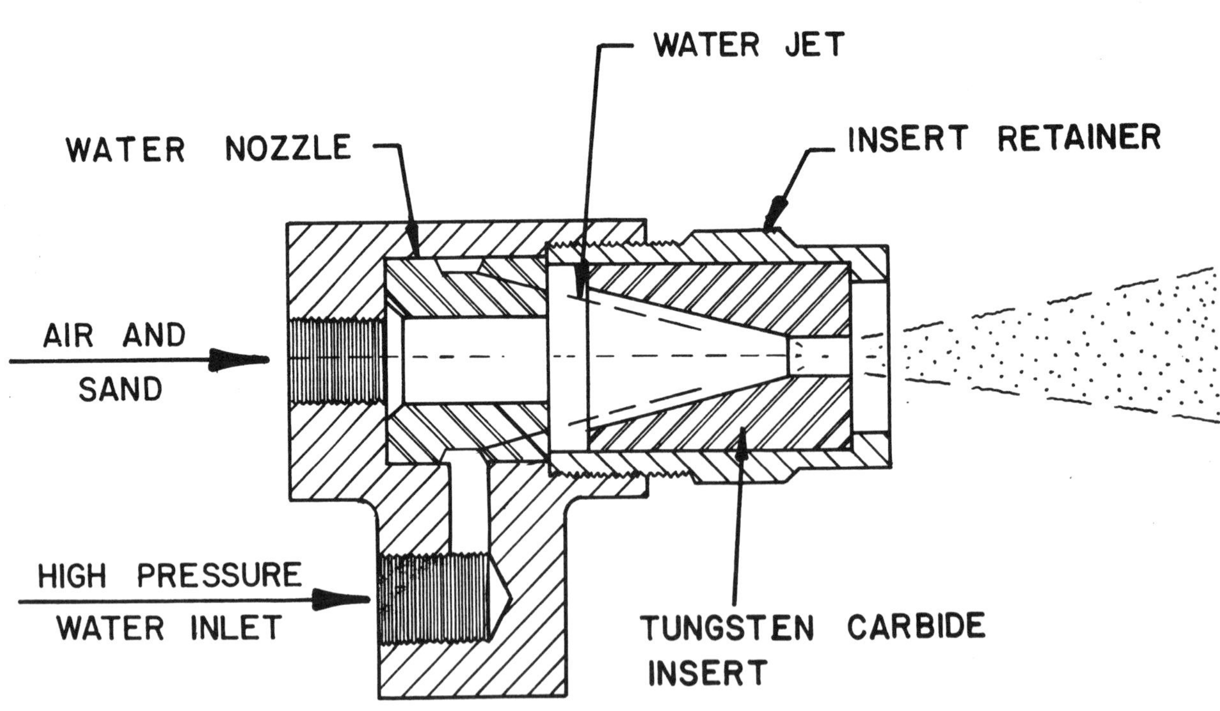

Figure 4. Details of the Abras-I-Jector.

Figure 5.

Test stand for glass cutting tests.

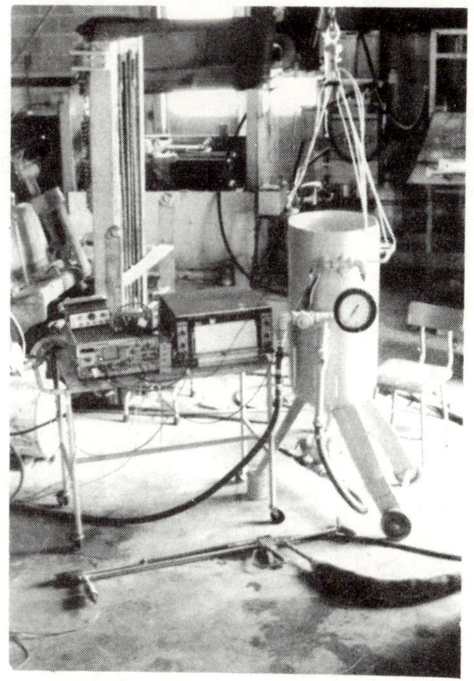

Figure 6.

Sand pot instrumentation.

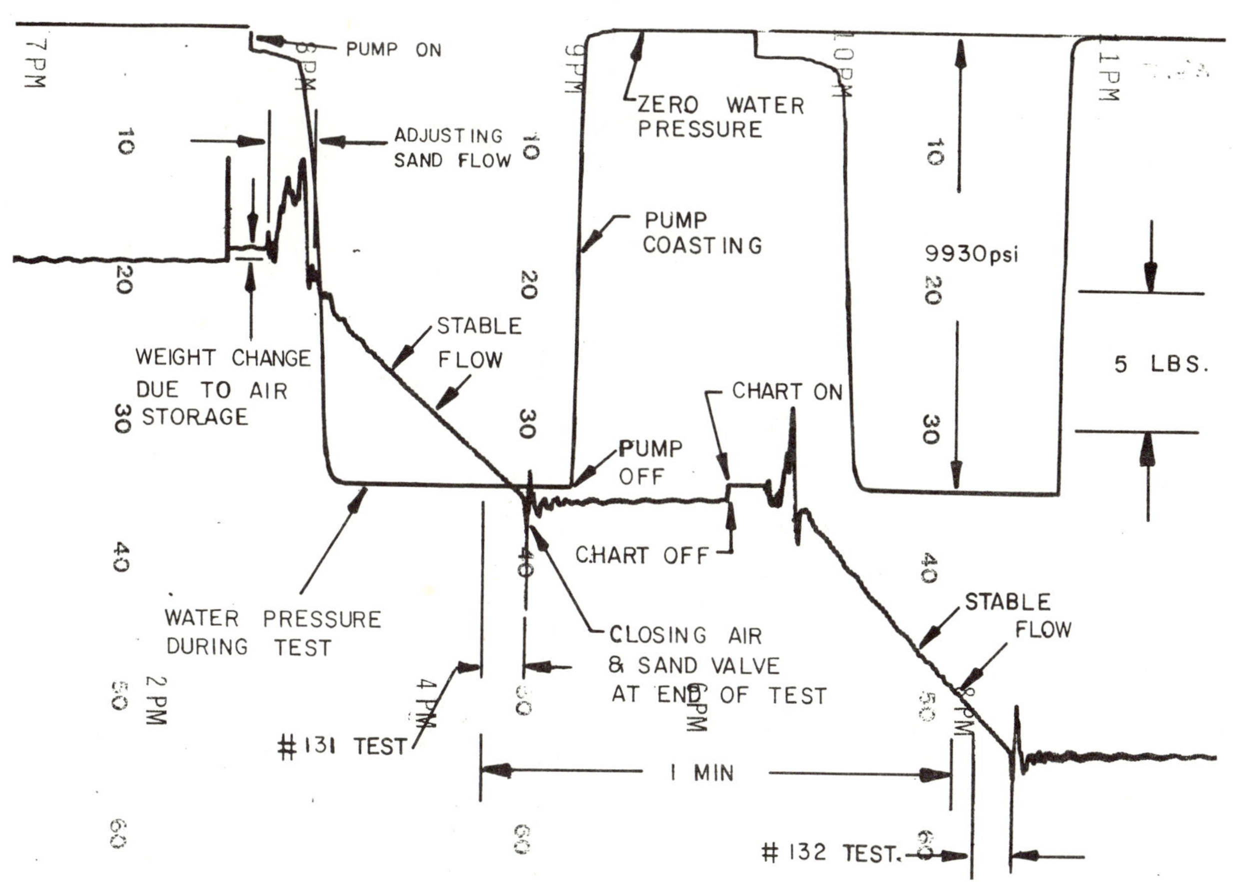

Figure 7. Typical cutting test record.

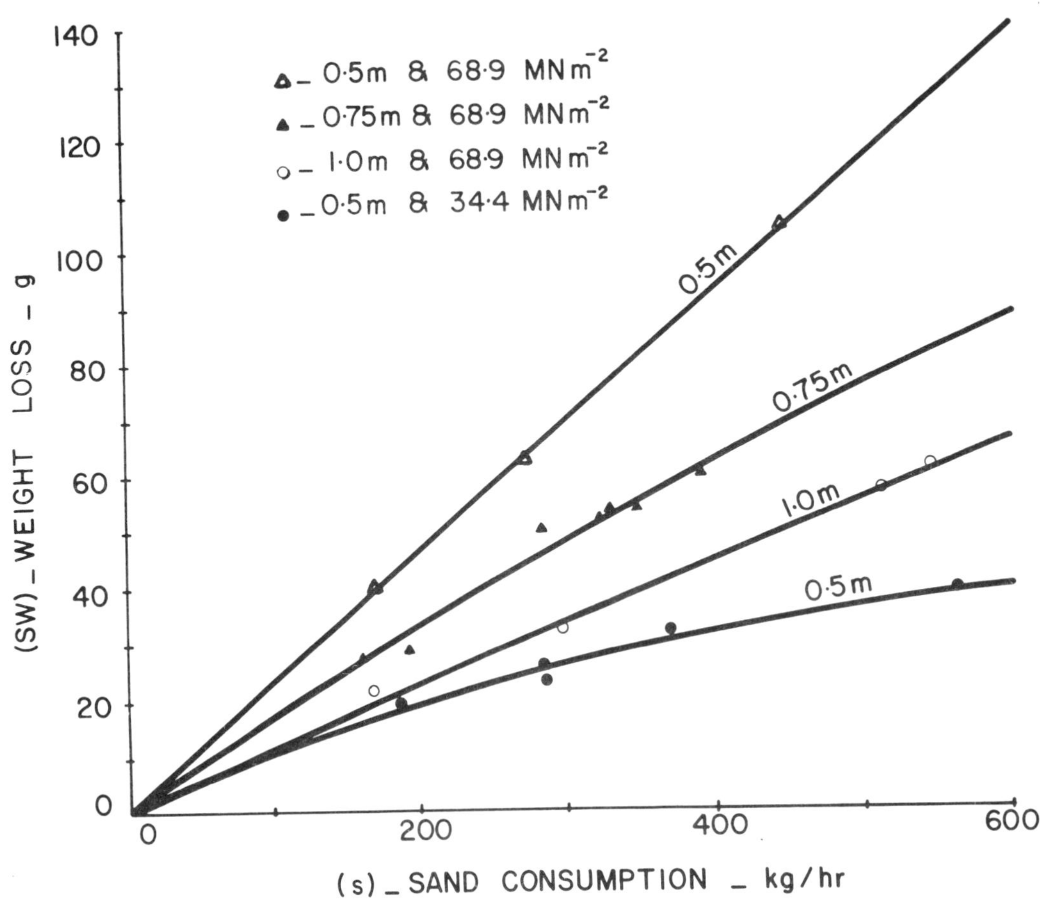

Figure 8. Weight loss for six jet nozzle with cone chamber.

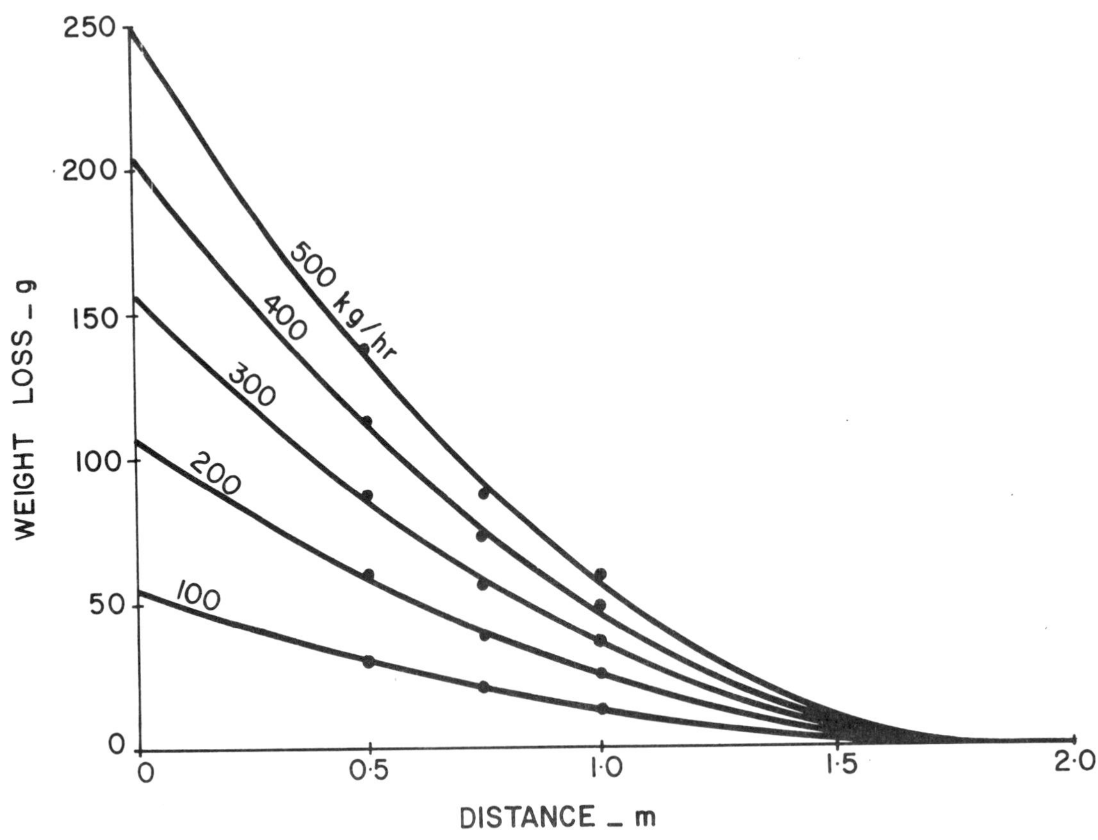

Figure 9. Weight loss for constant sand consumption.

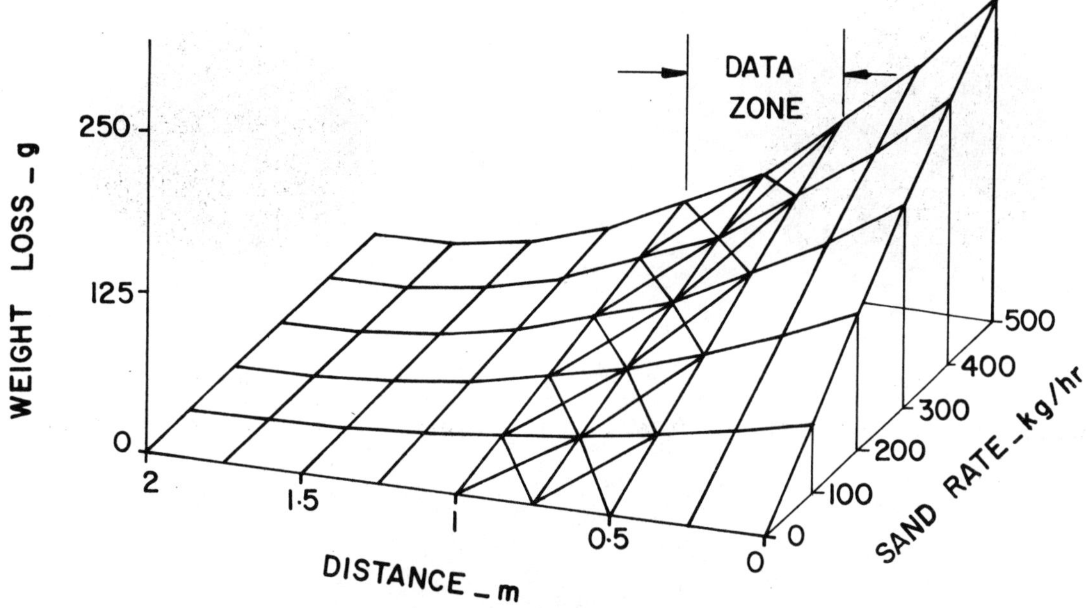

Figure 10. Weight loss for variation with distance and sand consumption.

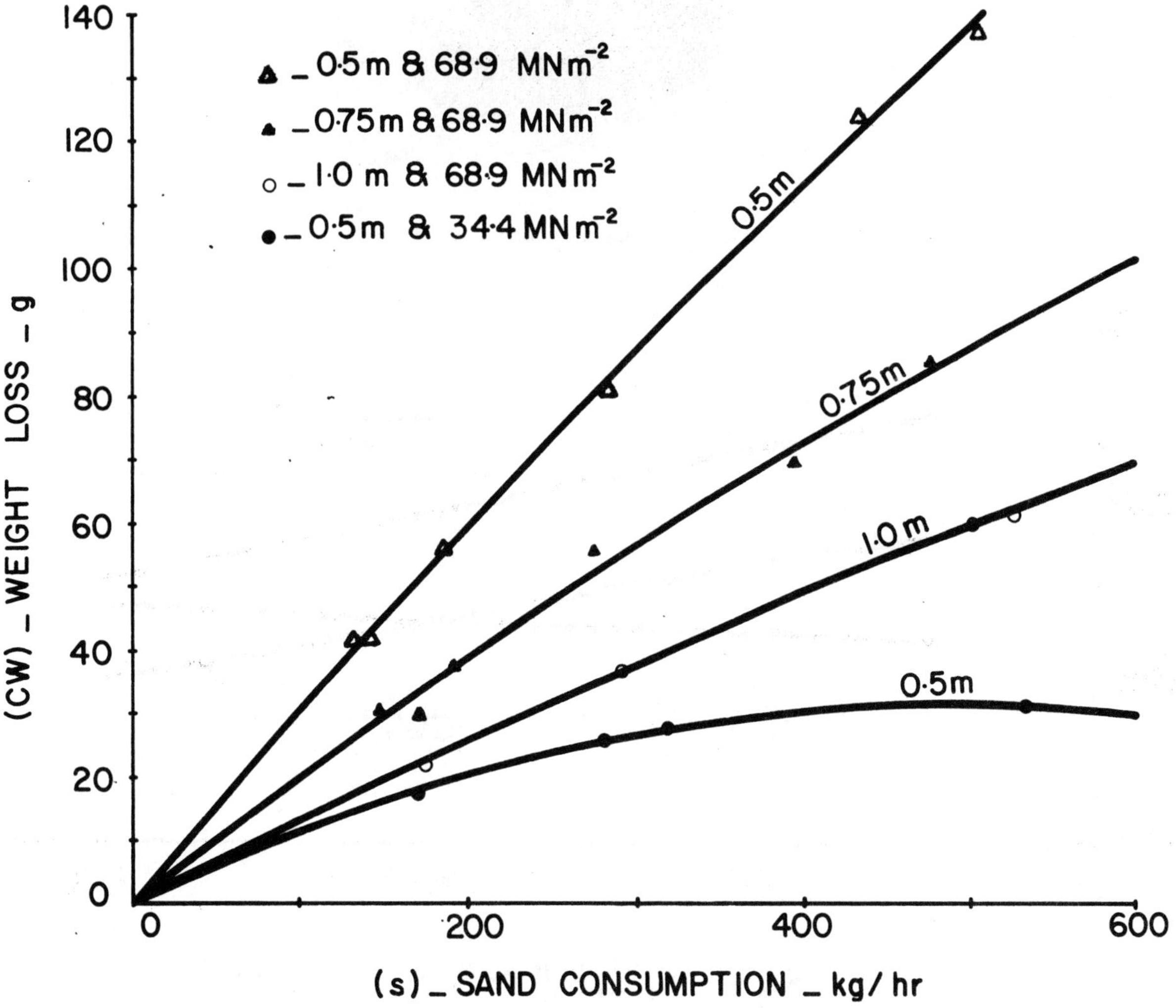

Figure 11. Weight loss for six jet nozzle without cone chamber.

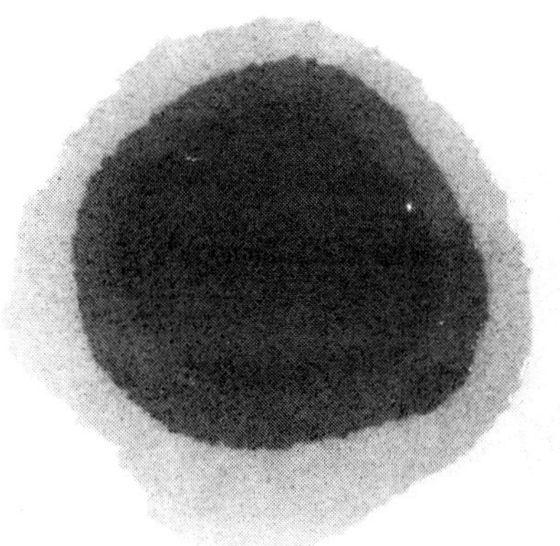

Figure 12.
Damage zone for sample #6 - 6 jets.

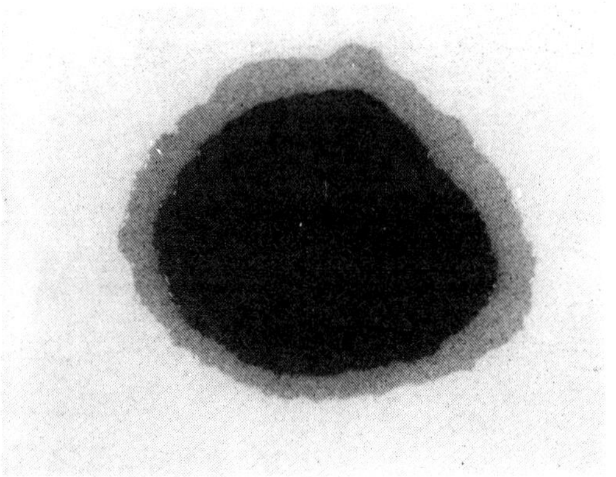

Figure 13.
Damage zone for sample #9 - 6 jets.

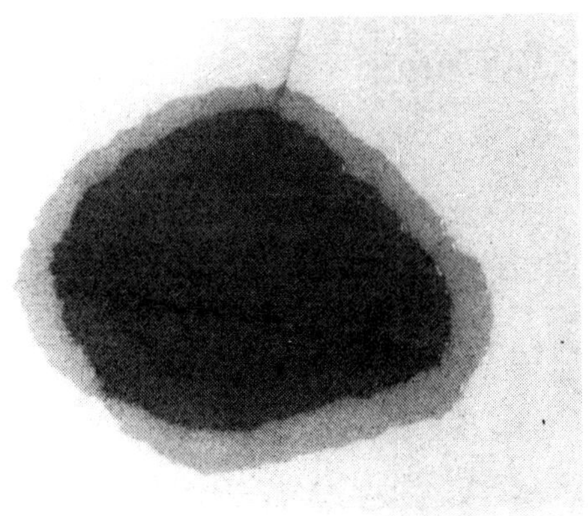

Figure 14. Damage zone for sample #16 - 6 jets.

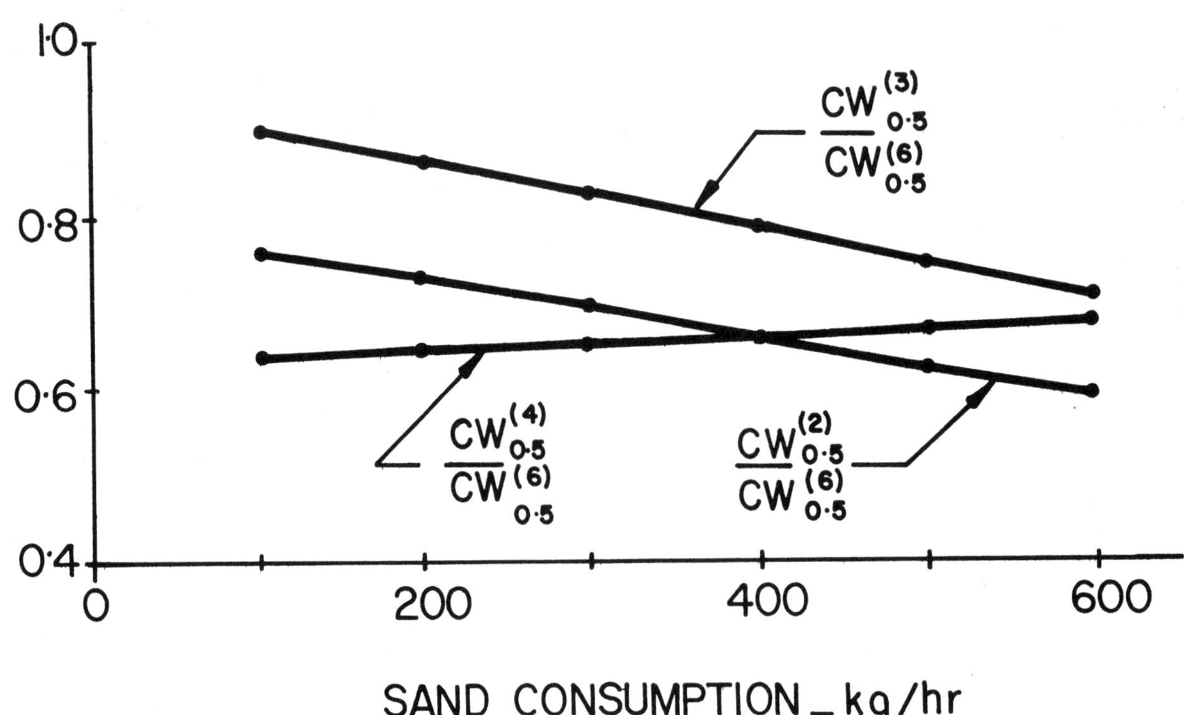

Figure 15. Cutting performance comparison at 0.5 m with cone.

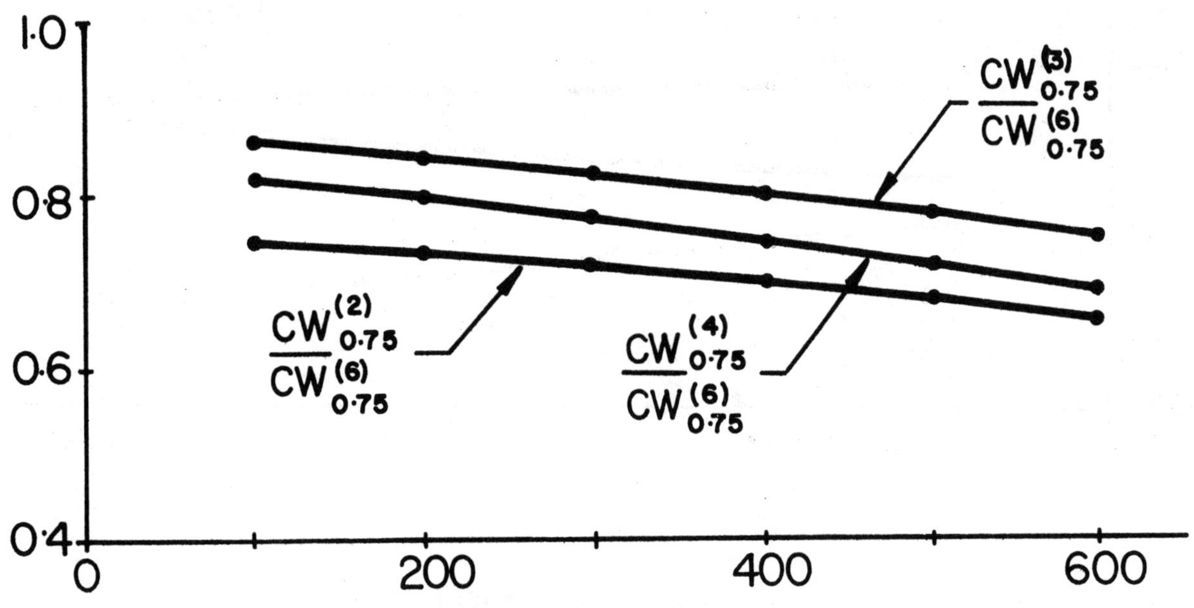

Figure 16. Cutting performance comparison at 0.75 m with cone.

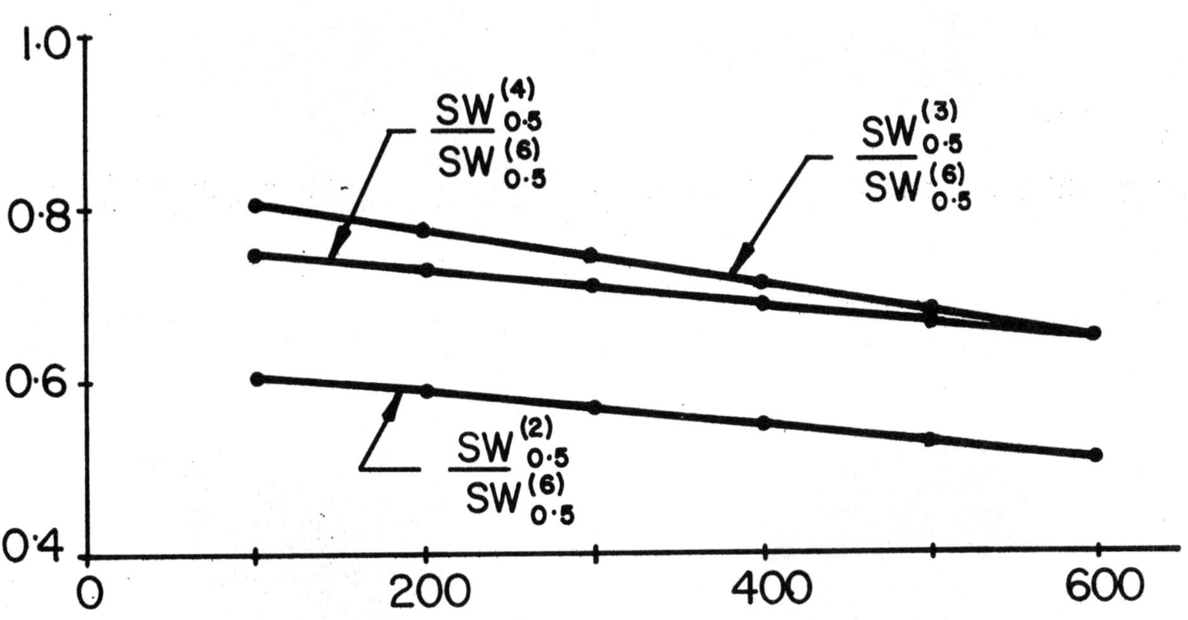

Figure 17. Cutting performance comparison at 0.5 m without cone.

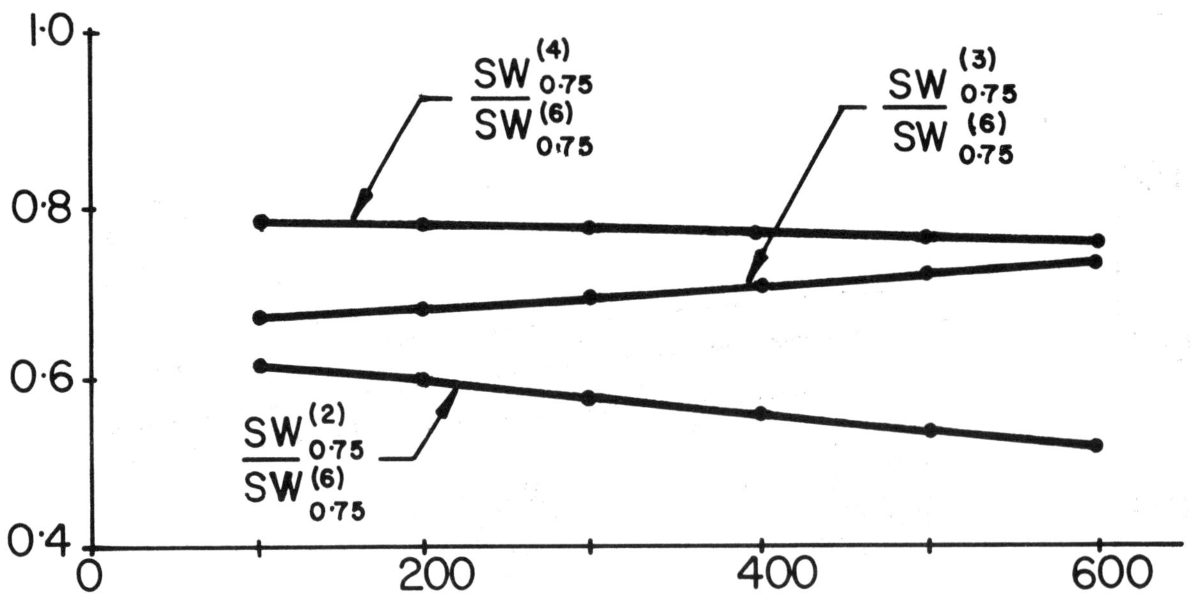

Figure 18. Cutting performance comparison at 0.75 m without cone.

Figure 19.

Damage zone for sample #67 - 2 jets.

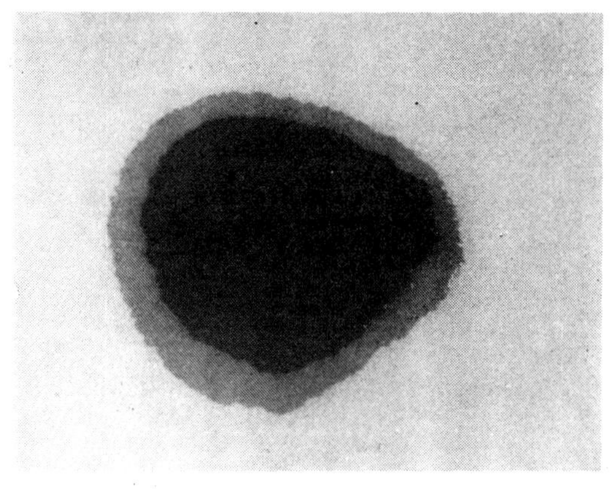

Figure 20.

Damage zone for sample #129 - 3 jets.

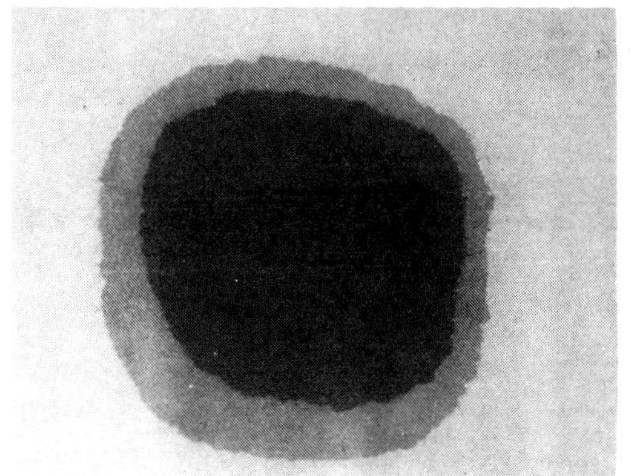

Figure 21.

Damage zone for sample #102 - 4 jets.

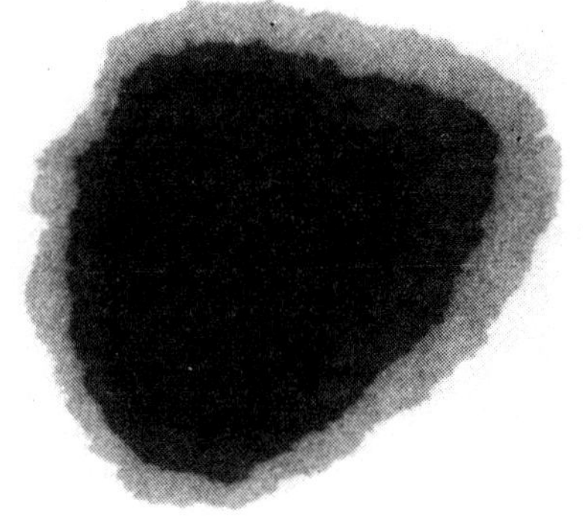

Figure 22.

Damage zone for sample #23 - 6 jets.

6th International Symposium on
Jet Cutting Technology
6-8, April, 1982

THE APPLICATION OF ABRASIVE JETS TO CONCRETE CUTTING

M. Hashish

Flow Industries Inc., U.S.A.

Summary

A study was undertaken to develop an abrasive jet cutting nozzle for concrete. This paper presents the experimental results on the effects of various abrasive jet cutting parameters. An optimization kerfing scheme is also presented. The investigation showed the great potential of abrasive jet cutting for a wide range of applications. For example, a 37.5 kW abrasive jet stream was shown to be capable of cutting concrete and steel reinforcing rebars to depths up to 380 mm without adjusting the original standoff distance at advance rates of up to 5 m h^{-1}.

Held at the University of Surrey, U.K.
Symposium organised and sponsored by
BHRA Fluid Engineering

NOMENCLATURE

G = garnet (the number following the variable represents mesh size)

N = number of passes

P = pressure

S.C. = silicon carbide (the number following the variable represents mesh size)

S.S. = silica sand (the number following the variable represents mesh size)

U = traverse rate

d_n = nozzle diameter

h = depth of cut

$\dot{m}_a$ = abrasive flow rate

1. INTRODUCTION

High-pressure waterjets show great potential for cutting and drilling minerals, rock, and concrete, especially when compared with conventional cutting and drilling tools which are inherently limited by their lack of both strength and resistance to abrasion. Such limitations are in direct contrast to waterjet cutting, in which the cutting rate is limited only by the available pumping capacities and there is no tool to wear out.

The Electric Power Research Institute (EPRI) and the Gas Research Institute (GRI) funded research programs at Flow Industries, Inc., to study methods of increasing waterjet cutting capabilities. Although the two programs were separate, both involved investigation into the use of high-pressure waterjets for concrete cutting, including deep trenching and cutting access manholes for underground utilities. This paper presents the results of both of these research programs.

Our approach to this research was to develop a high-pressure abrasive jet which could increase waterjet cutting capabilities for both deep and fast penetration in concrete. Abrasive jets are presently used in the sandblasting industry at relatively low pressures. A successful demonstration of the capabilities of abrasive jet cutting of concrete and steel reinforcing rebars will make it possible to extend this technology to a wide range of applications. In this paper, the results of our investigation on concrete cutting are presented, along with a performance comparison between abrasive jet and waterjet cutting.

2. WATERJET CONCRETE CUTTING RESEARCH

A summary of research on concrete slotting with waterjets is given in Table 1. From this survey the following conclusions can be made:

1. The power levels required for practicable concrete cutting systems using waterjets are generally higher than 150 kW.
2. The pressure required to cut concrete aggregates is in excess of 276 MPa.
3. The maximum depth for multipass surface cutting (where the jet stays on the surface) is 65 mm.
4. Deep kerfing up to 200 mm requires that the jet(s) enter the kerf by first cutting a wide slot. This has many disadvantages:
   a. The kerf should be wide enough to feed the nozzle into, which results in additional energy losses.
   b. It is difficult to exceed 125 mm in depth, as the slot becomes narrower with depth.
   c. The standoff control mechanism exhibits much wear as it moves within the kerf.
5. Waterjets may be superior to diamond saws in terms of slotting rates and productivity. However, specific energy values are much higher.
6. Waterjets are not capable of cutting steel reinforcing rebars.
7. Kerfing with waterjets is associated with spalling edges.

The strategies presented in the different publications listed in Table 1 (Refs. 1 to 7) for efficient cutting of concrete are somewhat similar. They all employ a multipass cutting mechanism at high traverse rates. For deep kerfing, either a rotating nozzle from which all the hydraulic power is delivered or a set of nozzles with narrow spacings is used to cut a wide slot to further allow the feed of the nozzle(s) head.

Other waterjet cutting research has focused on pulsed jets for the breaking and removal of pavements (Refs. 8 and 9). The availability of free surfaces proves to enhance the fracturing process. Also, large-diameter nozzles, 2.5 mm at 280 MPa, are needed. The different requirements for concrete cutting include:

1. Shallow slotting for controlling skidding due to hydroplaning on roads.
2. Deep kerfing for cutting manholes in the utility industry.
3. Removal of the top concrete layer for repairing roads.

4. Cutting and demolishing in the construction industry.
5. Slotting for laying cables.
6. Cutting highway concrete for expansion joints.
7. Trenching for the prevention of crack propagation.
8. Breaking pavement with pulsed jets for underground utility maintenance.

A performance comparison of waterjet cutting with other conventional methods, such as pneumatic hammers, rotary percussive drills, explosives, thermic lances, flame jets, plasma arcs, and diamond saws, is discussed by McCurrich and Browne (Ref. 1).

3. EXPERIMENTAL SETUP AND PROCEDURE

Very hard concrete blocks, 200x200x300 mm, were used in the tests. The concrete contained granite, river rock aggregates; the average size of these aggregates was 10 mm in diameter and the maximum size was 25 mm. These blocks were cast with a design compressive strength of 34.5 MPa about four years prior to these experiments. The weathering effect may have increased the compressive strength to higher values than 34.5 MPa. Steel reinforcing rods 10 mm in diameter were cast in some samples. A rotary table was used to traverse the concrete blocks up to 5.8 mm s^{-1}.

The parameters involved in the investigation were the abrasive material flow rate, the pressure, the traverse rate, and the number of passes. The nozzle diameter (0.635 mm) and standoff distance (7 mm) were held constant in all the tests. The abrasive flow rate during each test was measured by weighing the abrasives container before and after cutting. The depth of cut was measured after each test at at least five locations and averaged.

4. CONCRETE CUTTING RESULTS

The effects of each of the different parameters on the depth of cut are described in the following subsections.

4.1 Effect of abrasive material.

Fig. 1 shows the effect of the abrasive material on the depth of cut at two different traverse rates. The garnet abrasives proved to be more effective than other types. The silicon carbide abrasive is expensive, but it showed effectiveness in penetration. Silica sands are the least effective, but they may have potential when economic considerations must be made.

4.2 Effect of abrasive flow rate.

The depth of cut as a function of the abrasive flow rate is expected to have an optimum as shown in Fig. 2. The probable reason for this is discussed by Hashish (Ref. 10) for steel cutting. The optimum abrasive flow rate is about 68 g s^{-1}; however, an abrasive flow rate of 38 g s^{-1} is only 11 percent less effective. This consideration is important for an economic study of an abrasive jet cutting system if abrasives are not to be recycled. The optimum abrasive flow rate should also be a function of the nozzle size, pressure, and grit size. These factors control the momentum transfer between the two phases of the abrasive jet, which in turn determines the optimum abrasive flow rate.

4.3 Effect of pressure.

Fig. 3 shows the effect of pressure on the depth of cut achieved after three passes at 3.8 mm s^{-1}. The reason for cutting three passes was to obtain a more representative comparison for deep kerfing, where the standoff distance may limit the effect of the jet at low pressures. The dotted lines represent the results that should optimally be expected if the depth of cut is linearly proportional to the power. The most significant result of these tests is that pressures lower than that required for a pure waterjet can efficiently cut concrete, including the aggregate.

4.4 Effect of number of passes.

Random conditions of the abrasive type, size, and flow rate are plotted for the effect of the number of passes on the cumulative depth of cut in Fig. 4. The lowest curve shows the limited effect of waterjet cutting at the shown hydraulic parameters. This limiting value of depth (50 mm) will not improve by optimizing the traverse rate and the number of passes, as optimization can only reduce the specific energy but not increase the limiting depth. This maximum cumulative depth of cut for waterjets was also reported by Hamada et al. (Ref. 3) for the same type of concrete at more optimized conditions.

The top curve in Fig. 4 shows that abrasive jets can reach at least a 380-mm depth (the thickness of the concrete block for that test was 380 mm at an equivalent forward speed of 0.4 mm s^{-1}). This, of course, does not represent the optimum conditions, as will be discussed later, but rather shows the potential of abrasive jets. The slope of that curve, however, indicates that greater depths can be reached with more passes. A deeper penetration test was not conducted, as no sample thicker than 380 mm was available.

The middle five curves in the figure show similar trends to penetrate 300-mm-thick concrete blocks at different numbers of passes and abrasive flow parameters. The slopes of these lines show the linear (or close to linear) relationship between the depth of cut and the number of passes between 75- and 300-mm cumulative depths. The reason behind this behavior could be that the acceleration of the abrasive particles in the kerf, which acts to keep the jet coherent, overcomes the frictional drag of the kerf walls. For straight waterjet cutting, the jet will always lose effectiveness due to wall interaction and disintegration. The results shown in Fig. 4 indicate that the need to design a nozzle feed mechanism to enter the kerf is eliminated for abrasive jet cutting.

4.5 Effect of traverse rate.

Fig. 5a shows the effect of the traverse rate on a single pass depth of cut for two different cases. The relationship is quite linear between 2.5 and 5.0 mm s^{-1} traverse rates. The rate of increase of the depth of cut below this range is much higher. Fig. 5b shows the rate of surface area generation, which is the depth of cut (h) multiplied by the traverse rate (U), versus the traverse rate. The curves show an optimum traverse rate of at least 5.0 mm s^{-1} for a maximum surface area generation value. This criterion, however, may be misleading to use for deep kerfs, as the number of passes has to be included. For shallow kerfs, say up to 25 mm, the criterion of h x U maximization may be feasible. According to the effectiveness of abrasive jets, much higher values than those on Fig. 5b for the rate of surface area generation can be achieved for shallow cuts when only the top layer of concrete has to be removed or when anti-skid grooves are required, for example. An additional optimization procedure will be required, in general, to include all other factors, not only the traverse rate. The traverse rate effect at different numbers of passes is shown in Fig. 6. Optimization of the traverse rate and number of passes is discussed more thoroughly below.

4.6 Slotting optimization.

Full optimization of abrasive jet cutting is beyond the scope of this paper. However, an attempt is made here to provide some information which can be used as part of an optimization scheme that takes into account all economic, technical, and operational considerations. If our concern is to achieve a certain depth of penetration at a certain power level and operating pressure (a flexible hose limit, for example), optimization around the abrasive parameters, the traverse rate, and the number of passes will be required. At the optimum abrasive parameters, as can be determined from the ranges of the given parameters, a set of curves as shown in Fig. 7 can be generated. These curves represent the cumulative depth of cut as a function of the number of passes at a carefully selected set of traverse rates. If the slowest traverse is U_1, then other traverse rates can be determined from $U_n = U_1 \times n$, where n = 2, 3, 4, ..., N/2, with N as the maximum number of passes. For example, if the minimum traverse rate is 1.3 mm s^{-1} and the maximum number of passes for the investigation is 12, then a good set of traverse rates is 1.3, 2.6, 3.9, 5.2, and 6.5 mm s^{-1}.

This set may be reduced or increased depending on the trend of the results during the course of the experiment. Equal time curves, the dotted lines on Fig. 7, can then be used to determine the optimum traverse rate and number of passes for different target depths of cut. These lines--equal time or energy lines--satisfy the relationship U/N = C and their peaks indicate a minimum specific energy condition.

Fig. 7 shows that the optimum traverse rate for a 300-mm cumulative depth of cut is about 4 mm s^{-1}. The associated number of passes is 10. For shallower depths, the optimum traverse rate is reduced until a certain critical depth below which the optimum traverse rate increases. Fig. 8 shows a selection chart for the traverse rates deduced from Fig. 7 to achieve a certain required depth of cut. For depths below 100 mm, the traverse rate should be higher than 7.7 mm s^{-1}. From Fig. 7 it can be seen that the associated number of passes for a 100-mm depth is 3. These data, of course, are only valid at the selected power and abrasive parameters. Another useful plot, Fig. 9, shows the machine advance rate (U/N) versus the traverse rate U for a different required cumulative depth of cut. An advance rate of 4.6 m h^{-1} can be achieved to slot 200-mm-thick concrete with higher confidence at a 3.8 mm s^{-1} traverse rate and 3 passes than at lower values of traverse rate and fewer passes. This can be seen in Fig. 9, as this condition will actually result in a slightly deeper penetration than 200 mm. If the target cumulative depth of cut is only 100 mm, half of the previous case, the advance rate will be at least 14 m h^{-1}, which is approximately 3 times faster than the previous case. These advance rates are obtained with only a 37.5 kW jet stream. This indicates that two jets with output power of 37.5 kW each can be used to cut a 200-mm slot at an advance rate of 14 m h^{-1} by lowering the follower jet into the slot to only 80 mm. This requirement is not severe, as the kerf edges produced by abrasive jet cutting are straight and clean to at least a 120-mm depth, as will be seen later.

4.7 Enforcing steel rebar cutting.

Some concrete test samples contained four steel rebars, each 10 mm in diameter and 75 mm away from the parallel concrete surfaces. The top rebars were 80 mm below the exit of the abrasive jet, while the bottom rebars were 127 mm lower. The top rebars were cut simultaneously with the top concrete layer at a 1.3 mm s^{-1} traverse rate using garnet mesh No. 60-80 with a 37.5 kW jet stream. The lower rebars were only abraded 1 mm deep after 3 passes. Fig. 10 shows a picture of a rebar cut in concrete.

4.8 Kerf quality.

One of the most important advantages of concrete slotting with abrasive jets is the quality of the kerf. Fig. 11 shows typical quality cuts in concrete. The spalling effect of waterjet cutting is not observed with abrasive jet cutting. Also, the edges of the kerf are straight and parallel to the jet to at least a 120-mm depth. The roughness of the kerf will increase slightly for deeper penetrations. Fig. 12 shows a cross section along a 300-mm-deep kerf. The smoothness and straightness of the kerf along the top 120 mm will have a great advantage in allowing the jet to enter the kerf if deep kerfs at high advance rates are required. The smoothness of the edges, especially when cutting a brittle material such as concrete, suggests that the shear failure mechanism plays an important role in the complicated slotting mechanism. Direct impact on brittle materials always results in spalling and side cracks, as has been observed in cutting concrete, glass, and coal with waterjets. The increased roughness observed for depths more than 120 mm indicates that the shear mechanism is not the dominant factor beyond 120 mm and the relative smoothness of the kerf below this depth is due to the sandblasting effect of the backflow.

4. CONCLUSIONS

The results of this investigation into cutting concrete and steel reinforcing rebars with abrasive waterjets can be summarized as follows:

1. Abrasive waterjet cutting of concrete has great potential advantages for commercial applications in the near future.
2. The advantages of abrasive jet cutting of concrete are:
   a. High advance rates.

b. Low power requirements.
c. The capability of cutting steel reinforcing rebars.
d. A high-quality cut without spalling or chipping.
e. No need to adjust standoff distance for deep cutting.

3. For shallow penetration, up to 50 mm, the abrasive jet advance rate is at least 10 times faster than waterjet cutting using the same power.

4. The depth limitation of abrasive jets is at least 8 times more than waterjets without adjusting the standoff distance.

5. There exist optimum parameters for cutting concrete with abrasive jets, such as the abrasive material, size, and flow rate, the traverse rate, the number of passes, and the pressure, which will result in a minimum specific energy for slotting.

6. Further research is needed to fully optimize an abrasive jet concrete cutting system.

5. ACKNOWLEDGEMENT

The author is grateful to the Electric Power Research Institute (EPRI) and the Gas Research Institute (GRI) for providing funds for this research. The discussions and comments of Dr. J. Reichman, Dr. J. Cheung, and Mr. H. Massenburg were most helpful. The author also wishes to acknowledge the efforts of Mr. A. Deforrest, Mr. C. Anderson, Mr. D. Peecher, and Mr. B. Burwell during the experimental work.

6. REFERENCES

1. McCurrich, L. H., and Browne, R. D.: "Application of waterjet cutting technology to cement grouts and concrete." In: Proc. 1st Int. Symp. on Jet Cutting Technology (Coventry, U.K.: April 5-7, 1972) Cranfield, U.K., BHRA Fluid Engineering, 1972.

2. Olsen, J.: "Jet slotting of concrete." In: Proc. 2nd Int. Symp. on Jet Cutting Technology (Cambridge, U.K.: April 2-4, 1974) Cranfield, U.K., BHRA Fluid Engineering, 1974.

3. Hamada, H., Fukuda, T., and Sijoh, A.: "Basic study of concrete cutting by high pressure continuous jets." In: Proc. 2nd Int. Symp. on Jet Cutting Technology, (Cambridge, U.K.: April 2-4, 1974) Cranfield, U.K., BHRA Fluid Engineering, 1974.

4. Northworthy, A. G., Mohaupt, U. H., and Burns, D. J.: "Concrete slotting with continuous waterjets at pressures up to 483 MPa (70 ksi)." In: Proc. 2nd Int. Symp. on Jet Cutting Technology, (Cambridge, U.K.: April 2-4, 1974) Cranfield, U.K., BHRA Fluid Engineering, 1974.

5. Hilaris, J. A., and Labus, T. J.: "Highway maintenance application of jet cutting technology." In: Proc. 4th Int. Symp. on Jet Cutting Technology (Canterbury, U.K.: April 12-14, 1978) Cranfield, U.K., BHRA Fluid Engineering, 1978.

6. Reichman, J., Kirby, M., and Rodenbaugh, T.: "The development of a waterjet cutting system for trenching in concrete." In: Proc. 5th Int. Symp. on Jet Cutting Technology (Hanover, W. Ger.: June 2-4, 1980) Cranfield, U.K., BHRA Fluid Engineering, 1980.

7. Hilaris, J., and Bortz, A.: "Field study for highway maintenance application of jet cutting technology." In: Proc. 5th Int. Symp. on Jet Cutting Technology (Hanover, W. Ger.: June 2-4, 1980) Cranfield, U.K., BHRA Fluid Enginering, 1980.

8. Yie, G., Burns, D., and Mohaupt, U.: "Performance of high pressure pulsed waterjet device for fracturing concrete pavement." In: Proc. 4th Int. Symp. on Jet Cutting Technology (Canterbury, U.K.: April 12-14, 1978) Cranfield, U.K., BHRA Fluid Engineering, 1978.

9. Vallve, F., et al.: "Relationship between jet penetration in concrete and design parameters of a pulsed waterjet machine." In: Proc. 5th Int. Symp. on Jet Cutting Technology (Hanover, W. Ger.: June 2-4, 1980) Cranfield, U.K., BHRA Fluid Engineering, 1980.

10. Hashish, M.: "Steel cutting with abrasive waterjets." To be published in: Proc. 6th Int. Symp. on Jet Cutting Technology (Surrey, U.K.: April 5-8, 1982) Cranfield, U.K., BHRA Fluid Engineering, 1982.

Table 1. Summary of Research on Concrete Slotting with Continuous Waterjets

INVESTIGATOR	RESEARCH OBJECTIVES	FINDINGS OF RESEARCH
L.H. McCurrich and R.D. Browne Taylor Woodrow Construction Ltd. U.K. (Ref.1)	To Study the mechanisms by which concrete may be cut with waterjets.	For a practicable site tool, the following parameters are recommended to cut a 10 mm depth at 1 m s^{-1}: P = 380 MPa d_n = 1 mm Nozzle Power = 260 kW Pump Power = 370 kW
J. Olsen Flow Industries Kent, WA, USA (Ref. 2)	To explore the possibility of constructing a practical jet slotting system using a compact pumping system and a high-power system.	Two 50-mm-deep slots can be advanced at 1 to 2 km h^{-1} with a 600 kW pump. This is predicted from lab experiments on concrete cutting studying the different influencing parameters. Eighty shallow grooves 5 to 10 mm deep can be produced by the same power at a 1 to 2 km h^{-1} advance speed. The pressure used is 413 MPa.
H. Hamada et al. Kobe Steel Ltd. Japan (Ref. 3)	To determine the optimum operating conditions of a future practicable jet cutting machine.	The proposed specifications of a practicable waterjet machine for concrete cutting has the following parameters: P = 100 MPa, Nozzle Diam. = 2 mm, Jet Power = 140 kW The lowest specific energy observed is 1,000 J cm^{-2} for kerfing, traverse = 5 mm s^{-1}, h_{max} = 50 mm.
A.G. Northworthy et al. Univ. of Waterloo Canada (Ref. 4)	To study the effect of water-jet cutting parameters on concrete slotting for skid control grooves.	An equivalent forward speed of 200 mm s^{-1} that produces a depth of 8 mm will be achieved by: P = 483 MPa, U = 510 mm s^{-1}, d_n = 0.178 mm This represents an example and not the proposed system parameters.
J. Hilaris and T. Labus IIT, Chicago (Ref. 5)	Study to access the application of waterjets for highway pavement maintenance-Task (1).	The laboratory test results are as follows: P = 551 MPa, d_n = 0.6 mm, U = 380 mm s^{-1}, h = 20 mm P = 551 MPa, d_n = 0.6 mm, U = 380 mm s^{-1}, N = 8, h = 55 mm P = 551 MPa, hU = 7 × 10^3 mm^2 s^{-1}
J. Reichman et al. Flow Industries Kent, WA, USA (Ref. 6)	To develop a waterjet cutting system for trenching in concrete.	The optimum parameters were found to be: P = 300 MPa, U = 760 mm s^{-1} h = 100 mm (Using Single Jet) h = 125 mm (Using Three Spaced Jets) Flow Rate = 0.5 ℓ s^{-1}, Power = 186.5 kW (Intensifier)
J. Hilaris and A. Bortz IIT, Chicago (Ref. 7)	Study to access the application of waterjets for highway pavement maintenance-Task (2).	The optimum parameters found from the field tests are: P = 276.5 MPa, U = 50 mm s^{-1}, d_n = 2 × 0.5 mm (0.02) rpm = 600 rpm, power = 82 kW patch depth = 70 mm (ave.) Productivity Rate = 1.13 m^2 h^{-1}

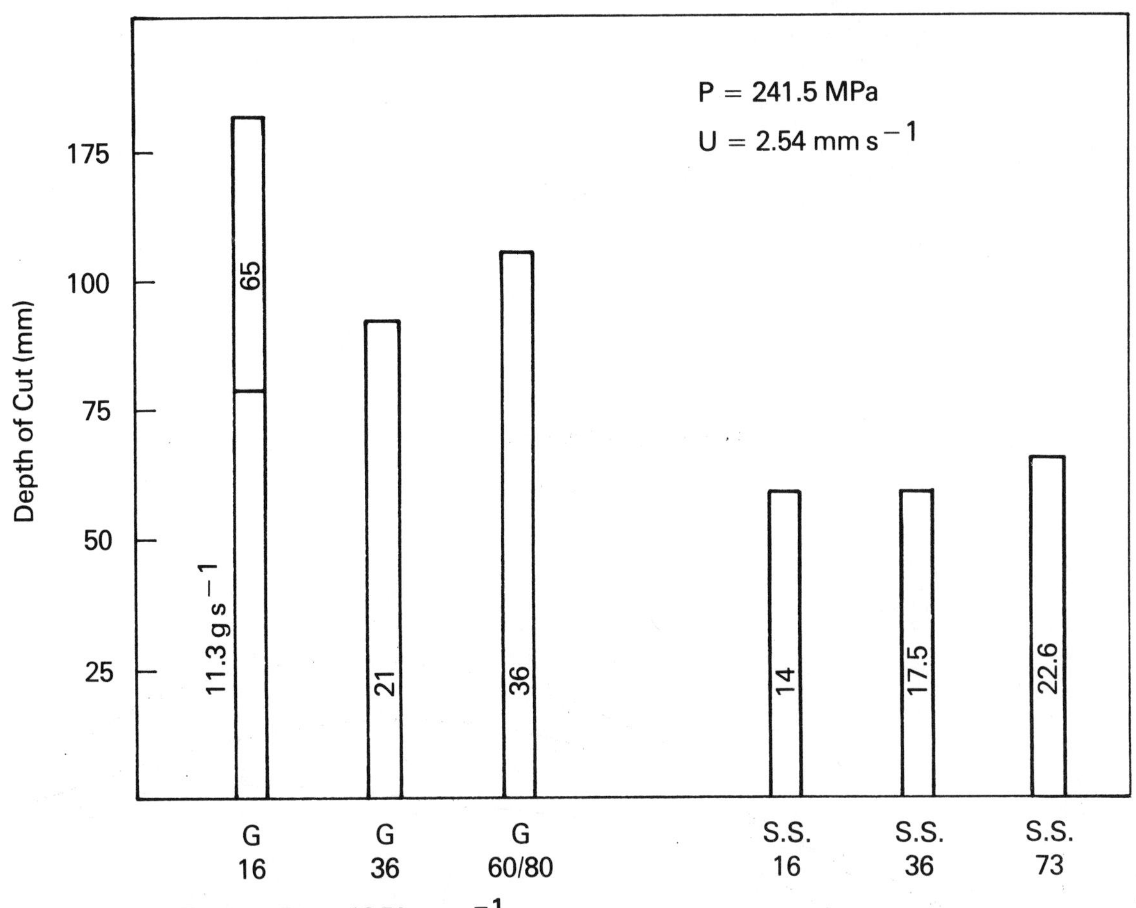

a. **Traverse Rate of 2.54 mm s^{-1}**

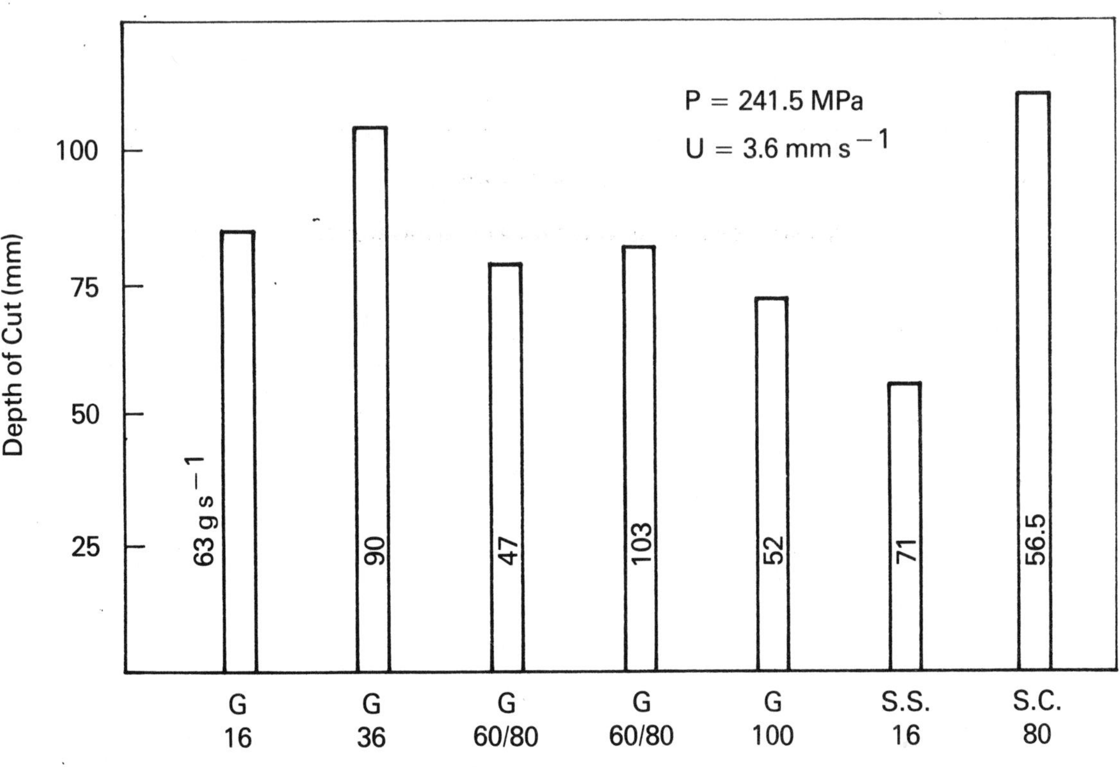

b. **Traverse Rate of 3.8 mm s^{-1}**

Figure 1. Effect of Abrasive Material on Depth of Cut

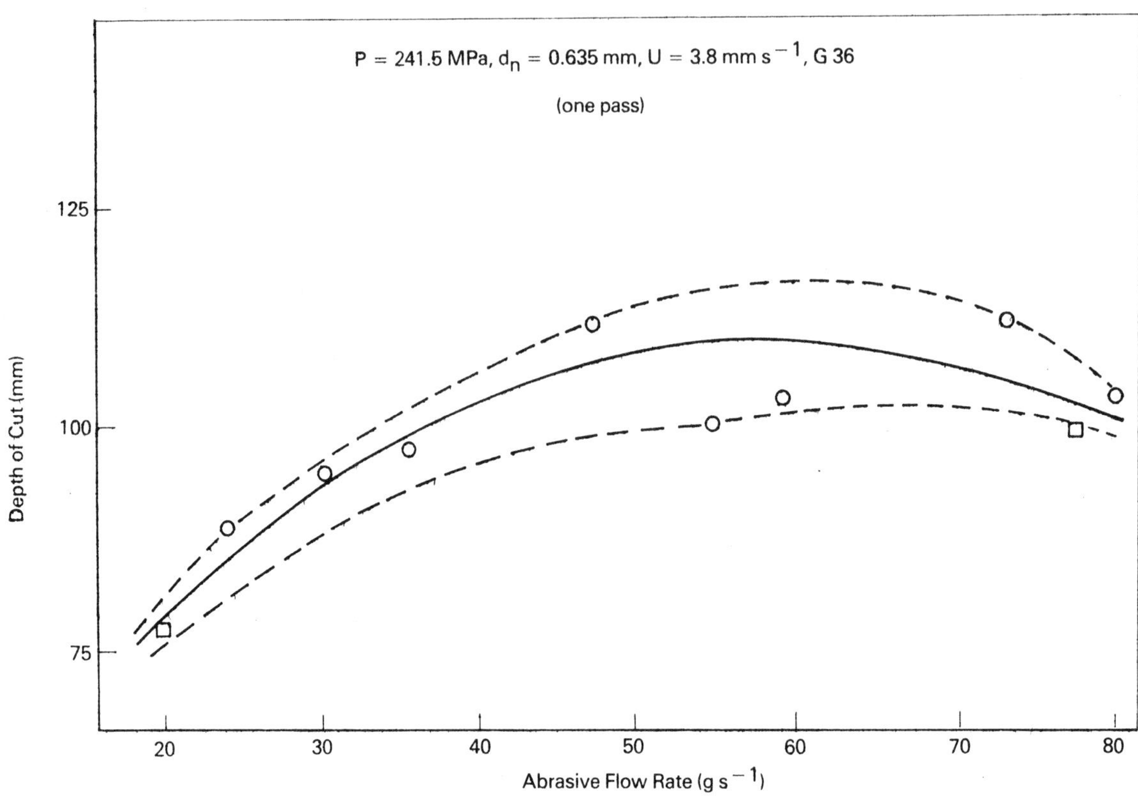

Figure 2. Effect of Abrasive Flow Rate on Depth of Cut

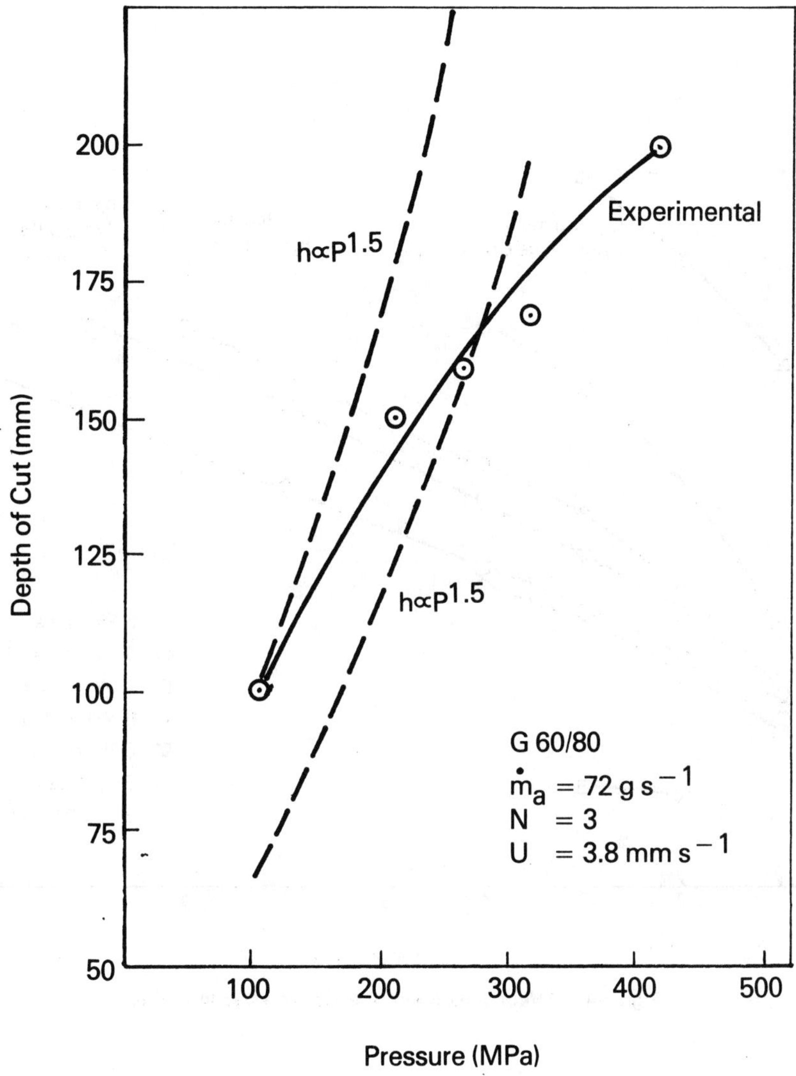

Figure 3. Effect of Pressure on Depth of Cut

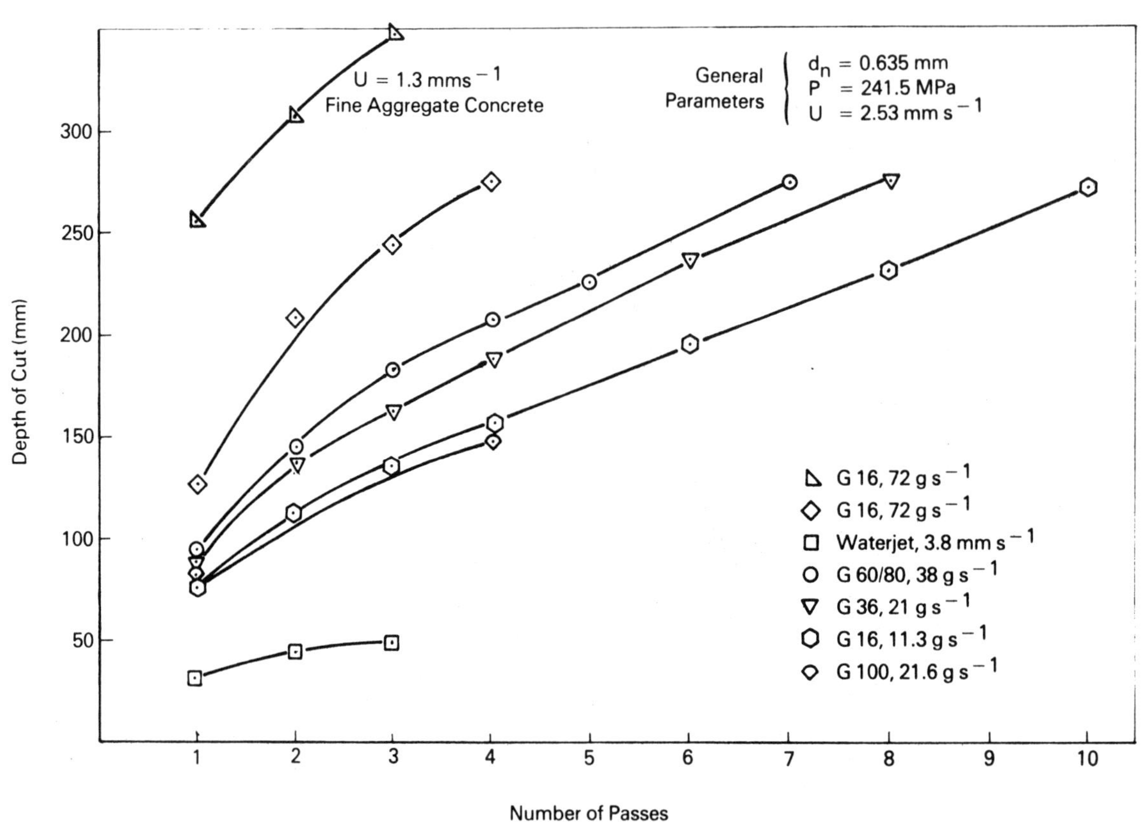

Figure 4. Effect of Number of Passes on Depth of Cut

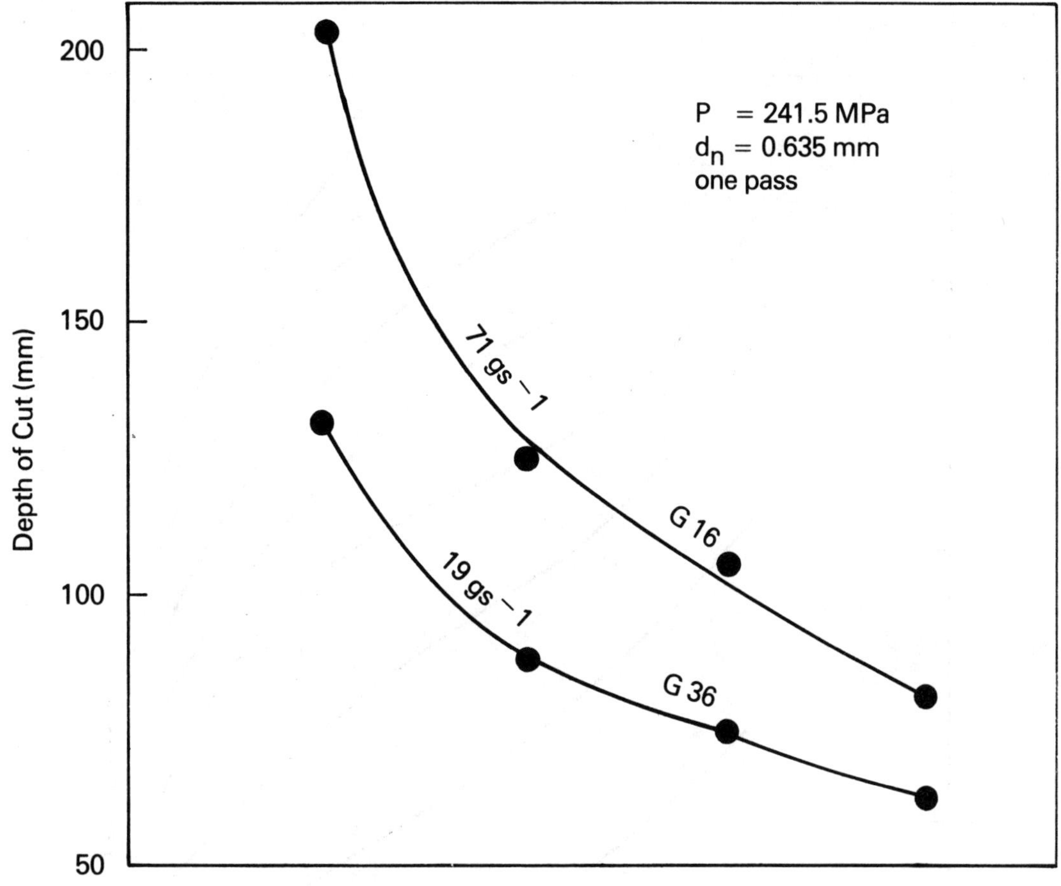

a. **Effect on Depth of Cut**

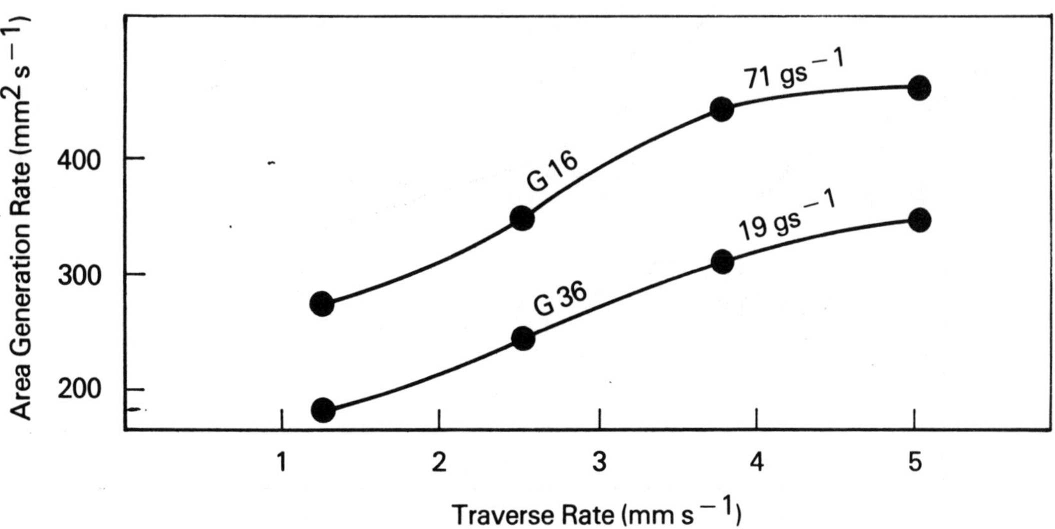

b. **Effect on Area Generation Rate**

Figure 5. Effect of Traverse Rate

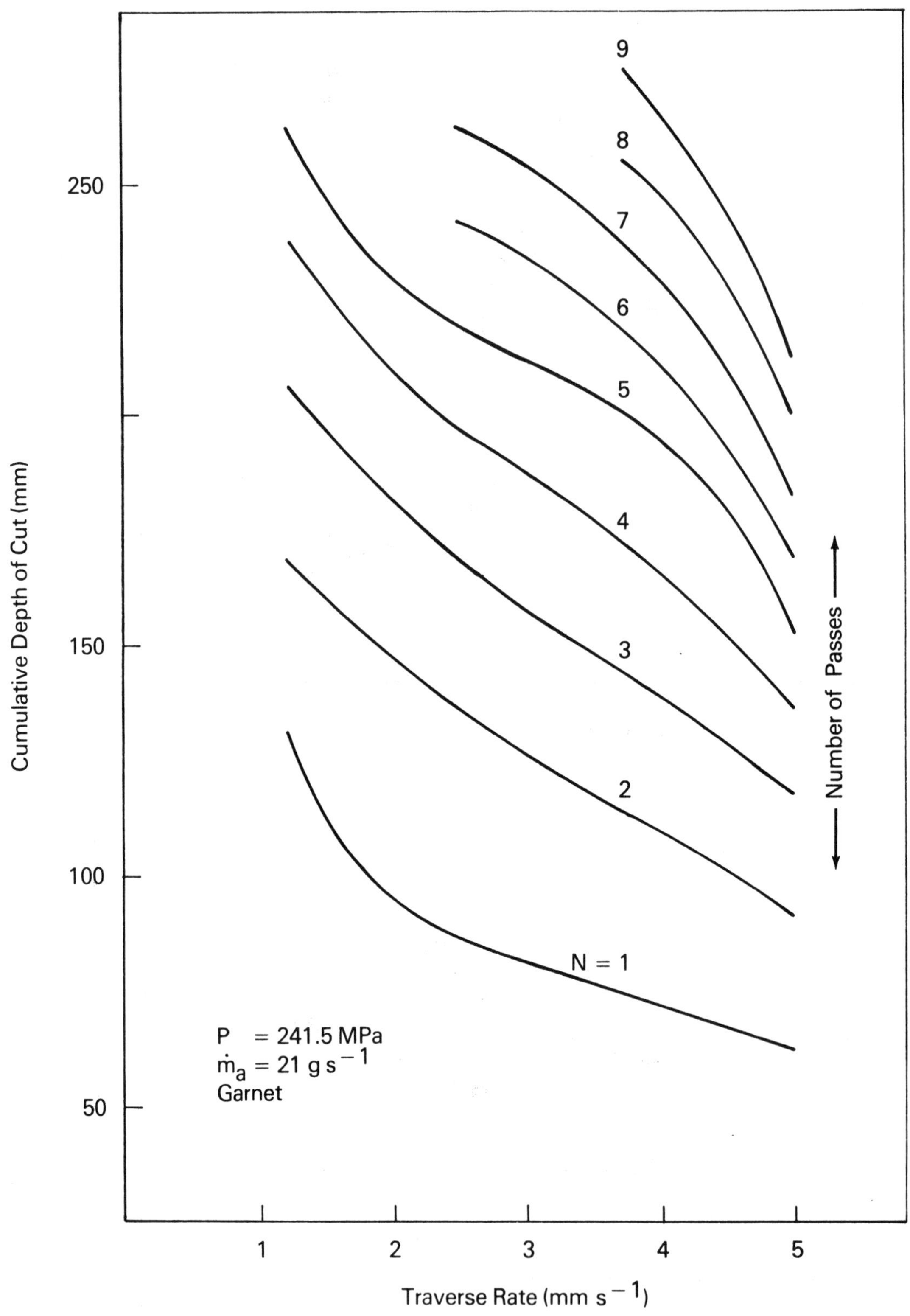

Figure 6. Effect of Traverse Rate at Different Numbers of Passes

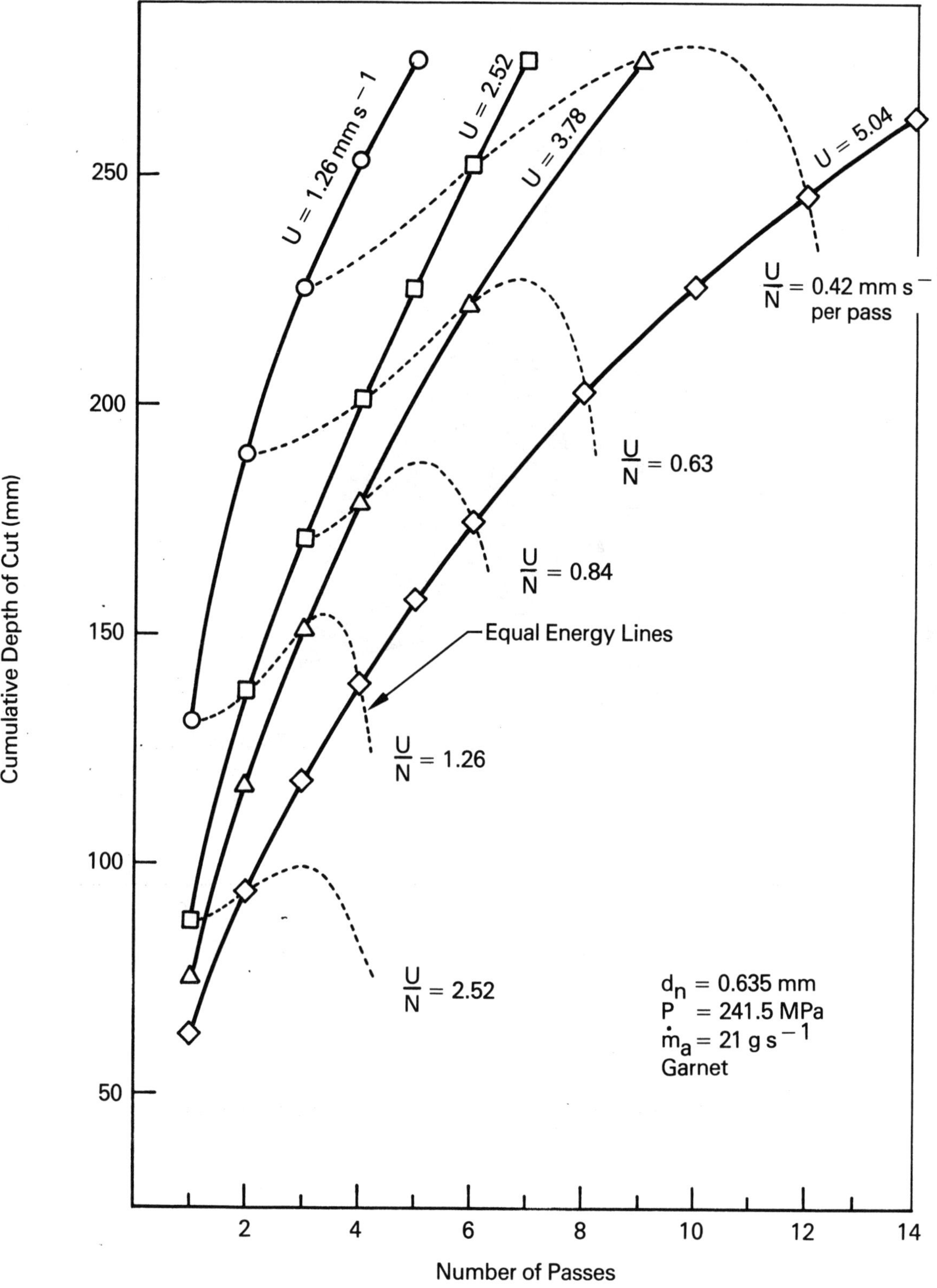

Figure 7. Effect of Number of Passes and Equal Energy Lines

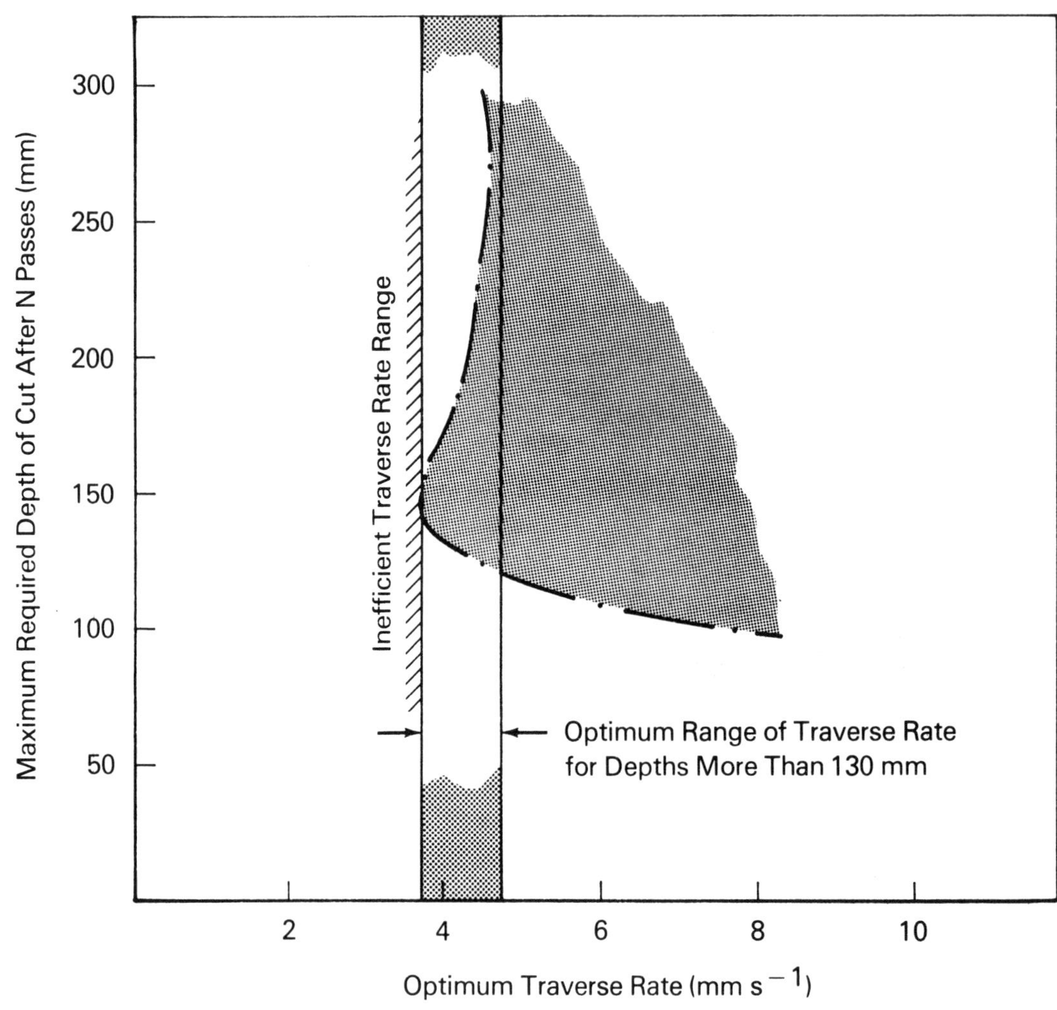

Figure 8. Traverse Rate Selection Chart for Concrete Cutting

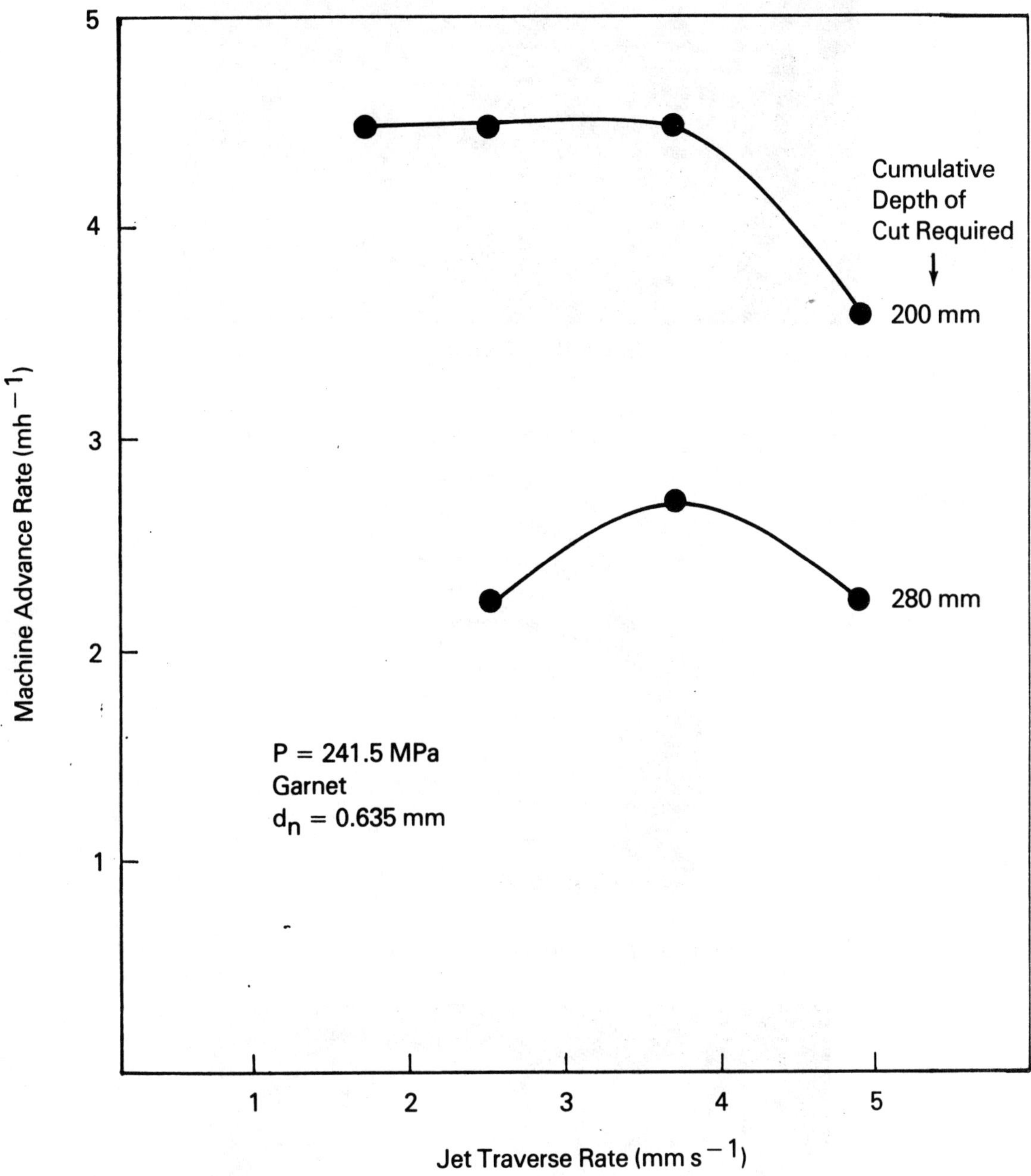

Figure 9. Machine Advance Rate for Target Depth of Cut

Figure 10. Rebar Cut

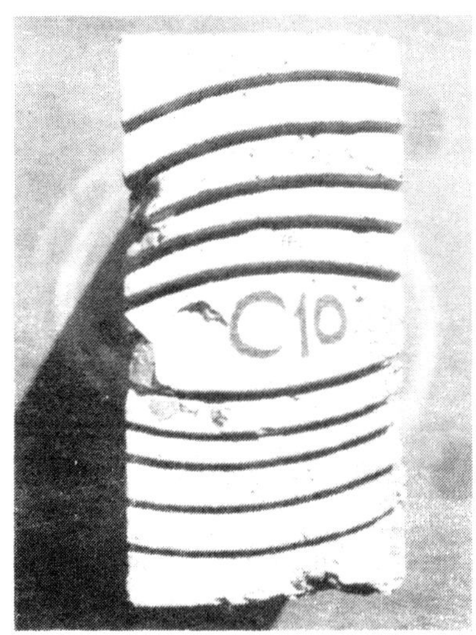

Figure 11. Concrete Cut (Top View)

Figure 12. Concrete Cut (Cross Section)

6th International Symposium on
Jet Cutting Technology
6-8, April, 1982

STEEL CUTTING WITH ABRASIVE WATERJETS

M. Hashish

Flow Industries Inc., U.S.A.

Summary

Steel cutting with high-velocity, abrasive waterjets is investigated. The effects of abrasive material, size, and flow rate, jet pressure, standoff distance, traverse rate, number of passes, and hydraulic power on the depth of cut are presented. The investigation shows the great potential of abrasive waterjets in various cutting applications. An example of the cutting results is the slicing of a 20 mm thick steel plate at a 3.3 mm s^{-1} traverse rate with a 37 kW hydraulic stream. Optimization of cutting parameters can further improve cutting results.

Held at the University of Surrey, U.K.
Symposium organised and sponsored by
BHRA Fluid Engineering

NOMENCLATURE

G = garnet (the number following the variable represents grit size)

P = pressure

U = traverse rate

V = slurry jet velocity

V_a = initial abrasive velocity

V_w = waterjet velocity

d_n = nozzle diameter

h = depth of cut

$\dot{m}_a$ = abrasive flow rate

$\dot{m}_w$ = water mass flow rate

1. INTRODUCTION

High-pressure waterjets are presently used in industrial applications to cut materials such as plastics, leathers, corrugated board, and asbestos. Waterjet cutting is also receiving increased attention as a viable method for mining mineral deposits and removing concrete for roadway repair. However, complications arise in fitting the jet(s) into a kerf for deep kerfing of concrete and rock. Also, limitations exist because the commercially available waterjet cutting devices are not capable of cutting metals, such as concrete steel rebars, metal laminates and foams, and advanced metal-reinforced composites. Although these limitations do not negate the many comparative advantages of waterjet cutting in many applications, they do indicate the need for significant advances in waterjet cutting technology.

The introduction of abrasives into the jet stream may be an excellent way to increase the cutting capabilities of high-pressure waterjets. Among the advantages of cutting with abrasive waterjets are:

1. The ability to cut almost any material.
2. Deep kerfing capability without entering the kerf.
3. No potential fire hazard is associated with the cutting process.
4. The equipment required is well within the present state of the art.
5. Low power levels.
6. Resultant high edge quality for metals, advanced composites, and rocks.
7. Easily adaptable to remote operations and control.
8. Not sensitive to the shape of cuts (contour cutting).
9. The cutting abrasives can be recycled.

This paper presents the results of research using high-velocity, abrasive waterjets to cut steel. The effects of various abrasive jet cutting parameters on the depth of cut in mild and stainless steels are presented.

2. LITERATURE REVIEW

The technique of employing abrasive streams for the removal of material has been in use for many years. Air-driven abrasive jets are presently used for surface finishing, for the removal of thin layers of materials, and in glass artwork. This technology is known as abrasive jet machining (AJM). The ultrasonic machining process is another existing example where abrasive grains in a liquid slurry are caused to impact a workpiece surface at ultrasonic frequencies, induced by a vibrating tool less than 1 mm away from the machined surface. Material removal rates in these processes are typically up to 1.5 $mm^3\ s^{-1}$ for steel and glass, and they require very low power levels and abrasive flow rates. Sandblasting techniques used for cleaning applications, on the other hand, require much higher air and/or water and abrasive flow rates. The patented devices shown in sketches a, b, and c in Table 1 are examples of the advanced techniques used in the sandblasting industry today for mixing abrasives and waterjets. Another technique, wet blasting, has recently developed to incorporate low- and medium-pressure waterjets to enhance the cleaning rates (see sketch d in Table 1).

Deep kerfing and cutting by abrasive jets is, however, a recent development in jet cutting technology. The oil industry was the first to employ abrasive mud jets to augment drill bits and increase their life by slotting formation rocks ahead of the roller bits. Several patented devices for oil well drilling are presented in sketches e, f, and g in Table 1; these devices are discussed in more detail by Maurer (Ref. 5). The high-pressure, abrasive-mud-jet drills can drill much faster than conventional drills. Severe equipment erosion and pumping problems, however, have terminated projects using this drilling technique.

New developments in jet cutting technology and the need for improved performances for cutting hard materials suggest that the abrasive waterjets are the solution. Cheung and Hurlburt (Ref. 6) and Hilaris and Labus (Ref. 7) studied the influence of various jet parameters for underwater cutting. The maximum depth of cut observed was below 2 mm, as shown on the upper graph in Fig. 1. Hilaris and Labus concluded that the abrasive jet cutting process is limited to thin sections with the parameters listed on the graph. A BHRA research team conducted studies for British Petroleum to design an abrasive waterjet cutting system for steel. The data disclosed in their annual report (Ref. 8) are brief but indicate a major success for a viable cutting system. The effect of pressure on depth of cut is shown on the lower graph in Fig. 1 for their steel cutting results.

3. ABRASIVE JET CUTTING APPROACHES

Abrasive waterjets can be classified according to the way the abrasive jet stream is formed. Basically, there are two ways to form an abrasive jet, depending on whether the abrasives are mixed upstream or downstream of the high-pressure nozzle. The advantages and disadvantages of each concept are listed in Table 2. For downstream mixing, further classifications can be made as shown in systems a, b, c, and d in Table 2. The general advantages of downstream mixing are (1) wear by abrasives is restricted to the mixing area and can be minimized, and (2) the system easily adapts with the state of the art of high-pressure pumps. Upstream mixing, on the other hand, is classified as either direct or indirect pumping, as shown in systems e and f in Table 2. The advantages of this approach lie in its more effective cutting capabilities and the compactness of the nozzle head for applications such as small-hole drilling.

The study described in this paper utilized the single-jet, side-feed, downstream-mixing approach (system a in Table 2). The objective was to generate a data matrix which shows the trends of the relative effects of different parameters. The results obtained, however, are not necessarily the maximum for such an approach. Optimization of the cutting results requires additional research.

4. EXPERIMENTAL SETUP

The experimental research described in this paper was performed to show the effects of various abrasive jet cutting parameters on the cutting results in mild and stainless steels. In this section, the abrasive jet parameters, the experimental setup and procedure, and the means of data collection and analysis are described.

4.1 Abrasive jet cutting parameters.

The parameters involved in the process of cutting with abrasive jets can be categorized as follows:

1. Hydraulic parameters
   - . Jet pressure
   - . Nozzle diameter
   - . Hydraulic power (dependent on jet pressure and nozzle diameter)
2. Abrasive parameters
   - . Type of abrasive
   - . Size of abrasive
   - . Flow rate of abrasive
   - . Type of feed (wet or dry)
   - . Pressure of abrasive slurry
3. Cutting parameters
   - . Traverse rate
   - . Standoff distance
   - . Number of passes
4. Cutting results
   - . Depth of cut
   - . Kerf width and quality

In this experiment, the depth of cut was taken as the dependent variable, and the independent variables, along with the range of values used, were as follows:

Pressure: up to 241.5 MPa with a nozzle diameter of 0.635 mm
Type and size of abrasives: wide range of commercially available materials
Abrasive flow rate: 7.57 to 75.7 g s^{-1}
Traverse rate: 1.27, 2.54, 3.81, and 5.08 mm s^{-1}
Standoff distance: up to 80 mm
Number of passes: 1 to 12

Each independent variable was incremented reasonably to provide the greatest possible number of observations. The pressure increment was 34.5 MPa; the abrasive flow rate increment was 7.5 g s^{-1}, the traverse rate increment was 1.27 mm s^{-1} and the number of passes increment was one pass. Data points between the increments were selected randomly to check their consistency with the rest of the data.

The parameters used as independent variables were varied according to the following procedure:

1. Vary abrasive type and size.
2. Vary abrasive flow rate.
3. Study effect of variations in the cutting parameters.
4. Study effect of pressure variations.
5. Conduct random parameter tests within the limit of parameters.

4.2 Experimental setup and procedure.

The equipment used included a high-pressure intensifier pump, which was used to generate the 0.635-mm-diameter jet. Pressure was limited to 241.5 MPa in most of the experiments. The feed table was derived by a linear actuator with its arm attached to the periphery of a rotating table. The location of the sample on the table determined the traverse rate.

Both mild and stainless steel cutting samples were used. The mild steel samples, 100x100x19 mm, were cut from the same plate to guarantee that physical and strength properties were the same from sample to sample. Some 15-5 PH stainless steel samples were used for cuts deeper than 19 mm, as observed at some low traverse rates and multipass cuttings.

The experimental test procedure used is given below. (The parameter variation procedure given in Section 4.1 falls within step 2 of this procedure.)

1. Weigh steel sample.
2. Perform a cutting test.
3. Reweigh steel sample and relate weight loss to average depth of cut.
4. Measure depth of cut in at least five locations.
5. Weigh abrasives before and after each cutting test to calculate abrasive flow rate.

4.3 Data collection and analysis.

After each cutting test, the depth of cut was measured at at least five locations along the kerf and averaged. Cuts with more than a 10-percent variation from the mean were repeated; this variation may result from an uneven flow of abrasives. The weight of material lost was also measured to check the mean depth of penetration (MDP). Weight loss per unit length data will be used for modeling the abrasive jet cutting process in future investigations.

The data collected were first plotted to check their trend and to determine if more data points had to be generated. Calculations using these data were conducted to determine optimum conditions. Two mixing chamber configurations were used, labeled mixing chambers (a) and (b). Care must be taken when comparing results as they are strongly dependent on the mixing chamber.

5. STEEL CUTTING RESULTS

The steel cutting results are presented in this section. The effects of each of the abrasive jet cutting parameters on the depth of cut are discussed.

5.1 Effect of abrasive material and size.

Different types of abrasive materials and sizes were used to select a suitable type for this comparative study. Garnet 100 grit was found suitable to offer reasonable cutting results and also satisfy other criteria, such as availability and ease of flow. Fig. 2 shows a magnified photograph of the garnet 100 grit abrasive particles. Other types of abrasives were found to be more effective but were not used due to their high cost.

5.2 Effect of abrasive flow rate.

Fig. 3 shows the linear relationship of the depth of cut versus the abrasive flow rate at random values of the other parameters. The linearity of this relationship should terminate at a certain critical flow rate. Beyond this critical flow rate, an optimum abrasive flow rate should exist at which the depth of cut peaks. A greater abrasive flow than the optimum will decrease the depth of cut, as the waterjet momentum loss rate exceeds the gain obtained by the increased number of impacts. The slopes of the lines on Fig. 3 should be functions of the hydraulic and abrasive parameters with increased slopes for higher pressures.

5.3 Effect of pressure.

Fig. 4 shows the linear relationship between the pressure, P, and the depth of cut, h, above 103.5 MPa. This implies that lower pressures, but higher than 103.5 MPa, are more energy efficient than higher pressures, greater than 241.5 MPa. This can be realized from the relationship of power, $\dot{E}$, and pressure: $\dot{E} \propto P^{1.5}$. Above 103.5 MPa, $h \propto P$ and not to $P^{1.5}$, which means that doubling the power does not double the depth of cut. Below 103.5 MPa, the divergence is greater as $h \propto P^{\beta}$ where $\beta < 1$. An optimum operating pressure should be determined to compromise between the rate of cutting and power requirements.

5.4 Effect of traverse rate.

Fig. 5 shows the relationship between the depth of cut and the traverse rate, U. The rate of the kerf area generation, h x U, is plotted versus U in Fig. 6, which shows that the optimum cutting speed is between 2.54 and 3.39 mm s^{-1} for the conditions listed on the graph. This optimum traverse rate should be a function of all the other parameters. The requirement to achieve a certain depth of cut at a certain power level requires optimization of the traverse rate and the number of passes, which also includes the effect of the standoff distance. The existence of critical low and high traverse rates, as observed in waterjet cutting (see Ref. 9), is not observed in the selected range of traverse rates in Fig. 6.

5.5 Effect of standoff distance.

Fig. 7 shows the effect of the standoff distance on the depth of cut. The rapid ineffectiveness of the jet with increased standoff distance can be explained by arguing that the effectiveness of the abrasive jet cutting results from the superimposed hydrodynamic pressure on the impacting particle. This reduces or eliminates the energy lost due to particles rebounding. If the particles are air driven, a large portion of the incident energy is reflected and the material removed is on the order of milligrams, as in the case of abrasive jet maching (AJM) for surface finishing and deburring. At large standoff distances, the liquid phase of the abrasive jet breaks up into droplets and the solid particles become freer to rebound upon impact, resulting in shallower penetration. The jet spreading and the associated reduction in local dynamic pressures explain the reduced effect of the abrasive jet with standoff distance before the jet breaks up. For multipass cutting, the jet is kept intact by the kerf produced from previous passes.

5.6 Effect of number of passes.

The effect of the number of passes was investigated for two ranges of cumulative depth. The first range was limited to 25.4 mm while the second range reached over 76.2 mm. Fig. 8 shows the trend for the first range. The increased slope of the curve suggests that the kerf acts as a local "mixing chamber" which tends to focus the abrasive jet stream for more effective cutting. Fig. 9 shows the trend for the second range (the greater cumulative depths at lower traverse rates). The top curve in Fig. 9 shows the trend for deeper kerfs, with a decreasing slope as in the case of cutting by waterjets. In this range, the effects of standoff distance and kerf frictional drag overcome the local focusing effect observed in the first range. The lower curve in Fig. 8 shows the two trends, as the cumulative depth of cut occurs in both ranges. The dotted lines in Fig. 9 represent equal energy lines, which suggests that slower traverses achieve deeper cuts than faster traverses with a greater number of passes. This may suggest the existence of an incubation period to initiate a kerf depth which starts to act as a local mixing chamber, resulting in an accelerated penetration rate. This incubation period should not be confused with that usually observed in erosion research (see Refs. 10 and 11), as the orders of magnitude are so far apart.

5.7 Effect of stream power and flow rate.

The parameter matrix shown in Table 3 was used to study the effect of the power level on the depth of cut. Within each power level, the nozzle diameter and pressure were changed to investigate the relative effects of pressure and flow rate. The last row in Table 3 shows the effects of flow rate at a fixed pressure (172.5 MPa). Fig. 10 shows the range of depths of cut obtained at two different power levels. The overlap shown in Fig. 10 indicates that high power levels are not necessarily better in achieving deeper cuts. The relationship between pressure and flow rate is controlled by the threshold values and material properties. A more detailed explanation of this observation is beyond the scope of this paper.

Fig. 11 shows the effect of the nozzle diameter (flow rate) on the depth of cut at a fixed pressure. The decreasing slope of the curve indicates that there is a saturation limit after which increased flow rates are not significant. This may confirm the previously mentioned assumption that the water acts only to superimpose high pressure on the cutting abrasives. At a high water flow rate, a larger portion of the abrasives will contribute to the depth of cut, as rebounding is reduced. Additional water flow rates over the saturation limit will then be insignificant. Another reason can be realized from the basic momentum theory: momentum transfer to the abrasives increases with an increasing water flow rate. The velocity of abrasive particles after mixing is

$$V = \frac{\dot{m}_w V_w + \dot{m}_a V_a}{\dot{m}_w + \dot{m}_a} .$$

According to this equation, the rate of increase of the velocity, V, is reduced by increasing the water flow rate, $\dot{m}_w$, for a fixed abrasive flow rate, $\dot{m}_a$. This is a similar qualitative trend as that shown in Fig. 11.

5.8 Kerf quality.

Fig. 12 shows typical top and side views of the shape of the cuts in a stainless steel sample. The top portion of the kerfs are generally smoother than the bottom portions, as is usually observed in straight waterjet cutting. However, the smoothness of the kerfs, are superior to conventional cutting methods. The wedge shape of the kerfs, usually seen in waterjet cutting of soft materials, is also observed for steel cutting with abrasive jets. However, the wedge angle is less in abrasive jet cuttng.

6. CONCLUSIONS

The results of this investigation of steel cutting with abrasive waterjets, as discussed in this paper, have led to the following conclusions:

1. Abrasive jets have great potential for steel cutting.
2. The depth of cut in steel varies linearly with the abrasive flow rate and pressure.
3. Low traverse rates are more efficient for deep cuts.
4. The optimum traverse rate for steel cutting lies between 3.33 and 4.17 mm s^{-1} for the minimum specific energy of surface generation.
5. The smaller the standoff distance, the deeper the cut for steel cutting by abrasive jets.
6. The kerf generated in steel acts as a local mixing chamber to accelerate deeper penetration.
7. High-quality kerfs are produced with abrasive waterjet cutting.

7. ACKNOWLEDGEMENTS

Thanks are due to Dr. J. Reichman and Dr. J. Cheung for their useful discussions and comments throughout this research program. The author also wishes to acknowledge the efforts of Mr. A. Deforrest, Mr. C. Anderson, Mr. D. Peecher, and Mr. B. Burwell for their help and participation during the experimental investigation.

8. REFERENCES

1. Maasberg, W., et al.: "Sand blasting apparatus." U.S. Patent 3,424,386, January 28, 1969.

2. Watson, J. D.: "Fluid-abrasive nozzle device." U.S. Patent 4,080,762, March 28, 1978.

3. Hart, B. E.: "Guns for forming jets of particulate material." U.S. Patent 3,972,150, August 3, 1976.

4. Easton, N.: "Wet abrasion blasting." U.S. Patent 4,125,969, November 21, 1978.

5. Maurer, W. C.: "Advanced drilling techniques." Houston, Texas, Maurer Engineering Inc., Technical Report No. TR79-1, January 2, 1979.

6. Cheung, J. B., and Hurlburt, G. H.: "Submerged waterjet cutting of concrete and granite." In: Proc. Third Int. Symp. on Jet Cutting Technology (Chicago: May 11-13, 1976), Cranfield, U.K., BHRA Fluid Engineering, 1976.

7. Hilaris, J., and Labus, T.: "Marine applications of high pressure water jets." ASTM, STP 664, 1977, pp. 582-596.

8. BHRA: "BHRA fluid engineering annual report." Cranfield, U.K., 1980, 12 pp.

9. Hashish, M.: "Critical and optimum traverse rates in jet cutting." In: Proc. of the First U.S. Waterjet Symposium, Golden, Colorado, Colorado School of Mines, April 7-9, 1981.

10. Springer, G.: "Erosion by liquid impact." New York, John Wiley & Sons, 1976.

11. Preece, C. (ed.): "Treatise on materials science and technology." New York, American Press, 1979.

Table 1. Patented Abrasive Jet Cutting Devices

INVESTIGATOR OR COMPANY	SKETCH OF CUTTING DEVICE	PARAMETERS/SPECIAL FEATURES	APPLICATION
W. Maasberg et al., 1969, for Woma-Apparatebau Wolfgang Maasberg & Co. G.m.b.H (Ref. 1)	a. **Sand Blasting Apparatus** 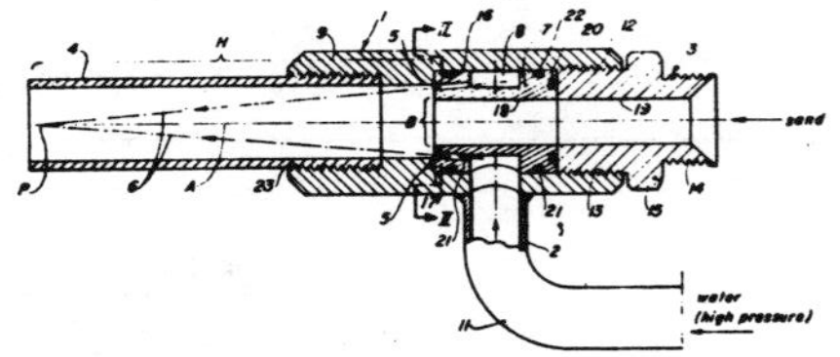	Pressure of 5.1 MPa. A number of convergent jets or a single annular jet.	Sand Blasting
J. D. Watson, 1978 (Ref. 2)	b. **Fluid-Abrasive Nozzle Device** 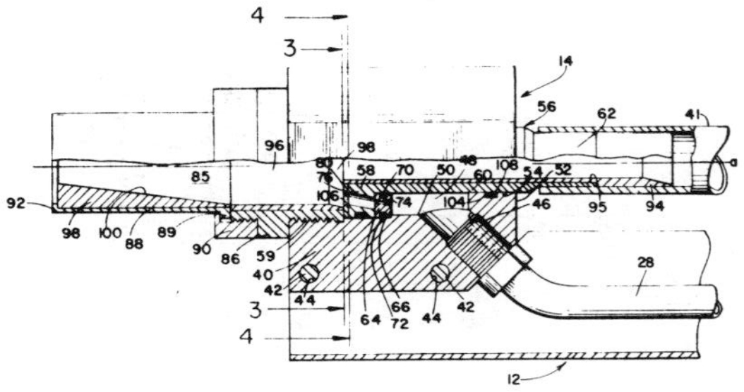	Patent claims applicability up to 207 MPa. Mixing efficiency is high. Spotless blasting. Suction feed. No material claims.	Sand Blasting
B. E. Hart, 1976 (Ref. 3)	c. **Guns for Forming Jets of Particulate Material** 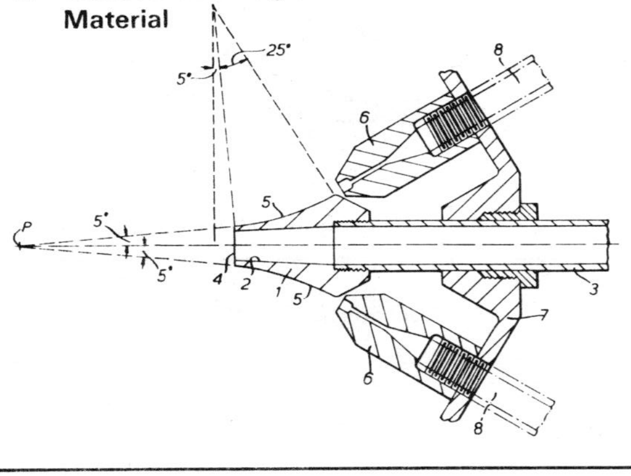	Slurry jet in the middle is mixed with waterjets from the sides to form a high-energy cleaning sheet of slurry.	Cleaning Buildings and Ship Hulls

Table 1. (Cont.)

INVESTIGATOR OR COMPANY	SKETCH OF CUTTING DEVICE	PARAMETERS/SPECIAL FEATURES	APPLICATION
N. Easton, 1978 (Ref. 4)	d. **Wet Abrasion Blasting** 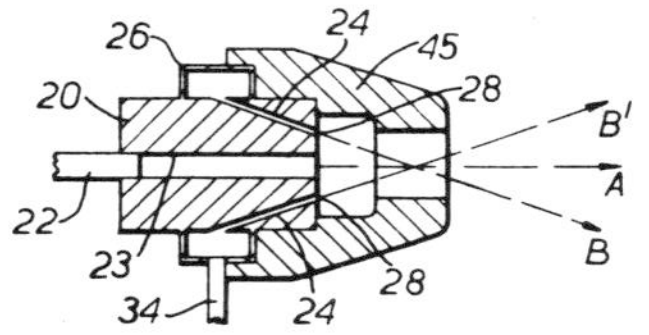	Particularly suitable when abrasive particulate material is soluble in the carrier liquid. 95% sodium silicate and 5% salt have been found effective in removing variety of soils from work surfaces.	Wet Abrasion Blasting
R. A. Bobo, 1963 (Ref. 5, p. 31)	e. **Abrasive Jet Drill** 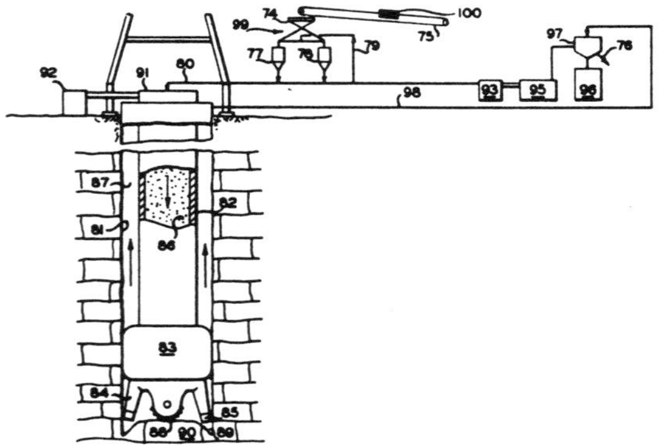	Abrasives are introduced into the mud line under pressure. Steel shot (1.6 to 3.2 mm diameter) was used at 890 N per barrel of mud or silica sand at 445 N per barrel.	Oil Well Drilling
R. J. Goodwin et al., 1968, for Gulf Oil (Ref. 5, p. 34)	f. **Abrasive Jet Nozzle** 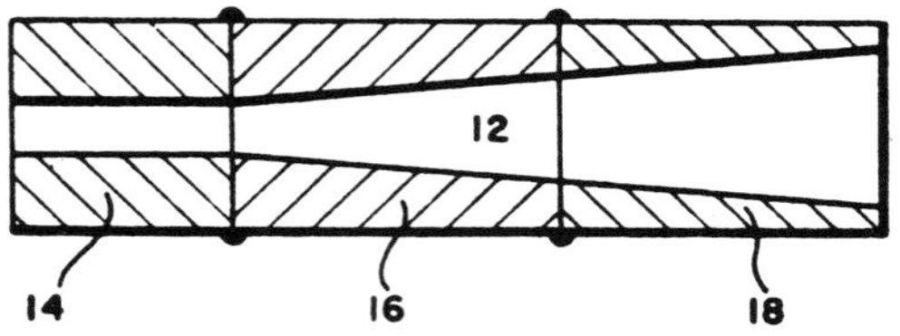	Sectional nozzle to adapt to impact and shear wear of abrasive flow.	Oil Well Drilling

Table 1. (Cont.)

INVESTIGATOR OR COMPANY	SKETCH OF CUTTING DEVICE	PARAMETERS/SPECIAL FEATURES	APPLICATION
M. R. J. Wylie, 1971, for Gulf Oil (Ref. 5, p. 38)	g. **Gulf High-Pressure Intensifier**	Abrasive mud containing steel shot was pumped at 69 MPa from 20 nozzles, each 3.1 mm in diameter. 20-40 ASTM mesh steel shot was used. About 3400 kW was applied, resulting in 170 kW per nozzle and an abrasive circulation rate of 20 million particles per second.	Oil Well Drilling

Table 2. Abrasive Jet Cutting Approaches

SYSTEM	SKETCH	ADVANTAGES	DISADVANTAGES
Single Jet Side Feed (Continuous Jet and Pulsed Jet)	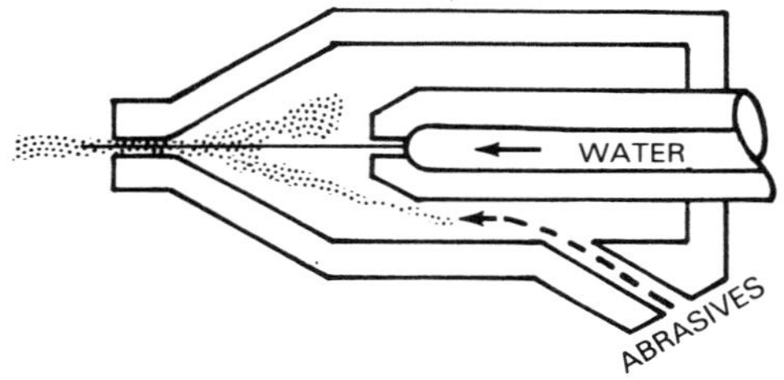 WATER ABRASIVES	Easy to machine. Size can be small. Cheap, conventional parts. Waterjet can be used without abrasives.	Inefficient mixing. Severe wear of exit section.
Multiple Jet Central Feed	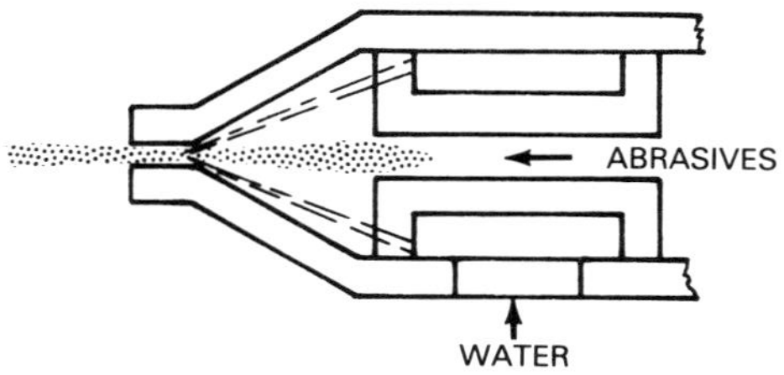 ABRASIVES WATER	Efficient mixing. Longer life of mixing section. Mixing chamber can be eliminated.	Difficult to machine. Bigger size. Standoff distance is great. Difficult maintenance
Annular Jet Central Feed	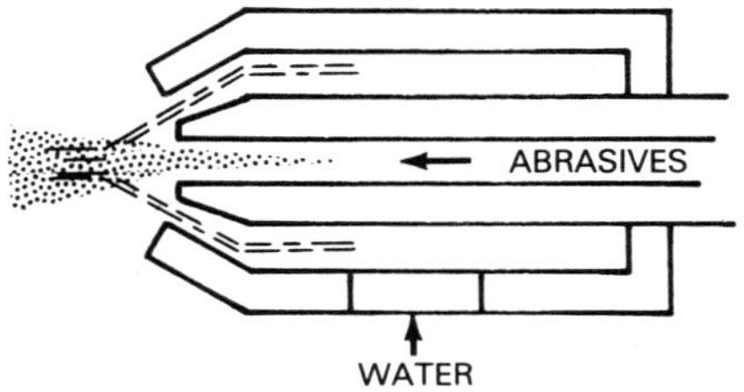 ABRASIVES WATER	Best mixing efficiency. Long life of mixing chamber if used. Size can be minimal.	Good only for large flow rates. Critical machining is needed. Possible wear of water exit annulus. High hydraulic losses.

Table 2. (Cont.)

SYSTEM	SKETCH	ADVANTAGES	DISADVANTAGES
Single Jet External Feed or Taped Abrasives	ABRASIVE JET ABRASIVE LAYER	Simple system. Waterjet can be used alone. No wear is involved. Good for cleaning.	Poor mixing. Limited to small abrasive consumption rates.
Direct Pumping	An abrasive slurry intensifier, such as Gulf Oil intensifier (sketch g in Table 1) and Halliburton mud pumps.	Most effective cutting. Minimum abrasive flow rate. Recyclable slurry.	Slurry should be stable. Severe nozzle wear. Limited to fine abrasives.
Indirect Pumping	HIGH PRESSURE SLURRY TANK WATER INTENSIFIER (PUMP)	Effective cutting. Use of available pumps.	Nozzle wear. Systems interface and control.

Table 3. Parameter Matrix Used to Study Hydraulic Power and Water Flow Rate Effects (Pressure Values Are Given in MPa)

Stream Power (kW) \ Nozzle Diam. (mm)	0.051	0.0635	0.076	0.089	0.102
26.1	255	190	150	123	103
37.3	324	241.5	190	155	131
Variable	172.5	172.5	172.5	172.5	172.5

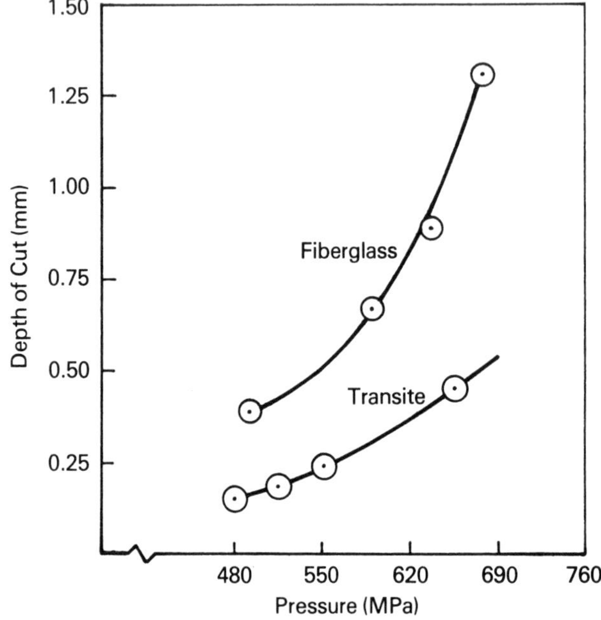

a. Underwater Cutting for Marine Cleaning Applications (Ref. 7)

Pressure Up to 690 MPa
Traverse Speed Up to 20 mm s^{-1}
Nozzle Diameter 0.4, 0.5 mm
Standoff Distance 127 mm
Cutting Rate 6.4 mm s^{-1}
1020 Steel Specimens
Fiberglass, Transite, and Sand as Abrasives
Optimum Jet Angle Was Found to Be 15°

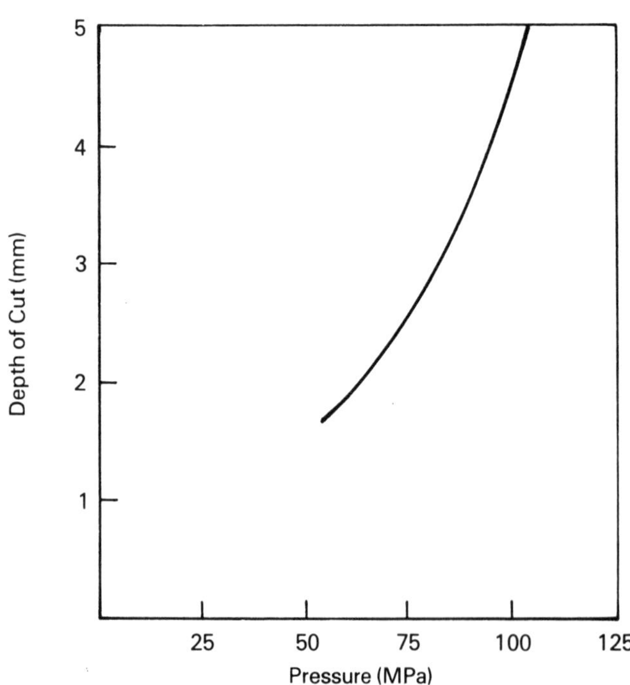

b. Steel Cutting (Ref. 8)

Pressure Up to 125 MPa
Traverse Speed Up to 5 mm s^{-1}
Mild Steel and Other Materials Were Tried
Cheap Throw-Away Abrasives Were Used as a Slurry
Cutting Head Was 150 m from the Pump

Figure 1. Effect of Pressure on Depth of Cut

200 x GARNET SAND, 150 GRIT

Figure 2. Magnified Garnet Particles (100 Grit)

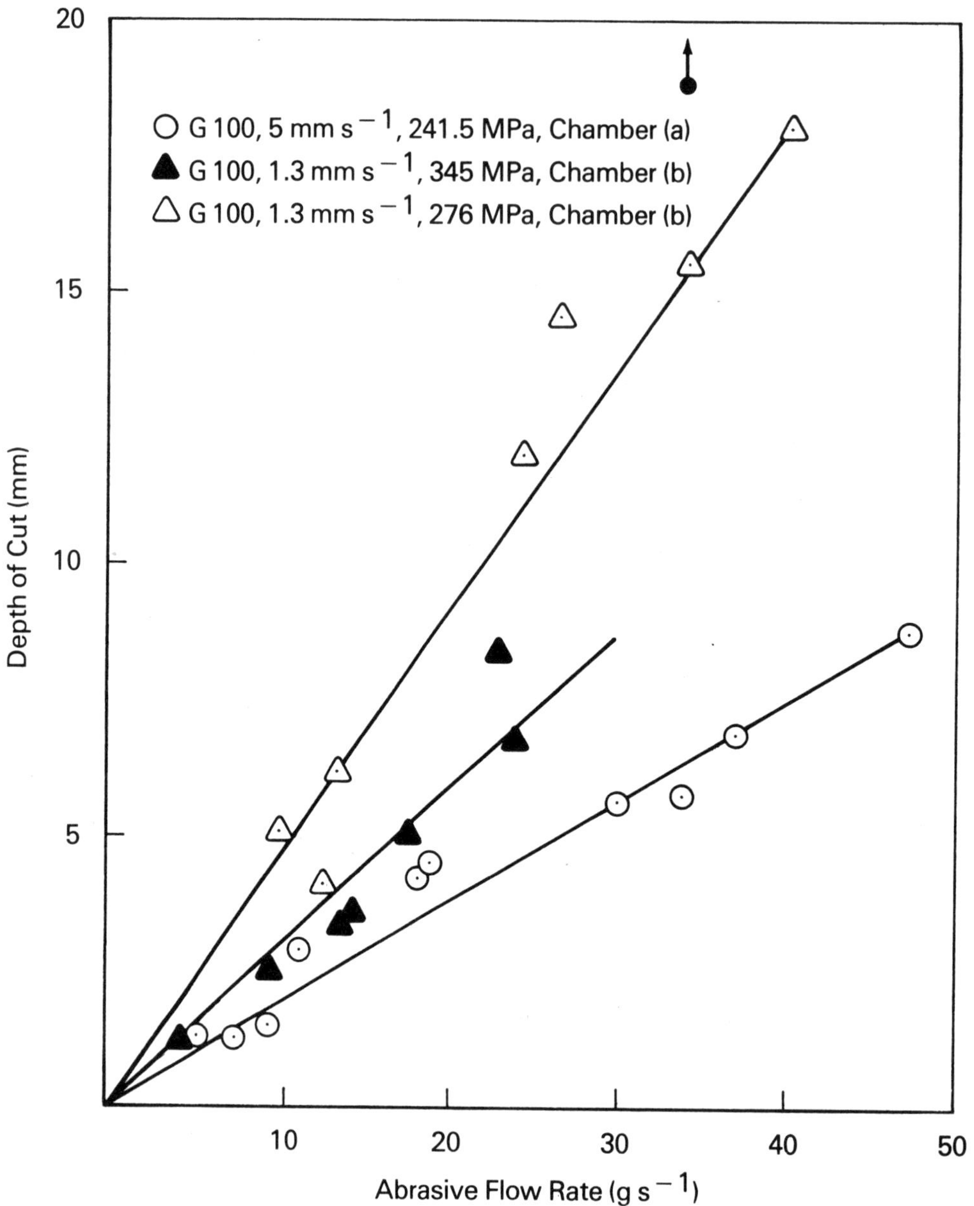

Figure 3. Effect of Abrasive Flow Rate on Depth of Cut

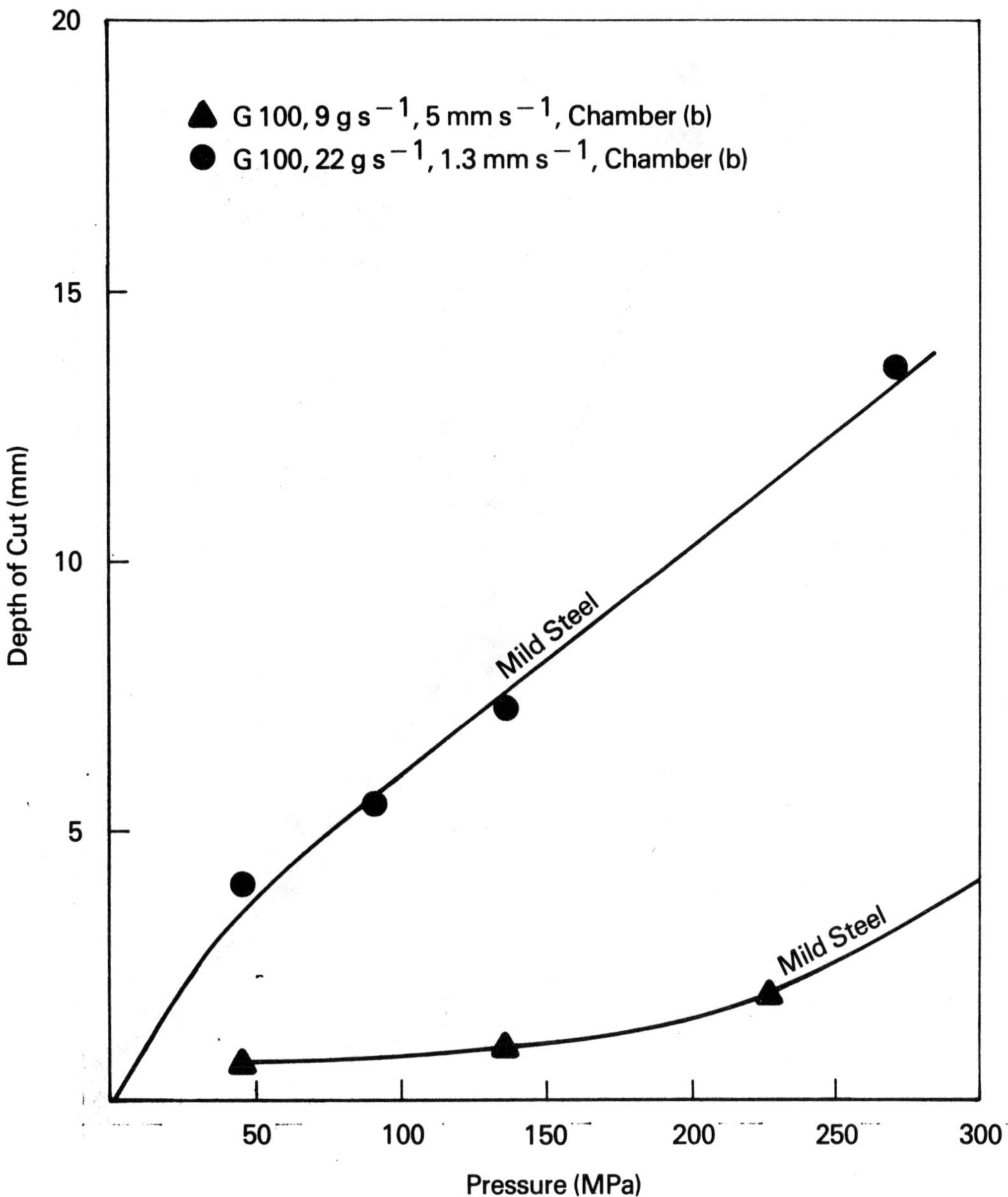

Figure 4. Effect of Pressure on Depth of Cut

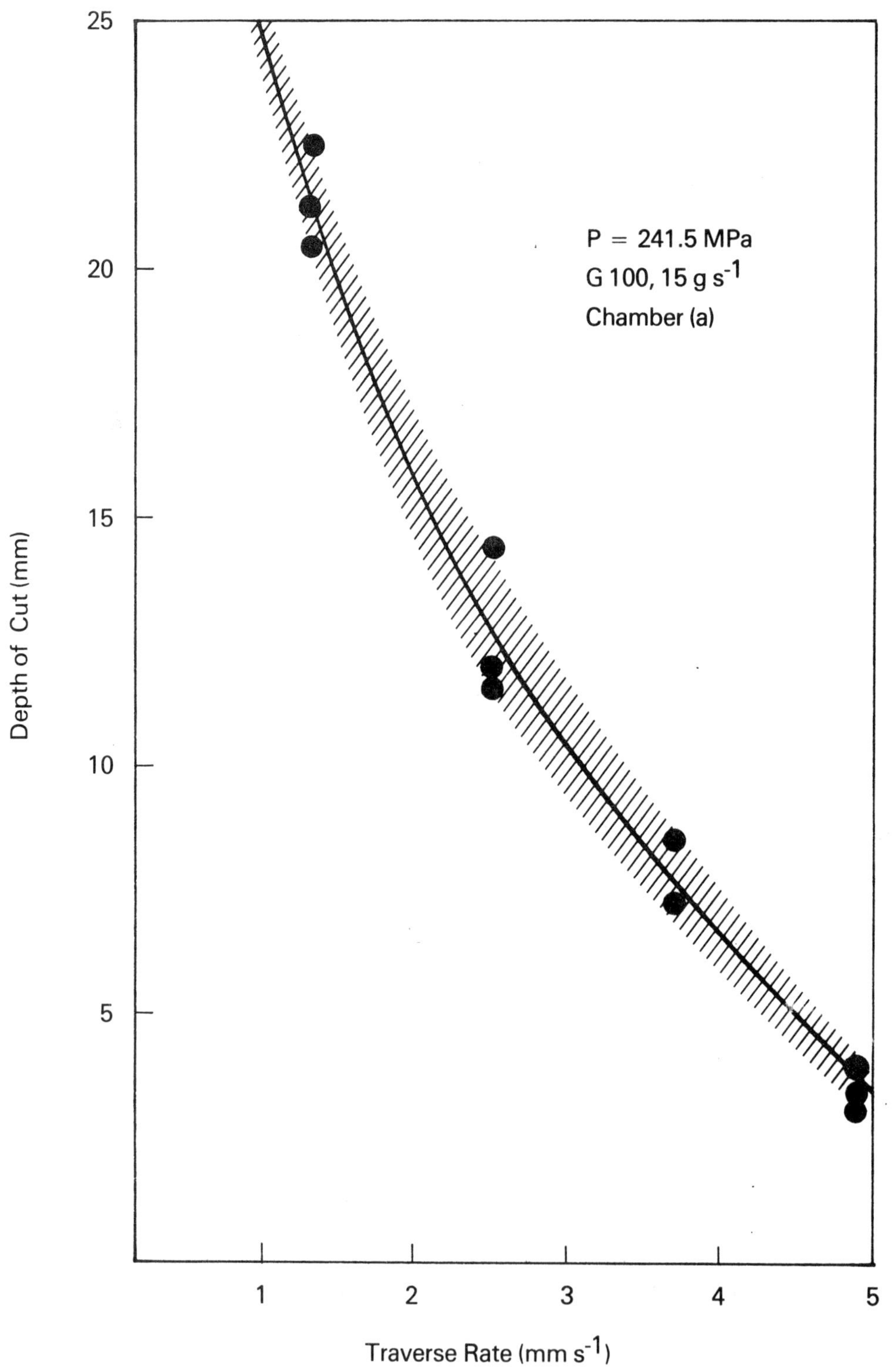

Figure 5. Effect of Traverse Rate on Depth of Cut

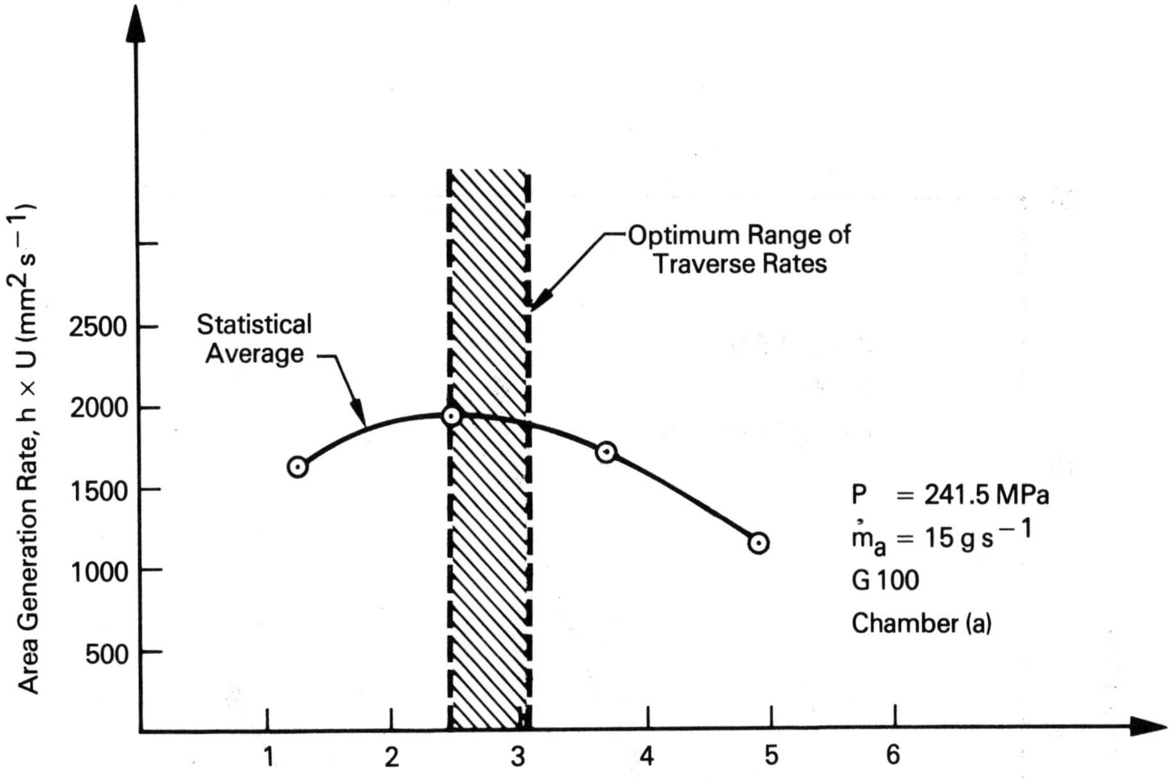

Figure 6. Effect of Traverse Rate on Area Generation Rate

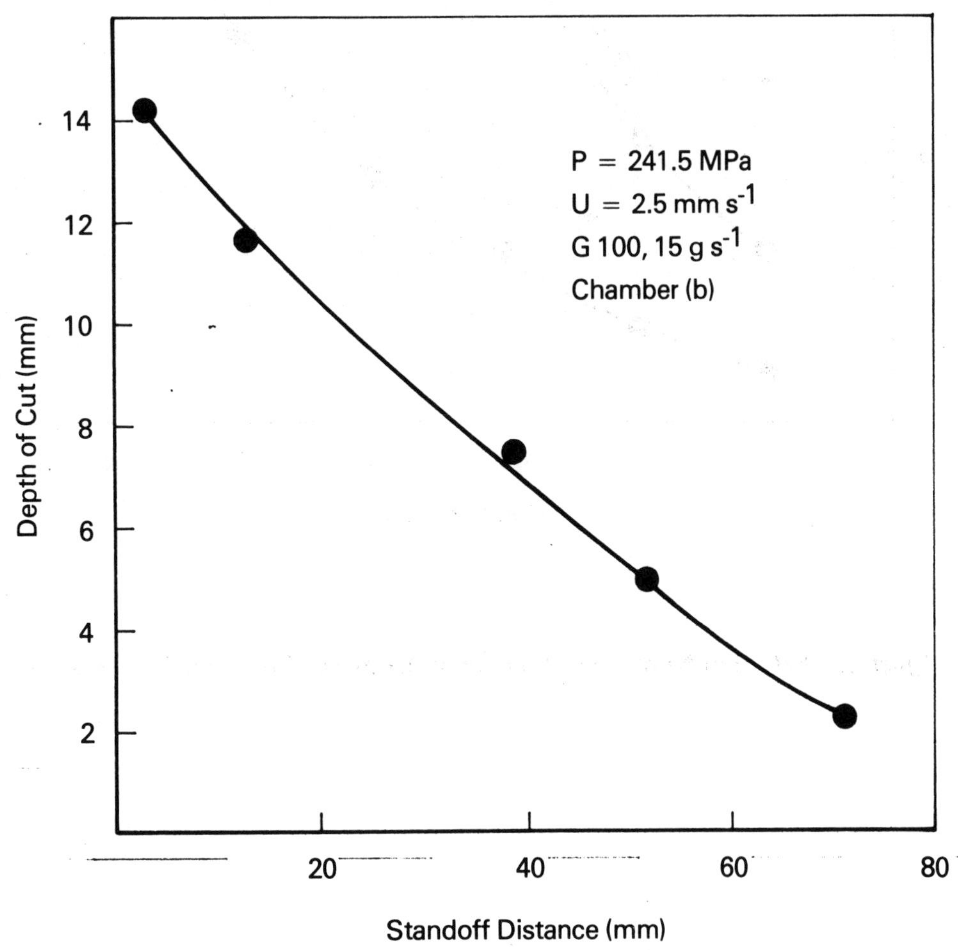

Figure 7. Effect of Standoff Distance on Depth of Cut

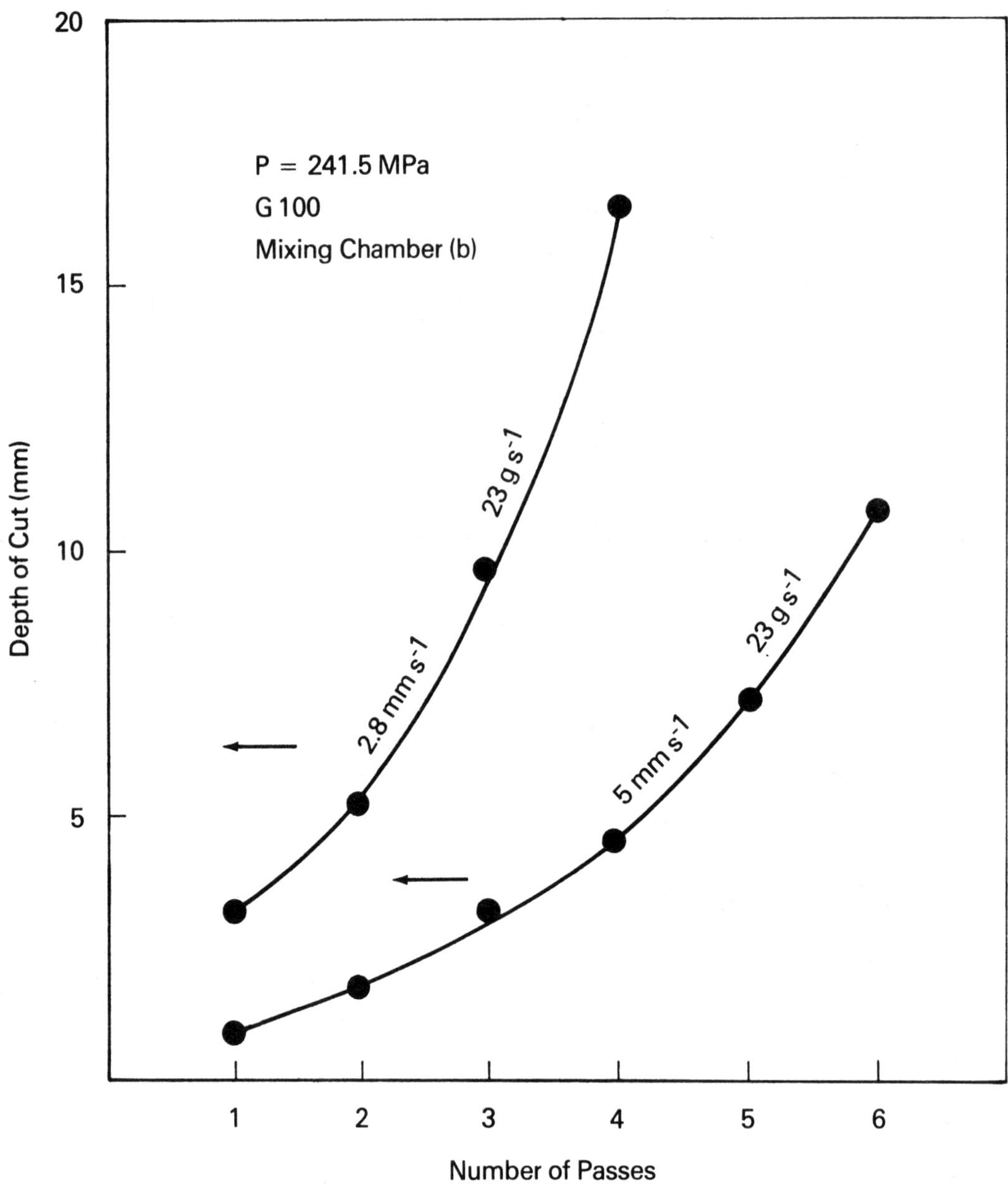

Figure 8. Effect of Number of Passes on Depth of Cut (High Traverse Rate)

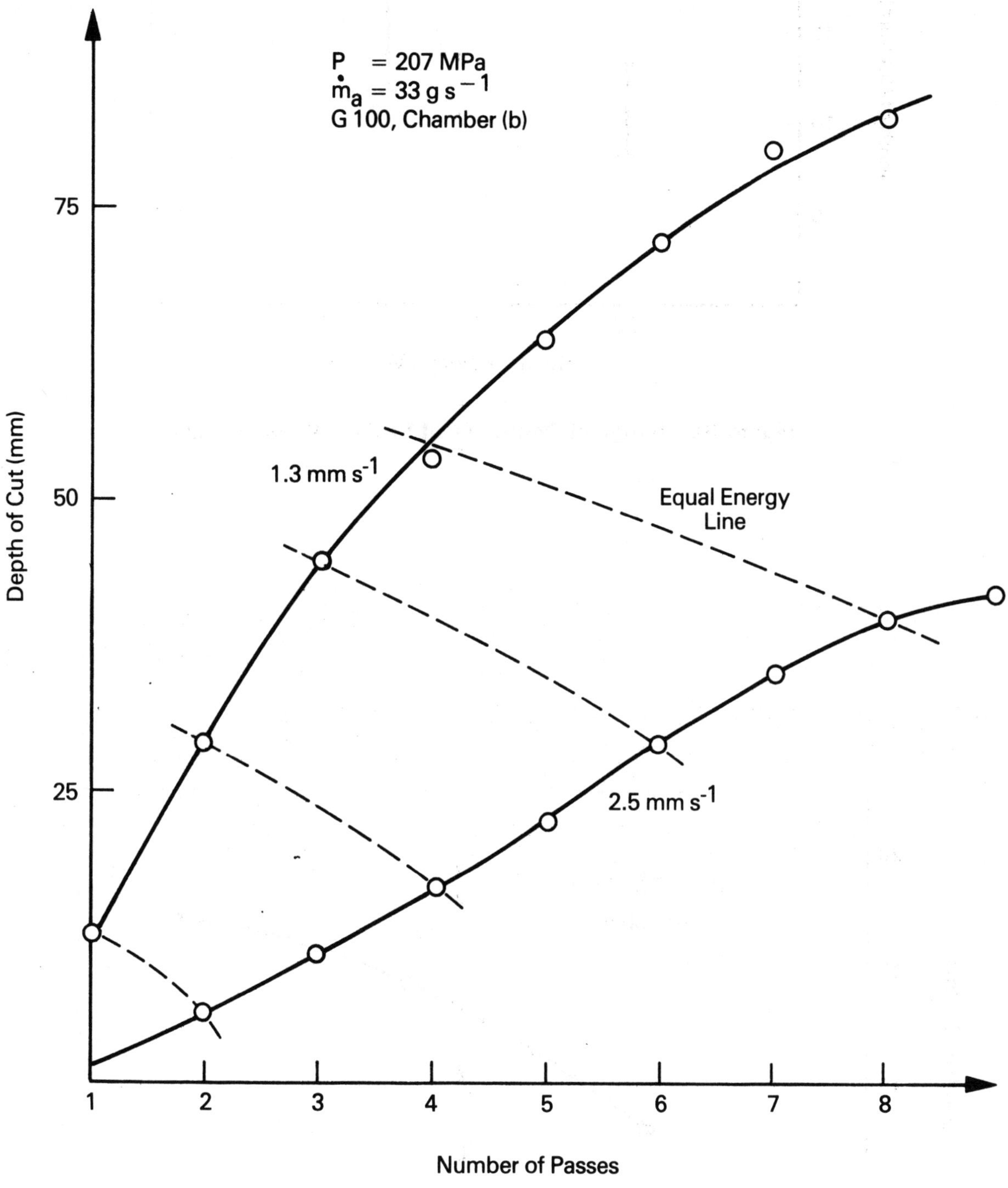

Figure 9. Effect of Number of Passes on Depth of Cut (Low Traverse Rate)

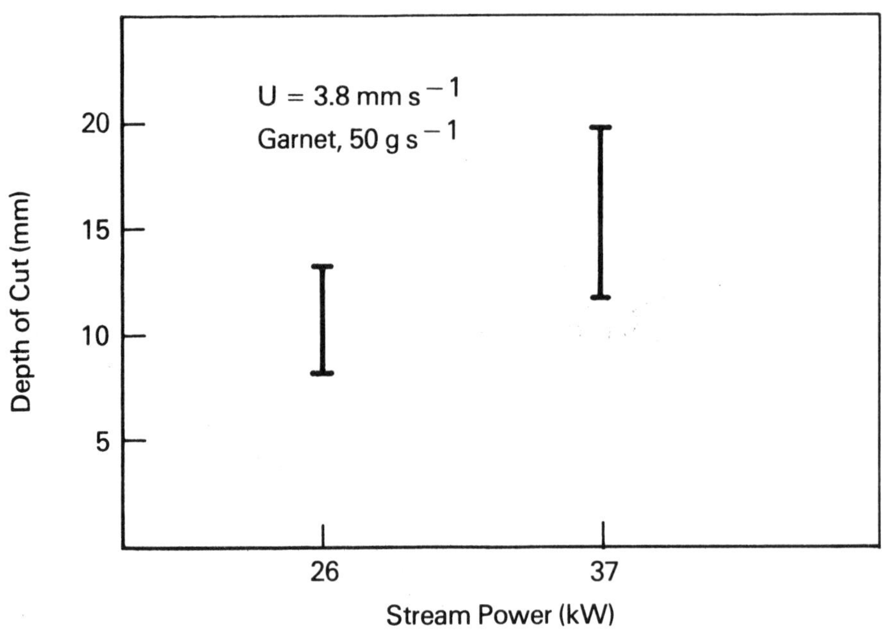

Figure 10. Range of Depth of Cut for Two Power Levels

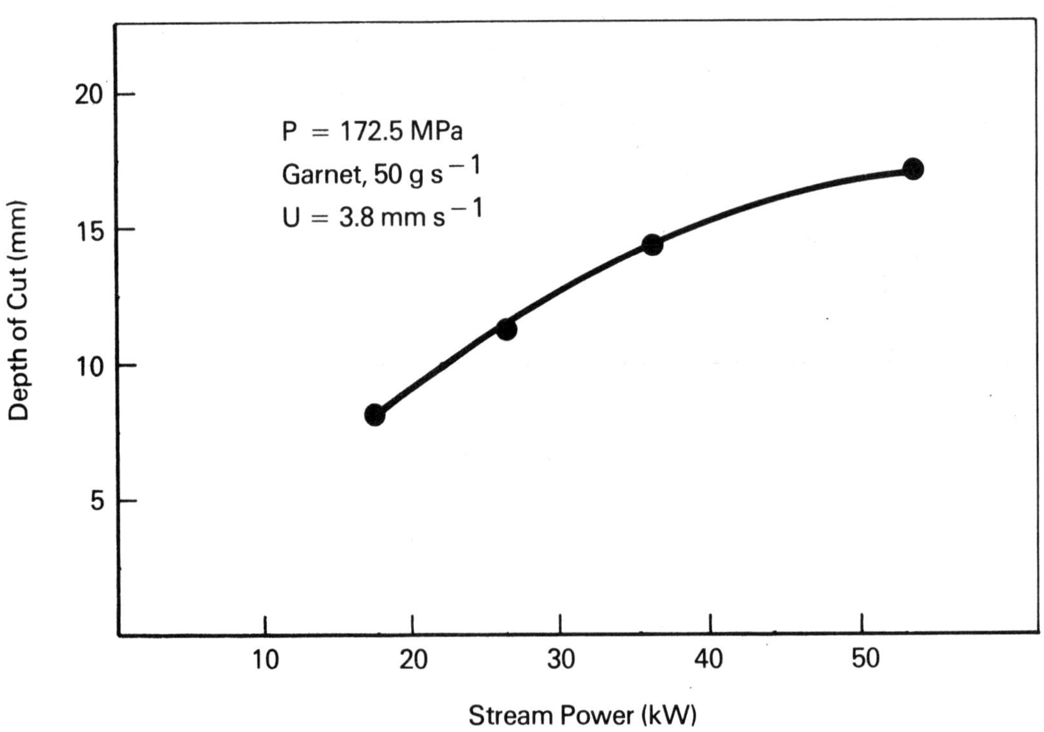

Figure 11. Effect of Flow Rate on Depth of Cut

Figure 12. Stainless Steel Sample Cut with Abrasive Waterjet

6th International Symposium on
Jet Cutting Technology
6-8, April, 1982

WATER/ABRASIVE JET CUTTING OF CONCRETE AND REINFORCED CONCRETE

R.E.P. Barton and D.H. Saunders

BHRA Fluid Engineering, U.K.

Summary

The use of water jets with entrained abrasives was investigated as a method of cutting high strength concrete. The method could be particularly valuable where access for more conventional methods is awkward, since the cutting head itself is small.

Trials on a number of specially prepared high strength concrete blocks are described. Blocks up to 700 mm deep were cut through using a cutting head developed at BHRA and operating at pressures up to 965 bar (14,000 p.s.i.). Some of the blocks included reinforcing steel (up to 32 mm dia.) which was also cut at the same time as the concrete, using the same equipment.

Held at the University of Surrey, U.K.
Symposium organised and sponsored by
BHRA Fluid Engineering

Introduction

There is to-day an ever increasing requirement to cut concrete both with and without reinforcement for a wide range of applications. These vary from relatively simple tasks such as removing a damaged section of concrete from a roadway, permitting a new section to be keyed into the old structure, to, at the other extreme, large scale demolition of, for example, a disused chemical plant.

The detailed requirements vary considerably between one application and another. For demolition work the basic requirement in most cases is for the structure to be broken up into manageable-size sections which can then be easily removed from the site. The tools that are presently available, including explosives and large breaking hammers, generally meet these requirements. However, there are circumstances, for example where damage to nearby buildings must be avoided, when the demolition must be carried out in a controlled manner. In these circumstances the requirements are similar to those when cutting.

The tools most commonly used for cutting operations are pneumatic drills and diamond tipped rotary saws or discs; these methods represent extremes in the quality of cut. Where a good quality cut is essential a rotary saw or disc would be used, and they are particularly worthwhile for long straight cuts. However, a limiting factor is the depth to which they can actually cut into concrete. This is determined by their physical size. Although a saw or disc will cut reinforcing bars, the operating time is greatly increased whilst at the same time the working life of the cutting blade is reduced. Pneumatic drills give a relatively poor quality cut and are not able to cut the reinforcing bars. They can also cause local damage in the area of cutting. The normal method used when cutting reinforced concrete with a pneumatic drill is to expose the reinforcing bars and then to use an additional method such as flame cutting or shearing to cut through them. There are a few applications where the use of these conventional cutting methods is restricted, for example, where there is limited access to the cutting area or where cutting is to take place in a flammable atmosphere.

In recent years plain water jets have been used to cut through small concrete sections. Water jetting heads do have the advantage of being small and can be readily manipulated in confined spaces. Typical operating pressures of up to 1000 bar have been used for this application. The action of a water jet is predominantly that of loosing the aggregate by washing out the softer matrix. This results in the cut obtained being of a very uneven nature with the reinforcing bars once again not being cut. For plain concrete a critical parameter seems to be the size of the jet which must be comparable with the size of aggregate. If the jet size is smaller than the aggregate will wedge in the kerf. With most other materials, the cutting performance in terms of the rate of cutting, increases markedly with pressure. However, it is doubtful if this would apply to high strength concrete, because to avoid excessive power consumption the nozzle size employed must be reduced which could in turn reduce the performance. Even if high pressures were available it would still not be possible to cut the steel reinforcing bars since pressure (Ref. 1) beyond the capability of any commercially available equipment would

Abrasive entraining water jetting devices have also been available for some time, and have been used in cleaning operations for removing hard surface deposits. A typical application is the removal of corrosion or paint from large surface areas. These devices are basically "jet pumps" using high velocity water jets to draw in abrasive laden air which is then mixed with the water before being discharged onto the surface to be cleaned. Devices of this type have also been used to give some improvement of the cutting performance of plain water jets on concrete. An advantage is that the steel reinforcing bars can be cut although the performance with respect to cutting rate is low.

BHRA studied the performance of a commercially available concrete cutting head for a steel cutting application. which led to the design and development of a new cutting head. (Ref. 2). Although this development was aimed at the specific application of cutting mild steel in an explosive environment, it was established that it could cut a wide range of target material (Ref. 2). One of the materials cut during trials was a section of reinforced concrete. The sample block of unknown strength was a 150 mm deep section with a 13 mm reinforcing cable in the centre.

The speed of cut was 13 mm per minute which although not giving a cut as smooth as that of a saw cut, was of a very good quality. It was observed that the individual pieces of aggregate had been cut.

The work described here was a limited programme of work sponsored by a client who had a requirement to cut high strength plain and reinforced concrete. The investigation was to check the performance of the cutting head whilst cutting through plain and reinforced blocks of concrete up to 700 mm deep. Although details of the actual application involved are confidential, the results obtained have been made available and are relevant to other concrete cutting situations.

Test Equipment

The test cuts were carried out in a BHRA facility (Fig. 1) which was purpose built to enable a wide variety of jetting applications, both cleaning and cutting, to be studied. The facility is equipped with a carriage which can be traversed at controlled speeds ranging from 1.0 mm/min to 100 metres /min and can support test pieces of up to about 200 Kg. The carriage is situated inside the enclosure about 3 metres long and can accommodate a wide range of test pieces from small metal castings to tube bundles and sections of marine structures. The enclosure can also be flooded to simulate underwater applications.

Water is supplied from a high pressure piston pump through a flexible hose to the cutting head, which was mounted in the jetting facility. This type of pump is already widely used by civil engineering contractors for cleaning and maintenance applications. The operating pressure of 690 bar (10,000 p.s.i.) selected for the majority of the test programme is commonly applied as a normal maximum for continuous operation of these jetting pumps. For some of the trials the pressure was raised to a maximum of 965 bar (14,000 p.s.i.).

A steady feed of dry abrasive was supplied to the cutting head by a gravity fed screw which discharged into the airstream being drawn into the head through a flexible hose. Copper slag abrasive was used throughout this investigation, at a fixed rate of 5.4 kg/min. This abrasive is widely used in abrasive jet cleaning applications and it is relatively cheap and readily available. The cutting head which was used for the trials was developed at BHRA for cutting mild steel in an offshore emergency situation and had also been used in a number of industrially sponsored cutting trials on a range of hard materials. (Ref. 2). Sample blocks of high strength concrete were specially prepared for the investigation. The strength of the sample blocks was measured in accordance to the relevant British Standard (Ref. 3), a strength of 45 N/mm^2 being recorded. The blocks were cast in two sizes to give depths of 400 mm and 700 mm, the overall dimensions of the blocks were selected to ensure that the maximum load capability of the carriage was not exceeded. Plain and reinforced blocks were made in each size.

Cutting Trials

The purpose of the investigation was to clearly demonstrate that concrete of this strength and depth could be successfully cut by this method. The time available to carry out this work was limited and because of this the test programme was concentrated on full depth trial cuts on a number of different sample blocks. Apart from some brief tests to indicate a satisfactory traverse speed, the operating parameters were selected from previous experience with other target materials. However where it became obvious that a change in operating parameter would improve the performance, changes were made but no attempt was made to determine the best value.

It was immediately obvious from the first trial cuts that an extremely low traverse speed would be required if the blocks were to be cut at a single pass. Consequently, it was decided to achieve the full depth cuts with a number of passes, and a nominal traverse speed of 25 mm/min was adopted. Measurements of the depth of cut varied across the width of the blocks, and measurements of minimum and maximum depth were taken. For trial cuts in plain concrete, the 'minimum' depth measurement in some cases only indicated the presence of a single piece of wedged aggregate, and was therefore not fully representative of the depth of cut achieved. For the reinforced blocks, however, the 'minimum' depth was frequently at or in line with one or more of the reinforcing bars.

Test Block No. 1 (Plain Concrete)

For this plain concrete block, 400 mm x 400 mm x 400 mm, the cutting head was set up at right angles to the block and at a stand-off distance of 13 mm, measured from the surface of the block, and an operating pressure of 690 bar.

Four passes of the jet were necessary for the jet to break through the block over part of the length of the cut. Depth measurements are shown in (Fig.2). As can be seen, the depth of cut advanced steadily with each pass, with no indication of any significant change in performance as the effective stand-off from the base of the cut increased over the 400 mm depth of the block.

Test Block No. 2 (Reinforced Concrete)

This block had the same overall dimensions as block no.1, but contained reinforcing bars as shown in Fig.3. The same operating parameters were applied and as before four passes were necessary for the jet to break through the block, including cutting through the four 25 mm reinforcing bars (Fig. 4). At this stage however, the minimum depth of cut was less than for block no.1. A fifth pass was carried out, which resulted in the block breaking in two. The cut block is shown in Fig. 5.

A brief single pass test cut on this block indicated that the depth of cut could be significantly increased if a larger stand-off distance was employed. This effect was not quantified, but it was decided to adopt this arrangement for further trial cuts.

Test Block No. 3 (Plain Concrete)

For this trial cut a deeper plain concrete block was used (300 mm x 300 mm x 700 mm deep). The stand-off distance was increased to 110 mm, with the operating pressure once again at 690 bar. Seven passes of the jet were necessary to break through the 700 mm depth of block (Fig. 6). In the early stages however, the rate of advance of the cut was faster than with the previous blocks, with a 'maximum' depth of 400 mm almost being achieved in only two passes. However, as the cut is deepened beyond this there does seem to be a definite reduction in cutting efficiency with a further five passes being required to complete the cut. The cut block is shown in Fig. 7.

Test Block No. 4 (Reinforced Concrete)

This block, Fig. 8, contained 32 mm diameter reinforcement bars evenly spaced throughout its 700 mm depth. The angle of attack of the jet was changed slightly to 70^0 so that the bars were not directly in line with the jet. The stand-off distance for the first pass was increased to 150 mm, but reduced to zero for the second and subsequent passes.

The cutting performance is given in Fig. 9. The rate of advance of the cut in this block is significantly slower than for the previous three blocks with six passes being required to achieve a 'maximum' depth of slot of 400 mm. The situation was complicated, however, because the jet appeared to have been deflected within the block (Fig. 10) by the combined effect of the steel bars and a particularly hard and stubborn piece of aggregate. An increase in operating pressure to 965 bar on the eighth and subsequent passes did help to straighten the cut (Fig. 11). From the tenth pass, the angle of attack of the jet was altered to 90^0, which immediately enabled the jet to break through the full depth of the block.

At this stage, the minimum depth of cut was about half the block depth, with four bars cut. Four further passes were made over the block,resulting in five bars completely cut, and the sixth partly cut (Fig. 12). In this condition, the block was still intact, and the sponsor of the work wished to retain it this way as a demonstration of the technique. Because further passes might have weakened the block, the trial was halted at this stage.

Discussion

This investigation has clearly shown that deep cuts can be made in high strength reinforced concrete with water/abrasive jets. Cuts of up to 700 mm deep have been made without the need to insert the cutting head into the slot. The fact that this has been possible without optimising the operating parameters of pressure, flow or abrasive type and feedrate is particularly significant. With the present equipment, it would appear that the rate of advance of the cut is somewhat reduced as the full 700 mm depth is approached. However, for depths of the order of 300 to 400 mm, there is no indication of any significant variation of the rate of advance with the depth of the cut. An important consideration is that the cutting head used was developed for cutting steel which, being a homogeneous material, will not react to the action of the jet in the same way as concrete. Further development of the head, together with some degree of optimisation of the operating conditions for each particular application, can be expected to give an improved cutting performance.

The test blocks were all made of the same concrete mix, with the same nominal compressive strength. In spite of this, it is obvious that there was considerable difference in the resistance of the four blocks to the water/abrasive jet. A difference of almost 2 to 1 was recorded between the 'maximum' depth measurements on the plain concrete for the same number of passes of the jet. The reason for this may be the non-homogeneous nature of concrete and, although a consistent measure of its strength can be obtained from a cube sample, the effect of a jet is very much more local. Because of this, compressive strength cannot be regarded as a reliable indication of the performance to be anticipated from a water/abrasive jet cutting system.

During the trials certain operating parameters were altered, particularly in the trial block 4, where stand-off distance, pressure and angle of attack were changed. From the cutting performance recorded (Fig. 9), there is no indication of significant changes in performance as a result of these changes. However, immediate effects were noted, with the cut being straightened when the pressure was increased, and the full depth cut being achieved with the altered angle. These effects might also be a direct result of the variability of the concrete.

This cutting method is based on a conventional water jetting pump. Pumps of this type are already widely used in the construction and other industries for surface cleaning and preparation, and the cutting head and its associated abrasive feed system can be considered as an add-on unit to an existing pump. Although in some situations, manipulation of the cutting head would be controlled automatically, in practice a hand held or hand controlled cutting head would allow a considerable degree of flexibility in operation. An experienced operator would be able to vary his operating conditions as necessary to take full advantage of any variability in the concrete.

Acknowledgements

The Authors wish to acknowledge the financial contributions to the work described here both of the Client and the Mechanical and Electrical Engineering Requirements Board of the Department of Industry. We would also like to thank our colleagues at BHRA especially Mr. R.M. Fairhurst who assisted throughout the trials.

References

1. Tmanska, O. Fryino, S. Shinohara, K. and Kawata, Y. : "Experimental study of machining characteristics by liquid jets of high power density up to $10^8 Wcm^{-2}$". Paper G3 p.p. G3-25 to G3-35, 1st Int. Symp. on Jet Cutting Tech. BHRA Fluid Engineering, Cranfield (April 1972).

2. Saunders, D.H. : "A safe method of cutting steel and rock in hazardous atmospheres". Paper to be presented at 6th Int. Symp. on Jet Cutting Tech. BHRA Fluid Engineering, Cranfield (April 1982).

3. British Standards Institution: B.S.I. 1881: Part 4: 1970."Methods of testing concrete for strength". British Standards Institution, 2, Park Street, London.

1. General view of test chamber.

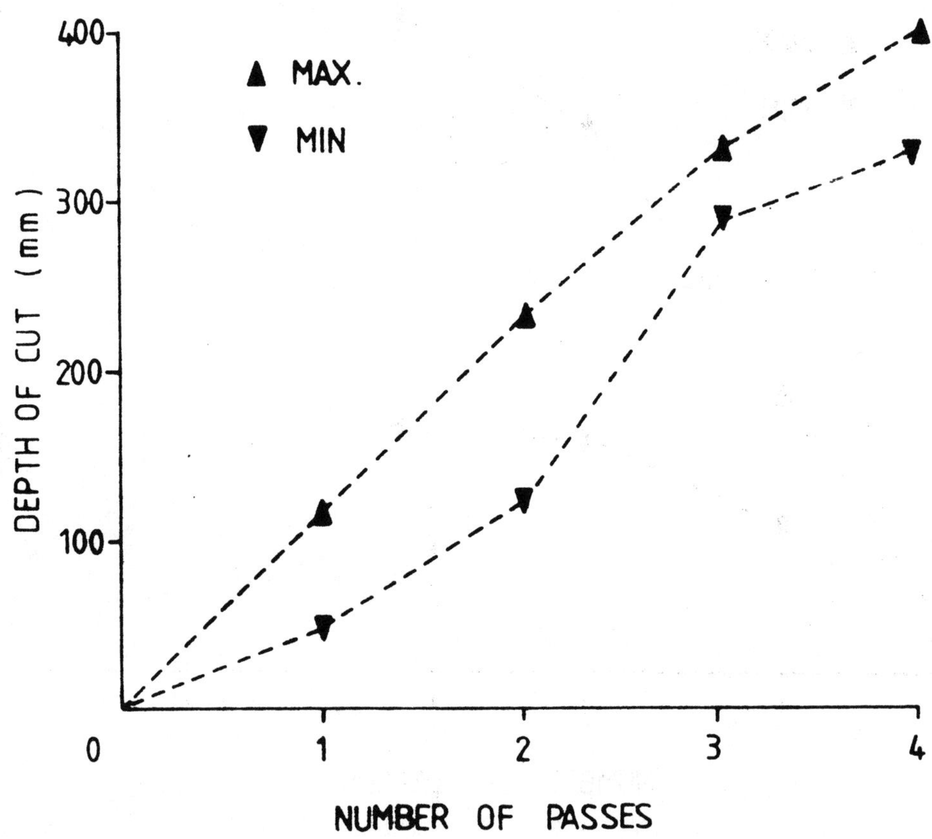

FIG.2. TEST BLOCK No.1

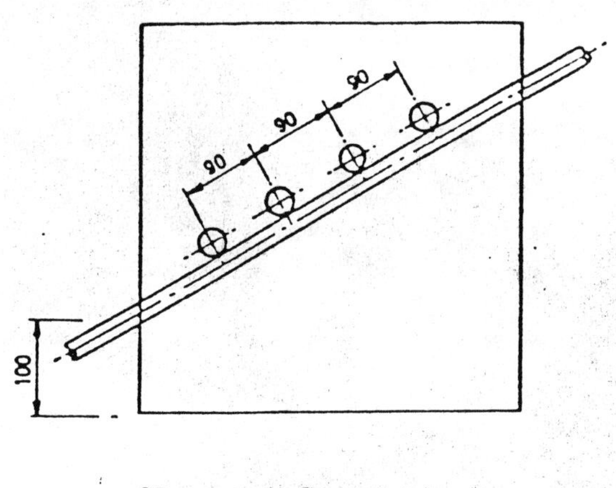

25 mm reinforcement

Fig.3. Specification of No.2 test block

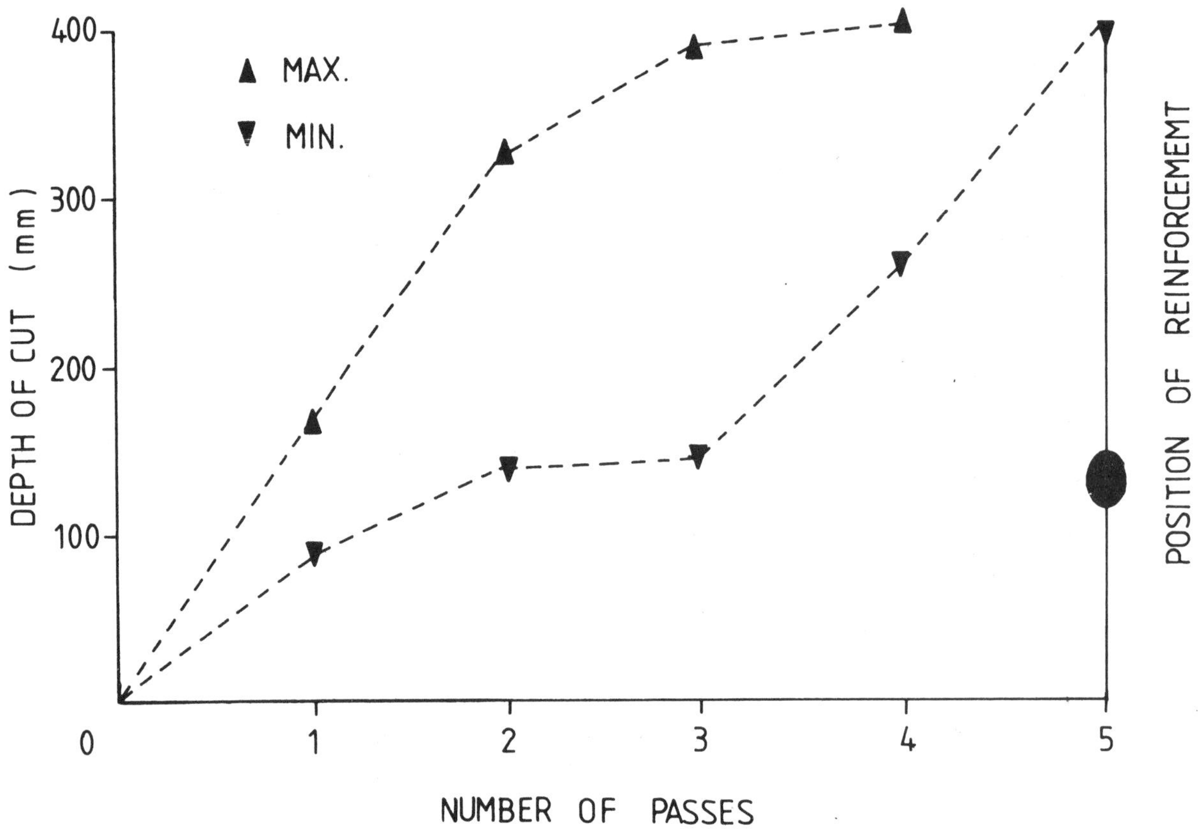

FIG.4. TEST BLOCK No.2

5. Photograph of trial cut in test block No. 2.

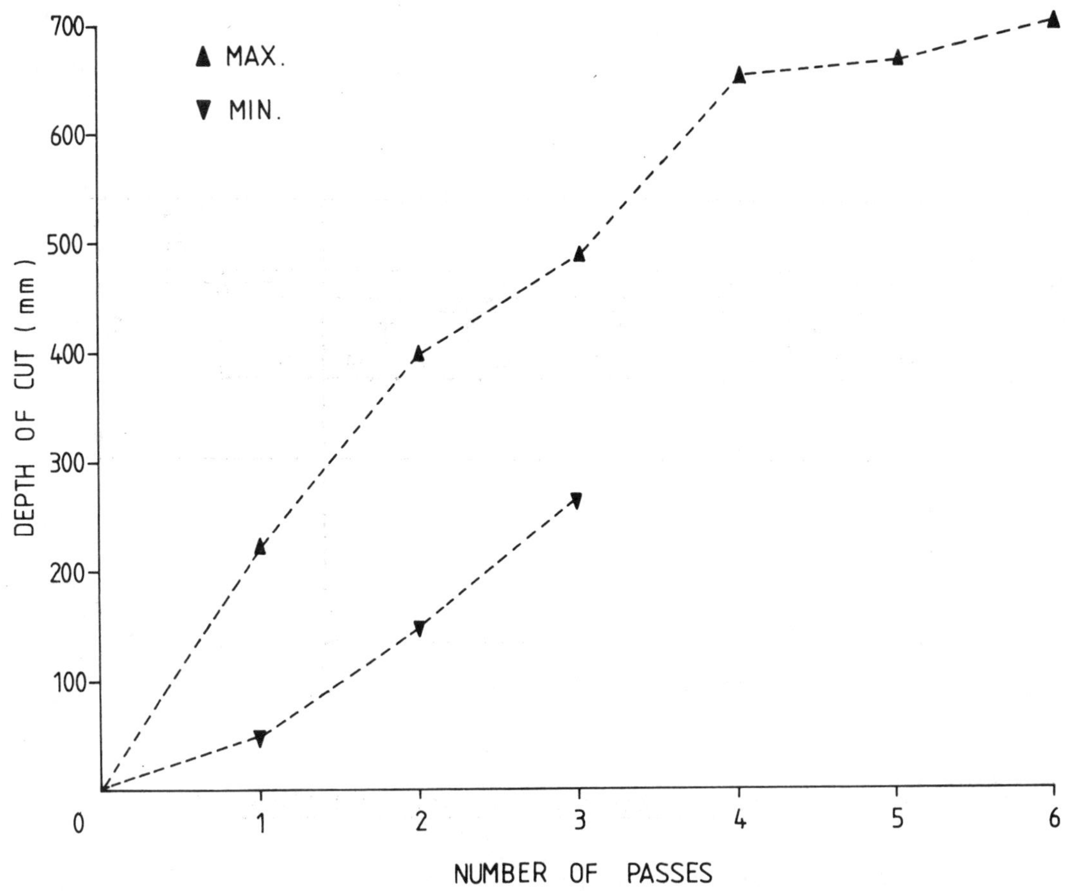

FIG.6. DEPTH OF CUT IN TEST BLOCK No. 3

7. Photograph of trial cut in test block No. 3.

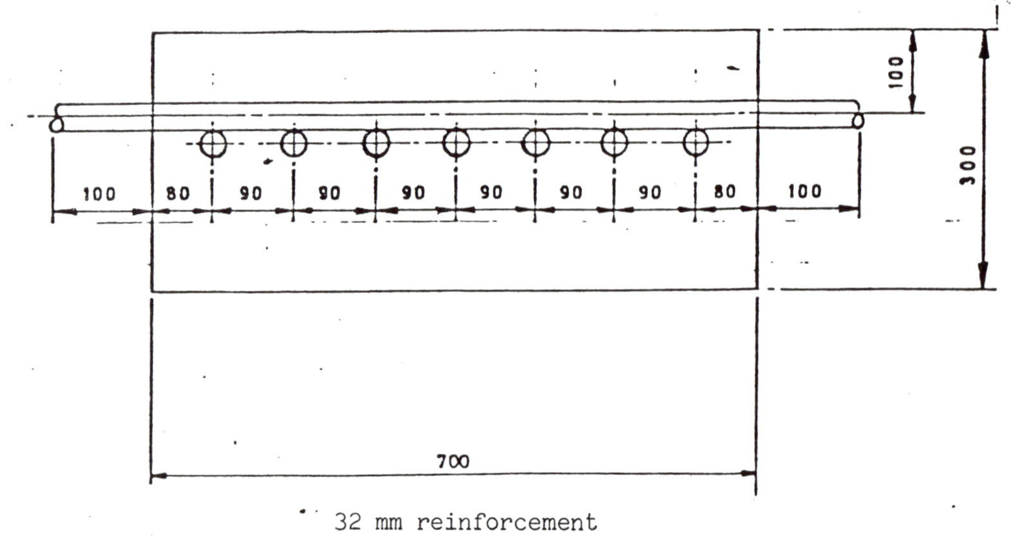

Fig.8. Specification of No.4 test block

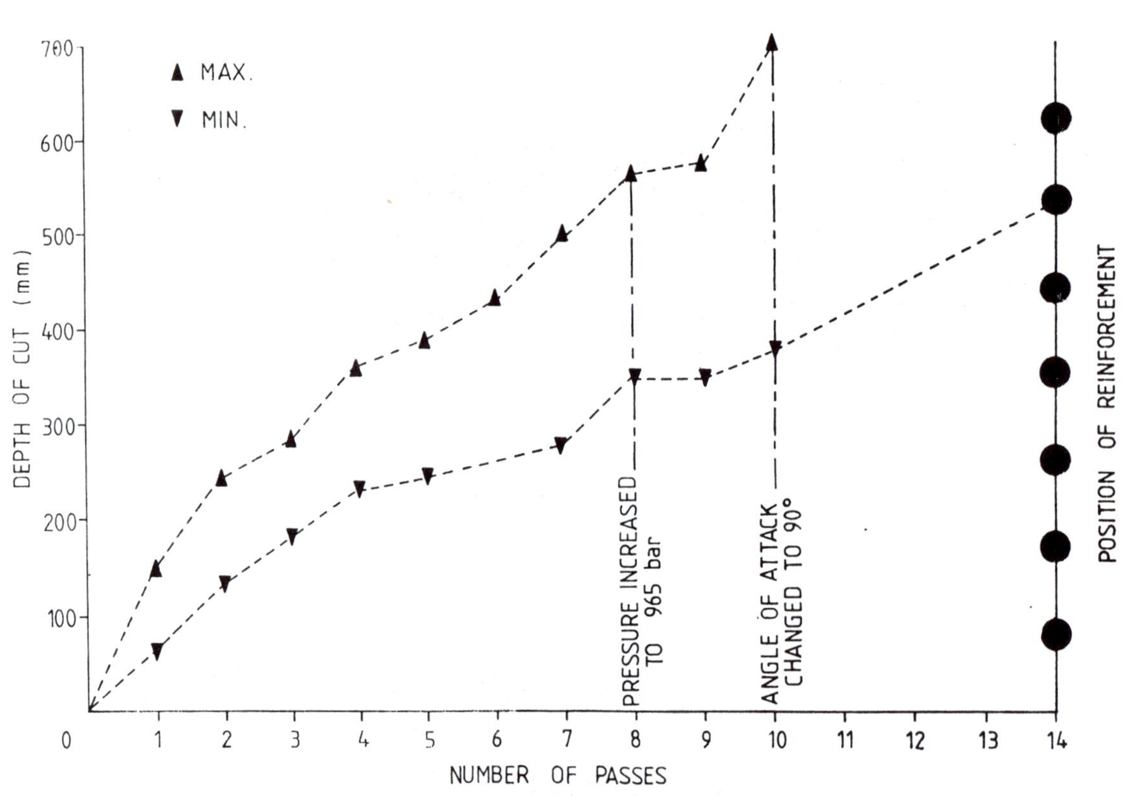

FIG. 9. DEPTH OF CUT IN TEST BLOCK No. 4

10. Photograph of test block No. 4 showing deflected cut.

11. Photograph of test block No. 4 showing straightened cut.

12. Photograph of completed trial cut in test block No. 4.

6th International Symposium on
Jet Cutting Technology
6-8, April, 1982

A SAFE METHOD OF CUTTING STEEL AND ROCK

D.H. Saunders

BHRA Fluid Engineering, U.K.

Summary

Water jet cutting has obvious advantages for use in hazardous flammable atmospheres, since the water provides continuous cooling to the cut. However, hard materials, such as steel, cannot readily be cut with water alone using currently available commercial equipment. This paper describes an investigation into the use of abrasive entraining jetting heads for this purpose.

The first part of the work was concerned with the use of conventional abrasive entraining water jet cleaning and cutting equipment. These tests, to determine cutting depth and speed, cover a range of operating pressures and flows as well as different abrasive types and grades. A satisfactory performance of cutting at 130 mm/min through 13 mm steel was obtained by using an expensive abrasive (silicon carbide). A jetting head was then developed at BHRA, which enables a similar performance to be obtained with a cheap throw-away abrasive (copper slag), and latest developments have increased this performance to 200 mm/min.

The paper includes results of trials carried out with the BHRA jetting head on a range of hard materials, including tests in an explosive atmosphere. In a special environmental chamber, cuts were made in steel and sandstone with atmospheres of methane/air and hydrogen/air, without causing ignition.

Held at the University of Surrey, U.K.
Symposium organised and sponsored by
BHRA Fluid Engineering

1. INTRODUCTION

With the increasing number and age of drilling and production platforms off-shore in the North Sea and elsewhere throughout the world, the need for specialised equipment to cut out damaged components and structural steel work has become increasingly important. Existing techniques, such as flame cutting, grinding discs, and high speed pneumatic saws, all have an inherent fire risk of differing orders of magnitude when used in an emergency situation. Water jets, even with abrasives, are potentially much safer, because any sparks that are formed can be satisfactorily drenched.

Previous work (Ref. 1) has indicated that, if water alone is used, pressures in excess of 10 kbar would be necessary to obtain a satisfactory rate of cut through mild steel. Abrasive assisted water jets have been used for several years for the cutting of concrete sections and large diameter concrete pipes. This includes the cutting of the reinforcing bars to be found within such sections. However, the technique has not in general been applied to the cutting of metals.

Primary considerations for the emergency cutting system were that it should be flexible in its use and capable of reliable operation in the off-shore environment. The use of flexible hose between the pump and the cutting head ensures that the cutting head can readily manoeuvred into position and enables the water pump to be sited a considerable distance from the working area. The operating pressure is limited because of this to a maximum of about 1000 bar. At this pressure, a conventional water jet cleaning pump can be employed. These pump units have a proven record of reliability and are already being widely used on off-shore structures for cleaning, both above and below water.

2. ABRASIVE JETTING HEADS

An abrasive jetting head is a device for entraining abrasive into a high velocity jet, or a number of jets in some cases. Trials on two different abrasive jet cutting heads are described here. The earlier work was carried out on a proprietory (Ref. 2) concrete cutting head used to cut steel. An alternative design of head, originating at BHRA, which is the subject of patent applications, was used in later work on a range of target materials. Both of these abrasive heads operate on the principal of a 'jet pump' (Fig. 1) by using a high velocity water jet to generate a region of low pressure in a mixing chamber into which an abrasive laden fluid is drawn. For most of the trials, air was used as the transporting fluid. However, for emergency cutting situations a water/abrasive slurry feed has the advantage that the abrasive is throughly wetted, and this condition was also tested with the BHRA design of cutting head.

3. TESTS ON PROPRIETORY HEAD

These tests were carried out to investigate the feasibility of using this type of equipment to cut steel. The head used was intended for concrete cutting at operating pressures up to about 1000 bar, although more commonly used at pressures around 550 bar. The tests were carried out in a purpose built chamber, so that all controls and personnel were remote from the working area.

Tests were carried out on 12.7 mm mild steel of 150 Vickers hardness. Each sample plate (460 mm x 150 mm) was used for a number of tests. These plates are shown in Fig. 2 . Water nozzle sizes used were 1.5 mm, 1.8 mm and 2.2 mm. The majority of tests were carried out using the 1.8 mm nozzle, with the 1.5 mm only being used for higher pressures. Pressures used were in the range 350 to 830 bar, this being the limit of the pumping equipment available for the tests. The cutting head was clamped to a carriage which was moved across the sample (Fig.3) by means of a variable speed drive. Abrasive was introduced into the air flow being drawn into the head from a variable feed screw which was gravity fed from a hopper.

The performance of the cutting head was measured under two different operating conditions. For a stationary head, the time taken to break through the sample plate is recorded in Table 1. For a moving head, the test samples were analysed in terms of depth of cut, with the mean of a number of depth measurements being recorded in Table 2. In order to limit the number of tests carried out, parameters were investigated individually, with satisfactory conditions carried forward to the next series of tests. A full depth cut was carried out occasionally to confirm and correlate results.

During the tests, problems were occasionally experienced with water leakage at the pump. This resulted in a reduced operating pressure, but did not otherwise affect the performance and it is noted in Table 2.

3.1 Stand off distance

Fixed conditions of abrasive (Mansel sand), pressure (690 bar), traverse speed (150 mm/min), and nozzle size (1.8 mm) were adopted for these tests. The stand off distance was measured as the gap between the end of the outlet tube and the test sample. The results, Fig. 4, show that the shorter the stand off, the deeper the cut. However, when using a stand off of 3.2 mm, abrasive particles were deflected back from the steel surface and caused severe erosion of the external surface of the cutting head. Consequently, a 6.4 mm stand off was adopted for further testing.

3.2 Abrasive feed rate

Tests were carried out using some of the abrasives at a fixed pressure, with varying rates of abrasive feed. A typical result for dry Mansel sand is shown in Fig. 5. The upper limit of this relationship was close to the maximum which the head could accept. Increasing the feed above this caused a blockage in the abrasive feed pipe. This maximum feed rate appeared to be determined by volume flow rather than mass flow of abrasive. Because of this, higher feed rates could be achieved with denser abrasives. Up to the maximum feed, the depth of cut appears to increase steadily with the feed rate. An approximate relationship of Depth of Cut $\propto$ (Feed Rate)$^{0.8}$ would seem to apply for this sand. A similar relationship can be applied for the other abrasives.

3.3 Angle of attack

This was studied with both a stationary and a moving head, using copper slag as the abrasive. An angled head was found to perform better in both situations. For a stationary head, because of the angle of attack, the effective thickness of the target is increased (to about 22 mm at 35^0). Even so, in all cases over the range tested, the drilling rate was found to be faster than at 90^0.

For a moving head, the direction of deviation relative to the direction of motion is important. A trailing angle (jet angled backwards from the direction of traverse) was found to improve performance (Fig.6). Although there was considerable scatter in the results, there did appear to be a general increase in performance for increasing angles of attack. An angle of attack of -45^0 was adopted for the majority of further tests.

3.4 Comparison of abrasives

Tests were carried out to determine the relative performance of the different abrasives using the maximum feed rate that could be consistently achieved without blockage. For a stationary head, the drilling rates are given in Fig. 7. Under the conditions of this test, two grades of silicon carbide gave the best results, with the copper slag and another grade of silicon carbide not far behind. These were considerably better than sand and the other abrasives used. Abrasive particle size also seems to be important, with the finer silicon carbide (80 grade) giving the best performance.

For a moving head, once again there is a wide scatter of the test results. However, similar relationships seem to apply between the various abrasives used, although for a -45^0 angle of attack there is some indication of a greater difference between copper slag and silicon carbide.

3.5 Operating pressure

It was not possible to maintain all other parameters constant while the operating pressure was changed. The maximum pressure from the pump could only be obtained with a smaller (1.5 mm) nozzle. With the 1.8 mm nozzle pressures in the range 350 to 740 bar were tested. However, at pressures below 690 bar, the feed rate of the abrasives had to be reduced to avoid blockage. The relationship obtained (Fig. 8) clearly indicates that depth of cut increased quite markedly with increasing pressure. However, when considered in terms of power consumption at the different pressure levels, the increase is not as great as it might at first appear.

3.6 Water nozzle size

A direct comparison of the performance at all three sizes is not possible. However, it would appear that the smaller the water nozzle the greater the depth of cut, and the best cutting performance (Test No.48) was obtained with the smallest (1.5 mm) nozzle. A possible explananation of this somewhat surprising result could be that the lower water flow from a smaller nozzle has increased the concentration of abrasive particles in the resultant jet. Another factor which may be particularly important is the size of the outlet tube relative to the nozzle. A significant improvement in performance was observed between Tests 45 and 46 for the same water nozzle when a partly worn outlet tube was fitted in place of a new one.

3.7 Full depth cuts

Full depth cuts were carried out in five tests. At an operating pressure of 690 bar and with copper slag abrasive a full depth (12.7 mm) cut was obtained at a traverse speed of 38 mm/min, with the cutting head both perpendicular and at -45^{0} (Tests 13 and 43). Multiple passes were also carried out at 76 mm/min with a complete cut being achieved after two passes (Test No.42), and at 150 mm/min, after four passes (Test No.40). A complete cut was also obtained in a single pass at 130 mm/min (Test No.48) at the somewhat higher pressure of 830 bar, with silicon carbide 80 as the abrasive.

4. ALTERNATIVE DESIGN OF CUTTING HEAD

The conclusion drawn from the trials on the proprietory head was that an acceptable speed of cut for an emergency cutting system could be achieved (>100 mm/min). However, to achieve this an expensive (silicon carbide) abrasive was required. In spite of this limitation, there was considerable interest in this method of cutting, not only for emergency use off-shore but also for other industrial requirements, which led to further developments.

It was obvious that the dimensions of the various parts of the head, both in absolute terms and also relative to other component parts, were important. An alternative cutting head was therefore constructed which, although operating on the same basic 'jet pump' principle, incorporates some novel features which form the basis of patent applications. Certain internal dimensions of this head were readily adjustable, so that some degree of 'optimisation' could be carried out.

For this method of cutting to become more widely applicable, a satisfactory performance with a relatively cheap abrasive is necessary. Because of this, the tests on the head have concentrated on abrasives such as sand and copper slag. For these tests, a new test facility was employed (Fig. 9), with the sample being moved at a controlled traverse speed past a stationary cutting head.

4.1 Cutting performance

The cutting performance of the head has been investigated on a number of different target materials as well as 12.7 mm mild steel plate. For the vast majority of requirements, water/abrasive cutting would be expensive compared with more conventional cutting tools. However, in addition to emergency situations, the higher operating costs can be justified in some circumstances where environmental or health and safety considerations are paramount, or where problems are experienced with the conventional cutting methods

The work referred to here has been carried out for industrial sponsors with particular cutting problems. Consequently, the results obtained are specific to the conditions and materials concerned, although they do give an indication of the possible range of application of water/abrasive cutting. Unless otherwise stated the abrasive used was copper slag and the operating pressure was 690 bar. Some examples of the performance achieved include,

a) <u>12.7 mm Mild Steel Plate</u> - Cut through at traverse speeds up to 114 mm/min. (Fig. 10).

b) <u>152 mm x 127 mm I Section Steel Beam</u> - Completely parted in a single pass (Ref. 3 and Fig. 10).

c) <u>10 mm Flash on Cast Iron</u> - Clean 'quality cut' removal of flash at 36 mm/min (Ref.4 and Fig.10).

d) Concrete - 65 mm deep slot in concrete containing aggregate up to 20 mm at 150 mm/min.
e) Reinforced Concrete - 150 mm deep section containing 13 mm aggregate and 13 mm reinforcing steel cable cut at 13 mm/min. Steel and individual aggregate pieces cleanly cut (Fig. 11).
f) High Strength Reinforced Concrete - Sections up to 700 mm deep cut through including steel reinforcing bars 32 mm diameter (Ref. 5).
g) Compressed Asbestos Insulation - Sections up to 76 mm cut through at 610 mm/min. (Fig. 12).
h) Hard Sandstone - 70 mm deep slot cut at 150 mm/min.
i) Slate - 200 mm thick section cut at 38 mm/min.
j) Slate (Decorative Panel) - 25 mm thick cut using sand to give good quality cut edge at 150 mm/min. (Ref. 6).
k) Sewer Lining Materials - A range of 13 mm thick sewer lining materials cut at 25 mm/min at a pressure of only 70 bar.

4.2 Cutting in Explosive Environments

Occasional sparks can be seen in a darkened test chamber when a dry feed of abrasive is used to cut steel. Although these sparks may not be sufficient to cause ignition of an explosive environment they do nevertheless give rise to some doubts. There is a distinct advantage in using a thoroughly wetted slurry fed abrasive,when sparks can no longer be observed. It has been found however that, although a similar cutting performance can be achieved, typically a higher feed rate of abrasive is required (about 20% more). The performance also does seem to be dependent on the concentration of abrasive in the slurry.

Cutting tests have been carried out in an explosive environment, cutting both steel and a sandstone that is known to cause ignition of a methane/air mixture when struck by a cutting pick. Tests were carried out with both dry and slurry fed abrasive and in explosive mixtures of hydrogen/air as well as methane/air, without causing an ignition during a total of 79 test cuts. (Ref. 7).

5. LATEST DEVELOPMENTS

A complete system, comprising an abrasive entraining head, a slurry abrasive feed system, and a flameproofed diesel engine driven water pump unit, is at present under construction for British Petroleum. This system is for installation on the emergency support vessel for use in the North Sea oil and gas fields.

Systems are also under development for other applications. Another system has been built at BHRA. This consists of an abrasive hopper, screw feed unit, and slurry pump, mounted on a separate skid. The slurry pump and screw feed are driven by air motors so that this unit as well as the cutting head can be used in explosive environments. The unit (Fig. 13) is used in conjunction with a standard water jetting pump.

Recent developments of the cutting head have enabled the cutting speed through 12.7 mm mild steel to be increased to 200 mm/min. As previously this has been achieved using copper slag abrasive and an operating pressure of only 690 bar.

6. DISCUSSION

The main conclusion to be drawn from this work is that water/abrasive jets operating at pressures below 1000 bar can provide the basis of a successful emergency steel cutting system. Cutting speeds in excess of 100 mm/min through 12.7 mm mild steel have been achieved using a low-cost abrasive without igniting a flammable atmosphere. With latest developments this cutting speed has been doubled for an operating pressure of only 690 bar.

There is also a great potential of application of this method of cutting to a wide range of materials, in spite of operating costs that are very much higher than for more conventional cutting methods, such as saws. In some circumstances, fixed installations can be envisaged, as for de-flashing castings and cutting asbestos, where health and safety considerations of dust and noise reduction can justify the extra costs. However, in many other circumstances it is the inherent flexibility of water jet cutting that is important, and the ease of manipulation of a relatively small cutting head gives this method an advantage.

Operation at higher pressures can give some improvement in efficiency. However, higher pressure pump or intensifier units are not widely used in industry, and a cutting system would therefore need to include the water pump as well. In many industries however, water jetting pump units capable of operating pressures up to 1000 bar are frequently used for cleaning, descaling and surface preparation, and have a proven record of reliability. In these circumstances, the cutting head and it's associated abrasive feed system can be considered as an add-on unit to an existing pump.

7. ACKNOWLEDGEMENTS

A large part of this work has been sponsored by a number of industrial and other organisations, including British Petroleum, the National Coal Board, and the National Research and Development Corporation. The support of the Mechanical Engineering and Machine Tool Requirements Board of the Department of Industry is also acknowledged.

The author would also like to thank his colleagues at BHRA and also staff at the Research and Laboratory Services Division of the Health and Safety Executive for their assistance in carrying out the work described.

REFERENCES

1) Imanaka, O. Fujino, S. Shinohara, K. and Kawata, Y. :"Experimental study of machining characteristics by liquid jets of high power density up to $10^8 Wcm^{-2}$". Paper G3 pp G3-25 to G3-35, 1st Int. Sym. on Jet Cutting Tech. BHRA Fluid Engineering, Cranfield. (April, 1972).

2) Part No. 010.1121. Page T-ZP-E-7.74, Accessories Catalogue, WOMA-Apparatebau, Duisburg.

3) Leeming, A.A. : "The feasibility of using high-pressure-water abrasive cutting equipment in coalmines". Confidential Internal Report No.81/4. N.C.B. Mining Research and Development Establishment, Burton-on-Trent (January 1981).

4) Griffiths, N.J. and Godding, R.G. : "A preliminary investigation into abrasive water jet cutting of cast iron". RR 1629,BHRA Fluid Engineering, Cranfield, (December 1980).

5) Barton, R.E.P. and Saunders, D.H. : "Water/Abrasive jet cutting of concrete and reinforced concrete". Paper to be presented at 6th Int. Symp. on Jet Cutting Tech. BHRA Fluid Engineering, Cranfield (April 1982).

6) Saunders, D.H. : "Water/Abrasive cutting of slate". Confidential report CR 1677, BHRA Fluid Engineering, Cranfield (March 1981).

7) Saunders, D.H. Griffiths, N.J. and Moodie, K. : "Water abrasive cutting in flammable atmospheres". RR 1608, BHRA Fluid Engineering, Cranfield. (June 1980).

TABLE 1.

Stationary Cutting Head Test Results
(12.7mm Mild Steel Plate)

ABRASIVE	WEIGHT OF ABRASIVE (kg/min).	PRESSURE (bar)	WATER NOZZLE DIAMETER (mm)	STAND OFF (mm)	IMPINGEMENT ANGLE (degrees)	TIME FOR PENETRATION (secs.)
Copper Slag	7.5	690	1.8	6.4	90	11.1
Copper Slag	7.5	690	1.8	6.4	90	11.9
Copper Slag	7.5	690	1.8	6.4	90	12.2
Copper Slag	7.5	690	1.8	6.4	45	9.6
Copper Slag	7.5	690	1.8	6.4	45	10.0
Copper Slag	7.5	690	1.8	6.4	45	9.4
Copper Slag	7.5	690	1.8	6.4	77	8.1
Copper Slag	7.5	690	1.8	6.4	77	8.6
Copper Slag	7.5	690	1.8	6.4	77	8.9
Copper Slag	7.5	690	1.8	6.4	45	9.6
Copper Slag	7.5	690	1.8	6.4	45	10.0
Copper Slag	7.5	690	1.8	6.4	35	9.0
Copper Slag	7.5	690	1.8	6.4	55	9.5
Copper Slag	7.5	690	1.8	6.4	45	9.5
Copper Slag	7.5	690	1.8	6.4	45	10.0
Mansel Sand	5.9	690	1.8	6.4	45	26.0
Mansel Sand	5.9	690	1.8	6.4	45	28.0
Mansel Sand	5.9	690	1.8	6.4	45	22.5
Olivine	6.3	690	1.8	6.4	45	18.5
Olivine	6.3	690	1.8	6.4	45	19.5
Aluminium Oxide 24-30	7.3	690	1.8	6.4	45	17.0
Aluminium Oxide 24-30	7.3	690	1.8	6.4	45	19.0
Aluminium Oxide 24-30	7.3	690	1.8	6.4	45	16.0
Silicon Carbide 80	6.3	690	1.8	6.4	45	8.5
Silicon Carbide 80	6.3	690	1.8	6.4	45	9.0
Silicon Carbide 46	6.3	690	1.8	6.4	45	9.0
Silicon Carbide 46	6.3	690	1.8	6.4	45	9.5
Silicon Carbide 16	6.3	690	1.8	6.4	45	11.0

TABLE 2.

Moving Cutting Head Test Results
(12.7mm Mild Steel Plate)

	ABRASIVE	WEIGHT OF ABRASIVE (kg/min).	PRESSURE (bar)	WATER NOZZLE DIAMETER (mm)	TRAVERSE RATE (mm/min)	DEPTH OF CUT (mm)	STAND OFF (mm)	IMPINGEMENT ANGLE (degrees)	COMMENTS
1.	Mansel Sand	5.9	690	1.8	150	0.8	25.4	90	
2.	Mansel Sand	5.9	690	1.8	150	1.0	19.0	90	
3.	Mansel Sand	5.9	690	1.8	150	1.0	12.7	90	
4.	Mansel Sand	5.9	690	1.8	150	1.1	6.4	90	
5.	Mansel Sand	5.9	690	1.8	150	1.2	3.2	90	
6.	Mansel Sand	3.5	690	1.8	150	0.7	6.4	90	
7.	Mansel Sand	2.5	690	1.8	150	0.6	6.4	90	
8.	Mansel Sand	5.9	690	1.8	150	1.2	6.4	90	
9.	Copper Slag	7.5	690	1.8	150	2.7	6.4	90	
10.	Copper Slag	7.5	690	1.8	150	2.3	6.4	90	
11.	Copper Slag	5.6	690	1.8	150	2.0	6.4	90	
12.	Copper Slag	3.7	690	1.8	150	1.8	6.4	90	
13.	Copper Slag	7.5	690	1.8	38	12.7	6.4	90	
14.	Aluminium Oxide 40-60	7.3	670	1.8	150	2.3	6.4	90	Slight leakage at pump
15.	Aluminium Oxide 40-60	7.3	670	1.8	150	2.6	6.4	90	" " " "
16.	Aluminium Oxide 40-60	3.6	670	1.8	150	0.7	6.4	90	" " " "
17.	Silicon Carbide 46	6.3	670	1.8	150	2.3	6.4	90	" " " "
18.	Silicon Carbide 46	3.2	670	1.8	150	1.3	6.4	90	" " "
19.	Aluminium Oxide 24-30	7.3	670	1.8	150	1.6	6.4	90	" " " "
20.	Silicon Carbide 80	6.3	670	1.8	150	2.3	6.4	90	" " " "
21.	Silicon Carbide 16	3.7	650	1.8	150	1.1	6.4	90	Leakage at pump
22.	Olivine	6.3	650	1.8	150	1.1	6.4	90	" " "
23.	Copper Slag	7.5	690	1.8	38	9.1	6.4	90	
24.	Copper Slag	7.5	690	1.8	150	2.9	6.4	-45	
25.	Copper Slag	7.5	690	1.8	150	1.9	6.4	-77	
26.	Copper Slag	7.5	690	1.8	150	1.8	6.4	+45	
27.	Copper Slag	7.5	690	1.8	150	2.8	6.4	-45	
28.	Copper Slag	7.5	690	1.8	150	2.5	6.4	-35	
29.	Copper Slag	7.5	690	1.8	150	2.4	6.4	-55	
30.	Copper Slag	5.0	350	1.8	150	0.8	6.4	-45	
31.	Copper Slag	6.2	480	1.8	150	1.5	6.4	-45	

TABLE 2 (Contd.)

Moving Cutting Head Test Results
(12.7mm Mild Steel Plate)

	ABRASIVE	WEIGHT OF ABRASIVE (kg/min).	PRESSURE (bar)	WATER NOZZLE DIAMETER (mm)	TRAVERSE RATE (mm/min)	DEPTH OF CUT (mm)	STAND OFF (mm)	IMPINGEMENT ANGLE (degrees)	COMMENTS
32.	Copper Slag	7.5	690	1.8	150	2.6	6.4	-45	
33.	Copper Slag	7.5	740	1.8	150	3.0	6.4	-45	
34.	Copper Slag	7.5	450	2.2	150	0.9	6.4	-45	
35	Mansel Sand	5.9	690	1.8	150	1.6	6.4	-45	
36	Olivine	6.3	690	1.8	150	1.5	6.4	-45	
37	Aluminium Oxide 24-30	7.3	690	1.8	400	1.1	6.4	-45	
38	Silicon Carbide 46	6.3	690	1.8	150	3.5	6.4	-45	
39	Silicon Carbide 16	6.3	690	1.8	150	2.7	6.4	-45	
40.	Copper Slag	7.5	690	1.8	150	12.7	6.4	-45	Four passes @ 150 (mm/min
41	Copper Slag	7.5	690	1.8	76	7.1	6.4	-45	
42	Copper Slag	7.5	690	1.8	76	12.7	6.4	-45	Two passes @ 76mm/min
43	Copper Slag	7.5	690	1.8	38	12.7	6.4	-45	
44.	Silicon Carbide 80	6.3	690	1.8	300	1.5	6.4	-45	
45.	Silicon Carbide 80	6.3	690	1.8	76	6.6	6.4	-45	New nozzle outlet tube
46.	Silicon Carbide 80	6.3	690	1.8	76	7.5	6.4	-45	Old outlet tube refitted
47	Silicon Carbide 80	6.3	830	1.5	150	7.3	6.4	-45	
48	Silicon Carbide 80	6.3	830	1.5	130	12.7	6.4	-45	
49.	Copper Slag	7.5	690	1.5	150	4.6	6.4	-45	
50	Copper Slag	7.5	760	1.5	150	4.8	6.4	-45	
51.	Copper Slag	7.5	790	1.5	150	5.0	6.4	-45	
52.	Copper Slag	7.5	790	1.5	130	6.2	6.4	-45	
53.	Silicon Carbide 46	4.2	690	1.5	150	3.2	6.4	-45	

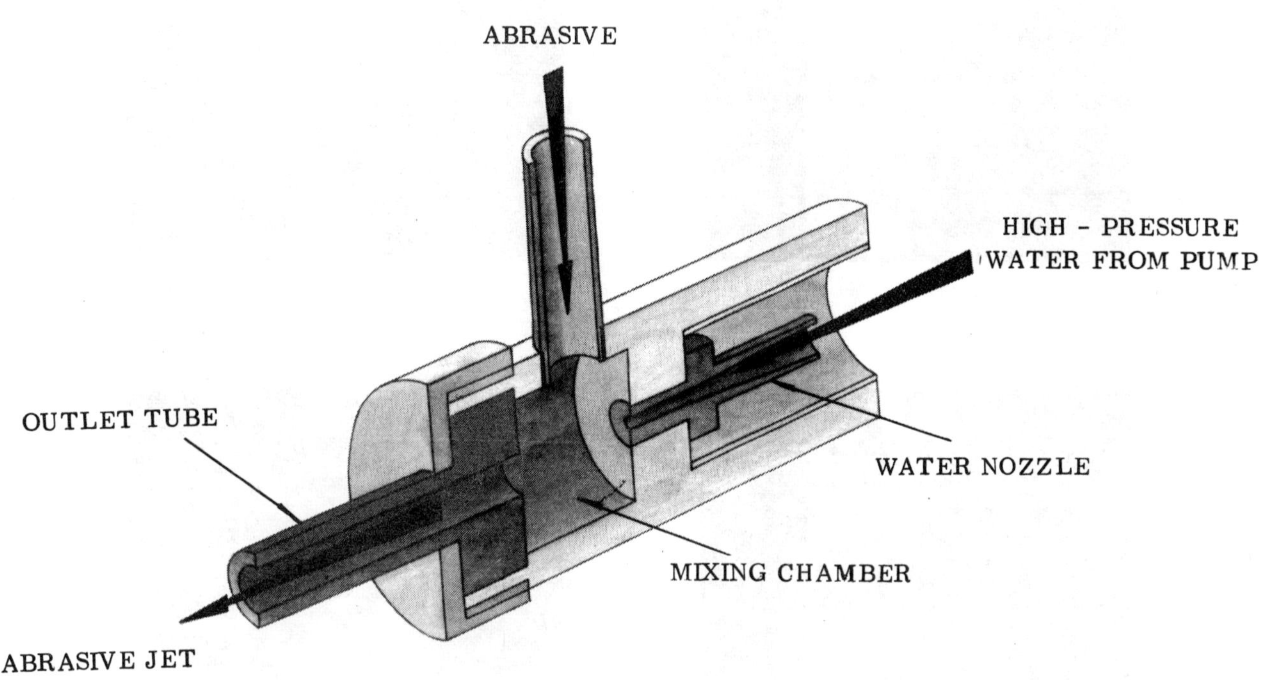

1) General layout of Abrasive Jet Cutting Head.

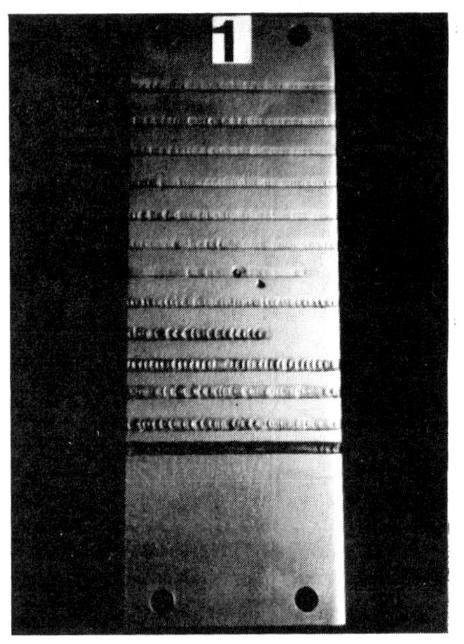

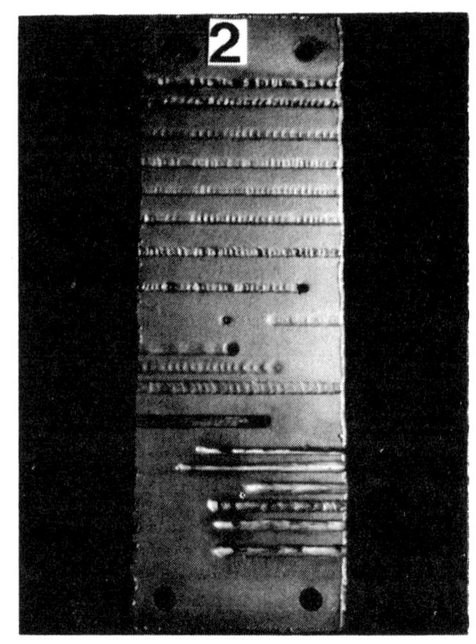

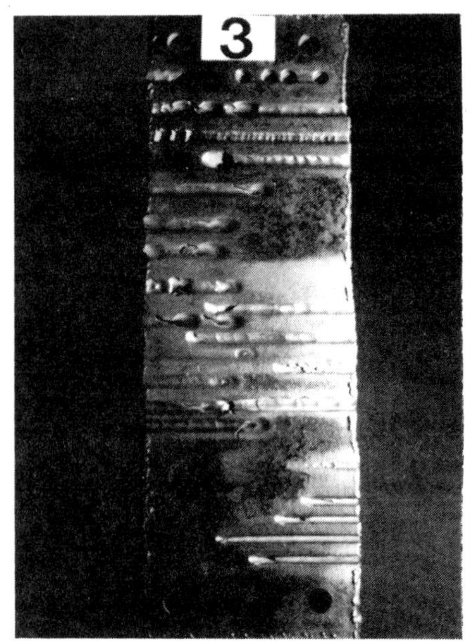

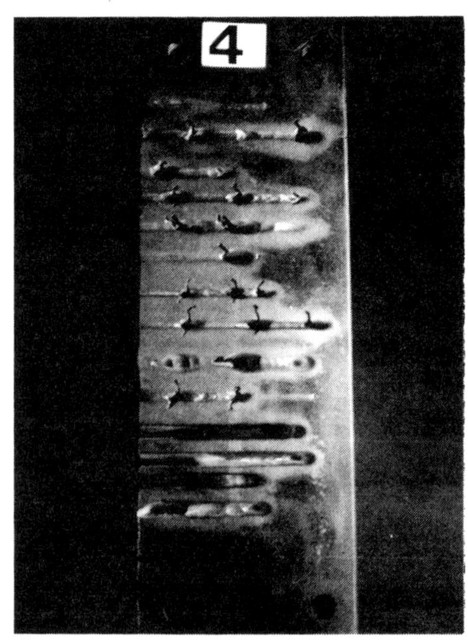

2) Mild Steel Sample Plates.

3) View of Cutting Head and Sample Plate.

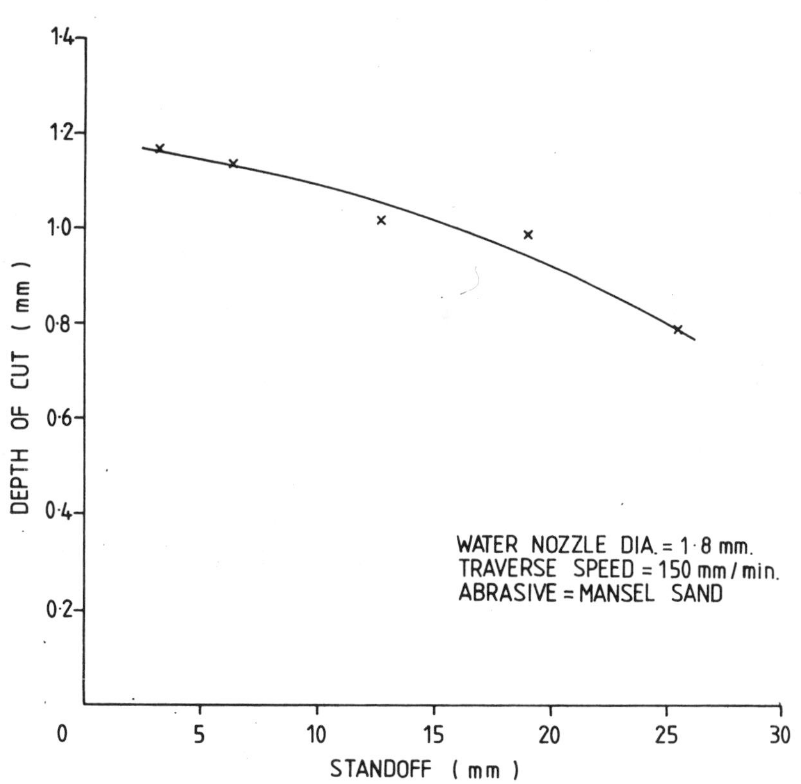

4) Variation of Cutting Performance with Stand Off Distance.

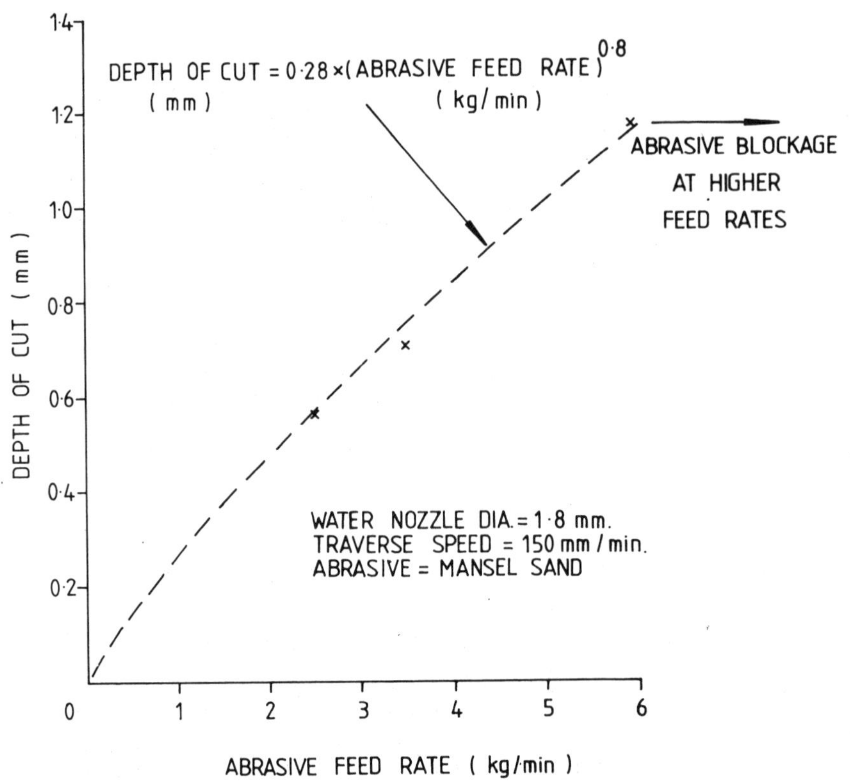

5) Variation of Cutting Performance with Abrasive Feed Rate.

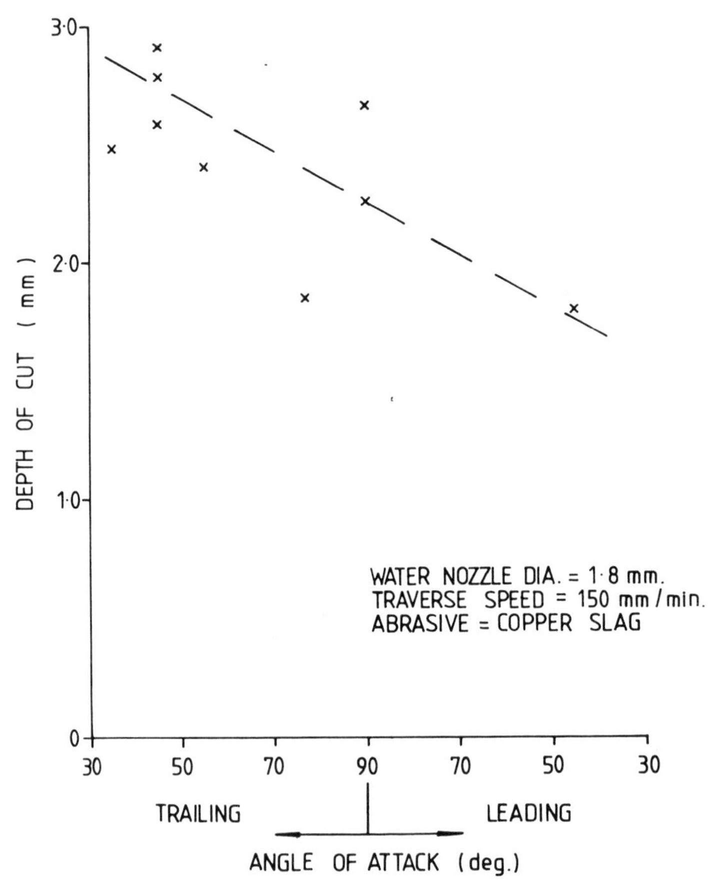

6) Effect of Angle of Attack.

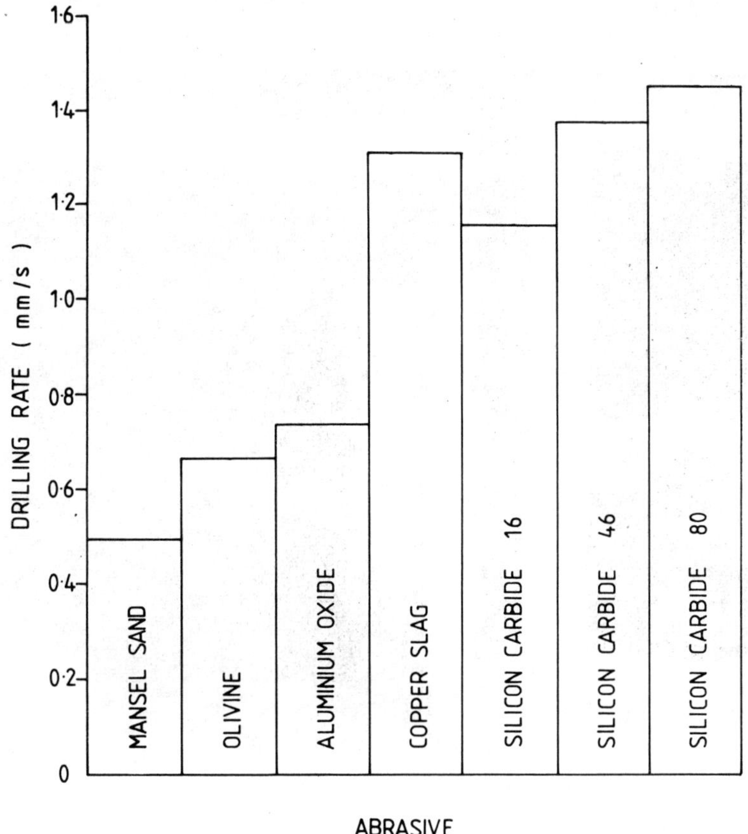

7) Comparison of Abrasives.

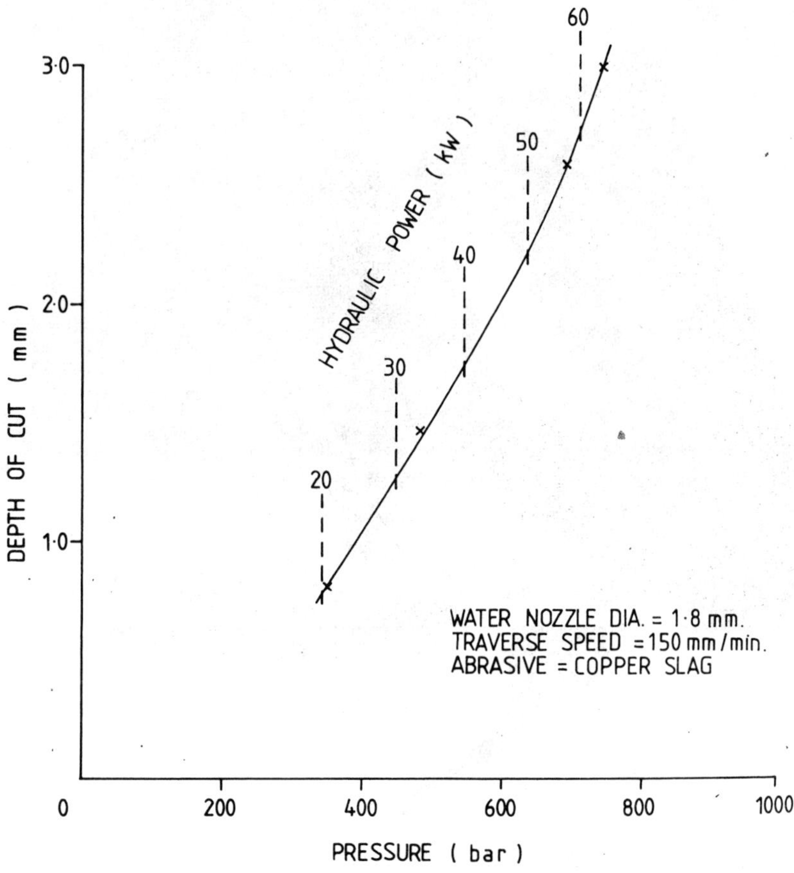

8) Relationship between Cutting Performance and Operating Pressure.

9) Photograph of BHRA Jetting Facility.

10) Mild Steel Plate, Steel I Beam, and Cast Iron cut by Water/Abrasive Jet.

11) Reinforced Concrete Section cut by Water/Abrasive Jet.

12) Range of Asbestos and Asbestos Cement Products cut by Water/Abrasive Jet.

13) Slurry Abrasive Feed System.

6th International Symposium on
Jet Cutting Technology
6-8, April, 1982

SAFETY ASPECTS OF JET CUTTING

D. Krüner, J. Wiedemeier

Industrial Inspection Board, F.R. Germany

H. Louis

University of Hanover, F.R. Germany

Summary

The field of jet cutting and how it is influenced by the safety regulation in order to give a high level of safety to users of these systems is examined from the West German view.

Firstly this paper attempts to highlight some of the risks of jet cutting systems which are not always appreciated.

Secondly this paper deals with safety regulations concerning the use of jet cutting systems that give more protection to the commercial user.

Held at the University of Surrey, U.K.
Symposium organised and sponsored by
BHRA Fluid Engineering

1. Introduction

Rapid developments in the field of jet cutting systems have led to many applications eg. precision cutting with very high pressures and the cleaning and removal of the most varied materials. Working with cutting systems as well as the jetting system itself, is not always intrinsically safe. Therefore careful analysis of the risks and planning of safety precautions is required to avoid these danger areas.

The majority of manufacturers of jet cutting systems agree with these criteria; nonetheless a few accidents, causing in some cases severe injury, should always remind us to think about the prevention of accidents.

From our point of view some safety regulations give the manufacturers a working basis for safe construction, and therefore for the prevention of accidents. On the other hand these obligations set the standards required by the administrative board.

2. The Danger Areas of Jet Cutting

While jet cutting is in itself an inherently safe method careful consideration should be given to the risks involved and the danger areas. With this method danger areas can be liquids compressed by pumps or intensifiers in conjunction with the stress levels of pressurised components.

In comparison with compressed air, compressed liquids, usually water, are often considered to be harmless. Whilst this statement is correct for lower pressures where the stored energy during compression ($\varkappa = 5.10^{-5}$ bar^{-1} in the case of water) is unimportant; at higher pressures in excess of 1000 bar the relationship changes; (Fig. 1) while the stored energy by compression per volume (dm^3) of water at 1000 bar pressure is comparable to that of air at 10 bar pressure, water at a pressure of 1200 bar is already comparable to air of 20 bar.

The elastic deformation of the loaded parts, eg. hoses, pipes, pumpheads, leads to an additional storage of elastic energy. In the case of a crack with small dimensions this storage of energy can give the precondition for a rapid crack propagation. Hence pressure in the system will not decrease spontaneously but in relation to the pressurised volumes with a delay.

Pressure fluctuations caused by the working mode of plunger pumps leads to the superimposition of cyclic stresses on pressurised parts of a jet cutting system. This cycle stressing is regarded as one reason for the limited durability of parts of the jet cutting system, eg. hoses, pipes, bolts or even compact parts. The limited durability itself is caused in the cyclic fatigue strength of the materials.

Above all very high cyclic stressing can appear on reaching resonant frequencies ie in a water column in pipes or in pressure accumulators.

When working with two or more pumps connected in parallel superimposition of pressure peaks may occur. The cyclic effect of dynamic loading is at this time in the German standard DIN 24295 "Safety requirements for pumps and pump systems for liquids", which gives a basis for the safe construction of pumps. Cyclic stresses are not taken directly into account.

This standard indirectly takes this into account by requiring a construction capable of withstanding a test pressure of 1.3 times max operating pressure. This safety

It is an operational requirement to guard moving machinery which in the case of jet cutting means guarding against the touching of the jet.

The standard DIN 31001 page 1 "Protectors (Schutzeinrichtungen)" gives clues for the safe dimension of protective distances (Fig. 4).

3. Some further safety regulations

The "Safety Regulations for Jet Blasters" defines specific requirements for jet cutting devices with pressures above 25 bar.

- Every device must have a trade mark on which the following specifications must be indelibly printed:

  Manufacturer or supplier, production number and type, year of manufacture, max. flow rate in l min^{-1} and related pressure and the max. permissible pressure.

- The device must have a safety valve which limits the excess pressure to no more than 10% (<100 bar 20%) of the max. permissible pressure.

  This valve must be secured against unauthorised alteration.

- The operating pressure must be indicated by a suitable pressure gauge.
- Manual lances or pistols must also have a trade mark which shows a sign of the producer and the maximum permissible pressure.
- Hoses for pressures of more than 10 bars are only allowed to be assembled by a skilled person. The hoses have to be proved with water pressure up to the test pressure. Hoses for pressures of more than 60 bars have to be marked with a sign of the person who has assembled it.
- An operational manual shall be supplied with every jet cutting device.
- The device must be inspected at intervals not less than every 12 months by a skilled person who has to record the results of the inspection. To this point the association of German machine building manufacturers (VDMA) has made a check-list, which gives good advices for the inspection: VDMA 24413, Dec. 1979 "Repeated inspection for protection of labour-jet cleaning devices" (Wiederholungsprüfung der Arbeitssicherheit-Hochdruckreiniger)

4. Protection of labour and security of devices in relation to the work of the authorities

In the field of security of devices and the protection of labour in Germany (FRG) three large organisations are working as authorities in law: the employers association as liability insurance (Berufsgenossenschaften), the mining offices (Bergämter) and the industrial inspection board (Staatl. Gewerbeaufsicht). In the case of devices eg. Jet cutting devices the demand for security is given by the "Law about technical working devices (Gesetz über technische Arbeitsmittel, Gerätesicherheitsgesetz)".

The protection of labour is also given by law, for example paragraphs 24a and 120a of the trade orders (Gewerbeordnung).

The employer has to protect the labour by organising the working process in a safe way, with safe machines and he must also protect the workers against risks for life and limb.

factor may (and only may) give sufficient safety against cyclic fatigue depending on the quality and type of the material used; experienced manufacturers of jet cutting equipment bear this in mind.
It would seem important that when using armoured hose an advantage can be derived from the date of production being marked on the hose for a periodical inspection or exchange. The "Safety-regulations for Jet-blasters (Richtlinien für Flüssigkeitsstrahler)", ZH1/406, 4/1980 of the employers association (Hauptverband der Berufsgenossenschaften) requires from a hose manufacturer the marking of hoses for pressures greater than 60 bars with the date of production.

Hour clocks, as fitted by some manufacturers, can enable the user to carry out periodic inspection and exchange wearing parts before cyclic limits are reached. An additional safety measure which was successfully approved is the use of a protective cover by a transparent safety hose with a thickness of some millimetres. This gives first a protection against wear caused by sharp edges, abrasive surfaces and second a shield should the pressure hose leak or burst.

Often a burst hose is far less harmful than a pin hole leak with its ensuing high velocity jet.

The nozzle carrier, be it a lance, pistol or spray bar, is subject to a reaction force given by F $\dot{m}v$ where F= reaction force, $\dot{m}$ = flow rate, and v = jet velocity. The reaction force must be controlled in a safe manner which is very important when working manually.

The wrong estimation of reaction forces has led to several accidents where the operator has been thrown off balance and injured.

Fig. 2 shows a typical situation where an operator is working in an unsafe manner. The "Safety Regulations for Jet Blasters" referred to above (5.4) require that an operator's working position has to be stable and safe and that the reaction force should not exceed 250N.

Only a person who is familiar with, and has been trained in the use and dangers of jetting equipment, shall be permitted to operate manually.

Problems can also arise in the jet shut-off mechanism. Examples have been recorded where shut-off devices, eg. triggers and foot-valves have failed to operate. Shut off devices must operate reliably. In other cases electrical shut-off devices have failed to operate when contacts have become wet. Generally speaking manual jetting devices must conform with "Safety Regulations for Jet-Blasters" (4.4.2.). This means that shut-off devices must have a "Deadman" action and levers must have a locking mechanism in the off position and be guarded.

In the application of water jets operating at pressures of more than hundred up to thousand's of bars the jet can be regarded as an effective tool, which is not inferior in comparison to other mechanical cutting tools.

In the case of the manual handling of jets or moving materials under the jet, the requirement for a close fitting guard is reasonable.

Here a jet like that in Fig. 3 used without any shield to protect the supporting fingers etc. cannot be a safe installation.

The regulations of these laws are complemented by implementing regulations and technical rules, which are for example safety rules (Unfallverhütungsvorschriften), guidelines of the association of engineers (VDI Richtlinien) or DIN standards. The "Law about technical devices" gives the requirement for a producer or an importer to build, sell or exhibit a device only if it can be used in a safe way. The device must be built with respect to the generally accepted technical rules.

Thinking about this, it must be right, that in Germany no device should be sold, exhibited or used which does not conform with the laws and general technical rules. That means for example that every jet cutting device must adhere to the safety regulations for jet blasters.

As a result in the years 1978 and 1979 the industrial inspection board was forced in three cases to prohibit the selling of jet cutting devices.

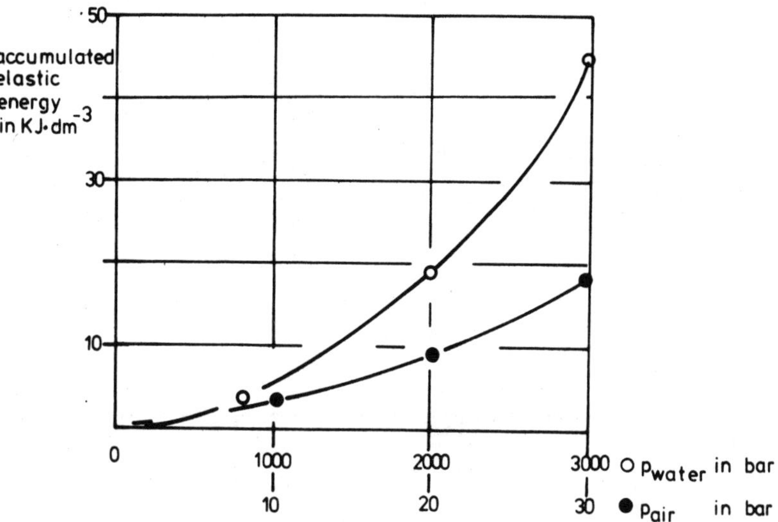

Fig. 1 Relation of the energy per volume between compressed water and air

Fig. 2 Cleaning with a nozzle armed lance an an unsafe position

Fig. 3 Jet-Cutting device without any shield to protect e.g. supporting fingers

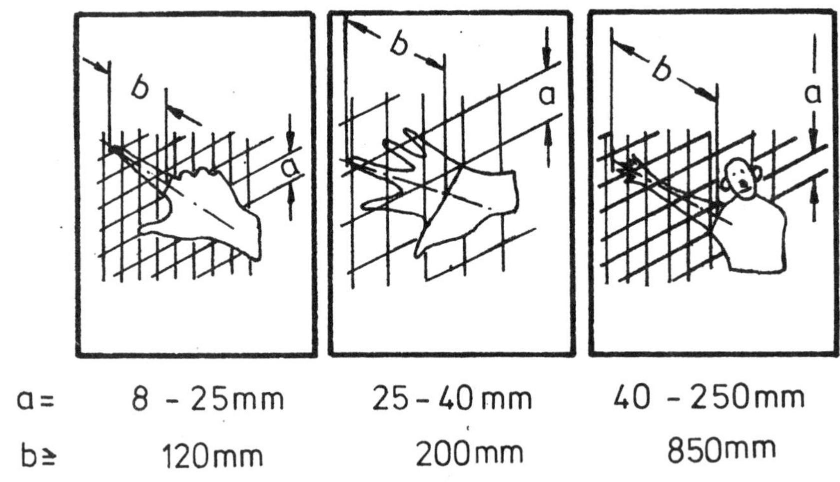

Fig. 4 Protective distances from the DIN standard 31oo1, page 1